Planning and Scheduling Optimization

Planning and Scheduling Optimization

Editors

Farouk Yalaoui
Taha Arbaoui
Yassine Ouazene

MDPI • Basel • Beijing • Wuhan • Barcelona • Belgrade • Manchester • Tokyo • Cluj • Tianjin

Editors
Farouk Yalaoui
University of Technology
of Troyes
France

Taha Arbaoui
University of Technology
of Troyes
France

Yassine Ouazene
University of Technology
of Troyes
France

Editorial Office
MDPI
St. Alban-Anlage 66
4052 Basel, Switzerland

This is a reprint of articles from the Special Issue published online in the open access journal *Applied Sciences* (ISSN 2076-3417) (available at: https://www.mdpi.com/journal/applsci/special_issues/PSO).

For citation purposes, cite each article independently as indicated on the article page online and as indicated below:

LastName, A.A.; LastName, B.B.; LastName, C.C. Article Title. *Journal Name* **Year**, *Volume Number*, Page Range.

ISBN 978-3-0365-2337-8 (Hbk)
ISBN 978-3-0365-2338-5 (PDF)

Contents

About the Editors

Farouk Yalaoui received a engineering degree in industrial engineering from the Polytechnic School of Algiers, Algeria, in 1995; a master's degree in industrial system engineering from the Polytechnic Institute of Lorraine, Nancy, France, in 1997; and a Ph.D. degree in production management from the University of Technology of Troyes (UTT), France, in 2000. He has been the Scientific Director of the Industrial Chair Connected Innovation, UTT, since 2016. He is currently a Full Professor with UTT, where he is also the Senior Vice President of Research. His research interests include scheduling problems, system design, operations research, modeling, the analysis and optimization of logistic and production systems, reliability and maintenance optimization, and optimization problems in general. He is author or coauthor of more than 440 contributions, publications, or communications, with 1 patent, 3 books, 3 edited books, 12 book chapters, and 80 articles in journals. He also published more than 270 articles in conference proceedings and presented 42 invited speeches (seminaries or conferences plenary sessions).

Taha Arbaoui holds an engineering degree in computer science from the Ecole nationale Supérieure d'Informatique of Algiers and a PhD in computer science from the University of Technology of Compiègne. Since 2015, he has been an assistant professor at the University of Technology of Troyes and a member of the Laboratory of Computer Science and Digital Society. He is also the head of industrial and foreign affairs of the Institute of Industries and Factories of the Future of Troyes (ISIFT). His research interests include developing optimization approaches for large-scale problems in scheduling, planning, and timetabling in various applications fields, such as industry 4.0, energy, and healthcare.

Yassine Ouazene received an engineering degree in industrial engineering from the Polytechnic School of Algiers, Algeria, in 2009, and Master's and Ph.D. degrees in systems optimization and safety from the University of Technology of Troyes, in 2010 and 2013, respectively. He is currently an Associate Professor at the Computer Science and Digital Society Laboratory, University of Technology of Troyes. He is author or coauthor of more than 50 scientific publications, including 15 articles from international journals. His research interests include production systems planning and scheduling, manufacturing system design, energy management issues in production systems, and optimization methods and algorithms applied for smart pricing.

Editorial

Planning and Scheduling Optimization

Yassine Ouazene [1,2,*], **Taha Arbaoui** [1] and **Farouk Yalaoui** [1,2]

1   Laboratoire Informatique et Société Numérique, Université de Technologie de Troyes, 12 rue Marie Curie, CS 42060, 10004 Troyes, France; taha.arbaoui@utt.fr (T.A.); farouk.yalaoui@utt.fr (F.Y.)
2   Chaire Connected Innovation, Université de Technologie de Troyes, 12 rue Marie Curie, CS 42060, 10004 Troyes, France
*   Correspondence: yassine.ouazene@utt.fr

Citation: Ouazene, Y.; Arbaoui, T.; Yalaoui, F. Planning and Scheduling Optimization. *Appl. Sci.* **2021**, *11*, 8980. https://doi.org/10.3390/app11198980

Received: 15 September 2021
Accepted: 24 September 2021
Published: 27 September 2021

Publisher's Note: MDPI stays neutral with regard to jurisdictional claims in published maps and institutional affiliations.

Optimizing the performance of services and industrial systems is a real lever for creating value for companies and society. However, this target requires significant research efforts. This optimization phase is one of the major steps that make it possible to guarantee optimized productivity while respecting numerous constraints, such as quality standards, safety of manufacturing processes, environmental impact and other economic constraints. These constraints are linked to the type of products, the type of manufacturing systems and their environment.

In most industries, the main objective is to guarantee an optimized use of workshops, plants and other logistics resources in order to be reactive to market changes and disturbances. From this perspective, industrial systems must be as flexible as possible in order to be able to adapt in terms of resources allocation, workshop reorganization, capacities allocation, etc.

This special issue is devoted to recent research and current developments on complex systems' planning and scheduling. This includes the principles and practice of the design, implementation and analysis of exact and approximate optimization methods to tackle the challenges of modern manufacturing, engineering and healthcare systems. The topics of interest of this issue include:

- Heuristic and metaheuristic algorithms for planning and scheduling problems in manufacturing, engineering and healthcare systems;
- Exact algorithms (brand and bounds, dynamic programming, etc.) for planning and scheduling problems;
- Planning and scheduling applications (timetabling, network routing, crew scheduling, production scheduling, resource-constrained project scheduling, etc.);
- Energy-efficient planning and scheduling problems;
- Intelligent optimization approaches for intelligent manufacturing systems;
- Artificial Intelligence and data analytics (manufacturing, services, healthcare, services and industries of the future, etc.).

In response to the call for papers, thirty-nine papers were submitted to this special issue among which sixteen were accepted for publication. One of the accepted works presents a systematic review and future directions for smart manufacturing scheduling approaches [1]. In this review, several papers published recently are analyzed to understand the requirements considered when developing scheduling solutions for manufacturing systems.

Considering healthcare problems, Huvent et al. [2] addresses the generation and development of a benchmark properly fitting different constraints of the Home Health Care (HHC) problem. Consequently, a generator is proposed dealing with all kinds of constraints, such as time window constraints, workload constraints, synchronization and precedence constraints. This generator allows researchers to validate and compare solving methods on a common dataset, regardless of confidentiality issues. The author validate their generator, firstly, by creating a common benchmark available for researchers and, secondly, by proposing a set of instances and a solving method based on an HHC problem found in the literature. In another work, Abderrabi et al. [3] study a real case of an optimization problem derived from a hospital supply chain. Their work focuses on developing operational decision support models and algorithms for the production process scheduling in a hospital catering. The addressed production system is represented as a flexible job shop system. A mathematical model and two metaheuristics for the production scheduling of multi-product and multi-stage food processes are developed to minimize the total flow time. These methods are tested, based on real data, for scheduling the operations of the food production at the hospital center of Troyes in France.

Some of the addressed papers are especially interested on the development of new theoretical properties regarding different contexts. In paper [4], the authors consider the problem of assigning non-preemptive jobs on identical parallel machines to optimize workload balancing criteria. Since workload balancing is an important practical issue for services and production systems to ensure an efficient use of resources, different measures of performance have been considered in the scheduling literature to characterize this problem. This study proposes a theoretical and computational analysis of these criteria. Based on these theoretical developments, the authors propose new mathematical formulations to provide optimal solutions to some unsolved instances in order to enhance the latest benchmark presented in the literature. In [5], a genetic crow search algorithm (GCSA) approach is proposed to deal with the operation sequencing problem in computer-aided process planning. The traditional CSA is improved by employing genetic strategies, such as tournament selection, three-string crossover, shift and resource mutation. Moreover, adaptive crossover and mutation probability coefficients are introduced to improve local and global search abilities of the GCSA. An operation precedence graph is adopted to represent precedence relationships among features. In [6], Oudani is interested in analyzing and solving the Intermodal Terminal Location Problem on incomplete networks. The problem is modelled as a mixed integer linear program. Then, a simulated annealing algorithm is introduced to tackle medium and large instances. The computational results show that the obtained solutions using simulated annealing are competitive and close to the exact solutions found by CPLEX solver for small and medium instances. The same developed algorithm outperforms the best-found solutions from the literature using heuristics for larger instances.

The study addressed by Wisittipanich et al. [7] focuses on a postman delivery routing problem of the Chiang Rai post office, located in the Chiang Rai province of Thailand. In this study, two metaheuristic methods—Particle Swarm Optimization (PSO) and Differential Evolution (DE)—are applied with particular solution representation to find delivery routings with minimum travel distances. The performances of PSO and DE are compared with those from current practice. The results show that PSO and DE clearly outperform the actual routing of the current practice in all the operational days examined. Moreover, DE performances are notably superior to those of PSO. Červeňanská et al. [8] present the results of the wider investigation of common priority rules combined with a discrete-event simulation model applied in a partially flexible job shop system in terms of several conflicting production performance indicators evaluated simultaneously. The results of the optimization experiments also indicate that the evaluation via applying multi-criteria optimization is relevant for identifying effective solutions in the design space when a specific priority rule is applied in the scheduling operations. In the same multiple-criteria optimization context, Lin and Chang [9] present an extended version of the Hadi-Vencheh model for multiple-criteria ABC inventory classification. The proposed model is based on the nonlinear weighted product method (WPM), which determines a common set of weights for all items. The proposed nonlinear WPM incorporates multiple criteria with different measured units without converting the performance of each inventory item, in terms of converting each criterion into a normalized attribute value, thereby providing an improvement over the model proposed by Hadi-Vencheh. This study mainly includes various criteria for an ABC classification and demonstrates an efficient algorithm for solving nonlinear programming problems, in which the feasible solution set does not have to be convex. The algorithm presented in this study substantially improves the solution efficiency of the canonical coordinates method (CCM) algorithm when applied to large-scale nonlinear programming problems.

The research work proposed by Klimek [10] presents the resource-constrained project scheduling problem with the discounted cash flow maximization criterion from the perspective of a contractor. Cash flows are considered as the contractor's expenses related to the execution of activities and client's payments after the completion of contractual stages. To solve the problem, the author develops dedicated techniques to generate solutions and a simulated annealing algorithm. Finally, the proposed procedures are examined using the Project Scheduling Library (PSPLIB). An experimental analysis identifies the efficient moves and techniques for creating solutions, that is backward scheduling with optimization of completion times of project stages and triple justification.

Viveros et al. [11] address the multi-level storage locations assignment problem for Stock-Keeping Units pallets, considering divisible locations in the first level to improve the picking operation and reduce the travel times associated with the routes of the cranes. A mathematical programming model is developed considering the objective of minimizing the total travel distance and, in the background, maximizing the use of storage capacity. To solve this complex problem, the authors consider its decomposition into four subproblems, which are solved sequentially. To evaluate the performance of the model, two analysis scenarios based on different storage strategies are proposed to evaluate both the entry and exit distance of pallets, as well as the cost associated with the movements.

In the book storage problem, the randomized storage strategy is known as a best practice for storing books of an online bookstore, it simplifies the order picking strategy as to retrieve books in purchase orders from closest locations of the warehouse. However, to be more responsive to customers, many distribution centers have adopted a just-in-time strategy leading to various value-added activities, such as kitting, labelling, product or order assembly, customized packaging, or palletization, all of which must be scheduled and integrated in the order-picking process; this is known as wave planning. Shiau and Huang [12] propose a wave-planning mathematical model by considering, simultaneously, (1) time window from master of schedule (MOS), (2) random storage stock-keeping units (SKUs) and (3) picker-to-order. A conceptual simulation, along with a simplified example for the proposed wave-planning algorithm, is examined to demonstrate the merits of the idea. The result shows that the wave-planning procedure can significantly reduce the waiting time for truck loading of packages and the time that packages are heaping in buffer areas.

The last four papers present different original real case studies in different sectors and different region in the world. In [13], the integration of simulation modeling and the response surface methodology (RSM) is presented to solve an order planning problem in the construction supply chain. The interactions of various factors are examined to observe their effects on key system measurements and a combination of factor levels is determined to achieve the optimal performance. RSM is applied to find the possible values of the optimal setting for system responses, which consists of three main steps: central composite design (CCD), Box–Behnken design (BBD) and a comparison of both designs. The model is tested with a realistic case study of a building material distributor in Vietnam to demonstrate its effectiveness. The developed framework could be applied as a useful reference for decision-makers, purchasing managers and warehouse managers to obtain the most suitable order policy for a robust order planning process. Mejía et al. [14] investigate the problem of sustainable rural supply and urban distribution of fresh food products in central Colombia. Paradoxically, while farmers in the countryside suffer from poverty due to the low profitability of the agricultural activity, inhabitants at urban centers pay high prices for fresh and nutritious foods. In this work, the authors propose a supply chain system and a business model based on food hubs located on existing (and often abandoned) public facilities in the central region of Colombia. They study a business strategy through a mathematical model which considers both the sustainable and efficient operation of the food hubs and better trading conditions for farmers. The authors also propose a variant of the competitive hub location problem adapted to this case study. They test the model under different scenarios, such as changes in the attractiveness parameters, operation costs and profit margins. The results suggest that, if hubs are able to attract farmers, the model can be both sustainable for the hub concessionaires and for the farmers. As claimed by Klement et al. [15], the management of industrial systems is conducted through different levels, ranging from the strategic (designing the system), to the tactical (planning the activities and assigning the resources) and the operational level (scheduling the activities). In their paper, the authors focus on the latter by considering a real-world scheduling problem from a plastic injection company, where the production process combines parallel machines and a set of resources. Two metaheuristics are tested and compared when used in the proposed scheduling approach: the stochastic descent and the simulated annealing. The method's performances are analyzed through an experimental study and the obtained results show that its outcomes outperform those of the scheduling policy conducted in a case-study company. Moreover, besides being able to solve large real-world problems in a reasonable amount of time, the proposed approach has a structure that makes it flexible and easily adaptable to several different planning and scheduling problems. The aim of the study presented by Ghaithan [16] is to develop an integrated mathematical model for the operation and maintenance planning for both the oil and gas supply chain. The utility of the proposed model is demonstrated using the Saudi Arabian hydrocarbon supply chain. In fact, the Hydrocarbon Supply Chain (HCSC) is a complex network that extends from oil and gas fields to demand nodes. Integrating operation and maintenance activities along this complex network are crucial since the hydrocarbon industry is the most influential sector in the world economy and any disruptions or variations in hydrocarbon product supply affect the whole world economy.

Author Contributions: The authors contributed equally to the different steps of this research. All authors have read and agreed to the published version of the manuscript.

Funding: This research received no external funding.

Institutional Review Board Statement: Not applicable.

Informed Consent Statement: Not applicable.

Data Availability Statement: Not applicable.

Acknowledgments: The authors would like to acknowledge the support from the Industrial Chair Connected-Innovation.

Conflicts of Interest: The authors declare no conflict of interest.

References

1. Alemão, D.; Rocha, A.D.; Barata, J. Smart Manufacturing Scheduling Approaches—Systematic Review and Future Directions. *Appl. Sci.* **2021**, *11*, 2186. [CrossRef]
2. Huvent, C.; Gagné, C.; Sioud, A. Randomized and Generated Instances Fitting with the Home Health Care Problem Subjected to Certain Constraints. *Appl. Sci.* **2021**, *11*, 3346. [CrossRef]
3. Abderrabi, F.; Godichaud, M.; Yalaoui, A.; Yalaoui, F.; Amodeo, L.; Qerimi, A.; Thivet, E. Flexible Job Shop Scheduling Problem with Sequence Dependent Setup Time and Job Splitting: Hospital Catering Case Study. *Appl. Sci.* **2021**, *11*, 1504. [CrossRef]
4. Ouazene, Y.; Nguyen, N.-Q.; Yalaoui, F. Workload Balancing on Identical Parallel Machines: Theoretical and Computational Analysis. *Appl. Sci.* **2021**, *11*, 3677. [CrossRef]
5. Djurdjev, M.; Cep, R.; Lukic, D.; Antic, A.; Popovic, B.; Milosevic, M. A Genetic Crow Search Algorithm for Optimization of Operation Sequencing in Process Planning. *Appl. Sci.* **2021**, *11*, 1981. [CrossRef]
6. Oudani, M. A Simulated Annealing Algorithm for Intermodal Transportation on Incomplete Networks. *Appl. Sci.* **2021**, *11*, 4467. [CrossRef]
7. Wisittipanich, W.; Phoungthong, K.; Srisuwannapa, C.; Baisukhan, A.; Wisittipanit, N. Performance Comparison between Particle Swarm Optimization and Differential Evolution Algorithms for Postman Delivery Routing Problem. *Appl. Sci.* **2021**, *11*, 2703. [CrossRef]

8. Červeňanská, Z.; Važan, P.; Juhás, M.; Juhásová, B. Multi-Criteria Optimization in Operations Scheduling Applying Selected Priority Rules. *Appl. Sci.* **2021**, *11*, 2783. [CrossRef]
9. Lin, P.-C.; Chang, H.-C. A Proposed Extended Version of the Hadi-Vencheh Model to Improve Multiple-Criteria ABC Inventory Classification. *Appl. Sci.* **2020**, *10*, 8233. [CrossRef]
10. Klimek, M. Financial Optimization of the Resource-Constrained Project Scheduling Problem with Milestones Payments. *Appl. Sci.* **2021**, *11*, 661. [CrossRef]
11. Viveros, P.; González, K.; Mena, R.; Kristjanpoller, F.; Robledo, J. Slotting Optimization Model for a Warehouse with Divisible First-Level Accommodation Locations. *Appl. Sci.* **2021**, *11*, 936. [CrossRef]
12. Shiau, J.-Y.; Huang, J.-A. Wave Planning for Cart Picking in a Randomized Storage Warehouse. *Appl. Sci.* **2020**, *10*, 8050. [CrossRef]
13. Wang, C.-N.; Nguyen, N.-A.-T.; Dang, T.-T. Solving Order Planning Problem Using a Heuristic Approach: The Case in a Building Material Distributor. *Appl. Sci.* **2020**, *10*, 8959. [CrossRef]
14. Mejía, G.; Granados-Rivera, D.; Jarrín, J.A.; Castellanos, A.; Mayorquín, N.; Molano, E. Strategic Supply Chain Planning for Food Hubs in Central Colombia: An Approach for Sustainable Food Supply and Distribution. *Appl. Sci.* **2021**, *11*, 1792. [CrossRef]
15. Klement, N.; Abdeljaouad, M.A.; Porto, L.; Silva, C. Lot-Sizing and Scheduling for the Plastic Injection Molding Industry—A Hybrid Optimization Approach. *Appl. Sci.* **2021**, *11*, 1202. [CrossRef]
16. Ghaithan, A.M. An Optimization Model for Operational Planning and Turnaround Maintenance Scheduling of Oil and Gas Supply Chain. *Appl. Sci.* **2020**, *10*, 7531. [CrossRef]

applied sciences

MDPI

Review

Smart Manufacturing Scheduling Approaches—Systematic Review and Future Directions

Duarte Alemão [1,2,*], **André Dionisio Rocha** [1,2] **and José Barata** [1,2]

[1] UNINOVA—Centre of Technology and Systems, FCT Campus, Monte de Caparica, 2829-516 Caparica, Portugal; andre.rocha@uninova.pt (A.D.R.); jab@uninova.pt (J.B.)
[2] Department of Electrical and Computer Engineering, Faculty of Sciences and Technology, NOVA University of Lisbon, 1099-085 Lisbon, Portugal
* Correspondence: d.alemao@uninova.pt

Abstract: The recent advances in technology and the demand for highly customized products have been forcing manufacturing companies to adapt and develop new solutions in order to become more dynamic and flexible to face the changing markets. Manufacturing scheduling plays a core role in this adaptation since it is crucial to ensure that all operations and processes are running on time in the factory. However, to develop robust scheduling solutions it is necessary to consider different requirements from the shopfloor, but it is not clear which constraints should be analyzed and most research studies end up considering very few of them. In this review article, several papers published in recent years were analyzed to understand how many and which requirements they consider when developing scheduling solutions for manufacturing systems. It is possible to understand that the majority of them are not able to be adapted to real systems since some core constraints are not even considered. Consequently, it is important to consider how manufacturing scheduling solutions can be structured to be adapted effortlessly for different manufacturing scenarios.

Keywords: manufacturing scheduling; smart manufacturing; intelligent manufacturing systems; scheduling requirements; cyber-physical production systems

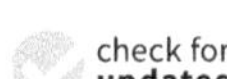

Citation: Alemão, D.; Rocha, A.D.; Barata, J. Smart Manufacturing Scheduling Approaches—Systematic Review and Future Directions. *Appl. Sci.* **2021**, *11*, 2186. https://doi.org/10.3390/app11052186

Academic Editor: Farouk Yalaoui

Received: 5 February 2021
Accepted: 22 February 2021
Published: 2 March 2021

Publisher's Note: MDPI stays neutral with regard to jurisdictional claims in published maps and institutional affiliations.

1. Introduction

Manufacturing has suffered deep changes over the past decades, mainly driven by the market trends that forced companies to move from traditional mass production lines to more dynamic and flexible manufacturing systems. The increasing demand for highly customized products with several variants led to smaller lot sizes, which requires companies to quickly adapt and adjust to new market opportunities in order to thrive in a very competitive world. Therefore, it is crucial that manufacturers develop approaches that allow for more dynamism, flexibility, and reconfigurability at the factory level.

The life cycles of product are becoming smaller and smaller, which leads to companies not adapting their production lines in time for new market opportunities, which takes time and is costly.

One of the biggest challenges of humankind has always been to maximize productive work in an efficient and effective way. To do so, it is of huge importance to plan a well-structured schedule with a detailed description of the tasks to execute, where they should be executed and when a task should be performed. This applies to areas such as transportation services, staff distribution, and, unquestionably, production systems.

Manufacturing scheduling has been studied for several decades and has been applied in industry in many different forms in an attempt to optimize the production processes and allocate resources as efficiently as possible. However, most of the scheduling has been carried out manually or using simple and limited programs that can barely improve the performance of the system. Until a few decades ago, mass production lines, producing huge lots of the same products always following the same method, were dominant; today that is not the case, and a better planning approach needs to be considered.

In recent years, new production paradigms have been proposed to support companies tackling this problem. These paradigms, such as lean production, agile manufacturing, or smart manufacturing, have been pushed and pulled by industry and academia, which contributes to huge advances in manufacturing. However, the gap between academia and industry is still huge and the link between both of them needs to be reinforced in order to achieve consistent and acceptable results. For instance, both parts should have a closer interaction which will provide more prosperous advances, since academia is often not aware of some manufacturing requirements, such as relevant production technical features, business environment, clients preferences, or societal requirements. On the other hand, companies that do not have a solid research department may not be aware of new technologies and processes being developed, which need to be strongly matured, mainly by the contribution of industrial partners.

Now, more than ever, there is an opportunity to implement robust and efficient schedule solutions, not only on the shopfloor but even along the value chain, since there is more information available than ever before. However, there is a big challenge to implement scheduling solutions in real manufacturing systems. Despite the required complexity of such implementation in the real world, there is not a reference guide in the context of smart manufacturing to assist in the implementation of these solutions.

Consequently, the authors see the importance of establishing a baseline that can serve as a starting point when developing manufacturing scheduling systems, with the objective of developing reliable solutions that can be applied to real manufacturing scenarios.

Thus, this work aims to identify, analyze, and point out the main trends regarding the adoption of industrial requirements or constraints which academia should focus more on in order to help developing scheduling solutions in smart manufacturing systems. These requirements were identified by analyzing some relevant studies in this area to understand which requirements are being considered when implementing manufacturing scheduling solutions. Furthermore, not only the requirements were analyzed but also the objective functions, i.e., the objective the solution is trying to optimize, which can be related to production efficiency, delivery time, energetic efficiency, and so on. Consequently, further research studies may focus on how to develop scheduling solutions based on some fundamental requirements identified in this study.

The rest of this document is composed of a brief overview of manufacturing scheduling and a description of the methodology adopted in this survey. Then, the main findings are introduced, highlighting the most common scheduling requirements and which research studies are considering them. After this, a discussion about the future of scheduling in smart manufacturing environment is presented. The document ends with a brief conclusion of the article.

2. Manufacturing Scheduling

The recent development and advances in technology as well as the market demand for highly customized and personalized products have been pushing manufacturing companies to develop new solutions to become more dynamic and flexible to face these emergent trends and the quickly changing markets.

Most of the existing production systems are based on automated systems built to achieve high performances and high delivery rates, coming from the second and third industrial revolutions, but have no capability regarding autonomy, adaptation, and flexibility. Consequently, a group of expert technicians is needed to solve a problem each time a disturbance occurs in the production line. In addition to these restrictions, the emergence of new manufacturing paradigms, the appearance of new technologies and processes, the cheaper development of IT infrastructures, and the emerging possibility of digitization, among other factors, led to a disruption in the industrial scene.

The fourth industrial revolution—under the abbreviation of I4.0 in Europe, industrial internet or smart manufacturing in the USA, smart factory in South Korea, and made in China 2025 in China [1–3]—is happening now but started with first steps several years ago. It makes use of different emerging technologies and paradigms such as AI, cyber-physical systems, Internet of Things (IoT), cloud computing, digital twin, agent-based systems, among others, and allows the development of more dynamic and agile approaches to improve the efficiency of manufacturing systems [4,5]. The capability to collect, store, and analyze data was hugely improved, which makes it possible for manufacturers to better understand their equipment, their products, their manufacturing processes, their customers, their workers, and even their competitors, which increases the smartness degree in manufacturing systems, and leads to a better interconnection between the different systems.

Smart manufacturing comprises both horizontal and vertical integrations. Horizontal integration connects the different players of the value chain along the entire product lifecycle, which allows the optimization of the production process from suppliers to manufacturers and end users. The digitization of data and processes allow the system to be shared, analyzed, and to dynamically adapt in real-time during the manufacturing process [6]. Vertical integration combines different hierarchical levels in the production process within the factory, from working stations and human workers on the shopfloor to software technology, such as manufacturing execution systems and marketing activities [1].

However, with all the recent developments in technology, one of the main challenges in production systems continues to be the development of scheduling solutions to deal with all the planning information as well as unpredictable events on distributed production processes [7]. These challenges occur not only in the development phase, due to the implementation complexity of the systems, but also in the design phase. Nevertheless, several research articles have discussed and proposed manufacturing scheduling approaches under the smart manufacturing umbrella, such as [5,8–14].

Scheduling has been largely applied in many different areas such as energy consumption [15,16], transportation [17], staff distribution [18], and manufacturing [19,20], among other areas, to help the industries to plan their activities. For each scheduling solution for different areas of application, specific algorithms and mathematical models should be developed, since it is not achievable to develop a one-fit-all solution [18].

In manufacturing, scheduling can be considered as a process of arranging, controlling, and optimizing work on the shopfloor [21]. Sometimes, parts need to wait too long on the shopfloor due to limited resources to manage them or due to weak planning of the system. Production scheduling aims to efficiently allocate the available resources and reach a predefined goal since scheduling is a process of optimizing work and time. A scheduling problem may be described as an environment composed of one or more machines, with specific characteristics, and a set of jobs (products with one or more operations that will be processed by the machines). The goal is to optimize an objective or group of objectives by assigning each job to a specific machine in a specific time in order to be processed, while conflicts between operations are avoided [22]. Succinctly, scheduling determines what is going to be carried out and where this will happen and with what resources.

Manufacturing processes can be very dynamic. Even in environments where the processes happening on the shopfloor are always the same and known in advance, they can be affected by one or another disturbance that forces all production to stop until the problem is solved. Although some years ago manufacturing systems were not ready for this change and were not efficient enough to deal with these disturbances, nowadays, manufacturing is becoming more adaptive, dynamic, and highly flexible to meet market requirements and adjusts to every change that may improve the process. This is even more important in the era of the emergence of mass individualization, where the disturbances in the production line can be even more significant. In order to minimize the unexpected events and improve the overall production performance, one of the key challenges is to develop reliable and robust scheduling solutions. I4.0 scheduling approaches should be designed to deal with these smart and dynamic manufacturing systems and their new technologies.

Although this has been studied for decades, complex and robust scheduling solutions are frequently disregarded in real manufacturing scenarios, where they are sometimes carried out manually, on data sheets, or on simple or limited software programs. These solutions frequently lead to significant errors since they do not consider the current status of the shopfloor and are not adaptive to different scenarios. Though, more robust solutions are not implemented mainly due to the complexity of implementing them in large-scale systems with real-time constraints, since it is considered to be a non-deterministic polynomial-time hard combinatorial optimization problem which is quite difficult to reach an optimal solution for with traditional optimization techniques [23]. However, scheduling optimization has direct impacts on the production efficiency, sustainability, and also on costs of manufacturing systems and must be developed to its full capabilities [5,24].

Most researchers assume some constraints, such as that resources are always available or that the processing time of a job is known in advance and remains constant during the entire process, but in real systems this is not always true.

Disturbances may occur during the production process, which lead to a rescheduling that should be performed as fast as possible. These disturbances can be the arrival of new orders, canceled orders, or machine breakdowns which lead to the machine's unavailability, or some emergency event [5,25]. Additionally, job processing times may increase over time, which is a situation knowing as *deterioration of resources* in scheduling problems [26], or even decrease when there is a learning factor or the workload can be reinforced [27]. Consequently, to adapt to the manufacturing system, it is vital that the scheduling process is dynamic and quick to avoid unnecessary system downtimes and costs.

In flexible and agile manufacturing environments, products can have several different feasible processing plans and most of the time it is very hard to find a good one for all the products. Production scheduling is a very important decision making in a factory and it can be a difficult problem depending on the number of calculations necessary to obtain a schedule that gives an optimal or near-optimal solution for the given objectives [19].

The production scheduling optimization problem may be decomposed into several categories, according to the factory type. There are several environments depending on the machine's layout and the flow of the products, which can mainly be divided into: flow shop, which is composed of a set of machines arranged in series, one after another, where the products follow the same execution order through all the machines [28]; job shop, which can be described as a set of machines that should process a set of different jobs, where each job is composed of a group of operations to be processed in a given order, so each product may have a different route [29]; open shop, composed of machines that can perform all operations and thus there are no fixed routes for each job, which consist of unordered operations that do not have precedence constraints [30]. Usually, the essence of these is that several jobs (products with one or more operations to be processed) are assigned to a set of machines at specific times, satisfying some constraints, while trying to minimize the makespan, i.e., the time between the moment that the first job started until the moment that the last job is finished, or optimize some other objective, such as the production due dates, or the number of finished products, or the load balancing, which refers to assigning the task among different resources equally to provide better quality service, in the case of human workers, and reduce idle times and work-in-process in the case of machines [23,26,31–33].

Moreover, to produce an optimized solution, restrictions regarding product parts, material availability, machines or work capacity, start and due dates, costs, distribution requirements, or setup efficiency conditions must be known [34].

3. Survey Approach

To identify and characterize tendencies on the application of scheduling concepts and approaches to the manufacturing area, this study was conducted following a systematic literature review (SLR) method to reach a systematic process that synthesizes research results [35]. The procedure is synthesized in Figure 1.

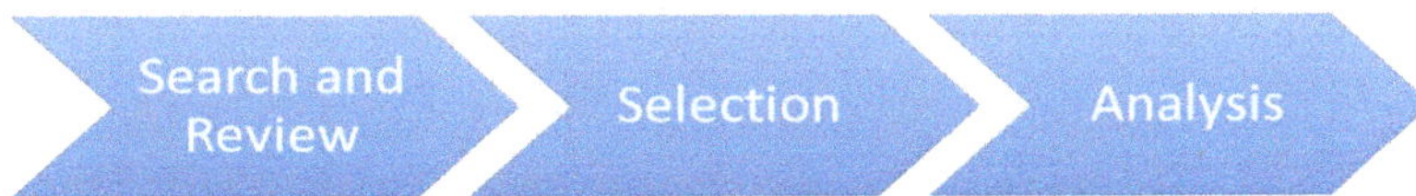

Figure 1. Procedure of the review.

An extended literature review was performed and the following research questions were formulated.

3.1. Research Questions

The research questions addressed by this work are:

RQ1 What are the constraints and targets of the manufacturing scheduling solutions found in the studied articles?

RQ2 How can scheduling systems be designed and developed so that they address different targets and requirements in the context of smart manufacturing?

3.2. Search Sources and Process

The first approach for finding publications related to the topic was a search on Google Scholar where it was possible to find a large number of articles related to manufacturing scheduling.

Then, for the search and selection of appropriate articles, three relevant indexed databases in manufacturing were used: Web of Science, Scopus, and IEEE Xplore. By restricting the search to these sources, some articles were automatically excluded; however, the papers indexed in these databases are expected to have more relevance, since these databases are broadly used by manufacturing and scheduling peers. Different terms were used individually or combined with others to come out with the current literature review. The most relevant terms are composed of "scheduling", "manufacturing scheduling", "production scheduling", "scheduling systems", "job shop scheduling", "dynamic scheduling", "industry 4.0", "industrie 4.0", "smart manufacturing", and "cyber-physical systems".

3.3. Inclusion and Exclusion Criteria

The main findings, i.e., the scheduling solutions pointed in Table 1, of this research are mainly focused on documents published during the 2013–2020 period to ensure that most information is up to date. Since the term "scheduling" is used in different applications (such as production, transportation, or staff allocation), it was necessary to exclude some papers that were not related to manufacturing. As a result, a total of 65 papers were kept as the foundation for this study.

4. Main Findings

In this section are presented numerous articles found in the literature that focus on different aspects of manufacturing scheduling. Contrary to traditional approaches that mostly use centralized manufacturing systems, underneath the smart manufacturing environment, most of the components are smart, autonomous, and dynamic, leading to a more intelligent and decentralized manufacturing system [36]. Consequently, a lot of data and different information need to be available in order to model and develop robust scheduling solutions.

One of the main findings is related to the innumerous different requirements that are considered among different solutions. A preliminary analysis of the selected literature allowed us to identify some requirements that may be crucial in real scenarios but are not always considered in the literature. The list of those requirements is presented next, and from here it was possible to build Table 1 and evaluate which studies consider each of these requirements.

4.1. Manufacturing Scheduling Requirements

Among these data, there are a lot of constraints and requirements that need to be considered before developing a production scheduling solution. Even though these requirements may be different for each particular case, some of them can be transversal to several real manufacturing scenarios. Although it is not possible to consider all of them, next are presented some relevant constraints and requirements.

4.1.1. Dynamic Environments

In most real-world manufacturing environments disturbances may occur over time, known as dynamic environments, and so it is important to be prepared to deal with them. This may mean that the optimal solution for the problem may also change. These disturbances may include the arrival of new jobs that need to be executed urgently, cancellation of jobs, changing of processing times, or machine availability, since machinery may be subject to maintenance operations or incur in breakdowns. Although the simplest way to cope with this problem may be to reschedule all the remaining jobs, as done by Tran and colleagues [37], it may be impractical from a temporal point of view. So, a possible solution is to the previous search space to improve the search after a change, by incorporating or removing the jobs in the previous schedule without affecting the other tasks. However, in extreme cases, reschedule could be the better option. However, in most real-world problems, changes are quite smooth, and so can be profitable to collect previous knowledge [38].

4.1.2. Flexibility

A flexible manufacturing system is able to produce different products by sharing tools. More factories are adopting flexible machines, which are able to perform more than one unique task; thus, a flexible scheduling solution should be adopted to those manufacturing systems. In flexible scheduling, an operation can be executed in more than one machine (routing flexibility) or each machine can be able to perform more than one operation by sharing resources (machine flexibility) [39,40].

4.1.3. Processing Times Variation

Most of the time, in the literature, task processing times are considered to be static, i.e., they are known in advance and do not change along the way. However, in real manufacturing systems, due to the most diverse situations, they may vary, mostly to increase over time. This situation can happen due to resource deterioration, a fault in the setup, or because of the surrounding conditions [26,41]. However, processing times may also be reduced, for example, by assigning more workforce to a task, which means that those time variations can sometimes be controlled [27]. This means that neglecting processing times may reflect in the actual production process.

4.1.4. Setup Times

Setup times encompass all the operations that are performed on the machines but are not related to the production process directly. It includes, for example, adding or removing product parts, calibration, machine cleaning, tests, etc. These operations can occur both before and/or after the processing of the task, and, depending on the industry, it may occur each time a new product needs to be processed. In a way to reduce setup costs, parts that need the same machine configuration may be scheduled one after the other, known as sequence-dependent setup problems [42]. However, most of the time these processes are not considered or are considered as part of the processing times of the product [27]. However, in real manufacturing systems, setup times can be a substantial part of the production time, and so should be managed wisely to generate correct information about the process. Otherwise, it can lead to incorrect information that can directly affect the production process.

4.1.5. Maintenance

Although often ignored in scheduling studies, maintenance activities play a crucial role in manufacturing systems, since they are a constant in real environments, either to prevent/avoid or to correct/recover failures. Even more, in the current globalized market where manufacturers focus more in reinforce the delivery reliability, sometimes at the cost of a good production and maintenance strategy [43]. Thus, maintenance activities are an important element to be considered when developing scheduling approaches, in order to have a more robust solution and achieve a better performance of the system [44,45]. Furthermore, by arranging maintenance operations strategically, companies may be able to pursue long-term competitiveness and sustainability, by providing better resources conditions, conservation, and functional life extension [46].

4.1.6. Precedence Activities

Even though precedence constraints have been studied for a few decades, it is still a very explored topic inside the research community. Even though it is usually assumed that every job or task to be processed is independent of any other, this is not always the case, since some jobs may be intermediates to other jobs. Thus, in the cases where there are precedence constraints, the first task of a certain job cannot start before all the tasks of its predecessors are finalized, as in the case of the assembly of two or more parts [47]. So, these constraints need to be known in advance to optimize the scheduling process.

4.1.7. Pre-Emption

In some cases, it may be necessary or desired that operations in jobs can be continued after a pause. This is known as pre-emption. On the other hand, when jobs cannot be interrupted, pre-emption is not allowed. Although pre-emption is rarely considered in the literature, in several scenarios, it may be needed, such as the arrival of new jobs with more importance than the ones being processed, which requires the machine to stop the operations. It can be beneficial to continue an operation in another machine or another time; unexpected cancellations by the clients might also require the stopping of production or even breakdowns in the machines as mentioned previously [27].

4.1.8. Release and Due Dates

Release dates (the time from when the job is available to be processed) and due dates (the time when the job must be completed) are other types of requirements or constraints present when developing scheduling solutions, although they can also be considered as part of the objectives of the scheduling. Sometimes, however, it may not be possible to complete all the jobs in the time interval between the release and due dates, but these times should not be disregarded [27]. It is important to respect the dates since the products need to be ready for delivery at some time [42] and it could be crucial to not overcome these dates. On the other hand, it could be important to not finish the products too soon as well, since this can lead to some wear in the parts or involve storage costs. Thus, when these dates are not respected, it may be necessary to apply penalties, both for early and late finishes [48,49].

4.1.9. Transportation

Product parts need to be moved inside the factory from one machine to another, or to the storage zone. This process can involve transportations through conveyors, robotic arms, automated guided vehicles, and many other solutions [27]. This means that first the product will not be immediately available in the next machine and a certain time is required to transport it, and second, the number of transporters is limited. So, they must be synchronized with the scheduling process along the chain or the parts need to wait for an available one, which may require to allocate each job to a transport vehicle and sequence the transport tasks to be executed by each vehicle [50]. However, transportation times are not often considered in the scheduling problems found in the literature, which can severely affect the scheduling performance.

4.1.10. Storage

Another constraint that is often ignored or considered to be infinite in the literature is the storage buffer. Products may need to be placed in storage, both during and/or at the end of the production process. Obviously, this space is not infinite, and full or poorly managed storage zones may imply additional problems in the production. Thus, storage buffers may be considered when developing scheduling approaches to have more realistic solutions and reduce unexpected problems [27].

4.1.11. Distributed Factories

Although scheduling systems are mostly associated with the scheduling process within the factory, the scheduling process of supply chains has evolved. This increased the complexity of the problem even more, since the products may be assigned to different factories, which may be distant from each other, and the transportation across these facilities needs to be considered [40,51].

4.1.12. Environmental Issues

During recent years, there has been an increase in concern about the negative environmental impact caused by the manufacturing environments. Since the population is increasing, it is quite natural that the energy consumption increases as well to respond to the demand for any type of goods. Nevertheless, to achieve sustainable development in order to reduce gas emissions or acidification, it is crucial to reduce the energy demand [24,52]. This can be done, for example, by creating energy-efficient machine tools and selecting appropriate tools when acquiring them, and assess a set of key performance indicators to support the design and selection of tools [53].

As stated by [24] in their research, "none of the IPCC (Intergovernmental Panel for Climate Change) reports identifying scheduling as either a method or an instrument to improve energy efficiency [...] scheduling is rarely considered as a suitable instrument to improve sustainability either in general or concerning energy efficiency in particular". However, scheduling can be an important tool to reduce the environmental impact and achieve sustainability, since it can inform what the best steps to improve and reduce the energetic consumption and costs during the manufacturing process are, such as machines' consumption or utilization of materials [54,55].

Based on the identified requirements, the next subsection will analyze the literature considering some important requirements and restrictions that are important to consider when designing scheduling approaches to be applied in manufacturing systems.

4.2. Existing Approaches

In this subsection are presented some studies found in the literature covering different characteristics of manufacturing scheduling. In Table 1, some requirements the authors consider to be important in manufacturing scenarios and that contribute to improving the development of scheduling solutions are presented, which can be found in the literature.

The following abbreviations were used:

ATCT—adjustment of total completion times;	MW—material wastage;
CJ—completed jobs;	Obj—objective function;
CW—cost of workers;	P—productivity;
DD—due date;	Pd—precedence;
DE—dynamic events;	PDC—total production and
DF—distributed factories;	distribution costs;
E—earliness;	PM—parallel machine;
EC—energy consumption;	Pr – pre-emption;

F—flexible shopfloor;
FS—flow shop;
I—industry-oriented;
JS—job shop;
LB—load balance;
M—makespan;
MAPE—mean absolute percentage error;
MC—manufacturing cost;
MDO—maximize delivery of orders;
MeanET—Mean earliness and tardiness;
MEP—maximize early production;
MET—sum of maximum earliness and tardiness;
MFT—mean flow time;
MOO—minimize overdue orders;
MSA—maximize system availability;
Mt – maintenance;
MtC—maintenance cost;

PT—processing time;
RD—release date;
RM—resources management;
S—storage;
Set—setup times;
SCT—sum of completion times;
SM—single machine;
ST—shop type;
Stab – stability;
T—tardiness;
TDR—tardiness delivery rate;
TET—total earliness and tardiness;
TFT—total flow time;
TT—transportation time;
TV—processing times variation;
TWM—total weighted makespan;
TWT—total weight tardiness.

Table 1. Aggregation of solutions based on scheduling type (column 2), requirements (columns 3–17), and objectives (last column).

Ref.	ST	I	DE	F	DF	TT	E	T	TV	Set	Mt	Pd	Pr	RD	DD	S	Obj
[56]	PM														✓		TWM
[57]	JS		✓	✓													M/EC
[40]	JS			✓	✓	✓											M
[58]	JS						✓	✓						✓	✓	✓	M/TWT/E/T/S
[59]	JS			✓						✓							M
[60]	JS		✓							✓							M
[49]	JS						✓	✓							✓		MET
[48]	FS							✓							✓		TWT
[51]	JS			✓	✓	✓											M
[44]	SM	✓	✓			✓				✓	✓						M
[13]	FS		✓					✓			✓	✓		✓			MC/RM/EC/TDR
[45]	SM		✓					✓			✓			✓			MC/T
[61]	FS/PM	✓		✓						✓							M/EC/MW
[62]	FS							✓									M/ATCT/PT
[63]	JS		✓	✓				✓			✓			✓			T/EC
[64]	JS		✓										✓				M/Stab
[65]	JS		✓					✓						✓			TWT
[25]	PM										✓			✓			M/SCT/T
[66]	SM		✓								✓						MtC
[41]	PM					✓	✓	✓			✓			✓			E/T/MtC
[67]	JS		✓		✓			✓						✓	✓	✓	M
[68]	JS	✓	✓										✓				MFT
[69]	FS		✓	✓				✓	✓	✓						✓	P/MDO/MOO/CW
[70]	JS		✓			✓	✓	✓						✓	✓		E/T/PT
[9]	FS		✓	✓				✓					✓				T/CJ/LB
[47]	JS		✓					✓						✓	✓		T
[71]	JS		✓	✓				✓						✓	✓		M/MFT/T
[72]	JS		✓									✓					M/MSA/EC
[73]	PM	✓	✓	✓							✓			✓			T
[74]	JS		✓					✓						✓	✓		M/TWT/MAPE

Table 1. *Cont.*

Ref.	ST	I	DE	F	DF	TT	E	T	TV	Set	Mt	Pd	Pr	RD	DD	S	Obj
[75]	PM									✓							M
[76]	JS																M
[77]	FS						✓	✓							✓		MET
[78]	JS			✓		✓										✓	M
[79]	FS			✓				✓							✓		EC/T
[80]	FS	✓										✓			✓	✓	MEP
[81]	JS			✓			✓	✓							✓		M/MeanET
[82]	JS		✓														M
[83]	JS		✓	✓													M/Stab
[84]	JS	✓	✓	✓													Stab/M/TFT/LB
[85]	JS			✓													M
[86]	SM						✓	✓							✓		TET
[87]	FS		✓	✓													M/EC
[88]	JS		✓	✓													M/LB
[89]	JS						✓	✓						✓	✓		M/TET
[90]	PM																M
[91]	JS			✓	✓				✓								M/LB/MFT
[26]	SM								✓								M
[31]	JS			✓	✓												M
[92]	JS	✓											✓				M
[55]	JS	✓															M/EC
[93]	FS			✓					✓								M
[94]	JS	✓		✓					✓								M/LB
[95]	JS			✓													M
[96]	JS			✓													M
[97]	FS		✓						✓						✓	✓	M
[8]	FS						✓	✓							✓		M
[98]	JS			✓	✓												EC/MC
[99]	JS			✓			✓	✓							✓		M/MeanET
[100]	JS	✓		✓			✓	✓							✓		MET/PDC
[101]	FS		✓	✓			✓	✓							✓		T
[37]	JS		✓					✓							✓		M/T
[102]	FS											✓					M
[42]	PM	✓					✓		✓						✓		T
[32]	PM			✓													LB

As [23] noticed in their research survey on AI strategies for resolving job shop scheduling problems, only a small percentage (8.06%) of the studied researchers published articles focused on solving real-life industrial problems, between 1997 and 2012. Additionally, [103] stated in their research survey study (focused on articles between 1990 and 2014) that most work was focused on testing the developed algorithms on benchmark instances, and just a fraction of the research has been applied to practical problem solutions as compared to pure research.

The data presented in Table 1 were processed using Microsoft Excel, taking advantage of the calculation and graphic tools features available.

From the 65 articles analyzed during this work (Table 1), it was possible to observe that only nearly 10% tried to solve the scheduling problem in real industrial scenarios. The focus on algorithm development is of huge importance and can contribute greatly to solve real problems. However, real scenarios have an entire set of conditions and circumstances that are not considered when algorithms are developed in the laboratory.

The current problem in obtaining feasible solutions for smart manufacturing scenarios is not related to technology by itself [104], as technology has evolved a lot during recent decades, it is related to managing all the actors and the connections between them, and the use of them to improve the industry. Thus, more effort should be dedicated to solving industry-oriented problems.

In addition to this, regarding the objectives, most researchers considered, i.e., tried to solve, single-objective problems, while some of them tried to optimize at least two or three objectives, as shown in Figure 2. In the case of Figure 2 a counting was performed for each number of objectives used in each article divided by the total number of articles. For example, when analyzing which articles considered two objectives, it is only necessary to count the number of articles with two objectives (16) and divide by the total number of articles (65), which is approximately 25%. For Figures 3 and 4 a similar approach was adopted.

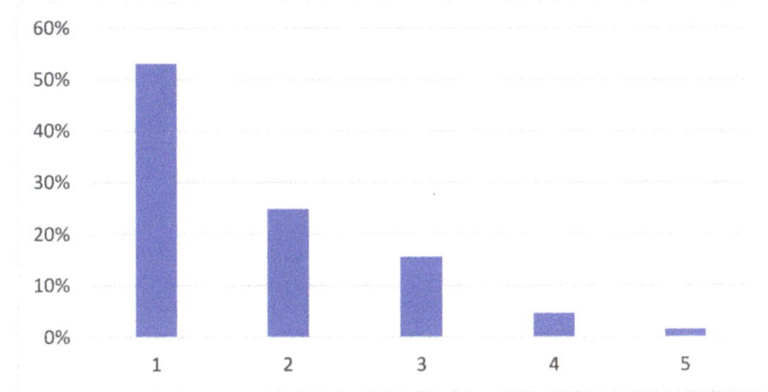

Figure 2. Number of objectives considered per study.

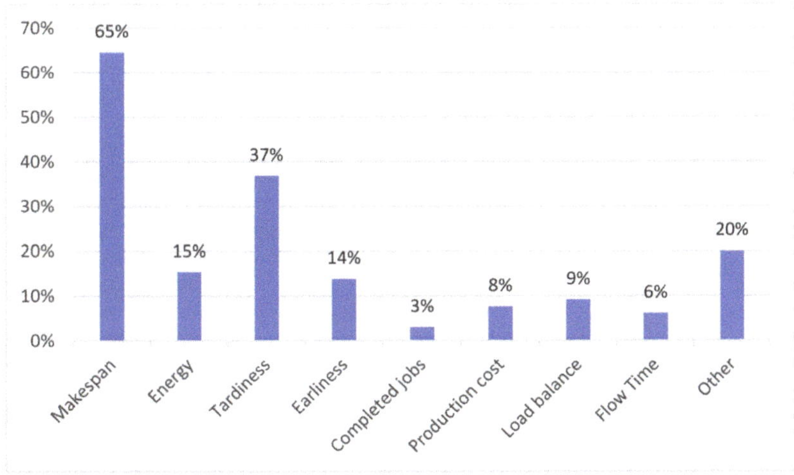

Figure 3. Percentage of articles considering each objective.

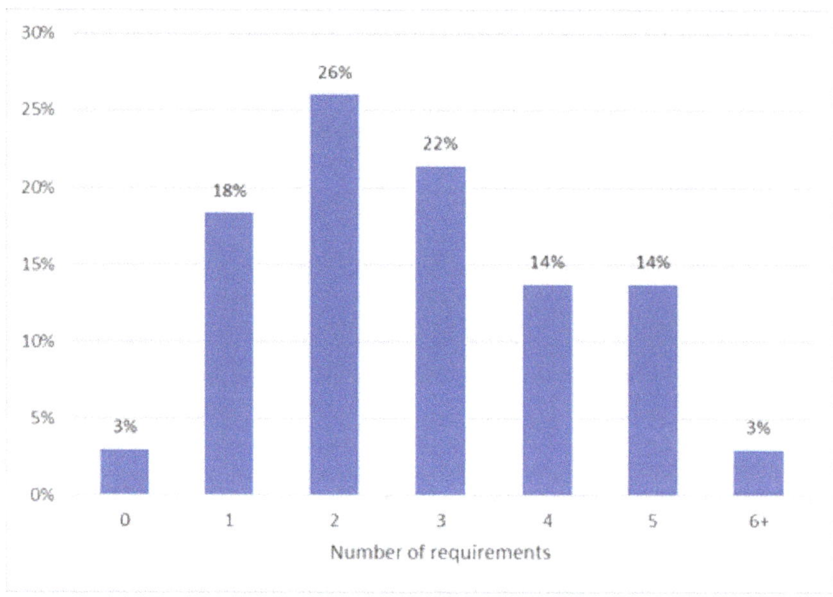

Figure 4. Number of requirements considered per study.

Within these objectives, most researchers, around 65%, focused on trying to optimize makespan, while more than a third try to deal with problems related to the tardiness of finished products, as it is possible to observe in Figure 3. The *Other* category is composed of features with weights of less than 2% each and so they were joined together. One outstanding point is that in the third position, 15% of the articles considered energy consumption issues, which makes clear that is a concern in the research community regarding environmental issues and sustainability. Nevertheless, although some authors use reference benchmarks for testing and comparison, it is not clear which techniques are superior to others for a specific problem, although they can be better than others in specific points. This leads to solutions that are good at beating benchmarks but are not able to be deployed in real manufacturing systems [103].

In addition to this, some problems arise when unrealistic assumptions are made since manufacturing environments are very dynamic and do not rely on static features. These assumptions can rapidly lead to unfeasible scheduling solutions which can be costly to the companies [27]. The most common assumptions are: all jobs and machines are available at starting time and release times are not considered; each machine can only execute one specific operation, which is not always true since some machines may have the flexibility to perform more than one operation, by changing tools, for example; the processing time of an operation is unchangeable, but processing times may change according to the conditions in a specific moment; machines never break and they are always available; setup times of any operation are sequence-independent and are included in the processing time, which is not always true and may compromise the entire schedule; pre-emption is not allowed; storage buffers are unlimited.

Aligned with this problem, the authors of [27] state that " . . . the intersection among three or more sets of constraints for any shop layout would return an empty set of references most of the time". This means that only a few studies contemplate more than three constraints. From Table 1, it is possible to observe that only about a third of the research studies considered more than three of the identified requirements in the same approach, as demonstrated in Figure 4.

It is also possible to observe that the most adopted requirements are flexibility, related to the ability of the shopfloor to adapt and adopt different features, and due dates, related with the date when each job shall be completed. After this comes the consideration of dynamic environments, where the arrival of new jobs, job cancellations or machine breakdowns during the execution process are considered and a reschedule may be necessary. Finally, the last consideration is tardiness, which evaluates the extent to which jobs are delayed from the initial deadline. All the other requirements are less expressive, being adopted in less than 20% of the articles.

This is a substantially small number considering all the requirements and constraints that can be found in manufacturing systems. Additionally, almost half of the studies considered two or three constraints. On the other side, two of the studied articles did not consider even one of the requirements mentioned in this study, as they only tried to optimize the objectives without any kind of limitation.

5. Scheduling in the Context of Smart Manufacturing and Next Steps

5.1. Gaps and Challenges

Sometimes it can look as researchers are facilitating when trying to narrow the bridge between academic studies and industry, as the published studies are often too simple, vague, and even have convenient restrictions and simplifications that do not reflect real industrial systems. This may be true. However, on the other hand, both computational and real-world complexity can difficult the problems solving. Additionally, most of the time companies are not willing to provide sensitive information that can be valuable for developing better manufacturing systems. These days, where data are becoming the core of industrial systems, it is crucial that data are provided to academia, so better solutions, not only for the shopfloor but for all value chains, may emerge. Nonetheless, academia should also make an effort to meet the industry's needs. Although there are plenty of scheduling studies in the literature, the research dealing with real-world problems is very uncommon. Likewise, most approaches do not allow solutions to be scalable and to be reused in some different scenarios.

Thus, the main challenges, and similarly the main gaps, in manufacturing scheduling research for smart manufacturing comes from the fact that most studies found in the literature, as mentioned before, are based on assumptions, related to the manufacturing environment, that are, commonly, not true and even naïve in real manufacturing systems. Furthermore, to cope with more realistic scenarios, more constraints and requirements need to be considered simultaneously to replicate as well as possible real manufacturing systems.

5.2. Human Factor

Based on the study presented, it is possible to identify that the objectives and requirements related to humans and the operators are not usually taken into account by manufacturing scheduling solutions. However, it is getting more and more usual for companies and society to explore how these new smart manufacturing solutions must deal with humans and how they can or cannot help humans within the industry. Hence, the authors believe that in the near future, it is necessary to explore new manufacturing scheduling approaches where these systems could use human-related aspects such as the available human resources or tiredness of the operators could be used to optimize the systems not only from the production or energy efficiency point of view but also taking into account these aspects to improve the integration of these systems and the operators, developing more harmonized workplaces.

The necessity to develop systems, including the human aspects, is critical for future research activities from the sustainability point of view. This aspect is one of the three pillars of sustainability. However, the addition of the human factors for manufacturing scheduling studies highly increases the studies' complexity since all humans are different and have different behaviors for the same situations. So, the authors believe that it is crucial to create interdisciplinary teams to explore this aspect because such systems' development will require analysis and understanding of elements that usually are far from the manufacturing scheduling experts' expertise. Simultaneously, this topic constitutes a relevant topic to be explored in the future due to the necessity for these new smart manufacturing-oriented solutions to be aware of the operators and their behaviors, taking that into account during the optimization process.

5.3. Opportunities Future Work

Nowadays, manufacturing systems are becoming enormous living robotic environments, and, in some cases, there is limited human presence. There are smart products, which constantly give feedback about their own status to the system, and smart resources, which can do the same and inform the system in real-time if some problem occurs in the production line. This intelligence within the system allows the different components to communicate and interact in order to reach common objectives. Nevertheless, on the other side, it also allows the extraction of more data and information from the system than ever before and transforms that information into useful knowledge. Knowledge can be used to improve multiple strands of the system, not only within the factory but also across the entire supply chain. This fact can immensely improve production scheduling systems, which are able to extract dynamically and in real-time more and more data from a vast variety of smart components, which can help to provide more robust and efficient scheduling solutions. However, in recent years there has been more and more interoperability between tools on shopfloors and existing legacy solutions are not able to account for this.

Although there is a huge complexity involved in developing smart manufacturing systems and there may be some restrictions regarding the process and memory power, nowadays the main limitations are not related to the hardware nor to the connection between entities since the emergence of cyber-physical systems has allowed all kind of entities on the shopfloor to be virtualized and connected together, thus permitting the easy execution of monitoring and controlling activities in the production line. One of the main problems is related to the linking between different types of data sources and how they need to cooperate to achieve better and more efficient performances. Consequently, it is necessary to go one step further and direct more efforts towards modeling, optimization, and standardization of manufacturing systems [105].

Therefore, to harmonize the designing and development of manufacturing systems, some reference architectures have emerged over recent years, such as ISA-95 [106], 5C architecture [107], Smart Grid Architecture Model—SGAM [108], Industrial Internet Reference Architecture—IIRA [109], or Reference Architectural Model for Industrie 4.0—RAMI4.0 [110]. These architectures aim to ensure a common comprehension, achieve standardization, enable semantic interoperability, and provide consistent operation models for the system. By adopting a reference architecture in manufacturing scheduling it will be possible to manage all the information coming from different sources in a consistent and homogeneous way and apply a core scheduling solution to different scenarios.

One of the possible ways to achieve this is to create an Asset Administration Shell (AAS) for scheduling and consequently for all other entities related to scheduling, under the RAMI4.0 architecture. An AAS consists of transforming physical components such as robots, machines, or devices and, similarly, intangible assets such as functions, plans, or an entire network into Industrie 4.0 components, which will then allow standardization as much as possible for solutions in engineering, operation, and management, and implementation of a heterogeneous communication structure in a smart manufacturing-oriented system. The AAS is the virtual representation of the asset that encompasses all the information and technical functionalities of the asset and manages communications with other Industrie 4.0 Components [110]. Through describing and modeling the asset components in smart manufacturing environments aligned with RAMI4.0 architecture, it will be possible to standardize and optimize the scheduling development process. In sequence, by clarifying which components should interact in the scheduling process and which data need to be available and flow within the system, scheduling designers and developers may be better prepared, and companies may benefit from the scheduling solutions created.

6. Conclusions

This work provides a literature review on smart manufacturing and, more specifically, on manufacturing scheduling. This is a very explored and discussed topic, which can hugely contribute to develop better manufacturing systems and improve the overall performance of those systems, regarding time, throughput, resources, or energetic optimization, among others. However, scheduling designing and development is not structured and harmonized between different entities, since it usually developed specifically for each case.

Consequently, the objective of this study is twofold. On one hand, the aim was to investigate which requirements and constraints are fundamental considerations when developing scheduling solutions for industrial scenarios. This was explored in subchapter 4.1 by analyzing some reference articles and books on the topic where requirements were used more commonly and their role in the scheduling process. On the other hand, this study analyzed several articles to identify the previous requirements they considered and which objectives they tried to optimize, which can render them robust approaches to be deployed in real manufacturing situations. The result of the last point came up as table containing all the 65 analyzed articles, pointing out which requirements they considered and the most common optimization objectives.

The results show that the vast majority of the articles only consider one or two objectives, mainly the makespan, which was found in around 65% of the articles, and tardiness related objectives, which were found in 37% of cases. It is also important to note that around 15% of the articles focused on solving energy consumption related issues, which shall be a major focus during these times of environmental awareness. Actually, environmental issues have been playing an increasingly important role in manufacturing scheduling, where studies are mainly making energy consumption assessment. However, in real manufacturing systems, depending on the case, it may be necessary to consider multiple optimization objectives. Furthermore, regarding the number of requirements considered in each study, it is possible to observe that most of the articles considered two or three requirements per case, while around 17% considered five or more constraints. It is important to understand that the more requirements considered, the more robust a scheduling solution can become, and then the more applicable it will be in real scenarios. As far as the authors know, this kind of study is not to be found in the literature.

In the analyzed literature, no study was found describing the components of manufacturing scheduling for reference architectures. Different approaches consider different constraints, and there is not a common and uniform way of developing scheduling solutions, even though there are common points in industrial systems that may be harmonized. Having a reference for scheduling designing and development can speed up the creation of scheduling solutions and make it easier to adapt these solutions to different scenarios. Knowing that scheduling can have a direct impact on production efficiency, sustainability, and costs of manufacturing systems, it is of huge interest to conduct in-depth research on how to model smart components to optimize scheduling approaches in smart manufacturing systems. Providing a structured model that includes relevant information about the inputs of the scheduling process and the desired outputs has great relevance for the design of scheduling solutions to cope with the smart manufacturing paradigm.

Funding: This work was supported in part by the FCT/MCTES project CESME—Collaborative & Evolvable Smart Manufacturing Ecosystem, funding PTDC/EEI-AUT/32410/2017.

Institutional Review Board Statement: Not applicable.

Informed Consent Statement: Not applicable.

Data Availability Statement: Data sharing not applicable.

Conflicts of Interest: The authors declare no conflict of interest.

References

1. Kagermann, H.; Wahlster, W.; Helbig, J. Recommendations for Implementing the Strategic Initiative INDUSTRIE 4.0—Securing the Future of German Manufacturing Industry. In *Final Report of the Industrie 4.0 Working Group*; Acatech—National Academy of Science and Engineering: Munich, Germany, 2013.
2. Li, Q.; Tang, Q.; Chan, I.; Wei, H.; Pu, Y.; Jiang, H.; Li, J.; Zhou, J. Smart manufacturing standardization: Architectures, reference models and standards framework. *Comput. Ind.* **2018**, *101*, 91–106. [CrossRef]
3. Wuest, T.; Weimer, D.; Irgens, C.; Thoben, K.-D. Machine learning in manufacturing: Advantages, challenges, and applications. *Prod. Manuf. Res.* **2016**, *4*, 23–45. [CrossRef]
4. Dionisio Rocha, A.; Peres, R.; Barata, J. An agent based monitoring architecture for plug and produce based manufacturing systems. In Proceedings of the 2015 IEEE 13th International Conference on Industrial Informatics (INDIN), Cambridge, UK, 22–24 July 2015; pp. 1318–1323. [CrossRef]
5. Zhang, J.; Ding, G.; Zou, Y.; Qin, S.; Fu, J. Review of job shop scheduling research and its new perspectives under Industry 4.0. *J. Intell. Manuf.* **2019**, *30*, 1809–1830. [CrossRef]
6. Stock, T.; Seliger, G. Opportunities of Sustainable Manufacturing in Industry 4.0. *Procedia CIRP* **2016**, *40*, 536–541. [CrossRef]
7. Rossit, D.A.; Tohmé, F.; Frutos, M. Industry 4.0: Smart Scheduling. *Int. J. Prod. Res.* **2019**, *57*, 3802–3813. [CrossRef]
8. Fu, Y.; Ding, J.; Wang, H.; Wang, J. Two-objective stochastic flow-shop scheduling with deteriorating and learning effect in Industry 4.0-based manufacturing system. *Appl. Soft Comput.* **2018**, *68*, 847–855. [CrossRef]
9. Ivanov, D.; Dolgui, A.; Sokolov, B.; Werner, F.; Ivanova, M. A dynamic model and an algorithm for short-term supply chain scheduling in the smart factory industry 4.0. *Int. J. Prod. Res.* **2016**, *54*, 386–402. [CrossRef]
10. Mourtzis, D.; Vlachou, E. A cloud-based cyber-physical system for adaptive shop-floor scheduling and condition-based maintenance. *J. Manuf. Syst.* **2018**, *47*, 179–198. [CrossRef]
11. Rossit, D.; Tohmé, F. Scheduling research contributions to Smart manufacturing. *Manuf. Lett.* **2018**, *15*, 111–114. [CrossRef]
12. Shiue, Y.-R.; Lee, K.-C.; Su, C.-T. Real-time scheduling for a smart factory using a reinforcement learning approach. *Comput. Ind. Eng.* **2018**. [CrossRef]
13. Zhang, Y.; Liu, S.; Liu, Y.; Yang, H.; Li, M.; Huisingh, D.; Wang, L. The 'Internet of Things' enabled real-time scheduling for remanufacturing of automobile engines. *J. Clean. Prod.* **2018**, *185*, 562–575. [CrossRef]
14. Zhong, R.Y.; Dai, Q.Y.; Qu, T.; Hu, G.J.; Huang, G.Q. RFID-enabled real-time manufacturing execution system for mass-customization production. *Robot. Comput. Integr. Manuf.* **2013**, *29*, 283–292. [CrossRef]
15. Lu, C.; Gao, L.; Li, X.; Pan, Q.; Wang, Q. Energy-efficient permutation flow shop scheduling problem using a hybrid multi-objective backtracking search algorithm. *J. Clean. Prod.* **2017**. [CrossRef]
16. Wang, J.; Tang, J.; Xue, G.; Yang, D. Towards Energy-efficient Task Scheduling on Smartphones in Mobile Crowd Sensing Systems. *Comput. Netw.* **2016**. [CrossRef]
17. Carosi, S.; Frangioni, A.; Galli, L.; Girardi, L.; Vallese, G. A Tool for Practical Integrated Time-Table Design and Vehicle Scheduling in Public Transport Systems. In *A View of Operations Research Applications in Italy*; Springer: Cham, Switzerland, 2019; pp. 207–217.
18. Ernst, A.T.; Jiang, H.; Krishnamoorthy, M.; Sier, D. Staff scheduling and rostering: A review of applications, methods and models. *Eur. J. Oper. Res.* **2004**, *153*, 3–27. [CrossRef]
19. Balin, S. Non-identical parallel machine scheduling using genetic algorithm. *Expert Syst. Appl.* **2011**, *38*, 6814–6821. [CrossRef]
20. Li, X.; Gao, L. An effective hybrid genetic algorithm and tabu search for flexible job shop scheduling problem. *Int. J. Prod. Econ.* **2016**, *174*, 93–110. [CrossRef]
21. Liu, Y.; Wang, L.; Wang, X.V.; Xu, X.; Zhang, L. Scheduling in cloud manufacturing: State-of-the-art and research challenges. *Int. J. Prod. Res.* **2018**, *7543*, 1–26. [CrossRef]
22. Nguyen, S.; Mei, Y.; Zhang, M. Genetic programming for production scheduling: A survey with a unified framework. *Complex Intell. Syst.* **2017**, *3*, 41–66. [CrossRef]
23. Çaliş, B.; Bulkan, S. A research survey: Review of AI solution strategies of job shop scheduling problem. *J. Intell. Manuf.* **2015**, *26*, 961–973. [CrossRef]
24. Gahm, C.; Denz, F.; Dirr, M.; Tuma, A. Energy-efficient scheduling in manufacturing companies: A review and research framework. *Eur. J. Oper. Res.* **2016**, *248*, 744–757. [CrossRef]
25. Yoo, J.; Lee, I.S. Parallel machine scheduling with maintenance activities. *Comput. Ind. Eng.* **2016**, *101*, 361–371. [CrossRef]
26. Chung, B.D.; Kim, B.S. A hybrid genetic algorithm with two-stage dispatching heuristic for a machine scheduling problem with step-deteriorating jobs and rate-modifying activities. *Comput. Ind. Eng.* **2016**, *98*, 113–124. [CrossRef]
27. Framinan, J.M.; Leisten, R.; Ruiz García, R. *Manufacturing Scheduling Systems*; Springer: London, UK, 2014; ISBN 978-1-4471-6271-1.
28. Pinedo, M.L. *Scheduling Theory, Algorithms, and Systems*; Springer International Publishing: Cham, Switzerland, 2016; ISBN 978-3-319-26578-0.
29. Mattfeld, D.C. Evolutionary search and the job shop: Investigations on genetic algorithms for production scheduling. In *Production and Logistic*; Springer: Hedelberg/Berlin, Germany, 1996. [CrossRef]
30. Werner, F. Genetic algorithms for shop scheduling problems: A survey. In *Heuristics: Theory and Applications*; Nova Science Publishers: Hauppauge, NY, USA, 2011; Volume 11, pp. 1–66. Available online: http://www.math.uni-magdeburg.de/~{}werner/preprints/p11-31.pdf (accessed on 28 September 2020).

31. Karimi, S.; Ardalan, Z.; Naderi, B.; Mohammadi, M. Scheduling flexible job-shops with transportation times: Mathematical models and a hybrid imperialist competitive algorithm. *Appl. Math. Model.* **2016**, *41*, 667–682. [CrossRef]
32. Ouazene, Y.; Yalaoui, F.; Chehade, H.; Yalaoui, A. Workload balancing in identical parallel machine scheduling using a mathematical programming method. *Int. J. Comput. Intell. Syst.* **2014**, *7*, 58–67. [CrossRef]
33. Ouazene, Y.; Hnaien, F.; Yalaoui, F.; Amodeo, L. The Joint Load Balancing and Parallel Machine Scheduling Problem. In *Operations Research Proceedings*; Springer: Berlin/Heidelberg, Germany, 2011; pp. 497–502.
34. Lin, L.; Hao, X.C.; Gen, M.; Jo, J.B. Network modeling and evolutionary optimization for scheduling in manufacturing. *J. Intell. Manuf.* **2012**, *23*, 2237–2253. [CrossRef]
35. Tranfield, D.; Denyer, D.; Smart, P. Towards a Methodology for Developing Evidence-Informed Management Knowledge by Means of Systematic Review. *Br. J. Manag.* **2003**, *14*, 207–222. [CrossRef]
36. Iwamura, K.; Sugimura, N. A study on real-time scheduling for autonomous distributed manufacturing systems. In Proceedings of the IEEE International Conference on Systems, Man and Cybernetics, Istanbul, Turkey, 10–13 October 2010; pp. 1352–1357. [CrossRef]
37. Tran, L.V.; Huynh, B.H.; Akhtar, H. Ant Colony Optimization Algorithm for Maintenance, Repair and Overhaul Scheduling Optimization in the Context of Industrie 4.0. *Appl. Sci.* **2019**, *9*, 4815. [CrossRef]
38. Jin, Y.; Branke, J. Evolutionary Optimization in Uncertain Environments—A Survey. *IEEE Trans. Evol. Comput.* **2005**, *9*, 303–317. [CrossRef]
39. Lei, H.; Xing, K.; Han, L.; Gao, Z. Hybrid heuristic search approach for deadlock-free scheduling of flexible manufacturing systems using Petri nets. *Appl. Soft Comput.* **2017**, *18*, 240–245. [CrossRef]
40. Lu, P.-H.; Wu, M.-C.; Tan, H.; Peng, Y.-H.; Chen, C.-F. A genetic algorithm embedded with a concise chromosome representation for distributed and flexible job-shop scheduling problems. *J. Intell. Manuf.* **2018**, *29*, 19–34. [CrossRef]
41. Zarook, Y.; Abedi, M. JIT-scheduling in unrelated parallel-machine environment with aging effect and multi-maintenance activities. *Int. J. Serv. Oper. Manag.* **2014**, *18*, 99. [CrossRef]
42. Klement, N.; Abdeljaouad, M.A.; Porto, L.; Silva, C. Lot-Sizing and Scheduling for the Plastic Injection Molding Industry—A Hybrid Optimization Approach. *Appl. Sci.* **2021**, *11*, 1202. [CrossRef]
43. Rivera-Gómez, H.; Montaño-Arango, O.; Corona-Armenta, J.; Garnica-González, J.; Ortega-Reyes, A.; Anaya-Fuentes, G. JIT Production Strategy and Maintenance for Quality Deteriorating Systems. *Appl. Sci.* **2019**, *9*, 1180. [CrossRef]
44. Kaplanoğlu, V. Multi-agent based approach for single machine scheduling with sequence-dependent setup times and machine maintenance. *Appl. Soft Comput. J.* **2014**, *23*, 165–179. [CrossRef]
45. Liu, Q.; Dong, M.; Chen, F.F. Single-machine-based joint optimization of predictive maintenance planning and production scheduling. *Robot. Comput. Integr. Manuf.* **2018**, *51*, 238–247. [CrossRef]
46. Holgado, M.; Macchi, M.; Evans, S. Exploring the impacts and contributions of maintenance function for sustainable manufacturing. *Int. J. Prod. Res.* **2020**, *58*, 7292–7310. [CrossRef]
47. Xiong, H.; Fan, H.; Jiang, G.; Li, G. A simulation-based study of dispatching rules in a dynamic job shop scheduling problem with batch release and extended technical precedence constraints. *Eur. J. Oper. Res.* **2017**, *257*, 13–24. [CrossRef]
48. Kuhpfahl, J.; Bierwirth, C. A study on local search neighborhoods for the job shop scheduling problem with total weighted tardiness objective. *Comput. Oper. Res.* **2016**, *66*, 44–57. [CrossRef]
49. Yazdani, M.; Aleti, A.; Khalili, S.M.; Jolai, F. Optimizing the sum of maximum earliness and tardiness of the job shop scheduling problem. *Comput. Ind. Eng.* **2017**, *107*, 12–24. [CrossRef]
50. Homayouni, S.M.; Fontes, D.B.M.M. Production and transport scheduling in flexible job shop manufacturing systems. *J. Glob. Optim.* **2021**. [CrossRef]
51. Chang, H.-C.; Liu, T.-K. Optimisation of distributed manufacturing flexible job shop scheduling by using hybrid genetic algorithms. *J. Intell. Manuf.* **2017**, *28*, 1973–1986. [CrossRef]
52. Alaouchiche, Y.; Ouazene, Y.; Yalaoui, F. Economic and Energetic Performance Evaluation of Unreliable Production Lines: An Integrated Analytical Approach. *IEEE Access* **2020**, *8*, 185330–185345. [CrossRef]
53. Tuo, J.; Liu, F.; Liu, P. Key performance indicators for assessing inherent energy performance of machine tools in industries. *Int. J. Prod. Res.* **2019**, *57*, 1811–1824. [CrossRef]
54. Helu, M.; Libes, D.; Lubell, J.; Lyons, K.; Morris, K.C. Enabling Smart Manufacturing Technologies for Decision-Making Support. In Proceedings of the ASME 2016 International Design Engineering Technical Conferences and Computers and Information in Engineering Conference, Charlotte, NC, USA, 21–24 August 2016; p. V01BT02A035. [CrossRef]
55. Salido, M.A.; Escamilla, J.; Barber, F.; Giret, A. Rescheduling in job-shop problems for sustainable manufacturing systems. *J. Clean. Prod.* **2017**, *162*, S121–S132. [CrossRef]
56. Lee, W.C.; Wang, J.Y.; Lin, M.C. A branch-and-bound algorithm for minimizing the total weighted completion time on parallel identical machines with two competing agents. *Knowl. Based Syst.* **2016**, *105*, 68–82. [CrossRef]
57. Chou, Y.C.; Cao, H.; Cheng, H.H. A bio-inspired mobile agent-based integrated system for flexible autonomic job shop scheduling. *J. Manuf. Syst.* **2013**, *32*, 752–763. [CrossRef]
58. Bürgy, R.; Bülbül, K. The job shop scheduling problem with convex costs. *Eur. J. Oper. Res.* **2018**, *268*, 82–100. [CrossRef]
59. Shen, L.; Dauzère-Pérès, S.; Neufeld, J.S. Solving the flexible job shop scheduling problem with sequence-dependent setup times. *Eur. J. Oper. Res.* **2018**, *265*, 503–516. [CrossRef]

60. Shahrabi, J.; Adibi, M.A.; Mahootchi, M. A reinforcement learning approach to parameter estimation in dynamic job shop scheduling. *Comput. Ind. Eng.* **2017**, *110*, 75–82. [CrossRef]
61. Zeng, Z.; Hong, M.; Man, Y.; Li, J.; Zhang, Y.; Liu, H. Multi-object optimization of flexible flow shop scheduling with batch process—Consideration total electricity consumption and material wastage. *J. Clean. Prod.* **2018**, *183*, 925–939. [CrossRef]
62. Mou, J.; Li, X.; Gao, L.; Yi, W. An effective L-MONG algorithm for solving multi-objective flow-shop inverse scheduling problems. *J. Intell. Manuf.* **2018**, *29*, 789–807. [CrossRef]
63. Alotaibi, A.; Lohse, N.; Vu, T.M. Dynamic Agent-based Bi-objective Robustness for Tardiness and Energy in a Dynamic Flexible Job Shop. *Procedia CIRP* **2016**, *57*, 728–733. [CrossRef]
64. Zhang, L.; Gao, L.; Li, X. A hybrid genetic algorithm and tabu search for a multi-objective dynamic job shop scheduling problem. *Int. J. Prod. Res.* **2013**, *51*, 3516–3531. [CrossRef]
65. Sobeyko, O.; Mönch, L. Heuristic approaches for scheduling jobs in large-scale flexible job shops. *Comput. Oper. Res.* **2016**, *68*, 97–109. [CrossRef]
66. Ladj, A.; Varnier, C.; Tayeb, F.B.S. IPro-GA: An integrated prognostic based GA for scheduling jobs and predictive maintenance in a single multifunctional machine. *IFAC-PapersOnLine* **2016**, *49*, 1821–1826. [CrossRef]
67. Liu, Y.; Wang, L.; Wang, Y.; Wang, X.V.; Zhang, L. Multi-agent-based scheduling in cloud manufacturing with dynamic task arrivals. *Procedia CIRP* **2018**, *72*, 953–960. [CrossRef]
68. Nikolakis, N.; Kousi, N.; Michalos, G.; Makris, S. Dynamic scheduling of shared human-robot manufacturing operations. *Procedia CIRP* **2018**, *72*, 9–14. [CrossRef]
69. Qu, S.; Wang, J.; Govil, S.; Leckie, J.O. Optimized Adaptive Scheduling of a Manufacturing Process System with Multi-skill Workforce and Multiple Machine Types: An Ontology-based, Multi-agent Reinforcement Learning Approach. *Procedia CIRP* **2016**, *57*, 55–60. [CrossRef]
70. Freitag, M.; Hildebrandt, T. Automatic design of scheduling rules for complex manufacturing systems by multi-objective simulation-based optimization. *CIRP Ann. Manuf. Technol.* **2016**, *65*, 433–436. [CrossRef]
71. Nie, L.; Gao, L.; Li, P.; Li, X. A GEP-based reactive scheduling policies constructing approach for dynamic flexible job shop scheduling problem with job release dates. *J. Intell. Manuf.* **2013**, *24*, 763–774. [CrossRef]
72. Mokhtari, H.; Hasani, A. An energy-efficient multi-objective optimization for flexible job-shop scheduling problem. *Comput. Chem. Eng.* **2017**, *104*, 339–352. [CrossRef]
73. Silva, C.; Klement, N.; Gibaru, O. A Generic Decision Support Tool for Lot-Sizing and Scheduling Problems with Setup and Due Dates. In *Closing the Gap Between Practice and Research in Industrial Engineering*; Springer: Cham, Switzerland, 2018; pp. 131–138. [CrossRef]
74. Nguyen, S.; Zhang, M.; Johnston, M.; Tan, K.C. Automatic design of scheduling policies for dynamic multi-objective job shop scheduling via cooperative coevolution genetic programming. *IEEE Trans. Evol. Comput.* **2014**, *18*, 193–208. [CrossRef]
75. Afzalirad, M.; Shafipour, M. Design of an efficient genetic algorithm for resource-constrained unrelated parallel machine scheduling problem with machine eligibility restrictions. *J. Intell. Manuf.* **2015**, *29*, 423–437. [CrossRef]
76. Asadzadeh, L. A local search genetic algorithm for the job shop scheduling problem with intelligent agents. *Comput. Ind. Eng.* **2015**, *85*, 376–383. [CrossRef]
77. Mokhtari, H.; Noroozi, A. An efficient chaotic based PSO for earliness/tardiness optimization in a batch processing flow shop scheduling problem. *J. Intell. Manuf.* **2018**, *29*, 1063–1081. [CrossRef]
78. Han, L.; Xing, K.; Chen, X.; Xiong, F. A Petri net-based particle swarm optimization approach for scheduling deadlock-prone flexible manufacturing systems. *J. Intell. Manuf.* **2015**, *29*, 1083–1096. [CrossRef]
79. Lei, D.; Gao, L.; Zheng, Y. A novel teaching-learning-based optimization algorithm for energy-efficient scheduling in hybrid flow shop. *IEEE Trans. Eng. Manag.* **2018**, *65*, 330–340. [CrossRef]
80. Azami, A.; Demirli, K.; Bhuiyan, N. Scheduling in aerospace composite manufacturing systems: A two-stage hybrid flow shop problem. *Int. J. Adv. Manuf. Technol.* **2018**, *95*, 3259–3274. [CrossRef]
81. Gao, K.Z.; Suganthan, P.N.; Pan, Q.K.; Chua, T.J.; Cai, T.X.; Chong, C.S. Discrete harmony search algorithm for flexible job shop scheduling problem with multiple objectives. *J. Intell. Manuf.* **2016**, *27*, 363–374. [CrossRef]
82. Jamili, A. Robust job shop scheduling problem: Mathematical models, exact and heuristic algorithms. *Expert Syst. Appl.* **2016**, *55*, 341–350. [CrossRef]
83. Ahmadi, E.; Zandieh, M.; Farrokh, M.; Emami, S.M. A multi objective optimization approach for flexible job shop scheduling problem under random machine breakdown by evolutionary algorithms. *Comput. Oper. Res.* **2016**, *73*, 56–66. [CrossRef]
84. Gao, K.; Yang, F.; Zhou, M.; Pan, Q.; Suganthan, P.N. Flexible Job-Shop Rescheduling for New Job Insertion by Using Discrete Jaya Algorithm. *IEEE Trans. Cybern.* **2019**, *49*, 1944–1955. [CrossRef]
85. Marzouki, B.; Belkahla Driss, O.; Ghédira, K. Multi Agent model based on Chemical Reaction Optimization with Greedy algorithm for Flexible Job shop Scheduling Problem. *Procedia Comput. Sci.* **2017**, *112*, 81–90. [CrossRef]
86. Parsa, N.R.; Karimi, B.; Husseini, S.M.M. Exact and heuristic algorithms for the just-in-time scheduling problem in a batch processing system. *Comput. Oper. Res.* **2017**, *80*, 173–183. [CrossRef]
87. Tang, D.; Dai, M.; Salido, M.A.; Giret, A. Energy-efficient dynamic scheduling for a flexible flow shop using an improved particle swarm optimization. *Comput. Ind.* **2016**, *81*, 82–95. [CrossRef]

88. Gao, K.Z.; Suganthan, P.N.; Pan, Q.K.; Tasgetiren, M.F.; Sadollah, A. Artificial bee colony algorithm for scheduling and rescheduling fuzzy flexible job shop problem with new job insertion. *Knowl. Based Syst.* **2016**, *109*, 1–16. [CrossRef]
89. Zhao, F.; Tang, J.; Wang, J.; Jonrinaldi. An improved particle swarm optimization with decline disturbance index (DDPSO) for multi-objective job-shop scheduling problem. *Comput. Oper. Res.* **2014**, *45*, 38–50. [CrossRef]
90. Costa, A.; Cappadonna, F.A.; Fichera, S. Minimizing the total completion time on a parallel machine system with tool changes. *Comput. Ind. Eng.* **2016**, *91*, 290–301. [CrossRef]
91. Petrović, M.; Vuković, N.; Mitić, M.; Miljković, Z. Integration of process planning and scheduling using chaotic particle swarm optimization algorithm. *Expert Syst. Appl.* **2016**, *64*, 569–588. [CrossRef]
92. Kundakcı, N.; Kulak, O. Hybrid genetic algorithms for minimizing makespan in dynamic job shop scheduling problem. *Comput. Ind. Eng.* **2016**, *96*, 31–51. [CrossRef]
93. Jung, S.; Woo, Y.B.; Kim, B.S. Two-stage assembly scheduling problem for processing products with dynamic component-sizes and a setup time. *Comput. Ind. Eng.* **2017**, *104*, 98–113. [CrossRef]
94. Gao, K.Z.; Suganthan, P.N.; Pan, Q.K.; Chua, T.J.; Chong, C.S.; Cai, T.X. An improved artificial bee colony algorithm for flexible job-shop scheduling problem with fuzzy processing time. *Expert Syst. Appl.* **2016**, *65*, 52–67. [CrossRef]
95. Wang, L.; Cai, J.; Li, M.; Liu, Z. Flexible Job Shop Scheduling Problem Using an Improved Ant Colony Optimization. *Sci. Program.* **2017**, *2017*. [CrossRef]
96. Xiong, W.; Fu, D. A new immune multi-agent system for the flexible job shop scheduling problem. *J. Intell. Manuf.* **2015**, *29*, 857–873. [CrossRef]
97. Helo, P.; Phuong, D.; Hao, Y. Cloud manufacturing—Scheduling as a service for sheet metal manufacturing. *Comput. Oper. Res.* **2018**, *110*, 1–12. [CrossRef]
98. Barak, S.; Moghdani, R.; Maghsoudlou, H. Energy-efficient multi-objective flexible manufacturing scheduling. *J. Clean. Prod.* **2020**, *283*, 124610. [CrossRef]
99. Zan, X.; Wu, Z.; Guo, C.; Yu, Z. A Pareto-based genetic algorithm for multi-objective scheduling of automated manufacturing systems. *Adv. Mech. Eng.* **2020**, *12*, 1–15. [CrossRef]
100. Mohammadi, S.; Al-e-Hashem, S.M.J.M.; Rekik, Y. An integrated production scheduling and delivery route planning with multi-purpose machines: A case study from a furniture manufacturing company. *Int. J. Prod. Econ.* **2020**, *219*, 347–359. [CrossRef]
101. Baxendale, M.; McGree, J.M.; Bellette, A.; Corry, P. Machine-based production scheduling for rotomoulded plastics manufacturing. *Int. J. Prod. Res.* **2020**, 1–18. [CrossRef]
102. Guo, J.; Shi, Y.; Chen, Z.; Yu, T.; Shirinzadeh, B.; Zhao, P. Improved SP-MCTS-Based Scheduling for Multi-Constraint Hybrid Flow Shop. *Appl. Sci.* **2020**, *10*, 6220. [CrossRef]
103. Chaudhry, I.A.; Khan, A.A. A research survey: Review of flexible job shop scheduling techniques. *Int. Trans. Oper. Res.* **2016**, *23*, 551–591. [CrossRef]
104. Ribeiro, L.; Member, S.; Bj, M. Transitioning From Standard Automation Solutions to Cyber-Physical Production Systems: An Assessment of Critical Conceptual and Technical Challenges. *IEEE Syst. J.* **2017**, *12*, 3816–3827. [CrossRef]
105. Kusiak, A. Smart manufacturing. *Int. J. Prod. Res.* **2018**, *56*, 508–517. [CrossRef]
106. International Society of Automation ISA 95. Available online: https://isa-95.com/ (accessed on 15 June 2020).
107. Lee, J.; Bagheri, B.; Kao, H.A. A Cyber-Physical Systems architecture for Industry 4.0-based manufacturing systems. *Manuf. Lett.* **2015**, *3*, 18–23. [CrossRef]
108. CEN, CENELEC, and ETSI. "Smart Grid Reference Architecture". 2012. Available online: ftp://ftp.cen.eu/EN/EuropeanStandardization/HotTopics/SmartGrids/Security.pdf (accessed on 18 June 2020).
109. Shi-Wan, L.; Bradford, M.; Jacques, D.; Graham, B.; Chigani, A.; Martin, R.; Murphy, B.; Crawford, M. *The Industrial Internet of Things Volume G1: Reference Architecture*; Industrial Internet Consortium White Paper; Version 1; Industrial Internet Consortium: Milford, MA, USA, 2017; p. 58.
110. ZVEI. "Communication in the Context of Industrie 4.0". 2019. Available online: https://www.zvei.org/fileadmin/user_upload/Presse_und_Medien/Publikationen/2019/Maerz/Communication_in_the_Context_of_Industrie_4.0/ZVEI_WP_Kommunikation_Industrie-4.0-Umfeld_ENGLISCH.pdf (accessed on 12 May 2020).

applied sciences

Article

Randomized and Generated Instances Fitting with the Home Health Care Problem Subjected to Certain Constraints

Colin Huvent *, Caroline Gagné * and Aymen Sioud *

Département d'Informatique et de Mathématique, Université du Québec à Chicoutimi, 555 Boulevard de Université, Chicoutimi, QC G7H 2B1, Canada
* Correspondence: colin.huvent1@uqac.ca (C.H.); caroline.gagne@uqac.ca (C.G.); aymen.sioud@uqac.ca (A.S.)

Abstract: Home Health Care (HHC) is a worldwide issue. It focuses on how medical and social organizations of different countries handle providing patients with health support at home. In most developed countries, reducing hospital cost constitutes a main objective. It is important to research the improvement of HHC logistics. This paper addressed the generation and development of a benchmark properly fitting different constraints of the HCC problem. Consequently, a generator was proposed dealing with all kinds of constraints such as time window constraints, workload constraints, synchronization, and precedence constraints. This generator allows researchers to validate and compare solving methods on a common dataset regardless of confidentiality issues. We validated our generator by firstly creating a common benchmark available for researchers and secondly by proposing a set of instances and a solving method based on an HHC problem found in the literature.

Keywords: smart health care systems; planning; scheduling; logistic systems; benchmark

Citation: Huvent, C.; Gagné, C.; Sioud, A. Randomized and Generated Instances Fitting with the Home Health Care Problem Subjected to Certain Constraints. *Appl. Sci.* **2021**, *11*, 3346. https://doi.org/10.3390/app11083346

Academic Editor: Farouk Yalaoui

Received: 26 February 2021
Accepted: 3 April 2021
Published: 8 April 2021

Publisher's Note: MDPI stays neutral with regard to jurisdictional claims in published maps and institutional affiliations.

1. Introduction

In most countries, the population of older adults is increasing. As people age, their health needs become more chronic and complex [1], which results in a rising demand for health care [2]. According to the Organisation for Economic Co-operation and Development (OECD), between 2010 and 2018, all developed countries registered an increase of 3% in their older population [3]. Furthermore, for most of those countries, a decrease in formal care leads to a substantial increase in demand for Home Health Care (HHC) services. In recent years, the Health Care (HC) industry has become one of the largest sectors of the economy in developed countries. In Europe, between 1% and 5% of the total public health budget is spent on HHC services [4].

In 2018, one-point-four million Canadians needed home care services [5] (4.1% of the total population). HHC services have emerged as a viable solution to meet the challenges of the aging population problem. HHC's aim is to improve patients' living conditions by providing them with medical, paramedical, and social services in their home. From an economic point of view, HHC decreases hospital congestion by avoiding or reducing hospitalizations, leading to both significant increases in quality of life for patients and relevant cost savings in the HC system [6,7].

Building a proper HHC plan is a complex task regarding the different constraints. Currently, operational planning is mostly done manually and often by an experienced senior caregiver [5,8–10]. According to the surveys of the HHC companies, each day, an HHC company carries out various logistics activities including the delivery of drugs or medical services from caregivers to patients and medical samples to the laboratory. Generally speaking, those activities can be considered as a Vehicle Routing Problem (VRP) or as a scheduling problem [11]. In both views, multiple constraints must be considered like workload constraints, interdependencies among services, etc. Interdependent services can be defined as having temporal dependencies between two services, or as synchronized

Appl. Sci. **2021**, *11*, 3346. https://doi.org/10.3390/app11083346　　

services [12] (also called shared services), or as services with priority constraints. Thus, it is possible for a single patient to need a pair of services that must be separated from a given time interval.

The HHC problem can be described regarding a short-, medium-, or long-term horizon, ranging from one day to a couple of months. Daily problems have received the most attention in the literature [13,14]. Usually, medium-term problems consider a planning horizon of a week. Another observation is that all constraints mentioned above are rarely met in the same problem. A set of constraints is chosen to reflect the scope of how the problem is treated. According to the review of Fikar and Hirsch [13], short-term problems usually consider time windows and qualification requirements, whereas medium-term problems deal with working time regulations and continuity of care considerations. However, interdependent and synchronized services are weakly discussed in the HHC literature. Finally, to establish an exhaustive HHC problem, many other constraints and aspects should be taken into account such as geographical constraints or legal working considerations.

The problem holds many constraints and parameters depending on the way it is handled. Due to medical confidentiality, researchers lack public benchmarks to validate and compare their models. Indeed, in the current literature, many studies cannot publish their datasets due to medical confidentiality. Some studies offer homemade datasets, but they are often developed to fit into the scope of the HHCRSP. These homemade datasets can rarely be used by other researchers. This article proposes a generator able to establish HHC models fitting most of the constraints and features linked to the HHC literature. The goal of this generator is to produce a large benchmark of the HHCRSP with different ways to tackle the HHCRSP by choosing a set of constraints regarding how researchers want to test and solve their HHCRSP. The organization of this article is as follows. Section 2 reviews related literature. Section 3 presents the HHC problem by defining the key aspects. The generator and its engine are detailed in Section 4. Section 5 shows an application of the generator by showing a solving method based on a skip descent algorithm. Finally, conclusions and some perspectives are provided in Section 6.

2. Literature Review

HHC-related models started with Begur et al. [15] in 1997. Currently, HHC problems are an extension of the VRP augmented by many side constraints that are specific to the HHC context linked to a scheduling problem. This new problem's formulation is called the Home Health Care Routing and Scheduling Problem (HHCRSP). To establish an HHCRSP, an extended definition of HHC is used, which includes various health-related parameters.

According to Lanzarone et al. [16], HHC deals with various challenges such as partitioning a territory, dimensioning human resources, and assigning and scheduling caregivers to patients or tasks. Mostly, the objective of a scheduling problem is to find an ideal pairing of caregivers and patients, whereas the routing problem aims to design caregivers' tours while minimizing the total travel time. The HHCRSP is an NP-hard problem [17], and due to this, it has received the most attention. The proposed models are evaluated on homemade datasets or derived from protected data [13]. Knowing that different sets of constraints and objectives can be considered, it is impossible to make a comparison between different solving methods. Indeed, in HHC, there is no commonly accepted benchmark due to medical confidentiality.

Bertels et al. [17] proposed an HHCRSP model formulated by taking into account workload balance for caregivers and introducing time windows. A time window represents a time interval in which something has to be started or can describe the working time or availability of someone. In 2013, Lui et al. [18] formulated an HHCRP with time window constraints. They also proposed a model with simultaneous delivery and pickup of drugs and biological samples. Mankowska et al. [12] proposed an HHCRSP as being multi-period and multi-service. The model takes into account the personal qualifications of the staff, possible interdependencies among services, and the individual requirements of the patients. Hiermann et al. [9] proposed a multimodal HHCRSP from an Austrian HHC provider based on multiple modes of transportation. Its aim was to determine efficient multimodal tours while considering staff and patient satisfaction. Liu et al. [19] studied scheduling and routing problems related to HHC with the consideration of lunch break requirements and working legal requirements. Frifita et al. [20] presented a model based on preferred time windows for patient and synchronization constraints. Among all these models, no comparison has so far been presented in the literature.

Each HHCRSP is distinguished regarding the planning horizon [14] and its composition in terms of strength (number of caregivers, patients) [21]. Some HC providers pursue continuity of care. Patients are assigned to only one caregiver named the reference caregiver, who follows the entire patient care pathway. A reference caregiver provides all the visits that are pertinent [22]. On another note, some HHCRSP definitions include patients' and caregivers' preferences [23,24], a minimal and maximal amount of working time for a caregiver, and time windows for patients. In the HHCRSP, some implicit parameters are not mentioned in the model such as the share of time when a caregiver is on the road or the capacity of the HC provider. The current literature suggests a rate between 15% and 20% for the share of traveling time [13]. The capacity of an HC provider is related to some constraints such as working legal requirements or the option for lunch breaks [10,13].

3. Home Health Care Problem Definition

HHC services, both at the international level and in Quebec, need tools to automatically and efficiently schedule all types of care. Finding a perfect schedule is a tough task due to the complexity and the repeating aspect of an HHC problem. HHC planning on a daily or a weekly horizon aims to assign and schedule all jobs or services required by patients to the proper caregivers.

In an HHC problem, caregivers are deeply involved. Indeed, identifying which caregiver may look after a new patient involves some considerations: the current workload of the caregiver, his/her qualifications, and geographic closeness. Interactions between the patient and the caregivers and temporal constraints such as care that needs to be performed at a specific time have to be taken into account.

An HHC problem can also be described as a problem in which patients have to be visited once or several times a week by caregivers. Cares services are delivered in a defined time during a visit. A large set of constraints has to be considered and satisfied as qualifications for each caregiver, working hours, geographical consistency, time windows, resting time and lunch breaks, working hours, workload, and precedence or synchronization constraints. Furthermore, compatibility constraints can be introduced. Compatibility takes wanted days, continuity of care, languages, and personal experience between a patient and a caregiver into account.

3.1. Jobs

Within this model, a patient has a list of needed jobs. The jobs $\mathcal{J}_\mathcal{P} = \{1, \ldots, J\}$ are care services that must be performed at patients' homes and have different natures: providing medication, performing an injection, changing a bandage, etc. A single patient may relate to one or more jobs. However, job $j \in \mathcal{J}_\mathcal{P}$ belongs to a unique patient. According to Mosquera et al. [21], a patient is mostly related to three jobs per day. A caregiver usually performs six jobs per day.

Each job has a priority and arduousness score [25,26]. The lowest score means that the job $j \in \mathcal{J}_\mathcal{P}$ has a low priority/arduousness. Sometimes, jobs have to be postponed until another day because no caregiver can perform the job on the current day or because of a lack of time. The priority scale is used to range jobs from basic to extremely important. Some jobs are more arduous than others (palliative care), and the caregiver can be affected by this job. These situations must be taken into account when building and scheduling the planning. The arduousness score is used to rate how much a job can affect a caregiver from low impact to deep impact. As an example and as illustrated in Figure 1, the first caregiver has three jobs scheduled. The first one has a low priority and a low arduousness level. The second job scheduled has a low priority and a high arduousness level. The third job scheduled has a medium priority and a high arduousness level. Taking care of how jobs are scheduled is important because as described in Figure 1, the first caregiver has to perform two jobs with the highest arduousness score, whereas the fourth caregiver only has jobs with a medium or low arduousness score.

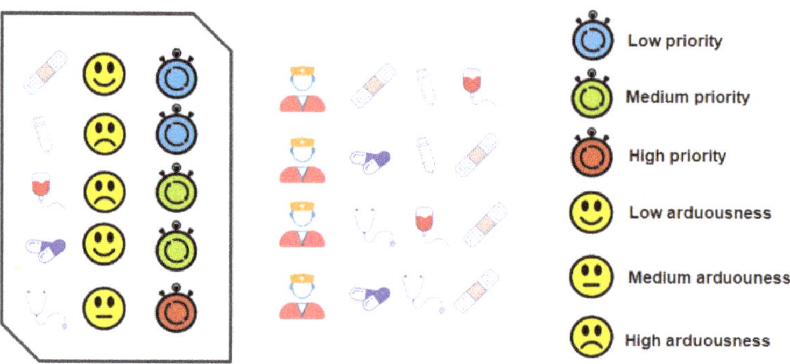

Figure 1. An example of priority and arduousness in an HCC scheduling problem.

Let $\mathcal{Q}$ be the set of qualifications required by a job. To perform a job j, a caregiver must have the required qualification $\mathcal{Q}_i \in \mathcal{Q}$, as illustrated in Figure 2. A job may need one or several qualifications, while a caregiver must have all qualifications required to perform the job. In Figure 2, Caregiver 2 is able to perform the second and third job because he/she has the required skills. As mentioned by Fikar [13], an HHC instance has about six different qualifications on average, which means that $card(\mathcal{Q}) \approx 6$. Moreover, as stated by Castillo-Salazar [27] in his review, on average, eighty percent of caregivers are able to perform any kind of job.

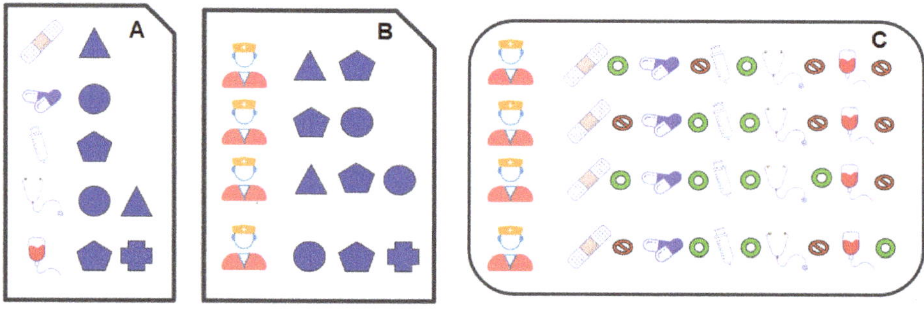

Figure 2. A qualification matrix (**C**) regarding job qualifications (**A**) and caregivers' qualifications (**B**).

Each job related to a patient has a given time window $[s_j, e_j]$ and a processing time dur_j. A patient can be an elderly person living at home and who has an agreement with an HHC center for his/her medical and social care (bathing, nursing, etc.). Depending on the care service, a time window has to be set up. The time window indicates when a job must start. For example, for a patient $p \in \mathcal{P}$ and a job $j \in \mathcal{J}$, a time window is set up to start at 2.30 P.M. and finish at 3.30 P.M. This means that a caregiver $c \in \mathcal{C}$ has to begin the job between 2.30 P.M. and 3.30 P.M.

Normally, each caregiver follows his/her own route. However, some patients require two caregivers to be present simultaneously (synchronization constraint) or in a given order (precedence constraint) to proceed to their jobs. If a job is synchronized, two caregivers are needed simultaneously to perform the job at the patient's home. Synchronized jobs are often met in the case of overweight patients or palliative care. The precedence constraint is met when, for example, medication has to be given by a caregiver before or after another task performed by another caregiver. Typically, ten to twenty percent of jobs in an HHC center follow such constraints. Moreover, other types of temporal and geographical constraints are considered in the literature; collecting biological samples is one of them. Indeed, in some jobs, a caregiver might need to collect a biological sample and bring it to a laboratory for testing. Thus, the location of an HHC center or medical laboratories involved in the care provided are among the parameters to be considered in the HHCRSP models.

Figure 3 sums up and exemplifies the characteristics of a job. First of all, a job is related to a patient. It has a priority and arduousness score indicating how the job must be scheduled. Then, a job requires skills to be performed. Each job has a fixed duration and must be started within a certain time frame on a specific day. Lastly, if a job may need another caregiver or produce a biological sample, synchronization or collection constraints are raised. Precedence constraints can also be raised if a job has to be performed before the current job.

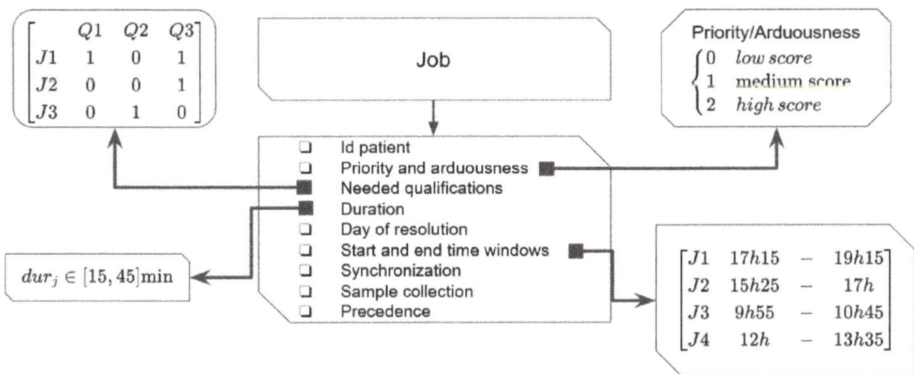

Figure 3. Description and summary of a job.

3.2. Caregivers

Let $\mathcal{C} = \{1, \ldots, C\}$ be a set of caregivers. As a reminder, the word caregiver is applied to anyone who provides personal care in his/her work, such as nurses or social workers. For each patient, there is at least one set of caregivers $\mathcal{P}_c \subset \mathcal{C}$ able to perform all jobs required by the patient at his/her home. All caregivers' tours can start at their own homes or at the HHC office and must finish at the HHC office. Traveled distances between patients' homes and caregivers' homes are calculated and stored in a specific matrix. For an HHC center, designing the optimal shifts for caregivers can be a difficult task. A shift can be consistent or mixed. In a consistent shift, there is only one type of caregiver (nurses only for example). However, a mixed shift has different types of caregivers (nurses and social workers work together with the patient). In both cases, caregivers are distinguished by their own qualifications and their labor agreement. Each caregiver $\mathcal{C}$ can accomplish a set of jobs regarding his/her set of qualifications $\mathcal{Q}_c \in \mathcal{Q}$. The labor agreement defines legal working hours. A caregiver can deliver a job only within his/her working hours. The lower and upper bound of availability $[s_{\mathcal{C}_i}, e_{\mathcal{C}_i}]$ represent the time frame in which a caregiver $\mathcal{C}$ is available during the day $\mathcal{D}_i$. As described in Figure 4, each caregiver has a different set of skills, and for each day of the planning, each caregiver has a different time frame for his/her legal working hours.

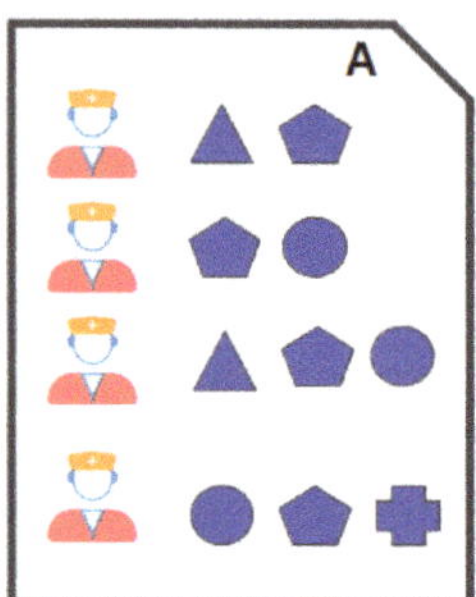 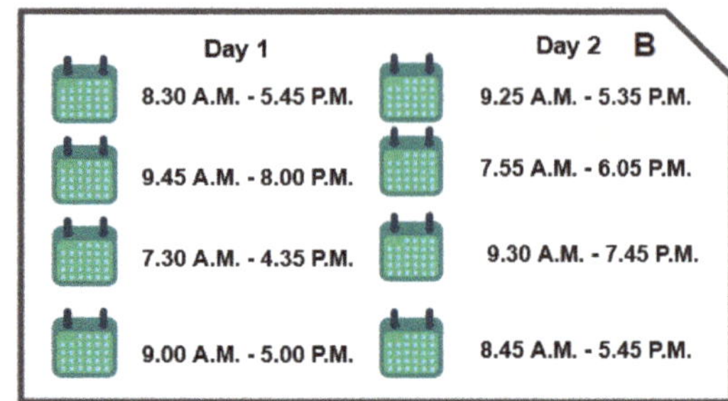

Figure 4. Caregiver qualifications' table (**A**); starting and ending working hours for each caregiver on a two-day basis (**B**).

The compatibility between a patient and a caregiver can be found in the literature as a constraint. They are many ways to tackle these compatibilities: caregivers can have a preference for some jobs, time windows, and patient relationship. A caregiver can choose simple jobs rather than difficult jobs such as palliative care. Caregivers might refuse to perform any of the patients' jobs for compatibility reasons. Two compatibility tables are stored to show how much a caregiver is compatible with a patient. The first compatibility table (A) in Figure 5 combines all hard constraints of a caregiver's/patient's compatibilities (e.g., if the caregiver has a dog allergy or spoken languages). The second compatibility table (B) in Figure 5 shows low constraints in a caregiver/patient relationship. A score of zero is the lowest feasible compatibility, and three is the highest. In the hard compatibility table, a score of four is only reachable if the caregiver is a reference caregiver of the current patient. As illustrated in Figure 5, each patient and caregiver has his/her own compatibility. For example, the first caregiver has a reference caregiver compatibility regarding the first patient, bu a medium compatibility for the same patient regarding the low compatibility table. Compatibilities are not mutual, which means that a caregiver can have a good compatibility with a patient, while this patient does not.

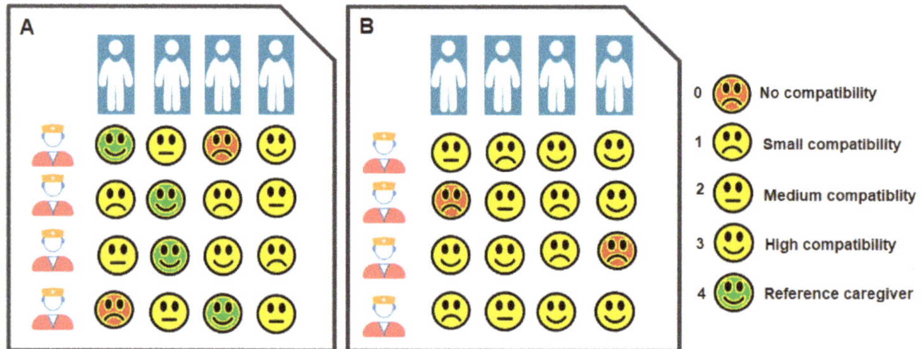

Figure 5. Illustration of the hard (**A**) and low (**B**) compatibility table.

Beyond daily planning, legal working requirements and the labor agreement are important to consider [28]. The caregiver can be full-time or part-time employee of the HHC provider. To do so, each caregiver has a contractual number of hours $\mathcal{H}_C$ per day. The number of daily hours that a caregiver is required to work must not exceed a threshold $[h_{min}, h_{max}]$. For example, $\mathcal{H}_C$ corresponds to 7.5h in the U.K. according to Akjiratikarl et al. [29] or 8h in Finland according to Bräysy and Gendreau [30]. This limit can also be set as a weekly availability [31] in which case, each caregiver is allowed to work a maximum amount of hours per week depending on his/her labor agreement. This is in addition to a daily working time regulation. This limit can be exceeded up to a maximum overtime per period. A break B of a duration dur_B can be planned for a caregiver $c \in \mathcal{C}$ for each day. This break can be taken at any time during the day. It is planned before or after a job.

Figure 6 describes the characteristics of a caregiver. Firstly, a caregiver possesses many skills due to his/her academic training. Secondly, a caregiver has the geographical coordinates of his/her current position and initial position. Compatibility tables are set up to reflect the caregiver/patient relationship. Legally, a caregiver has a working contract with the HHC center and may work each day during a fixed time frame for a minimum and maximum amount of hours.

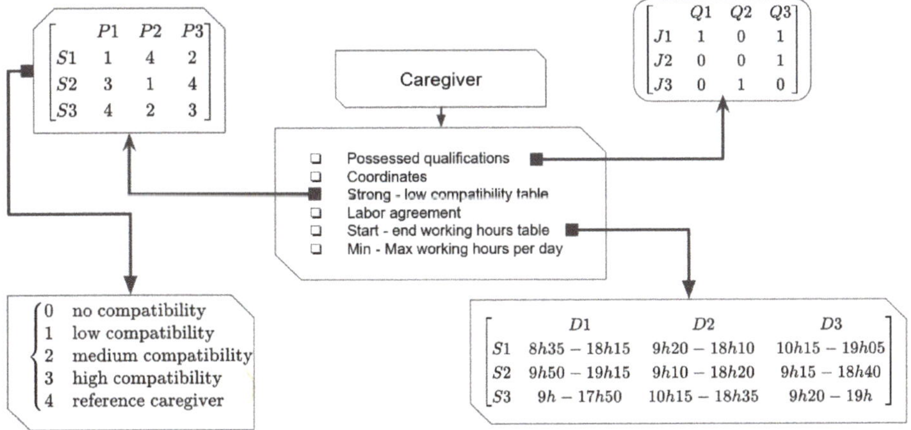

Figure 6. Description and summary of a caregiver.

3.3. Patients

Patients are key players in HHC or an HHCRSP. Let $\mathcal{P}$ be a set of patients scattered in a geographic area who need care services. For each $\mathcal{P}_i \in \mathcal{P}$, there is a subset $\mathcal{J}_{\mathcal{P}_i} \subseteq \mathcal{J}$ such that these jobs involve patient $\mathcal{P}_i$. Obviously, at least one job has to be scheduled in the given planning for each patient.

Compatibilities are often considered (language, patient/caregiver relationship, etc.). Patient compatibility is treated the same way as caregiver compatibility. Most studies assign a time interval to each patient in which he/she is available at home to receive care. This time interval, also called a time window, can be fixed or flexible. Fixed time windows force the caregiver to visit the patient in the time frame and to proceed with care, while flexible time windows allow the caregiver to delay care. Fixed time windows are met when the patient needs drug intake or medical care that cannot be delayed. As a result, for each patient $\mathcal{P}_i \in \mathcal{P}$, a time windows is defined as $[\alpha_i, \beta_i]$, where $\alpha_i \geq 0$ and $\beta_i \geq \alpha_i$ specify the earliest, respectively latest, start time for a visit.

The assignment constraints of an HHC service organization are largely related to continuity of care. According to the quality standards adopted by HHC providers, care services for patients can be provided under full, partial, or no continuity of care. Under full continuity of care, a patient is assigned to one and only one caregiver, who is responsible for the patient's care during his/her stay in the HHC service. On another note, partial continuity of care is usually observed in situations in which a patient needs more than one type of care. Then, for the most frequent type of care, a reference caregiver can be assigned (as a full continuity of care case). Usually, studies tend to deal with full continuity, but it is often more a criterion to maximize than a parameter of the HHCRSP model [23,29,30,32].

Figure 7 describes the characteristics of a patient. Firstly, a patient possesses a list of care services (jobs). The patient has the geographical coordinates (latitude/longitude) of his/her position. The distance between patients and the HHC center or laboratories is calculated and stored in a matrix. The compatibility table is set up to reflect the caregiver/patient relationship. Time windows are set up to show to the center when a patient is available to receive his/her care services.

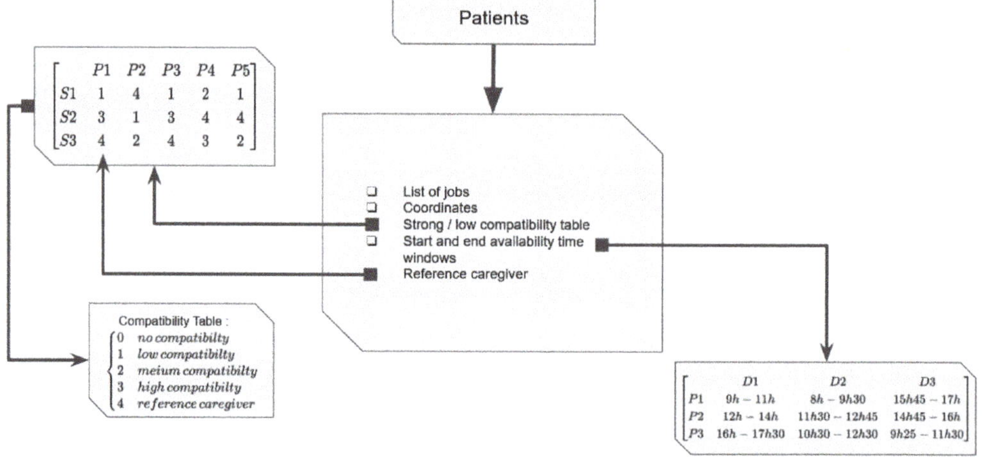

Figure 7. Description and summary of a patient.

3.4. Performance Measures: Common Objective Functions

In operational research, performance measures are represented by objective functions. Operational problems are often set up with many constraints and an objective function. As presented before, an HHCRSP can be tackled in different ways depending on the set of constraints. The objective function can also be elaborated differently depending on how the solution must be optimized. In the HHCRSP literature, different kinds of objective functions are introduced:

1. Minimizing the travel cost by optimizing the traveled distance [19,24,26].
2. Minimizing working hours for caregivers [10,29,33].
3. Minimizing overtime [16,25].
4. Continuity of care [24,34].
5. Preferences [35].

4. Overview and Functioning of the Generator

An HHC aims to increase the quality of life for patients and to improve the health system's efficiency. An HHC center deals with a wide range of patients suffering from various diseases. As described in the previous section, HHC problems are based on scheduling jobs linked to a routing problem. Generating a good reading of an HHCRSP is not an easy task. Listing and categorizing all different constraints and angles of an HHC problem are essential. Section 4.1 briefly illustrates the assumptions and constraints raised in the literature. In Section 4.2, a listing and description of different parameters found in the literature are proposed to fully describe the HHCRSP. The source code of the project is available at this address: https://drive.google.com/drive/folders/1NnZ8JwAtOOp4mm2 du1WjzKF0Dkoctekw?usp=sharing (accedessed date on 12 April 2021). This project was coded in C++ in Microsoft Visual Studio 2019. A readme file was implemented to explain how the generator works and describes the data provided by the generator. This file has an English version, as well as a French version.

4.1. Background Description of an HHCRSP

A typical HHCRSP consists of a set of caregivers starting and ending their work at an HC center. The goal is to find a valid plan for an n-day period for each caregiver. The resulting planning indicates which job should be carried out by which caregiver and when the job should start. Each caregiver has available time for carrying out patients' jobs with the possibility of working just a part of the day or the whole day. Consequently, jobs are assigned according to the feasible time of each caregiver. Depending on how geographical constraints are dealt with, caregivers use the same or various modes of transportation. Each job has specific qualifications or skills, and they are always considered as hard constraints. The preference relationship can be defined between patients and caregivers as a ratio or score between them. An HHC center can decide that patients should be visited by their preferred caregivers in order to ensure patients' satisfaction. Besides, when job schedules are made, legal constraints have to be respected as well. Usually, caregivers possess a full-time or part-time agreement. Then, for each caregiver, a maximum and a minimum working duration are set up by his/her own agreement. The working time of each caregiver starts from the beginning of the first scheduled job and ends at the completion of the last job. Other legal regulations can be considered such as breaks, especially lunch breaks. Breaks are flexible in the caregiver' schedule. Furthermore, the HHC company aims to improve the planning of caregivers' routes by balancing the working time among caregivers, in order to obtain a fair plan [11,35,36].

As a result, the HHC company seeks to optimize a multi-objective function. Usually, it aims at minimizing the total travel cost and the fixed cost of caregivers by minimizing the total overtime work (in hours) of caregivers and by limiting unscheduled visits [14,27]. Depending on how the HHCRSP is treated, different constraints can be raised, but some of them are always present, such as [13,27,35]:

- Only one HHC caregiver can visit a patient.
- Each patient must receive his/her requested HHC service provided by the appropriate caregiver.
- Each caregiver starts and ends at the HHC center.
- Each caregiver must leave the patient' home once he/she serves it.
- An unqualified caregiver cannot perform a qualified job.

However, some constraints are deeply linked to how the HHCRSP is tackled:

- The planning horizon is assumed to be n-days.
- Routing part: delivery and pick-up.
- Total working hours and route length limits for caregivers.
- Consistency of the working time with the possibility of a break between two successive jobs.
- The service time of a caregiver at a patient location.
- Patient time windows in the schedules of caregivers' routes.
- The feasibility of each job according to its time window.
- Patient/caregiver preference requirement.
- Continuity of care for all patients.
- Synchronization between two caregivers for a specific job.
- Biological sample production
- Precedence requirement.

4.2. Listing and Range of All Parameters

Each instance is described by a list of parameters. Some parameters are set up in the generator and can be only changed by modifying them in the source code. The configuration file stores a list of parameters. As presented in Figure 8, the generator creates a list of instances regarding a configuration. Every piece of information is reachable and editable.

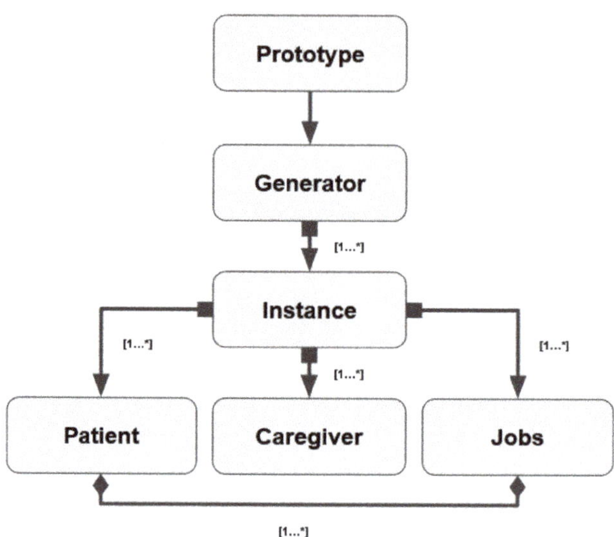

Figure 8. UML description of the generator. [1...*]: one occurrence to multiple occurrences.

4.2.1. Non-Editable Parameters

Some parameters are not editable in the configuration file. Firstly, as mentioned before, a caregiver is allowed to work a certain amount of time per day regarding his/her labor agreement. As noticed in the following Tables 1 and 2, the generator sets up the beginning of the day at 8 A.M. and the ending at 8 P.M. A caregiver with a full-time contract is allowed to work between 8 to 10 h a day, and a caregiver with a part-time contract can only work 4 to 6 h a day according to the literature [10,13,14,35,37,38]. Compatibilities between patients and caregivers are represented by a score as explained in the previous section. A low score means that the caregiver should not be assigned to this patient due to a low compatibility. The generator ranges compatibility between zero (lowest compatibility) and four (highest compatibility). A score of four is a special feature that allows the generator to set up a reference caregiver to a patient if the configuration file allows it.

Table 1. Range of non-editable parameters for caregivers.

Parameters	Range	References
Minimum and maximum working hours per day (full-time)	8–10 h	[10,13,14]
Minimum and maximum working hours per day (part-time)	4–6 h	[35,37,38]
Beginning and ending of the day	8 A.M.–8 P.M.	[23,39]
Compatibility score (lowest to highest)	0–4	[13,17,24,33]

Table 2. Range of non-editable parameters for patients.

Parameters	Range	References
Beginning and ending of the day	8 A.M.–8 P.M.	[23,39]
Range of the time windows	1–4 h	[6,13,25]
Compatibility score (lowest to highest)	0–3	[17,22,24,33]

The patient time frame availability on each day starts at 8 A.M. and finishes at 8 P.M. Furthermore, the compatibility scores are set up in the same way. However, as specified in Table 2, a patient has a time window. During this time window, our patient is available to receive his/her care. During the scheduling process, time windows for the patient can be considered as soft or hard constraints depending on how the HHC problem is tackled.

The generator sets up a range of 1 to 4 h, which means that a patient ks at least available for 1 h and at most 4 h to receive his/her care services.

Lastly, Table 3 sums up the different non-editable parameters for a job. A job has a processing time called duration in the generator. Duration can range from 30 min to an hour and a half. Some jobs in the instance need to be done, and some of them can be postponed to another day. A priority score is set up on all jobs to show how important the job is. Moreover, some jobs must be performed in a certain time frame. Usually, this time frame may not exceed one hour to two hours. Precedence constraints can be found in some particular jobs. The literature shows that the duration between two jobs related by a precedence constraint may not exceed two hours.

Table 3. Range of non-editable parameters for jobs.

Parameters	Range	References
Range of duration	30–90 min	[13,15,29]
Range of the time frame to proceed with the job	1–2 h	[9,13]
Range of the time frame between one job and another job (precedence)	2 h	[12,13,40]
Range of arduousness/priority score	0–3	[13,22]

4.2.2. Editable Parameters

Firstly, to setup up a benchmark correctly, the first parameter allows the generator to produce more than a unique instance following the current configuration. Some data are generated randomly such as the distance between patients or compatibilities. However, some can be similar between two instances such as the number of caregivers/patients or several proportions. The second parameter of an instance is the geographical boundary of the instance. Those coordinates enable the generator to set up different coordinates in the designated area for caregivers, patients, and depots. As explained earlier, distances are calculated by following a straight line between two patients/depots. Usually, studies [13,31,34] have been based on a daily planning horizon, but the third parameter allows the generator to set a different number of days in the planning horizon. Depending on how the HHCRSP is treated, the number of days can be one day for a short-term plan or two to seven days for a mid-term plan. Long-term planning is not taken into account due to its marked difference with short- and mid-term planning. The number of patients and caregivers has a huge impact on how jobs are generated. Indeed, several studies [6,23,25,39] concluded that a patient may have on average three to four jobs (cares) per day.

A caregiver is also able to deal with six to seven jobs per day depending on how much time a job needs to be performed properly [5,11,21,30]. Qualifications and skills are important to define different types of jobs in an instance [9,13,34,41]. The HHC literature has concluded the fact that all feasible jobs met in an HHC center can be categorized into six major qualifications on average. The HHC literature also presents two distinct ways to tackle the qualification constraint [9,16,22]. The first approach considers all skills as mutually independent. This approach is often met when the HHC center deals with a wide array of jobs (palliative care to daily care). The second approach is based on a pyramidal scheme. For example, a caregiver who possesses a specific qualification Q_i automatically possesses all qualifications at a lower level. The second approach is usually used in HHC problems with a huge VRP component [9,12,40]. The generator implements the first approach and uses independent qualifications.

To ensure continuity of care, studies in the literature implement a reference caregiver [42]. During the scheduling process, a reference caregiver is more likely to be assigned to his/her reference patient if it is possible to ensure a partial continuity of care. As explained in the previous section, caregivers can have different labor agreements (full-time or part-time), so the next parameter allows the generator to set up different proportions of part-time caregivers. This ratio can vary from a small portion to half of the number of caregivers. The traveled time parameter is an optional parameter. It measures how much time caregivers are on the road. On average, this time may be around 18% to 25% of their working time [13,33]. This parameter has a huge influence on how jobs are generated because the generator takes into account this percentage to create jobs in order to not overload the HHC center. The time traveled parameter is heavily linked to the next parameter: the occupation or saturation rate of instance. This parameter is the ratio between the number of working hours generated by all jobs and the number of working hours available in the HHC center. Finally, the generator can produce synchronized jobs, jobs with biological sample production, and jobs with precedence constraints. According to the HHC literature [13,32,38], currently, fifteen to twenty percent of care services are synchronized or have precedence constraints. As summed up in Table 4, those parameters design the configuration file and allow the generator to create different types of instances depending on those parameters.

Table 4. Range of all editable parameters.

	Parameters	Range	References
#1	Number of depots	1–3	[9,35]
#2	Number of days	1–7	[13,31,34]
#3	Number of patients	10–250	[11,21,28,39]
#4	Number of caregivers	2–25	[5,11,21,30,43]
#5	Number of different qualifications	5–7	[9,13,34,41]
#6	Reference caregiver	false-true	[42]
#7	Part-time/full-time proportion	5–50%	[16,34]
#8	Traveled time between patients (ratio)	15–25%	[13,33]
#9	Occupation/capacity of the instance (ratio)	50–100%	[33,35,43]
#10	Synchronization ratio (ratio)	10–25%	[13,32,38]
#11	Biological samples (ratio)	10%	[13,24,38]
#12	Precedence between jobs (ratio)	10–25%	[12,24,44]

5. A Descent Local Search Algorithm for an HHC Problem

In this section, we adopt a descent local search algorithm based on the work of Xio et al. [36], and we fix, as they did, the planning horizon to a daily one. The set of jobs demanded by patients is denoted $\mathcal{J}$. Each job $j \in \mathcal{J}$ is associated with a certain duration dur_j and a time frame $[\alpha_j, \beta_j]$. To reach another patient, a distance d_j (beeline) is set up, and a travel time t_j is calculated. Let $\mathcal{C}$ denote the set of caregivers. $\mathcal{C}$ includes both full-time and part-time caregivers. For each caregiver $c \in \mathcal{C}$, a time window corresponding to their available time is set up. Variables t_c^{begin} and t_c^{end} represent respectively the time when the caregiver c starts and ends his/her journey. The working time of a caregiver c is defined by t_c^{work}. It must not exceed the maximal working time planned by his/her labor agreement. We set p_i as a penalty that occurs when a job is not assigned. The objective function minimizes unassigned jobs and the daily operating cost of the HHC such as salaries and traveling cost by refunding the fuel cost.

The solving method was based on a descent algorithm. Firstly, the algorithm tries to reduce the bag of non-allocated jobs by switching jobs between caregiver and the bag until the job can insert itself into a caregiver schedule, as described in Figure 9. During this process, the qualification constraint must be respected, as well as the working balance, as much as possible. Compatibility constraints can be raised if necessary. Indeed, for example, a patient can have a low compatibility score with all caregivers. In Figure 9, for example, the algorithm tries neighborhood moves by inserting the first job of the bag in front of the second job of the second caregiver. At the same, it also tries to insert the second job of the second caregiver in front of the first job of the first caregiver. Lastly, the algorithm inserts the last job of the first caregiver at the end of the job list of the second caregiver to produce a correct solution. For more details about the algorithm, please refer to the main article written by Xio et al. [36]. Algorithm 1 presents a pseudo-code version of how the descent local search algorithm works regarding the different inputs.

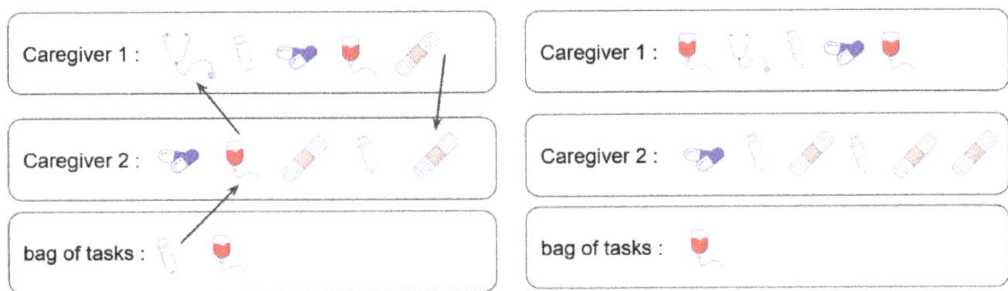

Figure 9. Bag of neighborhood moves.

Algorithm 1: Description of a descent local search algorithm.

Result: Optimal shifts for caregivers
Given a set of patients $\mathcal{P}$, a set of caregivers $\mathcal{C}$, and a set of jobs $\mathcal{J}$:
Create a feasible solution for all caregivers with a bag of jobs for unassigned jobs;
for *job in $\mathcal{J}$* **do**
    if *job can be affected* **then**
        | Assign it randomly to a caregiver with the right qualifications;
    else
        | Store it in the bag of jobs;
    end
end
while *the solution can be improved* **do**
    if *not an empty bag of jobs* **then**
        Try to insert a job in a caregiver shift;
        Fix the solution if the solution violates the constraints;
        if *job cannot be inserted* **then**
          | Do not try again to insert jobs from the bag of jobs.
        else
        end
    else
        Try to switch a job for one caregiver to another from a different caregiver to improve both shifts;
        Switch jobs in a caregiver shift to see if the shift can be improved;
    end
end

Then, the algorithm tries to balance working hours for all caregivers by minimizing their working hours per day and switching jobs between caregivers to see if a configuration results in a better solution than the current one. As described in Figure 10, two jobs can be switched between two caregivers depending on their qualifications and the patient-caregiver compatibility.

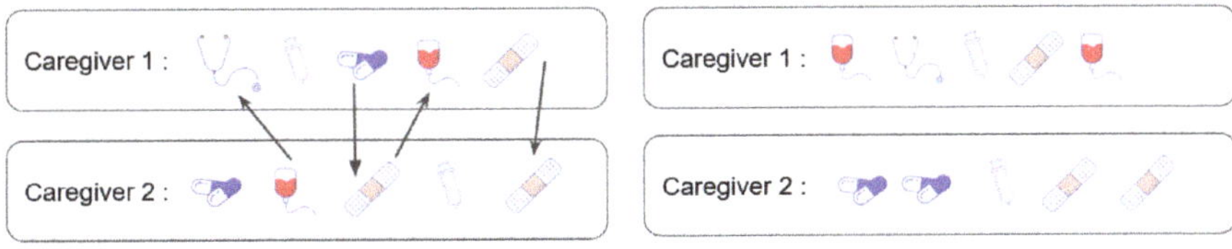

Figure 10. Caregiver neighborhood moves.

Finally, the solving method tends to improve each caregiver scheduled by trying to switch some jobs during the day. Some jobs are firstly scheduled because of priority or their large time windows, but sometimes, better configurations are found by switching them with other jobs, as described in Figure 11.

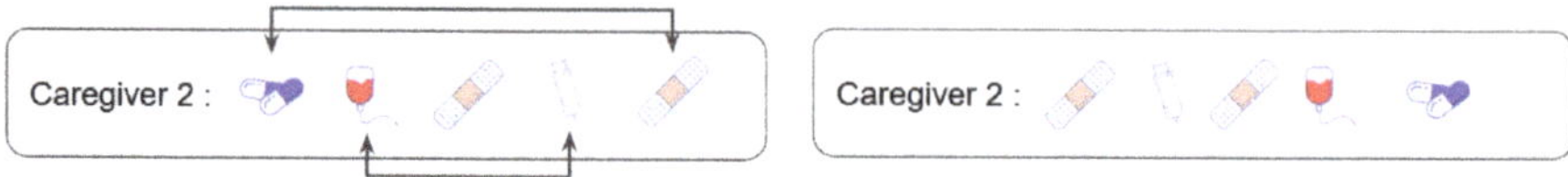

Figure 11. Patient neighborhood moves.

All empirical data were generated by our generator coded in C++ with Microsoft Visual Studio. This simulation was run on a computer with an i7-10700KF CPU 3.80 GHz. Ten instances were generated, and tests were carried out on a properly developed solving method explained before. For all tests, the objective function combined three parts of the costs: the salary of each caregiver calculated by hour. The payment was 20 CAD per hour; the fuel cost 1.50 CAD per kilometer, and each job postponed cost 100 CAD. To dispatch appropriate caregivers and to satisfy all patients, caregivers were not assigned to a patient if their compatibility scores were lower than one. If a patient did not have a compatibility score higher than one for all caregivers, he/she was assigned to the more available caregiver regarding his/her qualifications. The simulation was run several times before putting the results together. The postponed jobs and operating cost in Table 5 are the average results of the simulation. As mentioned before, ten instances were generated by the generator. The instances were designed to tackle the HHCRSP in the same way as Xio et al. [36] did. The configuration file was setup up with data from the article and can be found in the archive.

Table 5. Computational results.

Instance	Operational Results Postponed Jobs-Operating Cost
HCC-$p5\_c2$	0-501.05
HCC-$p10\_c2$	0-702.35
HCC-$p15\_c2$	1-1125.05
HCC-$p15\_c3$	0-1025.35
HCC-$p20\_c3$	0-1500.45
HCC-$p25\_c3$	2-2020.45
HCC-$p25\_c4$	0-2040.95
HCC-$p30\_c4$	0-2545.45
HCC-$p35\_c5$	2-3005.75
HCC-$p40\_c5$	7-5025.45

Table 5 shows the different sizes of the instances. The first column names the instance. Index p presents the number of patients and index c the number of caregivers that can be assigned for the different jobs. For example, $p5\_c2$ represents five patients and two caregivers. Then, we illustrate the results: Sub-column 1 represents the number of jobs postponed. The next sub-column presents the operating cost. Tables 6 and 7 show the result of two instances HHC$p10_c2$: the first column is the caregiver ID and the second column the number of jobs assigned to this caregiver. The next two columns present respectively the available time (when the caregiver starts and ends his/her journey) and the working time. The last column shows the schedule planning for each caregiver.

Table 6. Results of instance HHC$p10_c2_1$.

Caregiver	No. of Jobs	Available Time From–To	Working Time	Planning
1	12	8 h 30 min–16 h 30 min	432 min	3-1-2-5-7
2	14	9 h 30 min–17 h 30 min	454 min	4-9-8-0-6

Table 7. Results of instance HHC$p10_c2_2$.

Caregiver	No. of Jobs	Available Time From–To	Working Time	Planning
1	13	7 h 20 min–15 h 45 min	502 min	2-8-5-6-3
2	11	9 h 15 min–19 h 05 min	515 min	1-4-9-0-7

6. Conclusions

In this study, we introduced a common benchmark involving a multi-parameter generator for the HHC scheduling problem fitting a certain list of constraints found in the literature. First, we proposed a common definition of the HHCRSP dealing with an exhaustive list of constraints. These constraints can be attached to caregivers, patients, caregivers' actions, and the HHCRSP itself. Then, we described the different generator parameters with their ranges. We also introduced a short application of our generator based on a descent local search algorithm, which embedded adapted neighborhood moves for the HHCRSP. The main purpose of this contribution was to standardize testing and checking solving methods to allow researchers to compare their methods on a common benchmark with a robust and configurable generator. Our work will be directed toward the adaptation of some resolution methods in the future and especially for metaheuristics algorithms. As mentioned previously, this configurable generator allows researchers to work on a common benchmark to compare their solving methods without being affected by medical confidentiality. With these multiple parameters, the generator was able to reproduce a large type of HHCRSP with different constraints. A common benchmark was generated and stored in the archive and can be used be researchers to compare their methods or extract data to make their own dataset to verify their solving method.

Author Contributions: Conceptualization, C.H., C.G., and A.S.; formal analysis, C.H., C.G., and A.S.; funding acquisition, C.G.; methodology, C.H., C.G., and A.S.; project administration, C.G. and A.S.; software, C.H.; supervision, C.G. and A.S.; validation, C.H., C.G., and A.S.; writing—original draft, C.H.; writing—review and editing, C.G. and A.S. All authors read and agreed to the published version of the manuscript.

Funding: The financial support of the NSERC (Canada) and FUQAC made this research and its publication possible.

Institutional Review Board Statement: Not applicable for studies not involving humans or animals.

Informed Consent Statement: Not applicable for studies not involving humans or animals.

Conflicts of Interest: The authors declare no conflict of interest.

Abbreviations

The following abbreviations are used in this manuscript:

OECD	Organisation for Economic Co-operation and Development
HHC	Home Health Care
HHCRSP	Home Health Care Routing and Scheduling Problem
HHCRP	Home Health Care Routing Problem
VRP	Vehicle Routing Problem
HC	Home Care

References

1. Slade, S.; Shrich, A.; DiMillo, S. *Health Care for an Aging Population: A Study of How Physicians Care for Seniors in Canada*; The Royal College of Physicians and Surgeons of Canada: Ottawa, ON, Canada, 2019.
2. World Health Organization. *Global Strategy and Action Plan on Ageing and Health*; World Health Organization: Geneve, Switzerland, 2017.
3. OCDE and Union Européenne. *Health at a Glance: Europe 2018*; Éditions OCDE, Paris/Union Européenne: Brussels, Belgium, 2018.

4. Institute of Medicine and National Research Council. *The Future of Home Health Care: Workshop Summary*; Weisfeld, V., Lustig, T.A., Eds.; The National Academies Press: Washington, DC, USA, 2015.
5. Gilmour, H. Formal home care use in Canada. *Health Rep.* **2018**, *29*, 3–9. [PubMed]
6. Matta, A.; Chahed, S.; Sahin, E.; Dallery, Y. Modelling home care organisations from an operations management perspective. *Flex. Serv. Manuf. J.* **2014**, *26*, 295–319. [CrossRef]
7. Chahed, S.; Marcon, E.; Sahin, E.; Feillet, D.; Dallery, Y. Exploring new operational research opportunities within the home care context: The chemotherapy at home. *Health Care Manag. Sci.* **2009**, *12*, 179–191.
8. Eveborn, P.; Rönnqvist, M.; ; Einarsdóttir, H.; Eklund, M.; Lidén, K.; Almroth, M. Operations research improves quality and efficiency in home care. *Interfaces* **2009**, *39*, 18–34. [CrossRef]
9. Hiermann, G.; Prandtstetter, M.; Rendl, A.; Puchinger, J.; Raidl, G.R. Metaheuristics for solving a multimodal home-health care scheduling problem. *Cent. Eur. Oper. Res.* **2015**, *23*, 89–113. [CrossRef]
10. Nickel, S.; Schröder, M.; Steeg, J. Mid-term and short-term planning support for home health care services. *Eur. J. Oper. Res.* **2012**, *219*, 574–587. [CrossRef]
11. Yalcindag, S.; Matta, A.; Sahin, E. Human resource scheduling and routing problem in home health care context: A literature review. In Proceedings of the 37th Conference on Opisto Research Applied to Health Services: Cardiff, UK, 24–29 July 2011; pp. 8–22.
12. Mankowska, D.S.; Meisel, F.; Bierwirth, C. The home health care routing and scheduling problem with interdependent services. *Health Care Manag. Sci.* **2014**, *17*, 15–30. [CrossRef] [PubMed]
13. Fikar, C.; Hirsch, P. Home health care routing and scheduling: A review. *Comput. Oper. Res.* **2017**, *77*, 86–95. [CrossRef]
14. Cissé, M.; Yalçındağ, S.; Kergosien, Y.; Şahin, E.; Lenté, C.; Matta, A. OR problems related to Home Health Care: A review of relevant routing and scheduling problems. *Oper. Res. Health Care* **2017**, *13–14*, 1–22. [CrossRef]
15. Begur, S.V.; Miller, D.M.; Weaver, J.R. An integrated spatial dss for scheduling and routing home-health-care nurses. *Interfaces* **1997**, *27*, 35–48. [CrossRef]
16. Lanzarone, E.; Matta, A.; Sahin, E. Operations management applied to home care services: The problem of assigning human resources to patients. *IEEE Trans. Syst. Man Cybern. Part Syst. Hum.* **2012**, *42*, 1346–1363. [CrossRef]
17. Bertels, S.; Fahle, T. A hybrid setup for a hybrid scenario: Combining heuristics for the home health care problem. *Comput. Oper. Res.* **2006**, *33*, 2866–2890. [CrossRef]
18. Liu, R.; Xie, X.; Augusto, V.; Rodriguez, C. Heuristic algorithms for a vehicle routing problem with simultaneous delivery and pickup and time windows in home health care. *Eur. J. Oper. Res.* **2013**, *230*, 475–486. [CrossRef]
19. Liu, R.; Yuan, B.; Jiang, Z. Mathematical model and exact algorithm for the home care worker scheduling and routing problem with lunch break requirements. *Int. J. Prod.* **2017**, *55*, 558–575. [CrossRef]
20. Frifita, S.; Masmoudi, M.; Euchi, J. General variable neighborhood search for home health care routing and scheduling problem with time windows and synchronized visits. *Electr. Notes Discret. Math.* **2017**, *58*, 63–70. [CrossRef]
21. Mosquera, F. Flexible home care scheduling. *Omega* **2019**, *83*, 80–95. [CrossRef]
22. Lanzarone, E.; Matta, A. The nurse-to-patient assignment problem in Home Care services. In *Advanced Decision Making Methods Applied to Health Care*; Tànfani, E., Testi, A., Eds.; Springer: Berlin/Heidelberg, Germany, 2012; Volume 173, pp. 121–139.
23. Braekers, K.; Hartl, R.F.; Parragh, S.N.; Tricoire, T. A bi-objective home care scheduling problem: Analyzing the trade-off between costs and client inconvenience. *Eur. J. Oper. Res.* **2016**, *248*, 428–443. [CrossRef]
24. Rasmussen, M.S.; Justesen, T.; Dohn, A.; Larsen, J. The Home Care Crew Scheduling Problem: Preference-based visit clustering and temporal dependencies. *Eur. J. Oper. Res.* **2012**, *219*, 598–610. [CrossRef]
25. Sahin, E.; Matta, A. A contribution to operations management-related issues and models for home care structures. *Int. J. Logist. Res. Appl.* **2015**, *18*, 355–385. [CrossRef]
26. Hertz, A.; Lahrichi, N. A patient assignment algorithm for home care services. *J. Oper. Res. Soc.* **2009**, *60*, 481–495. [CrossRef]
27. Castillo-Salazar, J.A.; Landa-Silva, D.; Qu, R. Workforce scheduling and routing problems: Literature survey and computational study. *Ann. Oper. Res.* **2016**, *239*, 39–67. [CrossRef]
28. Wirnitzer, J.; Heckmann, I.; Meyer, A.; Nickel, S. Patient-based nurse rostering in home care. *Oper. Res. Health Care* **2016**, *8*, 91–102. [CrossRef]
29. Akjiratikarl, C.; Yenradee, P.; Drake, P.R. APSO-based algorithm for home care worker scheduling in the UK. *Comput. Ind. Eng.* **2007**, *53*, 559–583. [CrossRef]
30. Bräysy, O.; Gendreau, M. Vehicle routing problem with time windows, Part II: Metaheuristics. *Transp. Sci.* **2005**, *39*, 119–139. [CrossRef]
31. Duque, P.M.; Castro, M.; Sörensen, K.; Goos, P. OR Home care service planning. The case of Landelijke Thuiszorg. *Eur. J. Oper. Res.* **2015**, *243*, 292–301. [CrossRef]
32. Bredström, D.; Rönnqvist, M. Combined vehicle routing and scheduling with temporal precedence and synchronization constraints. *Eur. J. Oper. Res.* **2008**, *191*, 19–31. [CrossRef]
33. Ma, C.; Herrmann, L.; Miner, S.; Witkoski Stimpfel, A.; Squires, A. Home health care services to persons with dementia and language preference. *Geriatr. Nurs.* **2020**, *41*, 165–171. [CrossRef] [PubMed]
34. Di Mascolo, M.; Espinouse, M.-L.; El Hajri, Z. Planning in Home Health Care Structures: A literature review. *IFAC-PapersOnLine* **2017**, *50*, 4654–4659. [CrossRef]

35. Shi, Y.; Boudouh, T.; Grunder, O.; Wang, D. Modeling and solving simultaneous delivery and pick-up problem with stochastic travel and service times in home health care. *Expert Syst. Appl.* **2018**, *102*, 218–233. [CrossRef]
36. Xiao, L.; Dridi, M.; El Hassani, A.H. Mathematical Model for the Home Health Care Scheduling and Routing Problem with Flexible Lunch Break Requirements. *IFAC-PapersOnLine* **2018**, *51*, 334–339. [CrossRef]
37. Decerle, J.; Grunder, O.; El Hassani, A.H.; Barakat, O. A memetic algorithm for a home health care routing and scheduling problem. *Oper. Res. Health Care* **2018**, *16*, 59–71. [CrossRef]
38. Decerle, J.; Grunder, O.; El Hassani, A.H.; Barakat, O. A hybrid memetic-ant colony optimization algorithm for the home health care problem with time window, synchronization and working time balancing. *Swarm Evol. Comput.* **2019**, *16*, 171–183. [CrossRef]
39. Martin, E.; Cervantes, A.; Saez, Y.; Isasi, P. IACS-HCSP: Improved ant colony optimization for large-scale home care scheduling problems. *Expert Syst. Appl.* **2020**, *142*, 112994.
40. Pillac, V.; Gendreau, M.; Guéret, C.; Medaglia, A.L. A review of dynamic vehicle routing problems. *Eur. J. Oper. Res.* **2013**, *225*, 1–11. [CrossRef]
41. Nikzad, E.; Bashiri, M.; Abbasi, B. A matheuristic algorithm for stochastic home health care planning. *Eur. J. Oper. Res.* **2021**, *288*, 753–774.
42. Carello, G.; Lanzarone, E. A cardinality-constrained robust model for the assignment problem in Home Care services. *Eur. J. Oper. Res.* **2014**, *236*, 748–762. [CrossRef]
43. Leung, P.P.; Wu, C.H.; Kwong, C.K.; Ip, W.H.; Ching, W.K. Digitalisation for optimising nursing staff demand modelling and scheduling in nursing homes. *Technol. Forecast. Soc. Chang.* **2021**, *164*, 120512.
44. Nasir, J.A.; Kuo, Y. A decision support framework for home health care transportation with simultaneous multi-vehicle routing and staff scheduling synchronization. *Decis. Support Syst.* **2020**, *138*, 113361.

applied sciences

MDPI

Article

Flexible Job Shop Scheduling Problem with Sequence Dependent Setup Time and Job Splitting: Hospital Catering Case Study

Fatima Abderrabi [1,2,*], Matthieu Godichaud [1], Alice Yalaoui [1], Farouk Yalaoui [1], Lionel Amodeo [1], Ardian Qerimi [2] and Eric Thivet [3]

[1] Computer Science and Digital Society Laboratory, ICD, University of Technology of Troyes, 10300 Troyes, France; matthieu.godichaud@utt.fr (M.G.); alice.yalaoui@utt.fr (A.Y.); farouk.yalaoui@utt.fr (F.Y.); lionel.amodeo@utt.fr (L.A.)

[2] Hospital Center of Troyes, 10000 Troyes, France; ardian.qerimi@hcs-sante.fr

[3] University Hospital Center of Reims, 51100 Reims, France; ethivet@chu-reims.fr

[*] Correspondence: fatima.abderrabi@utt.fr or fatima.abderrabi@hcs-sante.fr

Abstract: This paper aims to study a real case of an optimization problem derived from a hospital supply chain. The present work focuses on developing operational decision support models and algorithms for production process scheduling in hospital catering. The addressed production system is considered as a flexible job shop system. The objective is to minimize the total flow time. A novel mathematical model and two metaheuristics for the production scheduling of multi-product and multi-stage food processes are developed. These methods have proven their effectiveness for the scheduling of operations of the food production processes and allowed significant improvements in the performance of the studied production system.

Keywords: hospital catering; production scheduling; flexible job shop problem; mathematical model; genetic algorithm; local search method; iterated local search algorithm

check for **updates**

Citation: Abderrabi, F.; Godichaud, M.; Yalaoui, A.; Yalaoui, F.; Amodeo, L.; Qerimi, A.; Thivet, E. Flexible Job Shop Scheduling Problem with Sequence Dependent Setup Time and Job Splitting: Hospital Catering Case Study. *Appl. Sci.* **2021**, *11*, 1504. https://doi.org/10.3390/app11041504

Received: 23 December 2020
Accepted: 1 February 2021
Published: 7 February 2021

Publisher's Note: MDPI stays neutral with regard to jurisdictional claims in published maps and institutional affiliations.

1. Introduction

Nowadays, hospital logistics has become an essential component of healthcare institutions. It allows the synchronization of all the flows inside a hospital to ensure the efficiency of the healthcare system. For many years, the management was commonly focused on improving the quality of medical care, while less attention was usually devoted to operation management. In recent years, the need for containing the costs while increasing the competitiveness along with the new national health service policies for hospital financing forced hospitals to necessarily improve their operational efficiency. It is in this context that the efficient use of resources and the research on optimal service stimulate logistical thinking in hospitals. The difficulties of optimizing flows have led managers to discover new avenues for rationalizing expenses and seeking refined solutions to these difficulties. In this context, optimized logistics solutions allow hospitals to improve inventory management, limit waste, and provide better inventory tracking and traceability of service products. On the other hand, the supply chain is a major source of costs, and its reorganization would make it possible to achieve crucial savings on all hospital expenses.

A hospital's logistics is part of its global performance, where the activities are organized and structured with the aim of patients' satisfaction in terms of quality, quantity, delay, safety, and low cost. The main purpose of this logistics is to control and optimize physical flows from suppliers to patients at the best cost that respects technical, economic, and regulatory conditions for optimal dispensing to patients. Hospital logistics is a complex process characterized by a diversity of needs, users, products, and distribution channels. The coordination of these activities requires logistical expertise that few institutions will be able to develop on their own. This has led researchers to focus for some years on the

45

management and optimization of the supply chain in hospitals. In this context and in order to improve the working conditions of the employees and their well-being, the hospital center of Troyes implements means to improve its daily efficiency. The hospital is carrying out a revision of its supply chain, which must notably consider the management of food flows within the hospital. In the present work, we focused particularly on the scheduling of the food manufacturing process in hospital catering, which is considered as a flexible job shop scheduling problem with a sequence-dependent setup time and job splitting by integrating specific industrial constraints.

The remainder of this paper is organized as follows: Section 2 presents the state of the art regarding the problem of food production process scheduling. The problem statement is defined in Section 3. In Section 4, the mathematical model developed for the studied problem is presented. Sections 5 and 6 present the genetic and the iterated local search algorithms specifically developed for the problem, the different elements of these metaheuristics, and the computational results. Finally, in the last section, an application to a real industrial case and the results obtained are presented.

2. Literature Review

The production scheduling problem in food industries belongs to a famous class of problems referred to as scheduling with sequence-dependent setups, which are well known to be NP-hard (Sun et al. [1]). In recent years, there has been great interest in the development of intelligent solutions for this problem in various fields of application. The promising results of scheduling methods, such as reduction of production costs, increased throughput and smoother operation of the production equipment, and improvement of working conditions and the well-being of employees, have stimulated considerable research efforts. The existing works in the literature are classified according to the number of products (single products or multiple products), the type of production system, and the expiration dates of products, which may be known or unknown.

Regarding works dealing with a single product, Entrup et al. [2] proposed three different mixed-integer linear programming for scheduling problems in the packing stage of stirred yogurt production in the fresh food industry. They accounted for shelf life issues and fermentation capacity limitations. Doganis and Sarimveis [3] proposed a model that aims to obtain the optimal production scheduling in a single yogurt production line. The model takes into account all the standard constraints encountered in production scheduling (material balances, inventory limitations, machinery capacity). It also considers special features that characterize yogurt production and that are limitations in production sequencing, mainly due to different fat contents and flavors of various products, as well as sequence-dependent setup times and costs. However, the model is limited to a single production line. In another study, Doganis and Sarimveis [4] presented a methodology for optimum scheduling of yogurt packaging lines that consist of multiple parallel machines. The methodology incorporates features that allow one to tackle industry-specific problems, such as multiple intermediate due dates, job mixing and splitting, product-specific machine speed, minimum and maximum lot size, and sequence-dependent changeover times and costs. However, the proposed mathematical model does not incorporate multi-stage production decisions and ignores some industry-specific characteristics, such as shelf life. Stefansdottir et al. [5] developed a generic optimization model for lot sizing and scheduling in the typical processing industry setting of flow shops. Sargut and Isık [6] presented a mathematical model for a dynamic economic lot sizing problem with a single machine for a single perishable item under production capacities. They also gave a dynamic programming-based heuristic for the solution of the overall problem.

Many studies have been carried out in the literature on the production scheduling of multi-product food processing industries. Akkerman and van Donk [7] developed a methodology for the analysis of the scheduling problems in food processing. This helps one to understand, describe, and structure scheduling problems in food processing and to evaluate the organizational structures and information flows related to scheduling. Smith, Daniels, and Larry [8] developed a general lot-sizing model for processing industries and applied their method to a representative situation of a food processing facility. Kopanos et al. [9] offered an efficient mathematical framework for detailed production scheduling in the food processing industry. Wauters et al. [10] introduced an integrated approach for production scheduling and demonstrated its applicability to the food processing industry. In this work, the scheduling had to deal with a combination of discrete, continuous, and batch processes, and it was complicated by particular characteristics. Acevedo-Ojeda et al. [11] presented a lot-sizing problem with a single machine that incorporated raw material perishability and analyzed how these considerations enforced specific constraints on a set of fundamental decisions, particularly for multi-level structures. Three variants of the two-level lot-sizing problem incorporating different types of raw-material perishability—fixed shelf life, functionality deterioration, and functionality–volume deterioration—were studied; the authors proposed mixed-integer programming formulations for each variant and performed computational experiments with sensitivity analyses. Copil et al. [12] considered a capacitated dynamic lot-sizing problem with parallel machines for the food industry, in which a given product produced during a specified time period is used to satisfy the related demand. Niaki et al. [13] addressed the integrated lot-sizing and scheduling problem of food production in batch manufacturing systems with multiple shared common resources and proposed a new mixed-integer linear programming formulation with multiple objective functions. Wei et al. [14] proposed a classical multi-level lot-sizing and flow-shop scheduling problem formulation to incorporate perishability issues.

Regarding the shelf life of products, Ahumada and Villalobos [15] reviewed models for the agri-food business, where products may be perishable or not, but their focus was on procurement and harvesting planning. The only goods they were interested in were crops. Sel et al. [16] introduced the planning and scheduling of decisions considering the shelf-life restrictions, product-dependent machine speeds, demand due dates, and regular and overtime working hours in the perishable supply chain. Arbib et al. [17] considered a three-dimensional matching model for perishable production scheduling, which was studied under two independent aspects: the relative perishability of products and the feasibility of launching/completion time. Basnet et al. [18] described an exact algorithm to solve a scheduling and sequencing problem in the same industry. Chen et al. [19] provided a review of literature on the integration of scheduling and lot sizing for perishable food products, and they categorized the papers by the characteristics of the lot sizing and scheduling that were included in their models, as well as the strategies used to model perishability.

In the present work, the studied production system is considered as a flexible job shop system. Since 1990, the flexible job shop scheduling problem (FJSP) has been extensively investigated by researchers. Liu and MacCarthy [20] discussed the problem in a flexible manufacturing system with transportation times and limited buffers. They developed a mixed-integer linear programming model and heuristics to minimize the makespan, the mean completion time, and the maximum tardiness. Guimaraes and Fernandes [21] proposed a genetic algorithm for the FJSP, where the objective function is used to minimize the makespan and the mean tardiness. Saidi-Mehrabad and Fattahi [22] took into account a special case of FJSP, where each operation could be assigned to one of two parallel machines. A tabu search algorithm for solving the sequencing and assignment problems was developed. The algorithm was compared to a branch and bound algorithm on several random test instances. Defersha and Chen [23] studied the FJSP with attached and detached setup times, machine release dates, and technological time lag constraints. For this problem, a mixed-integer linear programming model was proposed, and a parallel genetic algorithm using multiple threads was introduced. Mati et al. [24] proposed a genetic algorithm for an FJSP in which blocking constraints were taken into consideration. Bagheri and Zandieh [25] developed a variable neighborhood search algorithm. Mousakhani [26] presented a mixed-integer linear programming model and an iterated local search algorithm that minimize the total tardiness. Chaudhry and Khan [27] published a literature review on the methods used to solve the FJSP and the studied objective functions (Figure 1). In this bibliographic study, it was found that almost 59% of the papers used hybrid methods (Rajabinasab and Mansour [28], Geyik and Dosdogru [29], Zhou and Zeng [30]) or evolutionary algorithms. The minimization of the makespan turned out to be the most widely studied criterion. In 88 research papers (44.67%), the makespan was used as the sole objective function, while in another 78 papers (39.59%), makespan was used in combination with another objective function. However, the minimization of the total flow time, which is the target criterion in this work, has been studied very little in the literature. Table 1 represents some research works on flexible job shop problems with minimization of flow time. These works are classified according to the fields of its application in research or an industrial case, the resolution methods used, some workshop details, and the studied objective functions.

Table 1. Some research works on flexible job shop problems with flow time minimization.

Author	Year	Application	Algorithm and Workshop Details	Objective Function
The problem considered in this paper	2021	Research and industry	Mathematical model and metaheuristics for a flexible job shop problem with sequence-dependent setup time and job splitting	Minimum of total flow time
Gao et al.	2018	Research	Mathematical model and discrete Jaya algorithm	Minimum of makespan, total flow time, machine workload, and total machine workload
Gao et al.	2017	Research	Resolution approaches for uncertain resource assignment and job sequence in an automated flexible job shop	Minimum of makespan, number of late jobs, total flow time, and total weighted flow time
Nie et al.	2013	Research	Gene expression programming, dynamic flexible job shop problem with job release dates	Minimum of makespan, mean flow time, and mean tardiness
Doh et al.	2013	Research	Heuristic with priority rules	Minimum of makespan, total flow time, mean tardiness, number of tardy jobs, and max tardiness
Lee et al.	2012	Research	Heuristic	Minimum of makespan and mean flow time
Liu et al.	2009	Research	Multi-swarm particle swarm optimization	Minimum makespan of total flow time
Tanev et al.	2004	Industry	Genetic algorithm and prioritydispatching rules	Minimum ratio of tardy jobs, variance of the flow time, amount of mold changes, and maximum efficiency of machines

Finally, it is worth mentioning that, to the best of our knowledge, there are almost no studies addressing the problem of scheduling food production processes in hospital or collective catering. Most existing works in the literature on scheduling food production processes are from the food and dairy industries, where the production systems are parallel machine, flow shop, and single machine, in most cases. Moreover, in the majority of these works, the expiration date of products is unknown, while in this study, it is known. The works found in the literature cannot be adapted to the problem addressed in this study, since the production systems are different. In addition, several constraints specific to the considered problem have not been taken into account in the existing works, and they do not have the same objective of optimization.

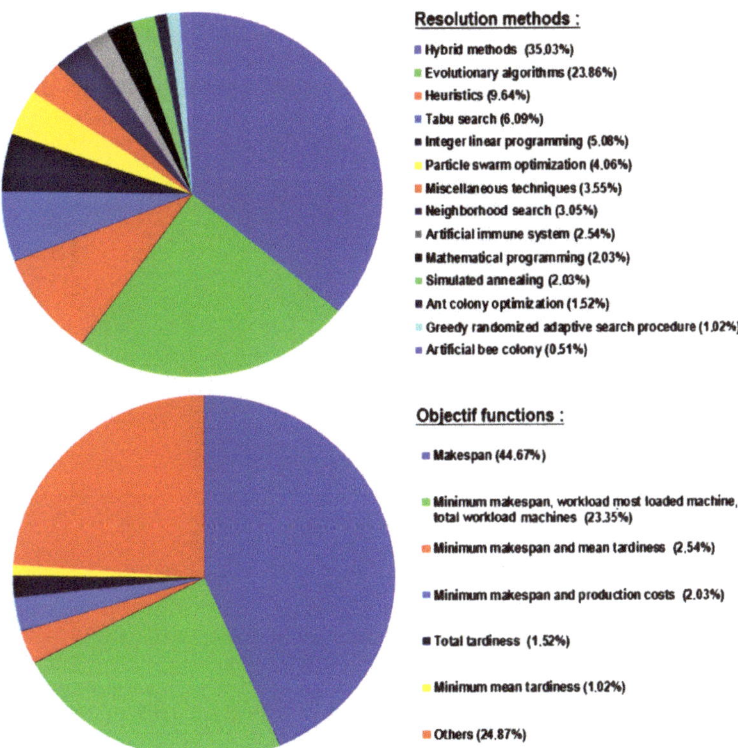

Figure 1. Bibliographical summary of the resolution methods used for flexible job shop problems and the objective functions studied in the literature: 191 papers between 1990 and 2014 (Chaudhry and Khan [27]).

3. Problem Description

The problem of food production process scheduling considered in this study aims to schedule the operations from the pre-treatment of raw materials to the stock of finished products of a meal manufacturing process in hospital catering or, more generally, in collective catering. Figure 2 represents the typical production areas and the different material resources available for realizing the operations of the meal-making process. This problem is considered as a flexible job shop problem with sequence-dependent setup time and job splitting by integrating specific industrial constraints, such as the days of pre-treatment and production of products, the product delivery times, and the amounts of human and material resources available, in addition to the different constraints described below.

The addressed problem is considered as a flexible job shop scheduling problem with sequence-dependent setup time, since each job has its own order of operations and each operation has to be assigned to one among a set of possible machines. This problem can be described by a set of N jobs, where each job i corresponds to the preparation of a dish characterized by a number of portions Q_i (quantity) and a set of operations J_i for the preparation of the dish (from raw material to finished product). It is worth highlighting that the dishes to be prepared do not have the same order of operating ranges (set of operations necessary for the preparation of the dish). For each job, there is a due date D_i to respect.

For each operation of an operating range, there is a set of material resources that are used to realize it, such as ovens, cooking pots, cookers, induction hobs, packaging machines, cooling cells (Figure 2). These material resources can be classified into three categories: M_1, M_2, and M_3. $M_1 \subset M$ is composed of material resources with a capacity of one portion and that can not process several jobs at the same time (material resources that can perform preprocessing and cold production operations). The set $M_2 \subset M$ represents the material resources with a capacity greater than one portion and that cannot process several jobs at the same time (ovens, etc.). $M_3 \subset M$ correspond to the set of material resources with a capacity greater than one portion and that can process several jobs at the same time (cooling cells).

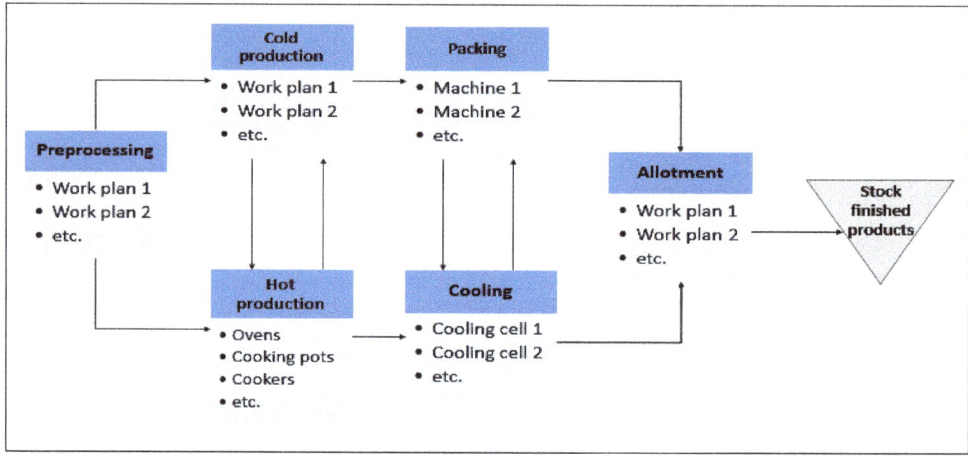

Figure 2. Representative scheme of the meal production process.

For each material resource, there is a setup time to take into account, which corresponds to the preparation time of the machine before carrying out an operation and the cleaning time of the machine between two consecutive operations. A time window of availability is known for each material resource.

Note that the corresponding machines may not be identical, involving different processing times according to the chosen machine. The setup times of machines are sequence dependent because they depends on the preceding operation on the same machine. The food production process scheduling involves two steps: (i) assignment of operations to machines and (ii) sequencing of operations on machines.

As mentioned previously, in order to respect the production capacities of material resources, a job can be split into smaller sub-lots in such a way that the operations of sub-lots of a job can be performed simultaneously on different machines. This strategy, which is useful when machine capacity does not allow the treatment of the whole job, also enables a more efficient processing scheme.

The criterion to minimize in the present study is the total flow time of jobs in the production system. The choice of this criterion is based on the fact that the respect of the cold chain at each stage of the product life cycle must be ensured. The aim is to constantly maintain a low temperature (positive or negative, depending on the product) to ensure the maintenance of all the food qualities (hygienic, nutritional, and gustatory).

4. Mathematical Model

In this section, a mixed-integer linear programming model is presented. This model formalizes and solves small-sized instances of the studied problem by using the Cplex solver. The solutions of the small-sized instances can be used to validate the efficiency of the developed metaheuristic methods.

4.1. Assumptions

The mathematical model for the scheduling of the food production process inherits its main assumptions from the standard flexible job shop scheduling problem with sequence-dependent setup times and additional features due to the job splitting:

- Jobs are independent of each other,
- A job can be split into sub-lots,
- The sub-lots of a job can be grouped on the machines to be treated at the same time,
- Each sub-lot of a job consists of a set of operations that must be processed consecutively (precedence constraints between operations of sub-lots of jobs),
- Each operation of a sub-lot has a given processing time,
- The preemption of operations of sub-lots of jobs is not allowed, i.e., operation processing on a machine cannot be interrupted,
- Each job has a given due date (finish date of production at latest),
- Sub-lot sizes (number of portions) are discrete,
- Sub-lots creation is consistent throughout the processing sequence, meaning that job splitting and sub-lot sizes remain constant for all operations,
- Machines are independent,
- A machine can process, at most, one operation of a job at the same time,
- The setup times of machines are dependent on the sequence of operations of sub-lots of jobs,
- Material resources have given availability time windows that must be taken into account.

Taking these assumptions into account, the objective is to find a schedule involving sub-lot assignments to machines and sub-lot sequencing for each machine in such a way that each job's demand is fulfilled, different constraints of the problem are respected, and the total flow time of jobs in the production system is minimized.

4.2. Notations

The definition of the developed mathematical model's parameters relies on the following sets and indexes:

- M: set of all material resources, where $m = |M|$.
- N: set of jobs (dishes to prepare), where $n = |N|$ and $\{0, n+1\}$ are two dummy jobs.
- J_i: set of operations of job $i \in N$, such that the operation $j \in J_i$ is done before the operation $j+1 \in J_i$ and $|J_0| = |J_{n+1}| = 1$.
- Q_i: number of portions (quantity) of job $i \in N$.
- q_i: number of portions in each sub-lot of job $i \in N$.
- L_i: set of sub-lots of job $i \in N$, with $|L_0| = |L_{n+1}| = 1$ and $l_i = |L_i|$ such that $l_i = \lceil \dfrac{Q_i}{q_i} \rceil$.
- d_i: due date of job $i \in N$.
- $M_{ij} \subset M$: set of material resources that can perform the operation $j \in J_i$ of job $i \in N$.
- R_k: maximum capacity in number of portions of the material resource $k \in M$.
- P'_{ijk}: unit processing time of operation $j \in J_i$ of job $i \in N$ on the material resource $k \in M_1$.
- P_{ijk}: processing time of operation $j \in J_i$ of job $i \in N$ on the material resource $k \in M_2 \cup M_3$.
- s_{ijhgk}: setup time of material resource $k \in M_{ij} \cap M_{hg}$ if the operation $j \in J_i$ of job $i \in N$ directly precedes the operation $g \in J_h$ of job $h \in N$ on the material resource $k \in M_{ij} \cap M_{hg}$.
- T_k: preparation time of the material resource k at the beginning of scheduling.
- E_k: cleaning time of the material resource k at the end of scheduling.
- $[A_k, Y_k]$: time window of availability of the material resource $k \in M$.
- B: big integer.

4.3. Decision Variables

- X_{iljk}: binary variable that equal to 1 if operation $j \in J_i$ of sub-lot $l \in L_i$ of job $i \in N$ is assigned to the material resource $k \in M_{ij}$, and 0 otherwise.
- $F_{iljhl'gk}$: binary variable that is equal to 1 if operation $j \in J_i$ of sub-lot $l \in L_i$ of job $i \in N$ directly precedes operation $g \in J_h$ of sub-lot $l' \in L_h$ of job $h \in N$ on the material resource $k \in M_{ij} \cap M_{hg}$, and 0 otherwise.
- $Z_{ill'jk}$: binary variable that is equal to 1 if operation $j \in J_i$ of sub-lot $l \in L_i$ of job $i \in N$ starts and finishes at the same time as the operation $j \in J_i$ of sub-lot $l' \in L_i$ of job $i \in N$ on the material resource $k \in M_{ij}$, and 0 otherwise.
- S_{iljk}: starting time of operation $j \in J_i$ of sub-lot $l \in L_i$ of job $i \in N$ on the material resource $k \in M_{ij}$.
- C_{iljk}: completion time of operation $j \in J_i$ of sub-lot $l \in L_i$ of job $i \in N$ on the material resource $k \in M_{ij}$.
- C_i: completion time of job $i \in N$.

4.4. Mathematical Model

In the literature, there exist different mathematical model formulations for the standard flexible job shop scheduling problem. The mathematical model proposed by Buddala and Mahapatra [31] takes into account constraints on processing times, precedence between operations, and machine capacity in number of operations processed at the same time, which corresponds with the studied problem. However, this model does not take into consideration the constraints on machine capacity in the number of portions, machine availability time windows, sequence-dependent setup time, due dates of jobs, splitting of jobs, and batching of sub-lots.

The developed mathematical model that integrates all these constraints is formulated as indicated through Equations (1) to (22):

$$Min \sum_{i \in N} C_i \tag{1}$$

$$C_i \geq \sum_{k \in M_{ij}} C_{iljk}, \ \forall \ i \in N,\, l \in L_i,\, j \in J_i \tag{2}$$

$$S_{iljk} + C_{iljk} \leq B * X_{iljk}, \ \forall \ i \in N, l \in L_i, j \in J_i, k \in M_{ij} \tag{3}$$

$$C_{iljk} - S_{iljk} \geq P_{ijk} - B * (1 - X_{iljk}), \ \forall \ i \in N, l \in L_i, j \in J_i, k \in M_2 \cup M_3 \tag{4}$$

$$C_{iljk} - S_{iljk} \geq P'_{ijk} * Q_i - B * (1 - X_{iljk}), \ \forall \ i \in N, l \in L_i, j \in J_i, \ k \in M_1 \tag{5}$$

$$S_{hl'gk} \geq C_{iljk} + s_{ijhgk} - B * (1 - F_{iljhl'gk}), \ \forall \ i \in N, l \in L_i, j \in J_i, h \in N, l' \in L_h, g \in J_h, \\ k \in M_{ij} \cap M_{hg} \backslash M_3 \tag{6}$$

$$S_{il'jk} \geq C_{iljk} + s_{ijijk} - B * (1 - F_{iljil'jk}) - B * Z_{ill'jk}, \ \forall \ i \in N, l, l' \in L_i, j \in J_i, k \in M_{ij} \cap M_{hg} \tag{7}$$

$$S_{iljk} - S_{il'jk} \leq B * (1 - Z_{ill'jk}), \ \forall \ i \in N, l, l' \in L_i, j \in J_i, k \in M_2 \cup M_3 \tag{8}$$

$$C_{iljk} - C_{il'jk} \leq B * (1 - Z_{ill'jk}), \quad \forall \ i \in N, l, l' \in L_i, j \in J_i, k \in M_2 \cup M_3 \tag{9}$$

$$\sum_{h \in N \setminus \{0\}} \sum_{l' \in L_h} \sum_{g \in J_h} F_{iljhl'gk} = 1, \quad \forall \ i \in N \setminus \{n+1\}, l \in L_i, j \in J_i, k \in M_{ij} \cap M_{hg} \tag{10}$$

$$\sum_{i \in N \setminus \{n+1\}} \sum_{l \in L_i} \sum_{j \in J_i} F_{iljhl'gk} = 1, \quad \forall \ h \in N \setminus \{0\}, l' \in L_h, g \in J_h, k \in M_{ij} \cap M_{hg} \tag{11}$$

$$\sum_{k \in M_{ij}} S_{iljk} - \sum_{k \in M_{ij-1}} C_{ilj-1k} \geq 0, \quad \forall \ i \in N, \ l \in L_i, \ j \in J_i \tag{12}$$

$$\sum_{k \in M_{ij}} X_{iljk} = 1, \quad \forall \ i \in N, \ l \in L_i, \ j \in J_i \tag{13}$$

$$C_i \leq d_i, \quad \forall \ i \in N \tag{14}$$

$$\sum_{l,l' \in L_i} q_i * Z_{ill'jk} \leq R_k, \quad \forall \ i \in N, \ j \in J_i, k \in M_2 \cup M_3 \tag{15}$$

$$S_{iljk} \geq A_k, \quad \forall \ i \in N, \ l \in L_i, \ j \in J_i, \ k \in M_{ij} \tag{16}$$

$$C_{iljk} \leq Y_k, \quad \forall \ i \in N, \ l \in L_i, \ j \in J_i, \ k \in M_{ij} \tag{17}$$

$$Z_{ill'jk} = 0, \quad \forall \ i \in N, l, \ l' \in L_i, \ j \in J_i, \ k \in M_1 \tag{18}$$

$$X_{iljk} \in \{0,1\}, \quad \forall \ i \in N, \ l \in L_i, \ j \in J_i, \ k \in M_{ij} \tag{19}$$

$$Z_{ill'jk} \in \{0,1\}, \quad \forall \ i \in N, l, l' \in L_i, j \in J_i, k \in M_{ij} \tag{20}$$

$$S_{iljk} \geq 0, C_{iljk} \geq 0, C_i \geq 0, \quad \forall \ i \in N, l \in L_i, j \in J_i, k \in M_{ij} \tag{21}$$

$$F_{iljhl'gk} \in \{0,1\}, \quad \forall \ i \in N, l \in L_i, j \in J_i, h \in N, l' \in L_h, g \in J_h, k \in M_{ij} \cap M_{hg}. \tag{22}$$

In the mathematical model presented previously, Equation (1) represents the objective function, which consists in minimizing the total flow time of jobs in the production system. It is defined as the sum of the completion times of all jobs since the release dates are equal to zero. The completion times of jobs are computed as the completion time of the last sub-lot derived from the considered job, as indicated in Equation (2). Note that, due to Equation (3), for given $i \in N, l \in L_i$, and $j \in J_i$, variables S_{iljk} and C_{iljk} are equal to zero if the machine $k \in M_{ij}$ is not chosen to realize the operation of the considered sub-lot. On the other hand, when X_{iljk} is equal to 1, Equations (4) and (5) activate the relationship between the starting time and completion time of an operation of a sub-lot. In this case, the processing time does not depend on the quantity of jobs for the material resources $M_2 \cup M_3$ (Equation 4), but it depends on the quantity of jobs for the material resources M_1 Equations (5) and (6) consider sequence-dependent setup times between the completion time and starting time of two operations of sub-lots that are processed on a machine one after another. Equation (7) disable Equation (6) if two different sub-lots of the same job are performed at the same time by the same material resource. Equations (8) and (9) require that if two operations of two different sub-lots of the same job are assigned at the same time to a material resource M_2 or M_3, they must have the same respective starting time and completion time. Equation (10) ensure that only one operation immediately follows the jth operation of sub-lot $l \in L_i$ of job $i \in N$ on machine $k \in M_{ij} \cap M_{hg}$, and Equation (11) guarantee that only one operation immediately precedes the gth operation of sub-lot $l' \in L_h$ of job $h \in N$ on machine $k \in M_{ij} \cap M_{hg}$. Equation (12) establishes the precedence constraints between two consecutive operations of the same sub-lot. Equation (13) enforce that each operation of each sub-lot should be assigned to exactly one machine among the possible ones. The respect for the due dates of jobs is modeled by Equation (14). Equation (15) ensure that the capacities of material resources in the number of portions are respected. The respect for time windows of availability of material resources is modeled by Equations (16) and (17). Finally, Equations (19)–(22) define the domain of the decision variables.

4.5. Computational Results of the Mathematical Model

The mathematical model presented previously was implemented in the Java programming language using the Cplex library. This mathematical model was tested on 150 instances of different types: real instances of the hospital center of Troyes (HCT), randomly generated instances of the HCT type, randomly generated instances, and instances adapted from the literature (Sriboonchandr et al. [32], Nouri et al. [33], Azzouz et al. [34], Mousakhani [26], Lee et al. [35], Bagheri and Zandieh [25], Pezzella et al. [36]). The details of the results of these different types of instances are presented in Section 6. The mathematical model was able to find solutions in less than three hours of execution for the small instances with less than eight jobs, 10 sub-lots, 39 operations, and 29 machines. The execution times of the mathematical model for these instances varied according to the number of jobs, sub-lots, and operations. The computational results of the proposed mathematical model for different types of instances and, specifically, for real instances of HCT show the limits of an exact resolution for the problem of scheduling of the food production process.

5. Resolution Methods

5.1. Genetic Algorithm

Solving the flexible job shop problem is known to be strongly NP-hard (Xia and Wu [37], Fattahi et al. [38]). The introduction of sequence-dependent setup time and job splitting complicates the already difficult flexible job shop problem. In order to solve this problem efficiently, we developed a hybrid method that combines a genetic algorithm and three local research methods. The elements of this hybrid method are presented in the following subsections.

5.1.1. Solution Representation

By solving a flexible job shop scheduling problem using a genetic algorithm, Kacem [39] used a solution representation by coding both the assignment and the sequencing of operations on different machines. A similar representation is used in this paper to solve the flexible job shop scheduling problem with job splitting by considering each sub-lot as a job. To illustrate this representation, let us consider a small example with three jobs and four machines. The numbers of operations of sub-lots for each job and all the machines eligible for each operation are given in Figure 3. A representation of the assignment of operations to machines and their sequencing is coded in a chromosome, as shown in Figure 3. In this chromosome, each gene is represented by a quadruple (i, l, j, k), designating the assignment of operation j of sub-lot l of job i to machine k. The sequence of genes in the chromosome represents the sequence of operations on machines. The chromosome decoding procedure is carried out by reading it from left to right. For example, the assignment and sequencing of operations on machine 1 can be decoded as follows: $(1, 1, 1, 1) \rightarrow (1, 2, 1, 1) \rightarrow (3, 1, 3, 1)$

$\rightarrow (3, 2, 3, 1) \rightarrow (3, 3, 3, 1)$. This information is obtained from genes 3, 8, 10, 16, and 17 in the chromosome, where $k = 1$. In this chromosome, for a given i and l, the gene (i, l, j, k) is always located on the right of all the other genes (i, l, j, k') with $j' < j$. This ensures that the precedence requirements of the operations of a particular sub-lot are not violated.

Jobs	Operations	Sub-lots	Set of eligible machines for operations		
			j1	j2	j3
1	3	3	{k1, k2}	{k3}	{k2, k4}
2	2	2	{k3, k4}	{k2}	
3	3	3	{k3}	{k2, k4}	{k1}

1	2	3	4	5	6	7	8	9	10	11	12	13	14	15	16	17	18	19	20	21	22	
3,1,1,3	3,3,1,3	1,1,1,1	1,3,1,2	3,1,2,4	3,2,1,3	3,3,2,4	1,2,1,1	1,1,2,3	3,1,3,1	1,2,2,3	1,1,3,2	3,2,2,2	2,1,1,3	1,2,3,4	3,2,3,1	3,3,3,1	2,2,1,4	1,3,2,3	1,3,3,2	2,2,2,2	2,1,2,2	i, l, j, k

i = job index, l = sub-lot index, j = operation index, k = machine index

Figure 3. Representation of the assignment of operations to machines and their sequencing.

5.1.2. Initial Population

The initial population plays an important role in the performance of genetic algorithms. An initial population with great diversity between solutions can avoid falling into a premature convergence or a local minimum. Different assignment and sequencing rules were used to achieve the diversity of solutions:

- Random assignment (RA): The operations are randomly assigned to machines.
- SPT assignment (SPTA): For each operation, the machine with a smaller processing time is selected to perform this operation.
- LPT assignment (LPTA): For each operation, the machine with a longer processing time is selected to perform this operation.
- Minimum machine workload assignment (MMWA): The operations are iteratively assigned to machines based on their processing times and machine workloads. The workload of a machine depends on its type. For the set of machines M_1, the workload is the sum of processing times of operations assigned to the machine. For the set of machines M_2 and M_3, the workload is the total machine occupation time. The procedure consists in finding the machine with the minimum workload for each operation.
- Random sequence (RS): This heuristic randomly orders the operations on each machine.
- SPT sequence (SPTS): Operations with the shortest processing time will be processed first.
- Most number of operations Remaining (MNOR): This heuristic consists of processing the operations of the sub-lot of the job that has the most operations remaining as a priority.
- Most work remaining (MWR): The operations that have the most remaining processing time will be prioritized for processed.

For the construction of the initial population using the heuristics described above, the procedure consists of randomly choosing an assignment and a sequencing heuristic and building solutions, such as the constraints of precedence between the operations of sub-lots of jobs, the due dates of jobs, the time windows of availability of material resources, and the respect for the production capacities of machines.

5.1.3. Fitness Evaluation

Algorithm 1 decodes the chromosomes to obtain the objective function value and a workable representation of the solutions.

Algorithm 1: Evaluation steps of the fitness function of a chromosome

---

- **Step 01:** Set $p = 1$.
- **Step 02:** Set i, l, j, k, the index values of the gene located at the P position of the chromosome.
- **Step 03:** Calculate the completion time C_{iljk}.

  - If operation j of sub-lot l of job i is the first operation assigned to the machine k and $j = 1$,

  $$S_{iljk} = A_k + T_k; \quad C_{iljk} = S_{iljk} + P_{ijk}; \quad C_i = Max\{C_i, C_{iljk}\}$$

  - If operation j of sub-lot l of job i is the first operation assigned to the machine k, $j > 1$, and the operation $j - 1$ is assigned to the machine k',

  $$S_{iljk} = Max\{A_k + T_k, C_{ilj-1k'}\}; \quad C_{iljk} = S_{iljk} + P_{ijk}; \quad C_i = Max\{C_i, C_{iljk}\}$$

  - If operation j' of sub-lot l' of job i' is the operation to be processed immediately before the operation j of sub-lot l of job i on the machine k and $j = 1$,

  $$S_{iljk} = s_{i'j'ijk} + C_{i'l'j'k}; \quad C_{iljk} = S_{iljk} + P_{ijk}; \quad C_i = Max\{C_i, C_{iljk}\}$$

  - If operation j of sub-lot l of job i and the operation j of sub-lot l' of job i are assigned to the machine k and $j = 1$,
    - If $(k \in M_2 \cup M_3$ and $2 * q_i <= R_k)$

    $$S_{iljk} = S_{il'jk}; \quad C_{iljk} = S_{iljk} + P_{ijk}; \quad C_i = Max\{C_i, C_{iljk}\}$$

    - If $(k \in M_2 \cup M_3$ et $2 * q_i > R_k)$ ou $(k \in M_1)$

    $$S_{iljk} = F_{il'jk}; \quad C_{iljk} = S_{iljk} + P_{ijk}; \quad C_i = Max\{C_i, C_{iljk}\}$$

  - If operation j' of sub-lot l' of job i' is the operation to be processed immediately before the operation j of sub-lot l of job i on the machine k, $j > 1$, and the operation $j - 1$ is assigned to the machine k',

  $$S_{iljk} = Max\{s_{i'j'ijk} + C_{i'l'j'k}, C_{ilj-1k'}\}; \quad C_{iljk} = S_{iljk} + P_{ijk}; \quad C_i = Max\{C_i, C_{iljk}\}$$

  - If operation j of sub-lot l of job i and the operation j of sub-lot l' of job i are assigned to the machine k, $j > 1$, and the operation $j - 1$ is assigned to the machine k',
    - If $(k \in M_2 \cup M_3$ and $2 * q_i <= R_k)$

    $$S_{iljk} = S_{il'jk}; \quad C_{iljk} = S_{iljk} + P_{ijk}; \quad C_i = Max\{C_i, C_{iljk}\}$$

    - If $(k \in M_2 \cup M_3$ and $2 * q_i > R_k)$ or $(k \in M_1)$

    $$S_{iljk} = Max\{C_{il'jk}, C_{ilj-1k'}\}; \quad C_{iljk} = S_{iljk} + P_{ijk}; \quad C_i = Max\{C_i, C_{iljk}\}$$

- **Step 04:** If p is less than the total number of operations of sub-lots of jobs, increment its value by 1 and go to Step 2; otherwise, go to Step 05.
- **Step 05:** Calculate the fitness function of the solution by using the equation $\sum_i C_i$.

It is worth noting that for a given i and l, the gene (i, l, j, k) is always located on the right of all the other genes (i, l, j', k') with $j' < j$. Based on this property, when the completion time of operation (i, l, j, k) on machine k is to be calculated, the completion time of operation $(i, l, j - 1, k')$ is already calculated and available, regardless of the machine to which this preceding operation is assigned. Moreover, the completion time of the operation (i', l', j', k) to be processed on machine k immediately before operation (i, l, j, k) is also calculated and available.

Genetic operators evolve the population to promising regions of the research space. The convergence behavior of the algorithm depends largely on them. These operators are generally categorized as selection, crossover, and mutation operators.

5.1.4. Selection Operator

The process of selecting two parents from the population for reproduction is called selection. The aim of the selection operator is to highlight individuals with the best fitness in hopes that their resulting offspring are more fit individuals. In the proposed genetic algorithm, we used the two-way tournament selection operator introduced in [40]. The selection operator is involved in holding competitions among randomly selected individuals and choosing the one with the best fitness (smallest total flow time). This individual is added to a mating population, which is used to form the next generation. Then, the individuals in the tournament are placed back in the current population, and the process is repeated until the number of individuals added to the mating population is equal to the population size.

5.1.5. Crossover Operator

After the selection of chromosomes for reproduction, a crossover operator is applied to combine the features of the parent chromosomes in order to produce a new child. The individuals in the mating population are randomly paired to create child individuals. For two parents, the algorithm arbitrarily selects one of the available crossover operators and applies it with a certain probability to create two child individuals by exchanging information contained in the parent chromosomes.

Different crossover operators are proposed and can be categorized as assignment or sequence crossover operators. The assignment crossover operators consist of generating offsprings by exchanging the assignment of operations to machines in the parent chromosomes. The role of sequence crossover operators is to produce two new offspring by exchanging partial information of the sequences of operations on machines in the parent chromosomes. The assignment and sequence crossover operators developed in the proposed genetic algorithm are: OMAC (operation to machine assignment crossover), JLOSC (job-level operation sequence crossover), and SLOSC (sub-lot-level operation sequence crossover).

From a given pair of parent chromosomes, OMAC creates two child chromosomes, preserving the sequence of operations in each parent chromosome and modifying the machines' affectation. The creation of a child chromosome using an OMAC crossover operator is illustrated in Figure 4. We assume that in the first step, the operations located in genes 1, 3, 4, 7, 12, 13, 14, and 15 are selected from parent 1 to change their assignments. Then, the second step is to copy all the genetic information of parent 1 to the offspring, except for the assignment information of the selected operations. The final step is to obtain the assignment information of the selected operations from parent 2 and then copy them to the corresponding genes in the offspring to produce a new child chromosome. The same process is repeated to create child 2 by starting to select the operations from the chromosome parent 2. Three versions of OMAC were developed to correspond to different ways of selecting the operations for which the assignment will be modified:

- OMAC 1: A set of operations of a sub-lot of jobs is chosen randomly.
- OMAC 2: A set of operations of sub-lots assigned to the loaded machines is selected.
- OMAC 3: A set of operations of sub-lots of jobs with larger completion times is chosen.

The objective of the crossover operator OMAC 2 is to balance the workload between the machines, whereas the goal of the crossover operator OMAC 3 is to reduce the ending dates of jobs with longer completion times.

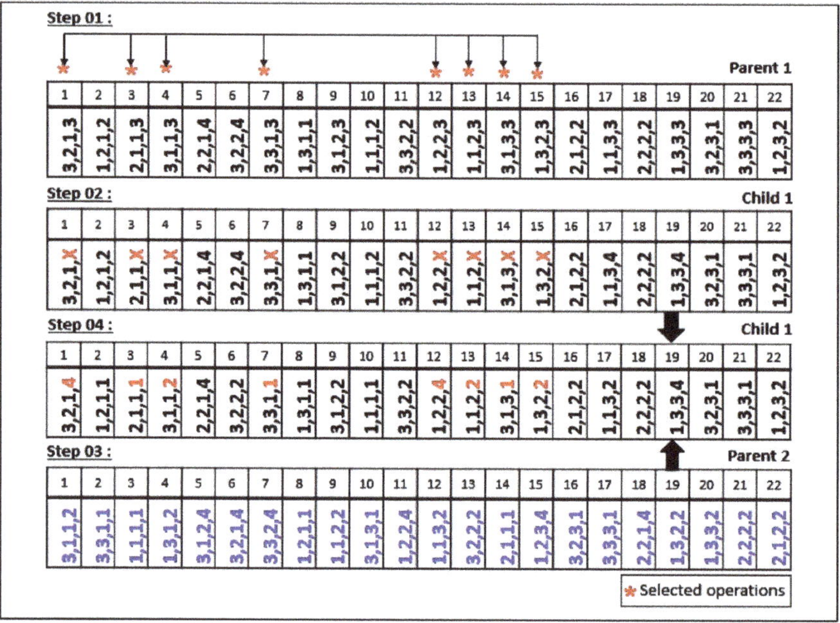

Figure 4. OMAC (operation to machine assignment crossover) operator.

The crossover sequencing operators JLOSC and SLOSC produce two new offspring by exchanging the sequencing information of parent chromosomes while keeping the assignment of machines in a gene unmodified. They are applied with given probabilities. The creation of the child chromosome by JLOSC and SLOSC with the preservation of the assignment information of parent chromosomes is illustrated in Figures 5 and 6. In the JLOSC crossover operator, the first step is to select a set of operations from parent 1. The next step is to copy all of the genetic information of the genes that are associated with the chosen operations to the offspring. The final step is to fill the empty genes in the offspring chromosome, while the machine assignment information of the empty genes is preserved from parent 1. This is done by copying the remaining operations from parent 2 to the empty genes, while the operations retain their appearance order from the second parent. The same procedure is repeated to create child 2 by starting to select the operations from parent 2. SLOSC is identical to JLOSC, with a difference in the second step of SLOSC: Only the genetic information of the chosen genes with the same sub-lot and job indexes is copied to the offspring. Two operators of each kind were developed (JLOSC1, JLOSC2, SLOSC1, and SLOSC2):

- JLOSC1 and SLOSC1: The operations are chosen randomly.
- JLOSC2 and SLOSC2: The operations are chosen according to specific rules that consist of choosing the operations of jobs with smaller completion times.

The SLOSC1 operator is similar to the JLOSC1 operator, but instead of keeping the unmodified operations of all the sub-lots, only the information from the sub-lot of the selected operation is kept and copied to the child chromosome.

The SLOSC2 crossover operator is similar to the JLOSC2 crossover operator except that in the first step, a set of sub-lots of jobs with smaller completion times is chosen. The goal of the JLOSC2 and SLOSC2 crossover operators is to keep the sequencing of operations of jobs and the sequencing of operations of sub-lots of jobs, respectively, with smaller completion times.

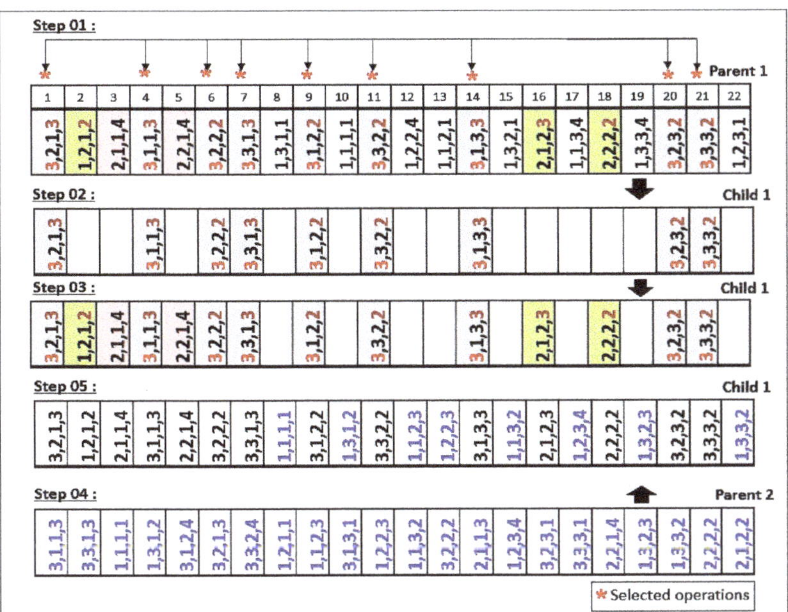

Figure 5. JLOSC2 (job-level operation sequence crossover 2) operator.

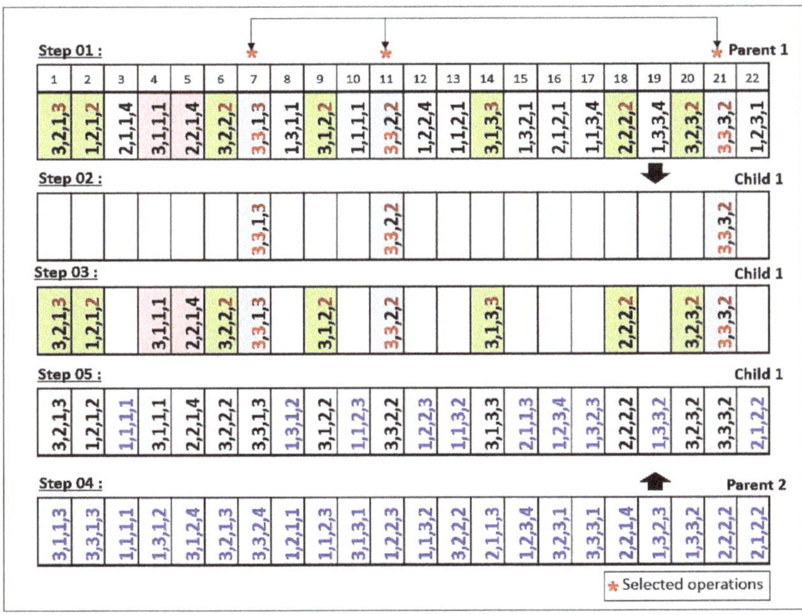

Figure 6. SLOSC2 (sub-lot-level operation sequence crossover 2) operator.

It is important to note that, after the application of any crossover operator, in the child chromosome, for a given i and l, the gene (i, l, j, k) is located after any gene (i, l, j', k'), where $j' < j$. This ensures that precedence constraints between operations of a particular sub-lot are not violated in the new created child chromosome.

5.1.6. Mutation Operator

The role of mutation operators is to prevent the algorithm from being trapped in a local optimum and to maintain genetic diversity in the population. The mutation operators are usually applied on chromosomes with given probabilities. Assignment and sequence operators also exist amongst the mutation operators. Assignment mutation operators change the assignment of operations to machines without changing the sequencing of these operations, and sequence operators only change the sequencing property of the chromosome undergoing the mutation, while the assignment property is preserved. The mutation operators used in the developed genetic algorithm are:

- ROAM (random operation assignment mutation): This operator is applied with a given probability on a set of operations of a given individual chromosome and randomly changes the assignment property of these operations to another machine.
- OSSM (operation sequence shift mutation): Whenever OSSM is applied on an individual, a set of operations is selected. Then, these operations are moved to other positions on the chromosome in such a way that no precedence constraint is violated.

5.1.7. Local Search Methods

Local Search through Gene Movement

This local search method improves the quality of any solution built after the application of the crossover and mutation operators. It is repeated as long as the quality of solutions obtained is not improved. The procedure for this method is illustrated through an example in Figure 7. The first step is to choose a set of operations of jobs with longer completion times; the goal is to reduce these completion times. In the second step, the procedure performs a search in the neighborhood of a solution by changing the positions of the operations chosen on the machines in order to improve the quality of this solution. This process is repeated until the iteration criteria are met. The steps of this local search method are as follows:

- Step 01: Choose a set of operations of jobs with larger completion times in the current chromosome.
- Step 02: Each operation chosen in Step 01 is positioned just before the previous operation assigned to the same machine (left movement) or just after (right movement) the following operation assigned to the same machine, while respecting the constraints of precedence between operations of sub-lots of jobs. This process allows one to lead research in the neighborhood of a solution by changing the sequencing of operations on the machines.
- Step 03: This process is repeated until the maximum number of iterations is not reached.

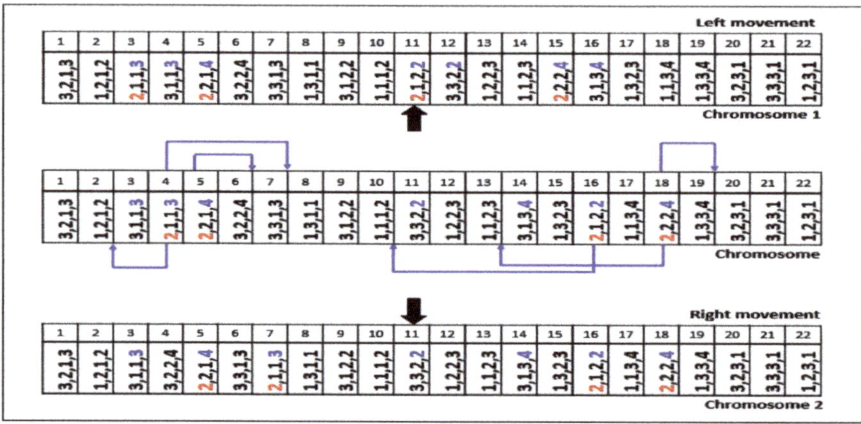

Figure 7. Local search procedure through gene movement.

Local Search by Grouping Sub-Lots

This local search method consists of grouping the operations of sub-lots of jobs on machines if the capacities of the machines allow the processing of several operations at the same time. This method can only be applied for machines such as ovens and cooling cells. The procedure for this local search method is illustrated through an example in Figure 8. For each gene, the sub-lot grouping procedure consists of browsing the chromosome from left to right. If the sub-lots of the same job as the considered gene are assigned to the same machine and the latter can process several sub-lots at the same time, all the genes of these sub-lots are repositioned one next to the other in the chromosome, which means that these sub-lots are treated at the same time by the same machine. The grouping of operations of sub-lots of jobs on machines must take into account the precedence between operations of these sub-lots.

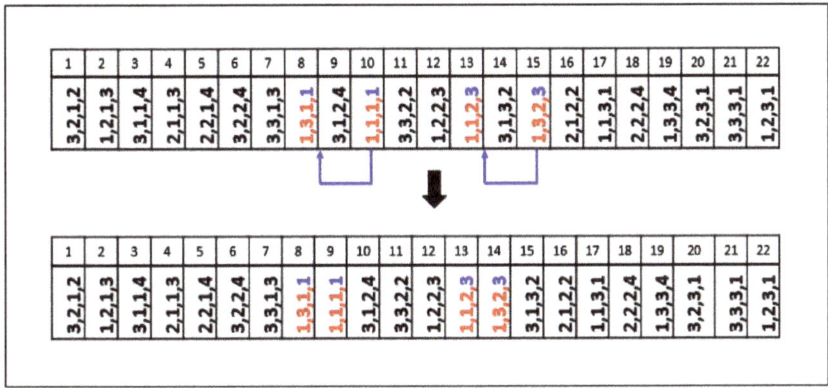

Figure 8. Local search procedure by grouping sub-lots.

Local Search through Intelligent Assignment of Operations

In this local search approach, a set of operations assigned to the most loaded machines is reassigned to the least loaded compatible ones.

Table 2 gives the common and different elements between the two developed algorithms. In this table, the heuristics for the generation of the initial population, crossover, and mutation operators of the two genetic algorithms are given. The two algorithms differ in the generation methods of the initial population and in the crossover operators. In both algorithms, several affectation and sequence heuristics were tested for the generation of the initial population. In Table 2, the heuristics allowing one to obtain the best solutions are given. The crossover operators of the first genetic algorithm are based on random choices of operations, while in the second algorithm, different operators that are specific and more adapted to the studied problem have been developed. Regarding the mutation operators, they are common between the two algorithms. The pseudocode of the generic genetic algorithm is given in Algorithm 2.

Table 2. Common and different elements between the two genetic algorithms.

	Initial Population	**Crossover Operators**	**Mutation Operators**
GA1	MMWA, SPT	OMAC1, JLOSC1, SLOSC1	ROAM, IOAM, OSSM
GA2	MMWA, MNOR	OMAC2, OMAC3, JLOSC2, SLOSC2	ROAM, IOAM, OSSM

Algorithm 2: Genetic algorithm

---

1. **Initialization:** The initial solutions are chosen using the heuristics described above ;
2. **Evaluation:** Evaluation of the initial solutions using the procedure of fitness function computation; ; nbIterations $\leftarrow$ 0; nbIndividuals $\leftarrow$ 0 ;
3. Application of the selection operator to choose the parent chromosomes to cross and mutate;
4. Randomly choose one of the crossover operators;
5. Application of the crossover operator chosen in 4;
6. Randomly choose one of the mutation operators ;
7. Application of the mutation operator chosen in 6;
8. Application of the local search by grouping sub-lots on child chromosomes obtained after crossover and mutation;
9. nbIndividuals $\leftarrow$ nbIndividuals + 1 ;
10. **If** the population size is reached, **then** go to 11, **else** go to 3 ;
11. Sorting solutions of the current population in ascending order of fitness functions ;
12. Construction of the new population with solutions of the previous population;
13. Application of the local search methods through gene movement and intelligent assignment of operations on the α best solutions of the new population;
14. nbIterations $\leftarrow$ nbIterations + 1 ;
15. **If** the maximum number of iterations is not reached, **then** go to 3, **else** go to 16 ;
16. **End of algorithm**

5.2. Iterated Local Search Algorithm

This section describes an iterated local search (ILS) algorithm for solving the studied problem. ILS is a simple, robust, and highly effective local search procedure that explores local optima in a given neighborhood (Lourenço et al. [41]).

The iterated local search algorithm starts from an initial solution and obtains a local optimum in its neighborhood through a local search procedure. To improve upon the current local optimum, ILS applies perturbation operators to this solution in order to move away from the current local optimum. An acceptance criterion is employed to determine which local optimum will become the current local optimum in the next iteration. The above process is repeated until a termination criterion is satisfied.

To generate the initial solutions of the iterative local search algorithms, the affectation and sequence heuristics for the generation of the initial population of the genetic algorithm are used, except that in the ILS algorithm, only one solution is generated.

The local search procedure is based on two operators: the affectation operator (AO) and sequence operator (SO). The affectation operator changes the assignment of a set of operations to the machines without changing their sequencing on the machines, while the sequence operator changes the sequencing of operations without changing their affectation to to machines.

Three affectation operators, AO1, AO2, and AO3, and three sequence operators, SO1, SO2, and SO3, were developed:

- AO1 and SO1: The operations are chosen randomly.
- AO2, AO3, SO2, and SO3: The operations are chosen according to rules developed to be specific to the studied problem.

The goal of the AO2 operator is to balance the workload between the machines, while AO3, SO2, and SO3 operators allow the reduction of the ending dates of jobs with larger completion times. The difference between these operators is in the choice of operations to change their assignments and sequences.

The first step of the SO operator consists of randomly choosing one of the sequence operators. Then, a set of operations in the current solution to be improved is chosen. The second step is to shift the selected operations to the left until the permutation of these operations with the previous operations assigned to the same machines is effective. If the left shift is blocked (precedence constraints are violated), then, from the solution obtained in the second step, the selected operations are shifted to the right until the permutation of these operations with the next operations assigned to the same machines is effective. This process is repeated until the maximum number of iterations is not reached.

In the AO operator, the first step is to choose one of the assignment operators. Then, a set of operations in the current solution to be improved is chosen. In the second step, the selected operations are assigned to another machine among a set of alternative machines. If the solution obtained after the reassignment of operations is not better than the current solution, then an SO sequence operator is applied on the selected operations with the new machine assignments obtained in the second step. This process is repeated for a given number of iterations.

The local search method can get stuck in a local optimum if the solution space is reduced. To fix this problem, a number of insertion moves to the current local optimum to obtain a perturbation solution are carried out. To determine the appropriate number of perturbations, different strategies have been tested. The best strategy, and that which is used in the algorithm, consists of performing a perturbation if the degree of similarity between the solutions is lower than a given threshold defined beforehand, which ensures a sufficient distance between solutions. The evaluation of the degree of similarity is based on the gaps between solutions. The similarity measures between solutions are evaluated to ensure the stability in the proposed solutions for the production scheduling.

The different steps of the iterative local search method in one iteration are illustrated through an example in Figure 9. The algorithm begins by generating an initial solution using one of the assignments and sequencing heuristics. Then, one of the SO sequence operators is applied. If no improvement of the solution is obtained after the application of the SO operator, the next step consists of applying an AO assignment operator on the operation chosen beforehand. If the application of the AO operator does not improve the initially generated solution, an SO operator with the new affectation obtained after the application of the AO operator is applied. If the solution is still not improved, a perturbation operator is applied on this solution, and the algorithm moves on to the next loop iteration. This operator consists of applying one of the randomly chosen sequencing or assignment operators. This process is repeated until the maximum number of perturbations is not reached.

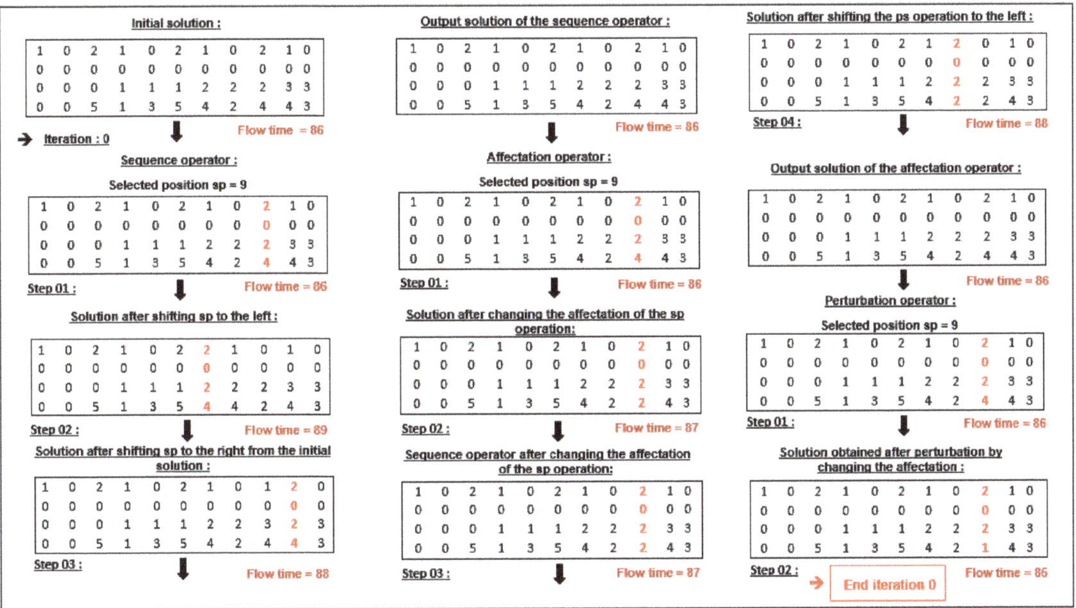

Figure 9. Steps of the iterative local search algorithm for one iteration.

Table 3 gives the common and different elements between the two developed algorithms. In this table, the heuristics for the generation of initial population, affectation, and sequence operators of the two iterated local search algorithms are given. The difference between the two algorithms lies in the heuristics used for the generation of initial solutions. In Table 3, the heuristics allowing one to find the best solutions are given. The affectation and sequence operators of the first iterated local search algorithm are based on random choices of operations, while in the second algorithm, the choice of these operations is based on specific rules more suited to the considered problem. The pseudocode of the generic iterated local search algorithm is given in Algorithm 3.

Algorithm 3: Iterative local search algorithm

---

1. **Initialization:** The initial solution is generated by using the previously described assignment and sequencing heuristics ;
   current solution ← initial solution ;
2. **Evaluation:** Evaluation of the initial solution ;
   SimilarDegree ← 0 ;
3. Application of the SO sequence operator;
4. **If** the solution is improved in 3, **then** ;
   current solution ← solution obtained in 3, **else** go to 5;
5. Application of the AO affectation operator ;
6. **If** the solution is improved in 5, **then**;
   current solution ← solution obtained in 5, **else** go to 7;
7. Application of the PO operator on the current solution ;
8. Update SimilarDegree by calculating the distance between the current solution and the improved one;
9. **If** the degree of similarity between the solutions obtained in 8 is less than a predefined threshold,;
   then go to 3, **else** go to 10 ;
10. **End of algorithm**

Table 3. Common and different elements between the two iterated local search algorithms.

	Initial Solution	**Sequence Operators**	**Affectation Operators**
ILS1	SPT, MNOR	SO1	AO1
ILS2	MMWA, MNOR	SO2, SO3	AO2, AO3

6. Computational Results of the Developed Algorithms

6.1. Data Generation

The genetic algorithms presented above were implemented using the Java programming language and were tested on 150 instances of different types: real instances of the HCT, randomly generated instances of the HCT type, randomly generated instances, and instances adapted from the literature (Sriboonchandr et al. [32], Nouri et al. [33], Azzouz et al. [34], Mousakhani [26], Lee et al. [35], Bagheri and Zandieh [25], Pezzella et al. [36]).

The real instances were built after collecting data for some examples of production days. The randomly generated instances of the HCT type were built using data (such as the number of machines, due dates of jobs, and time window of availability of machines) from the real instances of HCT. The other parameters were generated from uniform distributions, such as the parameters of the randomly generated instances. To adapt the instances from the literature to the studied problem, the missing data, such as the quantities of dishes, due dates of jobs, machine capacities, and time windows of availability of machines, were randomly generated according to uniform distributions and by taking into account the data on the processing times of operations and the setup times of machines given in these instances. The results of the developed resolution methods for these different categories of instances are presented below.

6.2. Optimization of the Metaheuristics' Parameters

To optimize the parameter values of the developed metaheuristics, the Taguchi method was used. The choice of these parameters has a significant effect on the efficiency of the metaheuristic algorithms. The parameter values that are needed for optimization act like controllable factors in the design of experiments. The aim is to find an optimal combination of the parameters such that the total flow time is minimized. To employ the Taguchi method, Minitab 19 was used; under the menu options, Stat-DOE-Taguchi Design Create Design was selected. For five levels of six factors, an L25 orthogonal array is used. This orthogonal array lists the different combinations of factors at different levels at which the response values of experiments have to be determined.

The parameters of the genetic algorithm are: population size (PS), maximum number of generations (MNG), crossover probability (Pc), mutation probability (Pm), number of elite individuals (NEI), and the number of individuals on which local research is applied (NILS). The parameters of the iterated local search algorithm are: maximum number of iterations of the sequence operator (MISO), maximum number of iterations of the affectation operator (MIAO), maximum number of perturbations of the perturbation operator (MPPO), and threshold distance between solutions to reiterate the steps of the iterative local search algorithm (DOS). The optimum values of the genetic algorithm and the iterated local search algorithm parameters are given in Table 4.

Table 4. Optimal parameters of the genetic and iterated local search algorithms.

	PS	MNG	NEI	NILS	Pc	Pm		DOS	MISO	MIAO	MPPO
GA 1	500	1000	50	50	0.8	0.2	ILS 1	0.05	100	100	50
GA 2	800	1200	80	80	0.8	0.2	ILS 2	0.07	200	200	100

6.3. Discussion of the Experimental Results

The optimization criteria of the developed metaheuristics are based on the quality of the solutions and rapidity. The quality and efficiency of the metaheuristics were proven by comparing the solutions obtained with these algorithms and the optimal solutions of the mathematical model. Regarding the rapidity, the computation times of the algorithms were compared with those of the mathematical model. From the results presented above (Tables 5–9), we remark that the two genetic algorithms are efficient in terms of quality and rapidity. The performance of the developed resolution methods depends on the types of instances and their sizes, and it also depends on the choice of heuristics for the generation of initial solutions. The first genetic algorithm is more efficient in terms of quality with the MMWA assignment heuristic and the SPT sequence heuristic, while the second genetic algorithm is more efficient with the MMWA and MNOR heuristics. For the iterated local search algorithms, the first algorithm gives better solutions with the SPT assignment heuristic and the MNOR sequence heuristic, while the second algorithm finds good solutions with the MMWA and MNOR heuristics. Regarding the rapidity, the four metaheuristics are more faster with the RA assignment heuristic and the RS sequence heuristic.

The results of the genetic algorithms for different types of instances (Tables 5–9) show that the developed metaheuristics are efficient relative to the quality of solutions. Table 5 represents the results of the different methods developed on a real instance with 82 jobs, 92 sub-lots, 370 operations, and 29 machines. From these instances, several sub-instances were built by increasing the number of jobs, sub-lots, and operations each time to see from what number of jobs, sub-lots, and operations the model is not able to find solutions. From Table 5, we observe that for the real instances of the HCT, the mathematical model quickly found solutions for the small instances with less than five dishes, five sub-lots, 24 operations, and 29 machines. It took a little longer for the instances with more than six dishes, six sub-lots, 29 operations, and 29 machines. However, for the instances with more than nine jobs, 44 operations, and 29 machines, after more than 3 h of execution, the mathematical model did not find solutions. For these instances, the two genetic algorithms succeeded in finding feasible solutions in very short computation times. For example, for the large instance with 82 jobs, 92 sub-lots, 370 operations, and 29 machines, the first genetic algorithm found a feasible solution that respected all the constraints after 5 min of execution.

Table 5. Computational results of the developed resolution methods for real instances of the hospital center of Troyes (HCT).

J	SL	O	M	F0 (h)	T0 (s)	F1 (h)	T1 (s)	F2 (h)	T2 (s)	F3 (h)	T3 (s)	F4 (h)	T4 (s)
2	2	10	29	15.3	2	15.3	0.1	15.3	0.3	15.3	0.01	15.3	0.08
3	3	15	29	24.4	4	24.4	0.2	24.4	0.5	24.4	0.04	24.4	0.14
4	4	20	29	32.8	120	32.8	0.3	32.8	0.8	32.8	0.06	32.8	0.18
5	5	24	29	39.9	240	39.9	0.4	39.9	1	39.9	0.1	39.9	0.28
6	6	29	29	49.2	1200	49.2	0.6	49.2	1.4	49.2	0.16	49.2	0.40
7	7	34	29	57.6	2700	57.6	0.8	57.6	1.8	57.6	0.26	57.6	0.54
8	10	39	29	69.4	9000	69.4	1	69.4	2	69.4	0.46	69.4	0.3
9	11	44	29	-	>10,800	77.7	1.5	77.7	3	77.7	0.71	77.7	1.2
10	12	48	29	-	>10,800	85.0	2	85.0	4	86.4	0.96	85.9	1.4
20	22	93	29	-	>10,800	162.4	30	162.4	60	166.5	5	165.8	11
30	32	138	29	-	>10,800	252.9	60	252.9	120	262.3	15	260.6	31
40	42	179	29	-	>10,800	339.7	90	339.7	180	354.9	23	352.3	49
50	58	227	29	-	>10,800	471.6	138	478.6	276	490.6	30	485.6	60
60	68	271	29	-	>10,800	590.4	180	593.9	360	624.4	43	616.7	87
70	78	315	29	-	>10,800	682.8	228	677.6	468	720.7	51	715.2	103
82	92	370	29	-	>10,800	798.3	300	788.6	600	846.5	70	840.8	141

J: number of jobs, SL: number of sub-lots, O: number of operations, M: number of machines, F0: total flow time of the mathematical model, T0: computational time of the mathematical model, F1: total flow time of GA 1, T1: computational time of GA 1, F2: total flow time of GA 2, T2: computational time of GA 2, F3: total flow time of ILS 1, T3: computational time of ILS 1, F4: total flow time of ILS 2, T4: computational time of ILS 2.

Tables 6–8 present the performances of the resolution methods in terms of quality and rapidity for randomly generated instances and a comparison between the solutions obtained by these different methods. From the results presented in these tables, we remark that the genetic algorithms found feasible solutions for the large-size instances in reasonable resolution times.

In Table 9, we observe that for the instances adapted from the literature, the genetic algorithms make it possible to improve the solutions of the instances adapted from the literature. For example, for the instance from Lee et al. [35], the genetic algorithms brought about an improvement of 3.92% over the solutions obtained with the methods proposed by Lee et al. [35].

Table 6. Computational results of the developed resolution methods for randomly generated instances of the HCT type.

J	SL	O	M	F0 (h)	T0 (s)	F1 (h)	T1 (s)	F2 (h)	T2 (s)	F3 (h)	T3 (s)	F4 (h)	T4 (s)
110	110	402	29	-	>10,800	986.7	480	986.2	720	1030.5	115	1018.4	225
100	100	342	29	-	>10,800	696.5	270	696.5	480	732.3	65	723.5	126
90	90	324	29	-	>10,800	704.7	240	705.1	360	740.2	58	732.5	112
80	80	286	29	-	>10,800	525.6	205	525.5	356	556.7	48	548.7	94
60	60	200	29	-	>10,800	248.8	140	248.5	210	262.4	34	258.6	65
50	50	173	29	-	>10,800	318.9	110	318.3	198	336.9	28	331.5	54
20	20	74	29	-	>10,800	158.6	60	158.2	108	165.3	15	163.9	28
15	15	59	29	-	> 10,800	421.5	52	421.1	92	432.1	12	428.8	22
10	10	31	29	-	>10,800	215.4	20	215.1	34	219.8	4	217.9	7
9	9	34	29	-	>10,800	368.8	15	368.1	25	376.5	3	374.1	5

Table 7. Computational results of the developed resolution methods for small randomly generated instances.

J	SL	O	M	F0 (h)	T0 (s)	F1 (h)	T1 (s)	F2 (h)	T2 (s)	F3 (h)	T3 (s)	F4 (h)	T4 (s)
3	12	11	6	117	16	117	3	117	6	123	0.1	122	0.4
3	8	13	6	85	24	85	5	85	10	89	0.2	88	0.8
3	10	8	6	45	10	45	1.5	45	3	47	0.06	47	0.1
3	11	11	6	55	20	55	4	55	8	58	0.2	57	1.0
3	12	12	6	117	104	117	24	117	48	120	1.6	119	3.8
3	11	13	6	78	30	78	6	78	12	81	0.4	80	1.2
3	10	12	6	88	172	88	42	88	82	92	3.6	91	8.8
3	10	13	6	103	240	103	56	103	112	106	7.6	105	13.6
3	10	10	6	76	28	76	6	76	12	80	0.5	79	1.7
3	8	9	6	44	24	44	5	44	10	45	0.3	45	1.4

Table 8. Computational results of the developed resolution methods for large randomly generated instances.

J	SL	O	M	F0 (h)	T0 (s)	F1 (h)	T1 (s)	F2 (h)	T2 (s)	F3 (h)	T3 (s)	F4 (h)	T4 (s)
10	10	132	6	-	>10,800	285.8	20	285.2	30	298.2	5	294.1	8
15	15	168	7	-	>10,800	302.4	24	302.0	35	312.5	7	308.7	12
20	20	146	8	-	>10,800	320.5	30	320.1	45	332.8	9	328.5	16
25	25	154	9	-	>10,800	332.6	35	332.0	52	345.0	11	340.2	20
30	30	151	10	-	>10,800	338.2	42	337.8	60	350.8	13	346.6	24
35	35	149	11	-	>10,800	354.3	50	353.8	75	368.4	15	363.1	28
40	40	179	12	-	>10,800	365.1	58	364.6	86	378.9	18	375.2	34
45	45	221	13	-	>10,800	381.8	70	381.0	102	396.3	22	392.7	42
50	50	191	14	-	>10,800	395.0	85	398.5	124	410.1	28	406.4	54
55	55	205	15	-	>10,800	412.7	108	418.2	158	428.6	34	424.5	65
60	60	195	16	-	>10,800	434.2	120	439.0	178	448.0	40	446.3	76
65	65	201	17	-	>10,800	454.3	138	448.5	202	472.1	46	468.9	88
70	70	256	18	-	>10,800	468.5	160	460.8	235	486.8	52	482.7	95
75	75	258	19	-	>10,800	485.6	184	492.1	272	504.6	60	501.4	116
80	80	300	20	-	>10,800	498.0	215	510.3	318	516.5	68	512.1	128

Table 9. Computational results of the developed resolution methods for instances adapted from the literature.

	J	SL	O	M	F0 (h)	T0 (s)	F1 (h)	T1 (s)	F2 (h)	T2 (s)	F3 (h)	T3 (s)	F4 (h)	T4 (s)
Azzouz et al.	3	3	11	3	74	8	74	2	74	4.2	78	0.5	76	1.05
Bagheri et al.	3	3	11	3	68	6	68	1.5	68	3.1	72	0.38	70	0.78
Pezzella et al.	3	3	10	4	33	7	33	1.8	33	3.8	35	0.46	34	0.95
Nouri et al.	3	3	9	5	31	10	31	2.8	31	5.6	33	0.82	32	1.42
Lee et al.	3	3	22	5	102	92	102	0.2	102	0.6	104	0.04	103	0.13
Mousakhani	4	4	12	3	101	9	101	2.5	101	5.2	106	0.64	104	1.35
Sriboonchandr et al.	4	4	14	5	53	12	53	3.2	53	6.5	56	1.2	55	1.48

By comparing the different resolution methods for all the tested instances, we observe that for some instances, the two genetic algorithms found the optimal solutions in a very short computation time compared to the mathematical model. For the instances for which the optimality was not reached, the gaps between the solutions obtained with the algorithms and the optimal solutions were very small. For the large instances for which the mathematical model failed to find solutions after more than three hours of execution, the genetic algorithms found feasible solutions within reasonable computation times.

Comparing the two genetic algorithms in terms of rapidity, the first algorithm is faster than the second algorithm, while comparing them in terms of stability both algorithms are stable for the real instances of HCT and randomly generated instances, while for the randomly generated instances of the HCT type, the second algorithm is more stable than the first algorithm. Comparing the two algorithms in terms of quality of solutions, the performance of the algorithms depends on the types of instances. For some instances, the first algorithm is better than the second algorithm, while for other instances, the second algorithm is more efficient than the first algorithm.

By comparing the results obtained with the genetic algorithms and the iterated local search methods for all the tested instances, we find that the ILSs are worse than the GAs in terms of the quality of solutions obtained. However, in terms of rapidity, the ILS methods are faster compared to the GAs.

7. Application to an Industrial Case

This work was carried out in collaboration with the hospital center of Troyes (HCT). To effectively meet the needs of patients and to improve working conditions and employees' well-being, the hospital center of Troyes implements important measures to improve its daily efficiency. It is in this context that this study of hospital catering activity optimization takes place.

The HCT is carrying out a project to revise its supply chain, which must, in particular, consider the management of food flows. The contribution of the present work is to determine the best plan for meeting customers' demands in terms of food flows and to propose ways to improve the well-being and working conditions of employees of the catering service of the hospital center of Troyes. The aim is to provide methods and tools for scheduling meal-making processes throughout the day.

The hospital center of Troyes is the main member of the South Champagne Hospitals group. Its logistic network (Figure 10) is composed of a set of customers whose meal needs are met from a warehouse, which is the central kitchen and a set of suppliers. These customers are hospitals, nursing homes, and psychiatric clinics. The central food production kitchen of the HCT produces, on average, 4800 meals per day. Its maximum daily production capacity is 5000 meals per day. The HCT catering service includes a production unit divided into several sectors, from the reception of raw materials to the dispatch of finished products and a coordination unit for menu management. In this study, we are particularly interested in the sectors from the pre-treatment of raw materials to the stock of finished products (Figure 11). This figure represents the production areas of the studied system and the different machines available for carrying out the operations of the meal-production process.

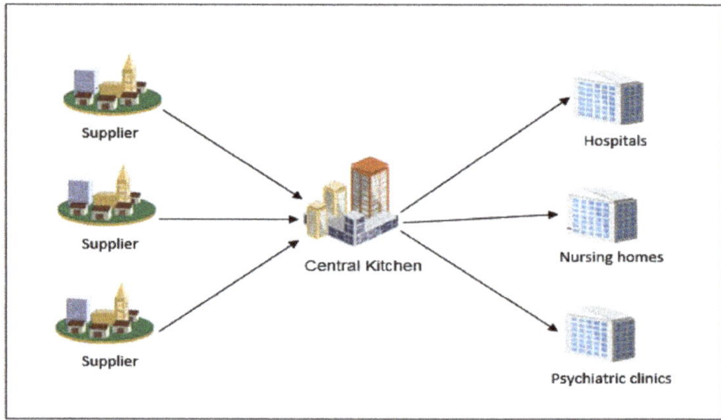

Figure 10. Logistics network of the hospital center of Troyes.

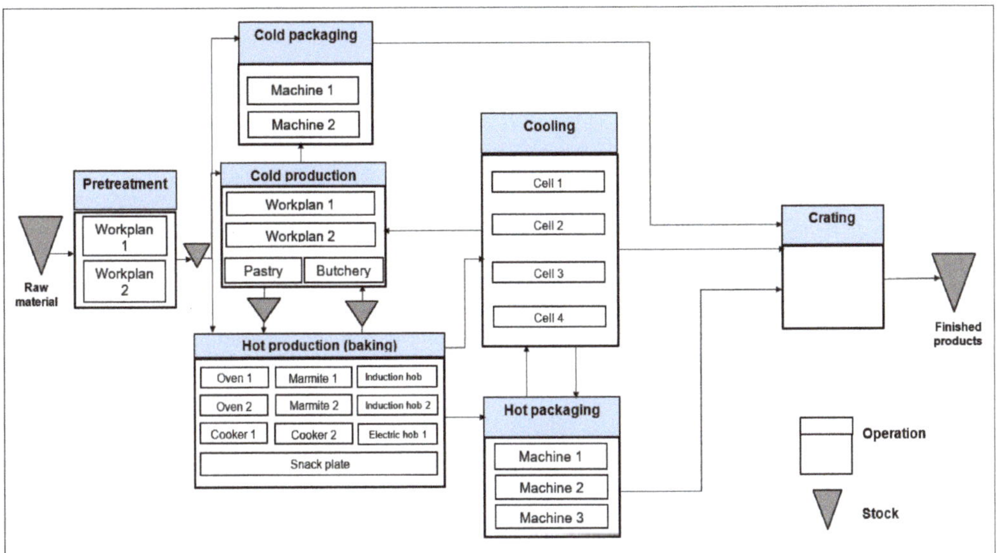

Figure 11. Representation of the studied production system.

Table 10 represents the results of the genetic algorithm for some examples of real production days with a comparison between the real solutions, as these production days were organized and the solutions were proposed by the genetic algorithm. In this table, for each instance, the number of dishes, the number of sub-lots of dishes, the total number of operations, the number of machines available, and the average number of meals produced per day are given. The performance indicators between solutions are based on the total flow time and the gaps between them. From these results, we remark that the gaps between the real solutions and those of the genetic algorithm are very important and significant. The performance of the genetic algorithm for these instances depends on their types and sizes. For example, for the instance with 62 dishes, 68 sub-lots, 218 operations, and 29 machines, we brought about a considerable improvement of 18.72% over the real solution, which shows the quality and performance of the developed algorithms.

Table 10. Comparison between the real and genetic algorithm solutions for some examples of production days.

	Instance 1	Instance 2	Instance 3	Instance 4
- Number of dishes	82	110	62	72
- Number of sub-lots of dishes	92	115	68	80
- Number of operations	370	392	218	328
- Number of material resources	29	29	29	29
- Average number of meals produced	4800	4800	4800	4800
- Real solutions	901.97 h	1062.66 h	278.23 h	784.84 h
- Genetic algorithm solutions	788.64 h	952.48 h	226.12 h	705.96 h
- Gaps between real and genetic algorithm solutions	−12.56%	−10.36%	−18.72%	−11.18%

8. Conclusions

The present article deals with the study of a new industrial problem. Different resolution methods for the scheduling of production processes in hospital catering were developed. A mathematical model integrating all the constraints of the studied problem was proposed. This model is an improvement of the standard flexible job shop scheduling problem with sequence-dependent setup times by integrating specific industrial constraints. An extensive study confirming the effectiveness of the developed model is presented. The computational results of the mathematical model for different types of instances show the limits of an exact resolution for the problem of scheduling production processes. To solve the large instances of the addressed problem, different metaheuristics were developed and tested on several types of instances. The computational results of these metaheuristics have proven their effectiveness and reliability for the scheduling of operations of the food production process and allowed significant improvements in the real current organization and system performance. As regards future research, this work is limited by the number of human resources available and does not take into account the possible hazards, such as the absence of staff, machinery breakdowns, and the unavailability of raw materials. Further research can be extended by considering these constraints. The present work also opens the way to other perspectives, such as the study of the production planning problem over several days, and our future work will focus on the development of resolution methods for this problem.

Author Contributions: Conceptualization, F.A., M.G. and A.Y.; Data curation, F.A.; Formal analysis, F.A.; Funding acquisition, F.Y. and L.A.; Investigation, F.A., M.G. and A.Y.; Methodology, F.A., M.G. and A.Y.; Project administration, M.G., A.Y., F.Y., L.A., A.Q. and E.T.; Software, F.A.; Supervision, M.G., A.Y. and A.Q.; Validation, M.G., A.Y., F.Y., L.A., A.Q. and E.T.; Visualization, F.A.; Writing—original draft, F.A.; Writing—review & editing, F.A. All authors have read and agreed to the published version of the manuscript.

Funding: This research was funded by the Hospital Center of Troyes.

Institutional Review Board Statement: Not applicable.

Informed Consent Statement: Not applicable.

Data Availability Statement: The data presented in this study are available on request from the corresponding author. The data are not publicly available due to confidentiality of the company.

Acknowledgments: The authors wish to thank the anonymous reviewers and the editorial board of the journal.

Conflicts of Interest: The authors declare no conflict of interest.

References

1. Sun, X.; Noble, J.S.; Klein, C.M. Single-machine scheduling with sequence dependent setup to minimize total weighted squared tardiness. *IIE Trans.* **1999**, *31*, 113–124. [CrossRef]
2. Lütkeentrup, M.; Günther, H.O.; Van Beek, P.; Grunow, M.; Seiler, T. Mixed-Integer Linear Programming approaches to shelf-life-integrated planning and scheduling in yoghurt production. *Int. J. Prod. Res.* **2005**, *43*, 5071–5100. [CrossRef]
3. Doganis, P.; Sarimveis, H. Optimal scheduling in a yogurt production line based on mixed integer linear programming. *J. Food Eng.* **2007**, *80*, 445–453. [CrossRef]
4. Doganis, P.; Sarimveis, H. Optimal production scheduling for the dairy industry. *Ann. Oper. Res.* **2008**, *159*, 315–331. [CrossRef]
5. Stefansdottir, B.; Grunow, M.; Akkerman, R. Classifying and modeling setups and cleanings in lot sizing and scheduling. *Eur. J. Oper. Res.* **2016**, *261*, 849–865. [CrossRef]
6. Sargut, F.Z.; Işık, G. Dynamic economic lot size model with perishable inventory and capacity constraints. *Appl. Math. Model.* **2017**, *48*, 806–820. [CrossRef]
7. Akkerman, R.; van Donk, D.P. Analyzing scheduling in the food-processing industry: Structure and tasks. *Cogn. Technol. Work.* **2009**, *11*, 215–226. [CrossRef]
8. Smith, Daniels, V.L.; Larry, P. A Model for Lot Sizing and Sequencing in Process Industries. *J. Prod. Res.* **1988**, *26*, 647–674. [CrossRef]
9. Kopanos, G.M.; Puigjaner, L.; Georgiadis, M.C. Efficient mathematical frameworks for detailed production scheduling in food processing industries. *Comput. Chem. Eng.* **2012**, *42*, 206–216. [CrossRef]
10. Wauters, T.; Verbeeck, K.; Verstraete, P.V.; Berghe, G.; De Causmaecker, P. Real-world production scheduling for the food industry: An integrated approach. *Eng. Appl. Artif. Intell.* **2012**, *25*, 222–228. [CrossRef]
11. Acevedo-Ojeda, A.; Contrerasa, I.; Chenb, M. Two-level lot-sizing with raw-material perishability and deterioration. *J. Oper. Res. Soc.* **2015**, *71*, 417–432. [CrossRef]

12. Copil, K.; Wörbelauer, M.; Meyr, H.; Tempelmeier, H. Simultaneous lotsizing and scheduling problems: A classification and review of models. *OR Spectr.* **2016**, *39*, 1–64. [CrossRef]

13. Niaki, M.K.; Nonino, F.; Komijan, A.R.; Dehghani, M. Food production in batch manufacturing systems with multiple shared-common resources: A scheduling model and its application in the yoghurt industry. *Int. J. Serv. Oper. Manag.* **2017**, *27*, 345. [CrossRef]

14. Wei, W.; Amorim, P.; Guimarães, L.; Almada-Lobo, B. Tackling perishability in multi-level process industries. *Int. J. Prod. Res.* **2018**, *57*, 5604–5623. [CrossRef]

15. Ahumada, O.; Villalobos, J.R. Application of planning models in the agri-food supply chain: A review. *Eur. J. Oper. Res.* **2009**, *196*, 1–20. [CrossRef]

16. Sel, C.; Bilgen, B.; Bloemhof-Ruwaard, J.M.; van der Vorst, J.G.A.J. Multi-bucket optimization for integrated planning and scheduling in the perishable dairy supply chain. *Comput. Chem. Eng.* **2015**, *77*, 59–73. [CrossRef]

17. Arbib, C.; Pacciarelli, D.; Smriglio, S. A three-dimensional matching model for perishable production scheduling. *Discret. Appl. Math.* **1999**, *92*, 1–15. [CrossRef]

18. Basnet, C.; Foulds, L.R.; Wilson, J.M. An exact algorithm for a milk tanker scheduling and sequencing problem. *Ann. Oper. Res.* **1999**, *86*, 559–568. [CrossRef]

19. Chen, S.; Berretta, R.; Clark, A.; Moscato, P. Lot Sizing and Scheduling for Perishable Food Products: A Review. *Ref. Modul. Food Sci.* **2019**. [CrossRef]

20. Liu, J.; MacCarthy, B.L. A global milp model for fms scheduling. *Eur. J. Oper. Res.* **1997**, *100*, 441–453. [CrossRef]

21. Guimaraes, K.F.; Fernes, M.A. An approach for flexible job-shop scheduling with separable sequence-dependent setup time. *Int. Conf. Syst.* **2006**, *5*, 3727–3731.

22. Saidi-Mehrabad, M.; Fattahi, P. Flexible job shop scheduling with tabu search algorithms. *Int. J. Adv. Manuf. Technol.* **2007**, *32*, 563–570. [CrossRef]

23. Defersha, F.M.; Chen, M. A parallel genetic algorithm for a flexible job-shop scheduling problem with sequence dependent setups. *Int. J. Adv. Manuf. Technol.* **2010**, *49*, 263–279. [CrossRef]

24. Mati, Y.; Lahlou, C.; Dauzère-Pérès, S. Modelling and solving a practical flexible job-shop scheduling problem with blocking constraints. *Int. J. Prod. Res.* **2011**, *49*, 2169–2182. [CrossRef]

25. Bagheri, A.; Zandieh, M. Bi-criteria flexible job-shop scheduling with sequence-dependent setup times Variable neighborhood search approach. *J. Manuf. Syst.* **2011**, *30*, 8–15. [CrossRef]

26. Mousakhani, M. Sequence-dependent setup time flexible job shop scheduling problem to minimize total tardiness. *Int. J. Prod. Res.* **2013**, *51*, 3476–3487. [CrossRef]

27. Chaudhry, I.A.; Khan, A.A. A research survey: Review of flexible job shop scheduling techniques. *Int. Trans. Oper. Res.* **2015**, *23*, 551–591. [CrossRef]

28. Rajabinasab, A.; Mansour, S. Dynamic flexible job shop scheduling with alternative process plans: An agent-based approach. *Int. J. Adv. Manuf. Technol.* **2010**, *54*, 1091–1107. [CrossRef]

29. Geyik, F.; Dosdogru, A. Process plan and part routing optimization in a dynamic flexible job shop scheduling environment: An optimization via simulation approach. *Neural Comput. Appl.* **2013**, *23*, 1631–1641. [CrossRef]

30. Zhou, D.; Zeng, L. A flexible job-shop scheduling method based on hybrid genetic annealing algorithm. *J. Inf. Comput. Sci.* **2013**, *10*, 5541–5549. [CrossRef]

31. Buddala, R.; Mahapatra, S.S. An integrated approach for scheduling flexible job-shop using teaching–learning-based optimization method. *J. Ind. Eng. Int.* **2018**, *15*, 181–192. [CrossRef]

32. Sriboonchandr, P.; Kriengkorakot, N.; Kriengkorakot, P. Improved Differential Evolution Algorithm for Flexible Job Shop Scheduling Problems. *Math. Comput. Appl.* **2019**, *24*, 80. [CrossRef]

33. Nouri, H.E.; Belkahla, D.O.; Ghédira, K. Solving the flexible job shop problem by hybrid metaheuristics-based multi-agent model. *J. Ind. Eng. Int.* **2018**, *14*, 1–14. [CrossRef]

34. Azzouz, A.; Ennigrou, M.; Ben Said, L. A hybrid algorithm for flexible job-shop scheduling problem with setup times. *Int. J. Prod. Manag. Eng.* **2017**, *5*, 23–30. [CrossRef]

35. Lee, S.; Moon, I.; Bae, H.; Kim, J. Flexible job-shop scheduling problems with 'AND'/'OR' precedence constraints. *Int. J. Prod. Res.* **2012**, *50*, 1979–2001. [CrossRef]

36. Pezzella, F.; Morganti, G.; Ciaschetti, G. A genetic algorithm for the Flexible Job-shop Scheduling Problem. *Comput. Oper. Res.* **2008**, *35*, 3202–3212. [CrossRef]

37. Xia, W.; Wu, Z. An effective hybrid optimization approach for multi-objective flexible job-shop scheduling problems. *Comput. Ind. Eng.* **2005**, *48*, 409–425. [CrossRef]

38. Fattahi, P.; Saidi Mehrabad, M.; Jolai, F. Mathematical modeling and heuristic approaches to flexible job shop scheduling problems. *J. Intell. Manuf.* **2007**, *18*, 331–342. [CrossRef]

39. Kacem, I. Genetic algorithm for the flexible jobshop scheduling problem. In Proceedings of the IEEE International Conference on Systems, Man and Cybernetics. Conference Theme-System Security and Assurance, Washington, DC, USA, 8 October 2003.

40. Goldberg, D. Genetic Algorithms in Search, Optimization, and Machine Learning. *Addion Wesley* **1989**, *1989*, 36.

41. Lourenço, H.R; Martin, O.C; Stutzle, T. Iterated local search. In *Handbook of Metaheuristics*; International Series in Operations Research & Management Science; Springer: Boston, MA, USA, 2003; Volume 57, pp. 320–353.

applied sciences

Article

Workload Balancing on Identical Parallel Machines: Theoretical and Computational Analysis

Yassine Ouazene *, Nhan-Quy Nguyen and Farouk Yalaoui

Laboratoire d'Optimisation des Systèmes Industriels, Chaire Connected Innovation, Université de Technologie de Troyes, 12 rue Marie Curie, CS 42060, 10004 Troyes, France; nhan_quy.nguyen@utt.fr (N.-Q.N.); farouk.yalaoui@utt.fr (F.Y.)
* Correspondence: yassine.ouazene@utt.fr

Abstract: This paper considers the problem of assigning nonpreemptive jobs on identical parallel machines to optimize workload balancing criteria. Since workload balancing is an important practical issue for services and production systems to ensure an efficient use of resources, different measures of performance have been considered in the scheduling literature to characterize this problem: maximum completion time, difference between maximum and minimum completion times and the Normalized Sum of Square for Workload Deviations. In this study, we propose a theoretical and computational analysis of these criteria. First, we prove that these criteria are equivalent in the case of identical jobs and in some particular cases. Then, we study the general version of the problem using jobs requiring different processing times and establish the theoretical relationship between the aforementioned criteria. Based on these theoretical developments, we propose new mathematical formulations to provide optimal solutions to some unsolved instances in order to enhance the latest benchmark presented in the literature.

Keywords: workload balancing; identical parallel machines; normalized sum of square for workload deviations; maximum completion time; minimum completion time; mathematical programming

Citation: Ouazene, Y.; Nguyen, N.-Q.; Yalaoui, F. Workload Balancing on Identical Parallel Machines: Theoretical and Computational Analysis. *Appl. Sci.* **2021**, *11*, 3677. https://doi.org/10.3390/app11083677

Academic Editor: José Firmino Aguilar Madeira

Received: 27 February 2021
Accepted: 12 April 2021
Published: 19 April 2021

Publisher's Note: MDPI stays neutral with regard to jurisdictional claims in published maps and institutional affiliations.

1. Introduction

Balancing workload among a set of resources or operators is an important issue in many services and production environments. For instance, in the manufacturing industry, balancing the workload among machines can reduce idle time. It also helps to manage bottlenecks in many production systems. This problem can be formally defined as the assignment of a set of jobs to identical parallel machines (or other types of resources) with the objective of balancing the total workload among those machines as equally as possible. Recently, this scheduling problem has found application in the Fog Computing empowered Internet of Things (IoT) [1]. This study proposes that a workload balancing scheme may reduce the latency of data flow, which is a key performance metric for IoT applications. In addition, workload balancing criteria have gained a lot of attention in key concepts of scheduling in Industry 4.0, such as of Flexible Manufacturing System [2] and Smart Manufacturing System [3]. Balancing the workload among a set of machines or human workers is also an important issue in many practical cases. For instance, assigning human resources is an important concern for many industries and organizations like hospitals, manufacturing firms, transportation companies, etc. Cuesta et al. [4] addressed a human resources assignment problem with the objectives of maximizing employee satisfaction and production rate on a real world case of a large production firm. Azmat et al. [5] presented different MIP models to solve workforce scheduling problems for a single-shift, taking into consideration constraints relative to labor regulations. Khouja and Conrad [6] addressed on a balancing problem encountered by a mail order firm. To handle order taking and billing, and to ease order tracking, customer groups should be assigned almost equally to the employees.

Many researchers, especially in the scheduling community, are interested in the workload balancing problem. They have addressed this problem in different ways: as a series of constraints or as an objective function. For example, Schwerdfeger and Walter [7] proposed an approached optimization method based on heuristics and a genetic algorithm in order to deal with semirelated parallel machines under sequence-dependent setups and load balancing constraints. Ouazene et al. [8] introduced a mixed integer-programming model and a genetic algorithm to minimize total tardiness and workload imbalance in a parallel machine-scheduling environment. In 2012, Keskinturk et al. [9] attempted to minimize the relative workload imbalance on parallel machines via two metaheuristic methods.

Cossari et al. [10] noted that there is no established performance measure in the literature to characterize the workload balancing problem. However, different criteria have been proposed in the literature. The first is the relative percentage of imbalances (RPI) introduced by Rajakumar et al. [11,12]. However, Walter and Lawrinenko [13] showed that this criterion is equivalent to maximum completion time minimization and does not provide perfectly balanced schedule.

Ho et al. [14] introduced a new workload balancing criterion, the so-called Normalized Sum of Squared Workload Deviations (NSSWD). They also analyzed some properties of the problem and introduced a heuristic algorithm to solve it. Later, Ouazene et al. [15] showed that the $NSSWD$ criterion is equivalent to $\sum_{m=1}^{M} C_m^2$. In other words, $NSSWD$ represents a normalized variant of the problem of minimizing $\sum_{m=1}^{M} C_m^2$. This problem was first treated by Chandra and Wong [16] in 1975, who analyzed the worst-case performance of Longest Processing Times rule, and derived its worst-case performance ratio, i.e., $\frac{25}{24}$.

Ho et al. [14] also stated that, in the case of identical parallel machines, a $NSSWD$-optimal schedule is necessarily a C_{max}-optimal schedule as well. However, [13] provided a counter-example for the case with more than two machines to prove that this result was wrong. They also carried out a computational study to give an overview on C_{max} as well as $NSSWD$ optimal schedules and empirically analyzed the correlation between both criteria. The numerical results revealed that the C_{max} criterion may be a good approximation for $NSSWD$ when the ratio $\frac{n}{m}$ is rather small. However, for some other instances, especially the largest ones, the C_{max}-optimal solution is not a good candidate solution for workload balancing.

Ouazene et al. [15] also recommended the minimization of the difference between the most and least loaded machine ($C_\Delta = C_{max} - C_{min}$) as a measure for workload balancing on identical parallel machines. They performed a computational study to show that, in general, a makespan optimal schedule is not optimal for C_Δ criterion.

One of the recent works dealing with this problem is that of Schwerdfeger and Walter [17]. The authors investigated the workload balancing problem on a parallel machines environment where the objective was to minimize the normalized sum of square for workload deviations. They proposed an exact subset sum-based algorithm to solve a three-machine case. In a follow-up paper, Schwerdfeger and R. Walter [18] provided a dynamic programming approach to solve the same problem. They also suggested two local search procedures to solve large instances and provided a complete and dedicated benchmark to the workload balancing problem.

This paper consists of a theoretical and computational analysis of the different criteria addressed in the literature to deal the workload balancing problem. More specifically, it aims to investigate the theoretical relationships between $P||C_{max}$, $P||NSSWD$ and $P||C_\Delta$ problems. As mentioned by Cossari et al. [10], while workload balancing is an important practical criterion given the need of production systems to efficiently use all of their resources, there is no established measure of performance in the scheduling literature that characterizes total workload balance. Different criteria have been addressed in the literature and some of them are trivial. That is why we present in this work a rigorous theoretical analysis of the most important criteria that have been considered for this problem. The main theoretical contribution of this paper is to establish the exact mathematical relationship between NSSWD and (C_Δ) as a measure for the workload balancing on identical parallel machines. Then, based on this theoretical analysis, new mathematical bounds are developed and incorporated into mathematical programming models to provide a number of optimal solutions to some unsolved instances in the benchmark proposed by Schwerdfeger and R. Walter [18].

The remainder of this paper is organized as follows. The next section introduces the problem considered in this study with the different notations. Section 3 presents the workload balancing problem using two cases: the case of identical jobs and the general case with jobs having different processing times. Section 4 introduces three mathematical formulations to the problem. Numerical tests are presented in Section 5. Finally, Section 6 summarizes the contribution of this paper.

2. Problem Description

2.1. Preliminary Notations and Assumptions

The problem considered in this paper is described as follows: a set of N independent jobs $J_1, J_2, \ldots, J_N$ are to be scheduled on M parallel machines. All machines or resources are identical, i.e., job j requires the same processing time on any machine. Each machine can process at most one job at a time, and each job must be processed without interruption by one of the M machines. All the jobs are assumed to be available at time zero and each job J_j has a deterministic integer processing time p_j. The setup times, if any, are included in the processing time.

To avoid trivialities, it is assumed that: $N > M \geq 2$.

A schedule is represented by an M-partition $S = S_1, S_2, \ldots, S_M$ of the set jobs J, where each S_m, represents the subset of jobs assigned to the machine m. The objective is to minimize the workload imbalance between the different machines.

Since the machines are identical, we can assume, without loss of generality, that the indexes of the machine are rearranged in nondecreasing order according to the completion times (or workloads) $C_1 = C_{min} \leq C_2 \leq \ldots \leq C_{M-1} \leq C_M = C_{max}$ as illustrated by Figure 1.

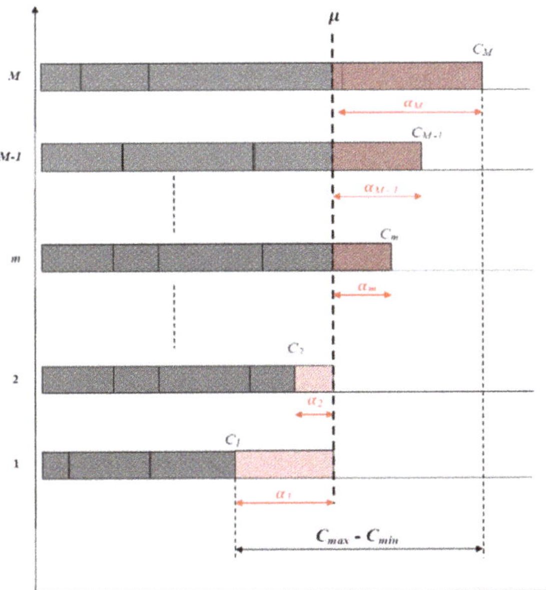

Figure 1. Illustration of the workload imbalance on identical parallel machines.

The different notations used in this paper are described below in Table 1.

Table 1. Nomenclature.

N	Total number of jobs
M	Total number of machine
m	Machine index
j	Job index
p_j	Processing time of job j
S_m	Set of jobs scheduled on the machine m
$C_m = \sum\limits_{j \in S_m} p_j$	Completion time or workload of machine m
C_{max}	Maximum of workloads
C_{min}	Minimum of workloads
$C_\Delta = C_{max} - C_{min}$	Difference between maximum and minimum workloads
$\mu = \frac{1}{M} \times \sum\limits_{m=1}^{(M)} C_m = \frac{1}{M} \times \sum\limits_{j=1}^{N} p_j$	Average workload
$NSSWD = \frac{1}{\mu} \times \sqrt{\sum\limits_{m=1}^{M} (C_m - \mu)^2}$	Normalized Sum of Square for Workload Deviations

2.2. Performance Criteria for Workload Balancing

In the case of identical parallel machines, a well-established criterion is the normalized sum of squared workload deviations ($NSSWD$). This criterion was introduced by Ho et al. [14]; it is based on the sum of squares principle known in measuring variability in statistics and serves as a precise measurement criterion. $NSSWD$ is defined as follows.

$$NSSWD = \frac{1}{\mu} \times \left[\sum_{m=1}^{M} (C_m - \mu)^2 \right]^{\frac{1}{2}} \tag{1}$$

where: $\mu = \frac{1}{M} \times \sum_{m=1}^{M} C_m = \frac{1}{M} \times \sum_{j=1}^{N} p_j$

The authors discussed some properties of the $NSSWD$ criterion and tackled the complexity of the problem. They also showed that a non makespan-optimal schedule can be improved in terms of $NSSWD$ by reducing its maximum machine completion time.

Cossari et al. [19] showed that: $NSSWD = M^{\frac{1}{2}} \times \dfrac{\sqrt{\frac{1}{M} \times \sum_{m=1}^{M} (C_m - \mu)^2}}{\mu}$, where $\sqrt{\frac{1}{M} \times \sum_{m=1}^{M} (C_m - \mu)^2}$ is the standard deviation and $\dfrac{\sqrt{\frac{1}{M} \times \sum_{m=1}^{M} (C_m - \mu)^2}}{\mu}$ is the coefficient of variation.

Hence, the authors concluded that the $NSSWD$ criterion is equivalent to the coefficient of variation and both two criteria may be used as performance measures.

In the particular case of identical parallel machines, $P||NSSWD$ is equivalent to $P||\sum_{m=1}^{M} C_m^2$ (cf. [15,17]). It makes no difference whether we minimize $NSSWD$ or $\sum_{m=1}^{M} C_m^2$.

3. Theoretical Analysis of Different Criteria

3.1. Case with Identical Jobs

This section introduces some of the properties of the different workload balancing performance measures and establishes equivalence between the following problems: $P|p_j = p|C_{max}$, $P|p_j = p|C_\Delta$ and $P|p_j = p|NSSWD$.

Proposition 1. *If any schedule of a parallel machines problem with N identical jobs and M machines, the problems $P|p_j = p|C_{max}$, $P|p_j = p|C_\Delta$ and $P|p_j = p|NSSWD$ are equivalent and trivial.*

Proof of Proposition 1. In any schedule of a parallel machines problem with N identical jobs and M machines, there is a set of l machines ($1 \leq l \leq M$) which have the same workload load $C_{max} = \left(\left\lfloor \frac{N}{M} \right\rfloor + 1 \right) \times p$ and $(M - l)$ other machines which must have the same workload $C_{max} = \left\lfloor \frac{N}{M} \right\rfloor \times p$, where $l = N - M \times \lfloor N/M \rfloor$.

We consider that all jobs have the same processing time: $\forall j = 1 \ldots N, p_j = p$, then we obtain the following cases.

Case 1: $\frac{N}{M} \equiv 0 \ (mod M)$

Then, the different workload measures can be expressed as follows.

$$\begin{cases} C_{max} = \frac{N}{M} \times p \\ C_{min} = \frac{N}{M} \times p \\ C_\Delta = 0 \\ NSSWD = 0 \end{cases} \tag{2}$$

Case 2: $\frac{N}{M} \equiv l \ (Mod M)$

Then, the different workload measures can be expressed as follows.

$$\begin{cases} C_{max} = \left\lfloor \frac{N}{M} \right\rfloor \times p + p \\ C_{min} = \left\lfloor \frac{N}{M} \right\rfloor \times p \\ C_\Delta = p \end{cases} \tag{3}$$

Concerning the $NSSWD$ calculation, let us consider: $\left\lfloor \frac{N}{M} \right\rfloor \times p = x$. Then, the average workload can be written as follows: $\mu = \frac{(M \times x) + (l \times p)}{M} = x + \frac{l}{M} \times p$. The average workload can be also expressed as: $\mu = \frac{1}{M} \times \sum_{m=1}^{M} C_m = \frac{N \times p}{M}$.

So, based on the $NSSWD$-criterion definition, we have:

$$NSSWD = \frac{1}{\mu} \times \left[\sum_{m=1}^{M} (C_m - \mu)^2 \right]^{\frac{1}{2}} = \frac{1}{\mu} \times \left[\sum_{m=1}^{l} (C_m - \mu)^2 + \sum_{m=l+1}^{M} (C_m - \mu)^2 \right]^{\frac{1}{2}}$$

$$= \frac{1}{\mu} \times \left[\sum_{m=1}^{l} (x + p_K - \mu)^2 + \sum_{m=l+1}^{M} (x - \mu)^2 \right]^{\frac{1}{2}}$$

$$= \frac{1}{\mu} \times \left[l \times (x + p_K - \mu)^2 + (M - l) \times (x - \mu)^2 \right]^{\frac{1}{2}}$$

$$= \frac{1}{\mu} \times \left[l \times \left(x + p_K - x - \frac{l}{M} \times p_K \right)^2 + (M - l) \times \left(x - x - \frac{l}{M} \times p_K \right)^2 \right]^{\frac{1}{2}}$$

$$= \frac{1}{\mu} \times \left[l \times \left(p_K - \frac{l}{M} \times p_K \right)^2 + (M - l) \times \left(-\frac{l}{M} \times p_K \right)^2 \right]^{\frac{1}{2}} \tag{4}$$

$$= \frac{1}{\mu} \times \left[l \times \left(\frac{M - l}{M} \right)^2 \times p_K^2 + (M - l) \times \left(\frac{l}{M} \right)^2 \times p_K^2 \right]^{\frac{1}{2}}$$

$$= \frac{1}{\mu} \times \left[\frac{l \times (M - l)^2 + (M - l) l^2}{M^2} \times p_K^2 \right]^{\frac{1}{2}}$$

$$= \frac{1}{\mu} \times \left[\frac{l \times (M - l) \times (M - l + l)}{M^2} \times p_K^2 \right]^{\frac{1}{2}}$$

$$= \frac{1}{\mu} \times \left[\frac{l \times (M - l)}{M} \times p_K^2 \right]^{\frac{1}{2}} = \frac{1}{\mu} \times \left[\frac{l \times (M - l)}{M} \right]^{\frac{1}{2}} \times p_K$$

$$= \frac{M}{\sum_{k=1}^{K} |S_k| \times p_k} \times \left[\frac{l \times (M - l)}{M} \right]^{\frac{1}{2}} \times p_K$$

$$= \frac{p_K}{\sum_{k=1}^{K} |S_k| \times p_k} \times [M \times (M - l) \times l]^{\frac{1}{2}}$$

Considering that $l = N - M \times \left\lfloor \frac{N}{M} \right\rfloor$, the Normalized Sum of Square for Workload Deviations can be expressed as follows:

$$
\begin{aligned}
NSSWD \quad &= \frac{[M \times (M-l) \times l]^{\frac{1}{2}}}{N} = \frac{[M \times (M-N+M \times \lfloor \frac{N}{M} \rfloor) \times (N-M \times \lfloor \frac{N}{M} \rfloor)]^{\frac{1}{2}}}{N} \\
&= \frac{[M \times (M \times (1+\lfloor \frac{N}{M} \rfloor)-N) \times (N-M \times \lfloor \frac{N}{M} \rfloor)]^{\frac{1}{2}}}{N}
\end{aligned}
\tag{5}
$$

Based on the demonstration detailed above, we can also conclude that in any parallel machine scheduling problem with N identical jobs and M machines, the Normalized Sum of Square for Workload Deviations ($NSSWD$)-criterion is constant and completely independent of the job processing times (rather, it depends only on the number of jobs and the number of machines). $\square$

3.2. Case with Different Blocks of Identical Jobs

We assume the particular case in which the N jobs can be partitioned into K sets (S_k) such as the jobs owned the same set have the same processing time: $\{1 \ldots, n\} = \cup_{k=1}^{K} S_k$ and $\forall j \in S_k, p_j = p_k$.

Lemma 1. *Considering the assumptions mentioned above and if we have $|S_k| \equiv 0 (mod\ M)$, $\forall k = 1 \ldots K$, then the Normalized Sum of Square for Workload Deviations (NSSWD) is equal to zero.*

Proof of Lemma 1. This proof is based on the preliminary results established in Proposition 1.

$$
\begin{aligned}
|S_k| \equiv 0 (mod\,M), \forall k = 1 \ldots K &\Longrightarrow \sum_{k=1}^{K} |S_k| \equiv 0 (mod\,M) \Longrightarrow N \equiv 0 (mod\,M) \\
&\Longrightarrow C_m = \sum_{k=1}^{K} \frac{|S_k|}{M} \times p_k, \forall m = 1 \ldots M
\end{aligned}
\tag{6}
$$

Finally, since all the machines have the same workload value, then the Normalized Sum of Square for Workload Deviations ($NSSWD$) is equal to zero. $\square$

Proposition 2 *Under the assumptions assumed above and if we have $|S_k| \equiv 0\ (mod\ M)$, $/\ \forall k = 1 \ldots K - 1$ and $|S_K| \equiv l (mod\ M)$, then the Normalized Sum of Square for Workload Deviations (NSSWD) is constant.*

Proof of Proposition 2. In the optimal schedule, there is a set of l machines ($1 \leq l \leq M$) which have the same workload load $C_{max} = \sum_{k=1}^{K-1} \frac{|S_k|}{M} \times p_k + p_K$ and $(M-l)$ other machines which must have the same workload $C_{min} = \sum_{k=1}^{K-1} \frac{|S_k|}{M} \times p_k$.

Let us consider: $\sum_{k=1}^{K-1} \frac{|S_k|}{M} \times p_k = x$. Then, the average workload can be written as follows: $\mu = \frac{(M \times x)+(l \times p_K)}{M} = x + \frac{l}{M} \times p$. It can be also expressed as: $\mu = \frac{1}{M} \times \sum_{m=1}^{M} C_m = \frac{\sum_{k=1}^{K} |S_k| \times p_k}{M}$.

So, based on the $NSSWD$-criterion definition, we have:

$$
\begin{aligned}
NSSWD \;&= \tfrac{1}{\mu} \times \left[\sum_{m=1}^{M} (C_m - \mu)^2 \right]^{\frac{1}{2}} = \tfrac{1}{\mu} \times \left[\sum_{m=1}^{l} (C_m - \mu)^2 + \sum_{m=l+1}^{M} (C_m - \mu)^2 \right]^{\frac{1}{2}} \\[4pt]
&= \tfrac{1}{\mu} \times \left[\sum_{m=1}^{l} (x + p_K - \mu)^2 + \sum_{m=l+1}^{M} (x - \mu)^2 \right]^{\frac{1}{2}} \\[4pt]
&= \tfrac{1}{\mu} \times \left[l \times (x + p_K - \mu)^2 + (M - l) \times (x - \mu)^2 \right]^{\frac{1}{2}} \\[4pt]
&= \tfrac{1}{\mu} \times \left[l \times \left(x + p_K - x - \tfrac{l}{M} \times p_K \right)^2 + (M - l) \times \left(x - x - \tfrac{l}{M} \times p_K \right)^2 \right]^{\frac{1}{2}} \\[4pt]
&= \tfrac{1}{\mu} \times \left[l \times \left(p_K - \tfrac{l}{M} \times p_K \right)^2 + (M - l) \times \left(-\tfrac{l}{M} \times p_K \right)^2 \right]^{\frac{1}{2}} \\[4pt]
&= \tfrac{1}{\mu} \times \left[l \times \left(\tfrac{M-l}{M} \right)^2 \times p_K^2 + (M - l) \times \left(\tfrac{l}{M} \right)^2 \times p_K^2 \right]^{\frac{1}{2}} \\[4pt]
&= \tfrac{1}{\mu} \times \left[\tfrac{l \times (M-l)^2 + (M-l)l^2}{M^2} \times p_K^2 \right]^{\frac{1}{2}} \\[4pt]
&= \tfrac{1}{\mu} \times \left[\tfrac{l \times (M-l) \times (M-l+l)}{M^2} \times p_K^2 \right]^{\frac{1}{2}} = \tfrac{1}{\mu} \times \left[\tfrac{l \times (M-l)}{M} \times p_K^2 \right]^{\frac{1}{2}} \\[4pt]
&= \tfrac{1}{\mu} \times \left[\tfrac{l \times (M-l)}{M} \right]^{\frac{1}{2}} \times p_K = \tfrac{M}{\sum_{k=1}^{K} |S_k| \times p_k} \times \left[\tfrac{l \times (M-l)}{M} \right]^{\frac{1}{2}} \times p_K \\[4pt]
&= \tfrac{p_K}{\sum_{k=1}^{K} |S_k| \times p_k} \times \left[M \times (M - l) \times l \right]^{\frac{1}{2}}
\end{aligned}
$$

$$(7)$$

Considering that $l = |S_K| - M \times \left\lfloor \tfrac{|S_K|}{M} \right\rfloor$, the Normalized Sum of Square for Workload Deviations can be expressed as follows:

$$
\begin{aligned}
NSSWD \;&= \tfrac{p_K}{\sum_{k=1}^{K} |S_k| \times p_k} \times \left[M \times (M - l) \times l \right]^{\frac{1}{2}} \\[4pt]
&= \tfrac{p_K}{\sum_{k=1}^{K} |S_k| \times p_k} \times \left[M \times \left(M - |S_K| + M \times \left\lfloor \tfrac{|S_K|}{M} \right\rfloor \right) \times \left(|S_K| - M \times \left\lfloor \tfrac{|S_K|}{M} \right\rfloor \right) \right]^{\frac{1}{2}} \\[4pt]
&= \tfrac{p_K}{\sum_{k=1}^{K} |S_k| \times p_k} \times \left[M \times \left(M\left(1 + \left\lfloor \tfrac{|S_K|}{M} \right\rfloor \right) - |S_K| \right) \times \left(|S_K| - M \times \left\lfloor \tfrac{|S_K|}{M} \right\rfloor \right) \right]^{\frac{1}{2}}
\end{aligned}
$$

$$(8)$$

$\square$

3.3. General Case with Different Jobs

In this section, we consider the workload balancing problem in its general version. The N jobs may have different processing times.

A strong correlation between the two criteria C_Δ and $NSSWD$ was suspected by different authors in the literature [13,14,19]. However, their different analyses were based on empirical study. As a major contribution of our paper, we establish the exact mathematical correlation between these two criteria.

3.4. Case with Two Machines

Proposition 3. *In the case of two parallel machines, the problems $P2||C_{max}$, $P2||C_\Delta$ and $P2||NSSWD$ are equivalent.*

Proof of Proposition 3. First, we establish that a C_{max}-optimal solution is also C_Δ-optimal one. As a preliminary result, we assume the well-established relationship: $\max\{a, b\} = \tfrac{1}{2} \times (a + b + |b - a|)$.

Considering this relationship and the definition of the maximum completion time, we can write:

$$C_{max} = \max\{C_1, C_2\} = \frac{C_1 + C_2}{2} + \frac{1}{2} \times (|C_2 - C_1|) = \mu + \frac{1}{2} \times C_\Delta \tag{9}$$

Since $\mu = \frac{C_1 + C_2}{2}$ is constant and $C_\Delta = C_2 - C_1$, it is clear that minimizing C_{max} implies the minimization of the difference C_Δ.

Then, we establish that a C_Δ-optimal solution is also $(NSSWD)$-optimal. Based on the Normalized Sum of Square for Workload Deviations definition, we can write:

$$
\begin{aligned}
NSSWD &= \frac{1}{\mu} \times \left[\sum_{m=1}^{2} (C_m - \mu)^2 \right]^{\frac{1}{2}} = \frac{1}{\mu} \times \left[(C_1 - \mu)^2 + (C_2 - \mu)^2 \right]^{\frac{1}{2}} \\
&= \frac{1}{\mu} \times \left[\left(C_1 - \frac{C_1+C_2}{2} \right)^2 + \left(C_2 - \frac{C_1+C_2}{2} \right)^2 \right]^{\frac{1}{2}} = \frac{1}{\mu} \times \left[\left(\frac{C_1-C_2}{2} \right)^2 + \left(\frac{C_2-C_1}{2} \right)^2 \right]^{\frac{1}{2}} \\
&= \frac{1}{\mu} \times \left[\tfrac{1}{2} (C_2 - C_1)^2 \right]^{\frac{1}{2}} = \frac{1}{\sqrt{2} \times \mu} \times (C_{max} - C_{min}) = \frac{1}{\sqrt{2} \times \mu} \times C_\Delta
\end{aligned}
\tag{10}
$$

Since μ is constant, minimizing (C_Δ) implies the minimization of $NSSWD$. $\square$

3.5. Case with More Than Two Machines

Proposition 4. *The normalized sum of the square for Workload Deviations criterion NSSWD can be bounded as follows:* $\dfrac{M}{\sqrt{2} \times \sum_{j=1}^{N} p_j} \times C_\Delta \leq NSSWD \leq \dfrac{M^{\frac{3}{2}}}{2 \times \sum_{j=1}^{N} p_j} \times C_\Delta.$

Proof of Proposition 4. As first step, we deal with the upper bound.

Before dealing with the demonstration, we remind the reader of the following result (see [20] for details) concerning the maximum value of the standard deviation measure. It is shown that if $a \leq c_m \leq b, \forall m$, then the maximum possible value of standard deviation occurs when half of the values are equal ta a and half of them are equal to b. That means:

$$C_{max} = \max\{C_1, C_2\} = \frac{C_1 + C_2}{2} + \frac{1}{2} \times (|C_2 - C_1|) = \mu + \frac{1}{2} \times C_\Delta \tag{11}$$

$$\max_{a \leq c_m \leq b} \left\{ \frac{1}{M} \times \left[\sum_{m=1}^{M} (C_m - \mu)^2 \right] \right\} = \frac{(b-a)^2}{4} \tag{12}$$

Based on the proposition cited above, we can establish that:

$$
\begin{aligned}
NSSWD &= \frac{1}{\mu} \times \left[\sum_{m=1}^{M} (C_m - \mu)^2 \right]^{\frac{1}{2}} = \frac{M^{\frac{1}{2}}}{\mu} \times \left[\frac{1}{M} \sum_{m=1}^{M} (C_m - \mu)^2 \right]^{\frac{1}{2}} \\
\Longrightarrow \max NSSWD &= \max \frac{M^{\frac{1}{2}}}{\mu} \times \left[\frac{1}{M} \sum_{m=1}^{M} (C_m - \mu)^2 \right]^{\frac{1}{2}} = \frac{M^{\frac{1}{2}}}{\mu} \times \left[\max \frac{1}{M} \sum_{m=1}^{M} (C_m - \mu)^2 \right]^{\frac{1}{2}} \\
\Longrightarrow \max NSSWD &= \frac{M^{\frac{1}{2}}}{\mu} \times \left[\frac{(C_{max} - C_{min})^2}{4} \right]^{\frac{1}{2}} = \frac{M^{\frac{1}{2}}}{\mu} \times \left[\frac{(C_{max} - C_{min})}{2} \right] \\
\Longrightarrow \max NSSWD &= \frac{M^{\frac{3}{2}}}{2 \times \sum_{j=1}^{N} p_j} \times (C_{max} - C_{min}) \\
\Longrightarrow NSSWD &\leq \frac{M^{\frac{3}{2}}}{2 \times \sum_{j=1}^{N} p_j} \times C_\Delta
\end{aligned}
\tag{13}
$$

The second step of demonstration deals with the lower bound.

$$NSSWD = \frac{1}{\mu} \times \left[\sum_{m=1}^{M} (C_m - \mu)^2 \right]^{\frac{1}{2}} = \frac{1}{\mu} \times \left[(C_1 - \mu)^2 + (C_M - \mu)^2 + \sum_{m=2}^{M-1} (C_m - \mu)^2 \right]^{\frac{1}{2}} \tag{14}$$

$$NSSWD \geq \frac{1}{\mu} \times \left[(C_1 - \mu)^2 + (C_M - \mu)^2 \right]^{\frac{1}{2}}$$

Or by considering the relationship: $a^2 + b^2 = \frac{1}{2}(a+b)^2 + \frac{1}{2}(a-b)^2$, we can conclude that:

$$
\begin{aligned}
NSSWD \;&\geq\; \frac{1}{\mu} \times \left[(C_1 - \mu)^2 + (C_M - \mu)^2 \right]^{\frac{1}{2}} \\
&\geq\; \frac{1}{\mu} \times \left[\frac{1}{2} \times (C_1 + C_M - 2 \times \mu)^2 + \frac{1}{2} \times (C_M - C_1)^2 \right]^{\frac{1}{2}} \geq \frac{1}{\mu} \times \left[\frac{1}{2} \times (C_M - C_1)^2 \right]^{\frac{1}{2}} \\
&\geq\; \frac{1}{\sqrt{2} \times \mu} \times (C_{max} - C_{min}) \\
NSSWD \;&\geq\; \frac{M}{\sqrt{2} \times \sum_{j=1}^{N} p_j} \times C_\Delta
\end{aligned}
\tag{15}
$$

$\square$

Remark 1. *In the case of $M = 2$, NSSWD and C_Δ criteria are linearly dependent. This confirms the equivalence between the $P2\|NSSWD$ and $P2\|C_\Delta$ problems established in the previous section. This relationship is given below.*

$$
\begin{aligned}
\frac{2}{\sqrt{2} \times \sum_{j=1}^{N} p_j} \times C_\Delta \;&\leq\; NSSWD \;\leq\; \frac{2^{\frac{3}{2}}}{2 \times \sum_{j=1}^{N} p_j} \times C_\Delta \\
\implies \frac{\sqrt{2}}{\sum_{j=1}^{N} p_j} \times C_\Delta \;&\leq\; NSSWD \;\leq\; \frac{\sqrt{2}}{\sum_{j=1}^{N} p_j} \times C_\Delta \\
\implies NSSWD \;&=\; \frac{\sqrt{2}}{\sum_{j=1}^{N} p_j} \times C_\Delta
\end{aligned}
\tag{16}
$$

Remark 2. *Based on Proposition 3, we can introduce $\frac{NSSWD}{C_\Delta}$ as a new criterion to deal with the workload imbalance minimization in an identical parallel machine problem. This criterion presents the advantage of being bounded, and these bounds depend on the configuration of the instance (number of machines and sum of processing times).*

$$
\begin{aligned}
\frac{M}{\sqrt{2} \times \sum_{j=1}^{N} p_j} \times C_\Delta \;&\leq\; NSSWD \;\leq\; \frac{M^{\frac{3}{2}}}{2 \times \sum_{j=1}^{N} p_j} \times C_\Delta \\
\implies \frac{NSSWD}{C_\Delta} \;&\in\; \left[\frac{M}{\sqrt{2} \times \sum_{j=1}^{N} p_j} , \frac{M^{\frac{3}{2}}}{2 \times \sum_{j=1}^{N} p_j} \right] \\
\implies \frac{NSSWD}{C_\Delta} \;&\in\; \left[\frac{\sqrt{2}}{2 \times \mu} , \frac{\sqrt{M}}{2 \times \mu} \right]
\end{aligned}
\tag{17}
$$

We can also notice an important asymptotic behavior of this criterion. In fact, for a fixed number of machines M, the criterion $\frac{NSSWD}{C_\Delta}$ decreases according to the sum of processing times and we have a perfect balanced solution when $\sum_{j=1}^{N} p_j \mapsto +\infty$.

$$
\lim_{\mu \mapsto +\infty} \frac{NSSWD}{C_\Delta} = 0
\tag{18}
$$

4. Mathematical Formulations

In this section, we introduce three formulation of the problem, namely mixed-integer quadratic programming (MIQ1), mixed-integer quadratic programming with linear cuts (MIQ2) and mixed-integer linear programming (MILP).

4.1. Mixed-Integer Quadratic Programming (MIQ1)

This formulation was first introduced by [17] to deal with the problem $P1 \mid NSSWD$. This formulation considers binary variables x_{jm}, which take the value 1 if job j is assigned to machine m and 0 otherwise. Equation (19) describes the objective function expressed as a quadratic function. Constraints (20) ensure that each job is assigned to exactly one machine. The machine completion times are determined by Equation (21). Finally, the domains of the binary variables are defined by Equation (22).

$$\text{Minimize } \frac{1}{\mu} \times \left[\sum_{m=1}^{M} (C_m - \mu)^2 \right]^{\frac{1}{2}} \tag{19}$$

Subject to

$$\sum_{m=1}^{M} x_{jm} = 1, \forall j = 1, .., N \tag{20}$$

$$C_m = \sum_{i=1}^{N} x_{im} p_i \forall m = 1, .., M \tag{21}$$

$$x_{jm} \in \{0, 1\} \forall j = 1, .., N \; \forall m = 1, .., M \tag{22}$$

The following two formulation represent a new contribution based on the theoretical analysis developed in Section 3. First, we incorporate the new bounds, established by Proposition 4, as new cats in the mixed-integer quadratic programming (MIQ1) proposed by [17] in order to reduce the solution space. Then, based on the same development we introduce a new mixed-integer linear programming (MILP) to solve the problem.

4.2. Mixed-Integer Quadratic Programming with Cuts (MIQ2)

This second formulation considers the same binary variables x_{jm}, which take the value 1 if job j is assigned to machine m and 0 otherwise. Equation (23) describes the objective function. Constraints (25) ensure that each job is assigned to exactly one machine. The machine completion times are determined by Equation (25). Constraints (26) and (27) define respectively the workload of the least and most loaded machine. Constraints (28) and (29) introduce the theoretical bounds established by Proposition 4 as new cats in the original model. Finally, the domains of the binary variables are defined by Equation (31).

$$\text{Minimize } z + \frac{1}{\mu} \times \left[\sum_{m=1}^{M} (C_m - \mu)^2 \right]^{\frac{1}{2}} \tag{23}$$

Subject to

$$\sum_{m=1}^{M} x_{jm} = 1, \forall j = 1, .., N \tag{24}$$

$$C_m = \sum_{i=1}^{N} x_{im} p_i, \forall m = 1, .., M \tag{25}$$

$$C_m \geq C_{min}, \forall m = 1 \ldots M \tag{26}$$

$$C_m \leq C_{max}, \forall m = 1 \ldots M \tag{27}$$

$$z \leq \frac{M^{\frac{3}{2}}}{2 \times \sum_{j=1}^{N} p_j} \times C_\Delta \tag{28}$$

$$z \geq \frac{M}{\sqrt{2} \times \sum_{j=1}^{N} p_j} \times C_\Delta \tag{29}$$

$$x_{im} \in \{0, 1\} \forall j = 1, .., N \; \forall m = 1, .., M \tag{30}$$

$$C_{max}, C_{min}, z \geq 0 \tag{31}$$

4.3. Mixed-Integer Linear Programming (MILP)

This third formulation addressed a linear version of the mixed integer programming model. It also considers the same binary assignment variables x_{jm}, which take a value of 1 if job j is assigned to machine m and 0 otherwise. Equation (32) describes the objective function represented by the variable z. Constraints (33) and (34) define respectively the workload of the least and most loaded machine. Constraints (35) and (36) introduce the theoretical bounds established by Proposition 4 as new cats in the original model. Constraints (37) ensure that each job is assigned to exactly one machine. Finally, the domains of the binary variables are defined by Equations (38) and (39).

$$\text{Minimize } z \tag{32}$$

Subject to

$$\sum_{j=1}^{N} x_{jm} \times p_j \geq C_{min}, \forall m = 1 \ldots M \tag{33}$$

$$\sum_{j=1}^{N} x_{jm} \times p_j \leq C_{max}, \forall m = 1 \ldots M \tag{34}$$

$$z \leq \frac{M^{\frac{3}{2}}}{2 \times \sum_{j=1}^{N} p_j} \times C_\Delta \tag{35}$$

$$z \geq \frac{M}{\sqrt{2} \times \sum_{j=1}^{N} p_j} \times C_\Delta \tag{36}$$

$$\sum_{m=1}^{M} x_{jm} = 1, \forall j = 1 \ldots N \tag{37}$$

$$x_{im} \in \{0,1\} \forall j = 1,..,N \; \forall m = 1,..,M \tag{38}$$

$$C_{max}, C_{min}, z \geq 0 \tag{39}$$

5. Computational Analysis

The objective of this computational study is to analyze the performance of three formulations, namely *MILP*, *MIQ1* and *MIQ2* under the same datasets used by Schwerdfeger and Walter [18]. Concretely, we would like to gauge the impact of the linearization to the solving of linear model. In addition, the numerical analysis would track the differences made by introducing the linear cuts onto the quadratic formulation.

5.1. Benchmarks

To conduct numerical tests, we used two datasets introduced by Schwerdfeger and Walter [18], namely *HGJ* and *DM*. The *HGJ* dataset was generated by [21], and includes 560 instances with the number of jobs varying from 10 to 200 and the number of machines from 3 to 80. For each couple (number of jobs, number of machines), Haouari et al. [21] generated 20 instances. The *DM* dataset was generated by Dell'Amico and Martello [22] with 1900 instances. The number of jobs was generated from 10 to 10000 and the number of machines from 3 to 15. Fifty instances are generated for each couple. The reader may refer to [18] for details regarding the generation of the datasets.

We use the solver ILOG IBM CPLEX 12.5, executed on an Intel(R) Xeon(R) Gold 5120 CPU @ 2.20 GHz, with 92 GB of RAM. The solving time limit was fixed to 300 s. For each problem size, we recorded the number of instances that had been solved to optimality and the average/maximum/minimum solving time.

5.2. Results

Performance criteria: The performance of each model is based on computational performance of the solver to reach optimal solutions. The indicators are, amongst the instances with the same size (M, N):

- *max time*: maximal solving time.
- *min time*: minimal solving time.
- *avg time*: average solving time.
- *opt*: number of instances which are solved to optimality within the solving time limit fixed to 300 s.

5.2.1. Dataset DM

Table 2 lists the numerical results from solving the DM dataset by using three models (MIL, MIQ1 and MIQ2). With the linear model MIL, CPLEX solved 500 out of 1900 to optimality, i.e., 26%. The quadratic model MIQ1 yields 1269 instances solved 67% to optimality. With linear cuts, the number of opt for MIQ2 is 717, which is 38% of the total instances. The best average solving times of CPLEX was given by MIQ1, followed by MIQ2 and lastly MILP. While combining the solutions found by all the models, we increased the number of instances solved optimally to 1370, which is 72% out of 1900 instances. For this dataset, [18] found 945 optimal solutions by solving the quadratic model by Solver Gurobi.

Table 2. Numerical results from *DM* Dataset.

M	N	Max Time			Min Time			Avg Time			Opt			
		MILP	MIQ1	MIQ2	MILP	MIQ1	MIQ2	MILP	MIQ1	MIQ2	MILP	MIQ1	MIQ2	All
3	10	0.3	0.3	0.2	0.0	0.1	0.1	0.2	0.1	0.1	50	50	50	50
3	25	300.3	300.2	0.5	0.1	0.0	0.1	63.2	6.2	0.3	46	49	50	50
3	50	321.5	0.4	300.1	0.1	0.1	0.1	182.4	0.2	12.4	20	50	48	50
3	100	317.8	0.5	307.9	0.1	0.1	0.1	188.3	0.2	18.9	19	50	47	50
3	250	309.3	0.3	305.3	0.2	0.0	0.1	170.7	0.2	31.0	22	50	45	50
3	500	307.6	0.3	303.9	0.1	0.1	0.1	194.9	0.2	31.2	18	50	45	50
3	1000	305.3	0.4	304.7	0.3	0.1	0.3	182.3	0.3	7.6	20	50	49	50
3	2500	304.5	0.9	304.7	0.3	0.3	0.6	163.9	0.4	33.1	23	50	45	50
3	5000	304.4	1.2	303.1	0.8	0.9	1.5	230.3	1.0	108.2	12	50	33	50
3	10,000	304.3	4.6	302.3	1.4	2.6	4.7	212.5	3.3	136.3	15	50	29	50
5	10	0.2	0.9	1.8	0.1	0.1	0.1	0.1	0.2	0.4	50	50	50	50
5	25	300.3	300.2	300.1	0.2	0.1	0.1	169.1	144.2	66.6	26	26	39	44
5	50	321.6	302.3	308.2	0.2	0.2	0.2	216.3	132.4	115.1	15	28	31	33
5	100	321.5	304.6	308.7	0.3	0.2	0.4	248.9	20.1	171.1	10	47	22	47
5	250	309.3	8.6	307.2	0.2	0.3	0.5	221.0	1.5	177.2	14	50	21	50
5	500	305.9	7.3	306.3	0.4	0.3	0.4	255.3	1.1	188.9	8	50	19	50
5	1000	304.2	1.1	305.9	0.4	0.3	0.6	255.0	0.5	189.4	8	50	19	50
5	2500	307.7	1.7	304.6	0.7	0.7	2.5	244.5	1.1	204.2	10	50	17	50
5	5000	305.0	4.2	302.9	1.3	1.7	4.0	263.5	2.4	213.2	8	50	16	50
5	10,000	307.2	8.4	300.4	2.6	5.4	10.4	249.2	7.0	208.9	10	50	18	50
10	25	300.4	305.0	304.8	0.4	300.0	300.0	66.0	300.2	300.1	44	0	0	44
10	50	327.0	300.2	300.1	72.6	3.5	9.1	294.7	282.5	282.8	3	3	3	3
10	100	319.0	304.2	304.3	1.2	0.7	3.7	283.5	220.0	283.0	4	14	5	14
10	250	307.2	304.0	305.0	8.2	3.4	129.1	292.6	184.0	298.8	2	22	1	22
10	500	309.8	303.9	304.4	11.4	14.4	12.1	293.2	174.0	291.2	2	23	2	23
10	1000	312.2	303.6	306.0	62.8	7.5	301.3	300.3	149.1	302.8	1	32	0	32
10	2500	308.8	227.3	300.3	302.2	3.1	32.4	304.2	59.6	294.9	0	50	1	50
10	5000	308.5	191.5	300.6	24.6	7.0	69.2	297.9	27.2	291.3	1	50	2	50
10	10,000	312.2	117.1	301.1	151.6	17.9	170.1	296.7	32.8	297.0	2	50	2	50
15	25	308.7	320.3	313.0	0.2	300.0	300.0	115.1	307.1	303.8	32	0	0	32
15	50	307.9	304.1	305.1	300.0	300.0	300.0	300.6	300.1	300.7	0	0	0	0
15	100	316.9	302.5	304.6	14.6	5.7	8.6	293.4	267.2	274.7	3	6	7	7
15	250	315.5	303.2	304.1	22.1	64.3	300.1	302.9	294.4	301.8	1	3	0	3

Table 2. *Cont.*

M	N	Max Time			Min Time			Avg Time			Opt			
		MILP	MIQ1	MIQ2	MILP	MIQ1	MIQ2	MILP	MIQ1	MIQ2	MILP	MIQ1	MIQ2	All
15	500	313.1	304.9	306.0	43.3	24.4	80.6	303.2	270.3	297.6	1	7	1	7
15	1000	323.9	304.1	303.0	305.5	108.0	300.1	313.0	229.6	300.6	0	22	0	22
15	2500	313.6	300.3	300.6	303.3	283.1	300.2	306.8	299.9	300.3	0	1	0	1
15	5000	315.5	300.6	301.0	300.4	72.8	300.5	305.5	285.5	300.6	0	6	0	6
15	10,000	301.6	301.1	302.0	300.5	28.6	300.7	300.9	193.0	301.1	0	30	0	30
Avg/ Total		-	-	-	-	-	-	228.5	110.5	190.4	500	1269	717	1370
											26%	67%	38%	72%

Solving difficult instances of DM dataset (10, 25), (15, 25) and (15, 50): While the MIQ1 had the best performance overall, it encountered difficulties when dealing with the difficult-to-solve instances: (10, 25), (15, 25) and (15, 50). The same phenomenon was also frequently observed in the literature, especially in the most recent work of [18]. Schwerdfeger and Walter [18] did not find an optimal solution for all of the mentioned combinations. Interestingly, the linear formulation performed very well in those combinations: 44 (88%) instances solved to optimality for the problem size of (10, 25), and 32 (64%) instances solved to optimality for the problem size of (15, 25). The configuration (15, 50) remains problematic for our three models.

5.2.2. Dataset HGJ

Table 3 lists the numerical results from solving the HGJ dataset using three models (MILP, MIQ1 and MIQ2). For this dataset, the gap of the performance between the three models is slightly smaller than that of the *DM* dataset. Accordingly, out of 560 instances, the percentage of instances solved optimally was 52% for MILP, 55% for MIQ1 and 51% for MIQ2. Altogether, 389 out of 560 instances were solved optimally, which is equivalent to 61%. Again, in the configuration whereby one model has some difficulties in finding a solution, the others can solve those instances more easily. For this dataset, the authors of [18] found 37 optimal solutions by solving the quadratic model by Gurobi and 45 optimal solutions using a local search.

Table 3. Numerical results from the *HGJ* Dataset.

M	N	Max Time			Min Time			Avg Time			Opt			
		MILP	MIQ1	MIQ2	MILP	MIQ1	MIQ2	MILP	MIQ1	MIQ2	MILP	MIQ1	MIQ2	All
3	10	0.2	0.1	0.2	0.1	0.1	0.1	0.1	0.1	0.1	20	20	20	20
3	12	0.3	0.2	0.2	0.1	0.1	0.1	0.1	0.1	0.2	20	20	20	20
3	15	0.8	0.5	0.4	0.1	0.1	0.1	0.3	0.3	0.2	20	20	20	20
3	20	6.7	0.4	300.1	0.1	0.1	0.2	3.1	0.3	28.8	20	20	19	20
3	50	300.5	0.2	304.1	0.1	0.1	42.0	225.2	0.1	276.3	5	20	3	20
3	100	300.7	0.4	305.5	0.1	0.1	0.9	225.3	0.2	243.9	5	20	4	20
4	10	0.2	0.4	0.7	0.1	0.1	0.2	0.1	0.2	0.2	20	20	20	20
4	12	0.2	0.4	0.2	0.1	0.1	0.2	0.1	0.2	0.2	20	20	20	20
4	15	1.1	0.9	38.9	0.2	0.1	0.2	0.6	0.5	7.9	20	20	20	20
5	10	0.2	0.9	0.5	0.1	0.2	0.2	0.1	0.2	0.2	20	20	20	20
5	12	0.3	25.3	12.2	0.1	0.4	0.7	0.2	2.0	3.3	20	20	20	20
5	15	0.9	2.0	5.2	0.2	0.2	0.4	0.5	0.8	1.8	20	20	20	20
5	20	110.9	143.0	300.1	0.4	0.2	0.5	11.8	28.8	105.5	20	20	13	20
5	50	313.5	300.1	304.2	0.5	0.4	300.0	260.5	210.2	301.1	3	6	0	6
5	100	316.9	306.1	302.0	0.5	0.8	300.1	279.3	33.0	300.2	2	20	0	20
8	20	1.6	300.1	314.6	0.2	0.3	300.0	0.7	96.2	304.0	20	15	0	20
15	20	2.3	325.0	0.5	0.4	309.4	0.1	0.9	314.9	0.4	20	0	20	20
15	50	300.2	300.3	300.1	300.0	25.4	0.1	300.1	272.9	15.4	0	3	19	19

Table 3. *Cont.*

M	N	Max Time			Min Time			Avg Time			Opt			
		MILP	MIQ1	MIQ2	MILP	MIQ1	MIQ2	MILP	MIQ1	MIQ2	MILP	MIQ1	MIQ2	All
15	100	315.8	300.1	309.4	168.9	281.4	0.1	303.0	299.1	46.5	1	1	17	18
16	40	300.2	306.5	304.2	7.0	7.8	300.0	181.8	273.6	300.7	10	2	0	10
20	50	312.0	305.9	300.1	29.4	300.0	0.5	245.7	302.8	150.4	5	0	10	12
24	60	309.5	305.2	305.0	172.8	28.2	300.0	297.8	288.8	301.8	1	1	0	2
28	70	308.4	305.0	303.0	300.0	300.1	300.1	302.9	302.1	301.3	0	0	0	0
32	80	306.9	304.9	304.4	300.1	53.6	300.1	302.8	288.9	301.3	0	1	0	1
36	90	307.7	304.0	303.3	300.1	300.1	300.1	302.9	301.1	300.5	0	0	0	0
40	100	306.3	304.1	304.2	300.1	300.1	144.4	304.0	300.5	293.5	0	0	1	1
60	150	305.6	303.4	300.2	300.1	300.1	300.2	303.6	301.0	300.2	0	0	0	0
80	200	306.1	300.3	300.2	300.1	300.2	300.1	303.5	300.3	300.2	0	0	0	0
Avg/ Total		-	-	-	-	-	-	148.5	140.0	149.5	292	309	286	389
											52%	55%	51%	69%

Solving difficult instance of HGJ dataset (15, 20): we observed the same phenomenon in the DM dataset. In configuration (15, 20), the authors of [18] did not find an optimal solution; the quadratic model MIQ1 does not yield any optimal solution either. However, with a linear cut, MILP and MIQ2 found 20 optimal solutions out of 20 instances.

6. Conclusions and Perspectives

Workload balancing problems among a set of identical machines or resources is a challenge in many organizations and production systems. Different criteria have been addressed in the literature as performance measures to tackle this problem. The minimization of the difference between the most and least loaded machine (C_Δ) and the Normalized Sum of Square for Workload Deviations ($NSSWD$) are the most important and well-established ones. To the best of our knowledge, the most important works addressed in the literature have considered empirical approaches to analyze this problem.

In this paper, we proposed a rigorous theoretical analysis of the workload balancing problem on identical parallel machines and established the mathematical correlation between the two problems: $P||C_\Delta$ and $P||NSSWD$.

Based on these new theoretical proprieties, three mathematical formulations were tested using the latest benchmark described in the literature. The combination of these three models allowed us to provide optimal solutions for some unsolved instances in this benchmark.

Recently, Asadollahi-Yazdi et al. [23] discussed the concept of Industry 4.0 and its effect on the industrial world and global society. This study provided a comprehensive overview of industrial revolutions, manufacturing production plans over time, as well as a detailed analysis of Industry 4.0 and its features. One of the important concerns addressed in this paper was the necessity of integrating human factors into existing scheduling techniques and information systems in the production control of real factories. Workload balancing was identified as one of these criteria with which to assess and measure the balance or "equity" of workload assignments. We see an interesting direction for future research, which consists of generalizing the obtained results to establish a performance measure of workload equity in different, real cases, taking into account the perspective of the industry of the future.

Author Contributions: Y.O., N.-Q.N. and F.Y. contributed equally to the different steps of this research. All authors have read and agreed to the published version of the manuscript.

Funding: This research was funded by the Industrial Chair Connected-Innovation and the FEDER (Le Fonds Européen de Développement Régional).

Institutional Review Board Statement: Not applicable.

Informed Consent Statement: Not applicable.

Data Availability Statement: All the data have been presented in the manuscript. The codes developed for the current study are available from the corresponding author if requested.

Acknowledgments: The authors gratefully acknowledge the Industrial Chair Connected-Innovation and its industrial partners (https://chaire-connected-innovation.fr (accessed on 1 February 2021)), as well as FEDER (Le Fonds Européen de Développement Régional) for their financial support.

Conflicts of Interest: The authors declare no conflict of interest.

References

1. Fan, Q.; Ansari, N. Towards Workload Balancing in Fog Computing Empowered IoT. *IEEE Trans. Netw. Sci. Eng.* **2020**, *7*, 253–262. [CrossRef]
2. Florescu, A.; Barabas, S.A. Modeling and Simulation of a Flexible Manufacturing System—A Basic Component of Industry 4.0. *Appl. Sci.* **2020**, *10*, 8300. [CrossRef]
3. Li, D.; Tang, H.; Wang, S.; Liu, C. A big data enabled load-balancing control for smart manufacturing of Industry 4.0. *Cluster Comput.* **2017**, *20*, 1855–1864. [CrossRef]
4. Asensio-Cuesta, S.; Diego-Mas, J.A.; Canós-Darós, L.; Andrés-Romano, C. A genetic algorithm for the design of job rotation schedules considering ergonomic and competence criteria. *Int. J. Adv. Manuf. Technol.* **2012**, *60*, 1161–1174. [CrossRef]
5. Azmat, C.S.; Hürlimann, T.; Widmer, M. Mixed integer programming to schedule a single-shift workforce under annualized hours. *Ann. Oper. Res.* **2004**, *128*, 199–215. [CrossRef]
6. Khouja, M.; Conrad, R. Balancing the assignment of customer groups among employees: Zero-one goal programming and heuristic approaches. *Int. J. Oper. Prod. Manag.* **1995**, *15*, 76–85. [CrossRef]
7. Yildirim, M.B.; Duman, E.; Krishnan, K.; Senniappan, K. Parallel machine scheduling with load balancing and sequence dependent setups. *Int. J. Oper. Res.* **2007**, *4*, 1–8.
8. Ouazene, Y.; Hnaien, F.; Yalaoui, F.; Amodeo, L. The Joint Load Balancing and Parallel Machine Scheduling Problem. In *Operations Research Proceedings 2010*; Hu, B., Morasch, K., Pickl, S., Siegle, M., Eds.; Springer: Berlin/Heidelberg, Germany, 2011; pp. 497–502.
9. Keskinturk, T.; Yildirim, M.B.; Barut, M. An ant colony optimization algorithm for load balancing in parallel machines with sequence-dependent setup times. *Comput. Oper. Res.* **2012**, *39*, 1225–1235. [CrossRef]
10. Cossari, A.; Hob, J.C.; Paletta, G.; Ruiz-Torres, A.J. Minimizing workload balancing criteria on identical parallel machines. *J. Ind. Prod. Eng.* **2013**, *30*, 160–172. [CrossRef]
11. Rajakumar, S.; Arunachalam, V.P.; Selladurai, V. Workflow balancing strategies in parallel machine scheduling. *Int. J. Adv. Manuf. Technol.* **2004**, *23*, 366–374. [CrossRef]
12. Rajakumar, S.; Arunachalam, V.P.; Selladurai, V. Workflow balancing in parallel machines through genetic algorithm. *Int. J. Adv. Manuf. Technol.* **2007**, *33*, 1212–1221. [CrossRef]
13. Walter, R.; Lawrinenko, A. A note on minimizing the normalized sum of squared workload deviations on m parallel processors. *Comput. Ind. Eng.* **2014**, *75*, 257–259. [CrossRef]
14. Ho, J.C.; Tseng, T.L.; Ruiz-Torres, A.J.; López, F.J. Minimizing the normalized sum of square for workload deviations on m parallel processors. *Comput. Ind. Eng.* **2009**, *56*, 186–192. [CrossRef]
15. Ouazene, Y.; Yalaoui, F.; Chehade, H.; Yalaoui, A. Workload balancing in identical parallel machine scheduling using a mathematical programming method. *Int. J. Comput. Intell. Syst.* **2014**, *7*, 58–67. [CrossRef]
16. Chandra, A.K.; Wong, C.K. Worst-Case Analysis of a Placement Algorithm Related to Storage Allocation. *SIAM J. Comput.* **1975**, *4*, 249–263. [CrossRef]
17. Schwerdfeger, S.; Walter, R. A fast and effective subset sum based improvement procedure for workload balancing on identical parallel machines. *Comput. Oper. Res.* **2016**, *73*, 84–91. [CrossRef]
18. Schwerdfeger, S.; Walter, R. Improved algorithms to minimize workload balancing criteria on identical parallel machines. *Comput. Oper. Res.* **2018**, *93*, 123–134. [CrossRef]
19. Cossari, A.; Ho, J.C.; Paletta, G.; Ruiz-Torres, A.J. A new heuristic for workload balancing on identical parallel machines and a statistical perspective on the workload balancing criteria. *Comput. Oper. Res.* **2012**, *39*, 1382–1393. [CrossRef]
20. Al-Saleh, M.F.; Yousif, A.E. Properties of the standard deviation that are rarely mentioned in classrooms. *Austrian J. Stat.* **2009**, *38*, 193–202. [CrossRef]
21. Haouari, M.; Gharbi, A.; Jemmali, M. Tight bounds for the identical parallel machine scheduling problem. *Int. Trans. Oper. Res.* **2006**, *13*, 529–548. [CrossRef]
22. Dell'Amico, M.; Martello, S. Optimal Scheduling of Tasks on Identical Parallel Processors. *ORSA J. Comput.* **1995**, *7*, 191–200. [CrossRef]
23. Asadollahi-Yazdi, E.; Couzon, P.; Nguyen, N.Q.; Ouazene, Y.; Yalaoui, F. Industry 4.0: Revolution or Evolution? *Am. J. Oper. Res.* **2020**, *10*, 241–268. [CrossRef]

Article

A Genetic Crow Search Algorithm for Optimization of Operation Sequencing in Process Planning

Mica Djurdjev [1], Robert Cep [2], Dejan Lukic [3], Aco Antic [3], Branislav Popovic [4] and Mijodrag Milosevic [3,*]

[1] Department of Mechanical Engineering, Technical Faculty "Mihajlo Pupin", University of Novi Sad, 23000 Zrenjanin, Serbia; mica.djurdjev@tfzr.rs
[2] Department of Machining, Assembly and Engineering Metrology, Faculty of Mechanical Engineering, Technical University of Ostrava, 70800 Ostrava, Czech Republic; robert.cep@vsb.cz
[3] Department of Production Engineering, Faculty of Technical Sciences, University of Novi Sad, 21000 Novi Sad, Serbia; lukicd@uns.ac.rs (D.L.); antica@uns.ac.rs (A.A.)
[4] Department for Power, Electronic and Telecommunication Engineering, Faculty of Technical Sciences, University of Novi Sad, 21000 Novi Sad, Serbia; bpopovic@uns.ac.rs
* Correspondence: mido@uns.ac.rs; Tel.: +381-21-485-2346

Abstract: Computer-aided process planning represents the main link between computer-aided design and computer-aided manufacturing. One of the crucial tasks in computer-aided process planning is an operation sequencing problem. In order to find the optimal process plan, operation sequencing problem is formulated as an NP hard combinatorial problem. To solve this problem, a novel genetic crow search approach (GCSA) is proposed in this paper. The traditional CSA is improved by employing genetic strategies such as tournament selection, three-string crossover, shift and resource mutation. Moreover, adaptive crossover and mutation probability coefficients were introduced to improve local and global search abilities of the GCSA. Operation precedence graph is adopted to represent precedence relationships among features and vector representation is used to manipulate the data in the Matlab environment. A new nearest mechanism strategy is added to ensure that elements of machines, tools and tool approach direction (TAD) vectors are integer values. Repair strategy to handle precedence constraints is adopted after initialization and shift mutation steps. Minimization of total production cost is used as the optimization criterion to evaluate process plans. To verify the performance of the GCSA, two case studies with different dimensions are carried out and comparisons with traditional and some modern algorithms from the literature are discussed. The results show that the GCSA performs well for operation sequencing problem in computer-aided process planning.

Keywords: crow search; genetic algorithm; process planning; operation sequencing; precedence constraints

check for
updates

Citation: Djurdjev, M.; Cep, R.; Lukic, D.; Antic, A.; Popovic, B.; Milosevic, M. A Genetic Crow Search Algorithm for Optimization of Operation Sequencing in Process Planning. *Appl. Sci.* **2021**, *11*, 1981. https://doi.org/10.3390/app11051981

Academic Editors: Farouk Yalaoui, Taha Arbaoui and Yassine Ouazene

Received: 8 February 2021
Accepted: 16 February 2021
Published: 24 February 2021

Publisher's Note: MDPI stays neutral with regard to jurisdictional claims in published maps and institutional affiliations.

1. Introduction

According to the Society of Manufacturing Engineers, process planning is considered as a systematic definition of methods for economical and productive manufacturing. In Xu et al. [1] it is considered that computer-aided process planning (CAPP), as a crucial component of todays advanced manufacturing systems, can be imagined as the "bridge" between the product design (CAD) and product manufacturing (CAM). CAPP system development has shown to be a very complex task, assuming today's high diversity of products and complexity of planning numerous activities within a process plan in different manufacturing conditions. The consequence reflects the fact that CAPP systems are at the lower level of implementation in industry and therefore represent a bottleneck in integrated production environment, as stated by Lukić et al. [2].

89

As reported in Denkena et al. [3], the main activities of process planning, whether computer-aided or manual, are the following: receiving order and design details, selecting raw material and shape, selecting process technology, determining process order, preparing a process plan (including activities such as the selection of machining operations, sequencing of machining operations, selection of cutting tools, determination of setup requirements, calculations of cutting parameters, selection and design of jigs and fixtures, planning tool paths, estimating processing setup costs and times) and generating planning output (CNC programs, routing sheets, operation sheets etc.). The aim of these stages of process planning is to transform the part design (drawing of a part or a product) into a finished physical part or product in a cost-effective and competitive way.

Among the mentioned activities, the sequencing of machining operations, i.e., the operation sequencing, has been considered as one of the most complex optimization tasks in scientific community. Two main steps of the operation sequencing task are taken into account: (1) selecting the most suitable machining resources such as machines, tools and tool approach directions based on the specific machining features that have to be machined and (2) finding the sequence of all machining operations for a considered part with respect to precedence relationships among operations and features. Such tasks, when handled simultaneously, form a process plan which is evaluated using certain optimization criteria, such as the minimization of production time or minimization of production cost.

According to many authors such as Duo et al. [4], Falih and Shammari [5], Gao et al. [6], Petrović et al. [7] and Huang et al. [8], the operation sequencing is classified as an NP hard optimization problem due to computational ineffectiveness when dealing with these problems using conventional non-heuristic techniques. Milošević et al. [9] reported that metaheuristic algorithms have proven to be robust and efficient methods for these challenges and are therefore employed to find near-optimal solutions in an acceptable time period. Many intelligent metaheuristics have been applied in operation sequencing and process planning optimization in the last few decades. Among them, we can primarily mention traditional optimization approaches, such as genetic algorithms, ant colony optimization, particle swarm optimization or simulated annealing. Nowadays, novel modifications in terms of constraint handling mechanisms, representation of individual solutions and different techniques for improving local and/or global search abilities are still being proposed. In the next section, we will highlight the most recent advances in this area of research.

2. Related Work

Wang et al. [10] proposed a hybrid particle swarm-based method to solve the process planning problem. A novel representation scheme besides two local search algorithms were introduced to improve solutions in each generation. In work by Huang et al. [11], a hybrid approach based on operation precedence graph and genetic algorithm was developed. Topologic sorting algorithm was used to ensure feasibility of solutions and a modified crossover operator with two mutation strategies were adopted. A socio-politically inspired metaheuristic called imperialist competitive algorithm for process planning optimization was proposed by Lian et al. [12]. Various flexibilities were taken into account and the steps such as assimilation, position exchange, imperialist competition and elimination were proceeded to solve the NP hard problem. Liu et al. [13] developed an ant colony optimization algorithm reinforced with the constraint matrix and the state matrix to check the state of the operations and range of operation candidates. A honey bees mating algorithm was implemented by Wen et al. [14]. The solution encoding, crossover operator and local search strategies have been employed to verify the performance of this metaheuristic. Wang et al. [15] presented a two-stage ant colony optimization algorithm to minimize the total production cost. In this approach, a directed graph is used to represent the process planning problem. In the first stage, the operation nodes are selected and the graph is reduced to a simple graph. Later, in the second stage, the selected operations are used to create the sequence and generate process plans. Petrovic et al. [7] utilized particle swarm optimization and chaos theory to solve the flexible process planning problem. Various flexibilities were represented using AND/OR networks and the total production time and cost were used to develop the mathematical model for the minimization of objective function. Hu et al. [16] proposed an ant colony optimization algorithm-based approach to solve the operation sequencing problem in CAPP. Precedence and clustering constraint relationships are taken into account to ensure the feasibility of process plans. Global search ability is enhanced using adaptive updating method and local search mechanism. A hybrid genetic algorithm and simulated annealing approach to optimize the production cost of a complicated part in a dynamic workshop environment was proposed by Huang et al. [8]. The directed graph was used to model precedence relationships among machining operations and the graph search algorithms were embedded into framework of the optimization system. Su et al. [17] presented precedence constrained operation sequencing problem as a mixed integer programming model and incorporated an edge selection based strategy in the genetic algorithm to solve the problem. The novel encoding strategy assures the feasibility of solutions in the initialization stage with acceptable probability, and therefore improves the GA's convergence efficiency. Dou et al. [4] developed a feasible sequence-oriented discrete particle swarm optimization algorithm that is also directed toward generating feasible operation sequences in the initial stage of the algorithm. A crossover-based mechanism is adopted to evolve the particles in discrete feasible solution space. On the other hand, fragment mutation, as well as greedy and uniform mutations, are used to alter the feasible sequence and machining resources respectively. The adaptive mutation probability is used to improve the exploration ability of the approach. A new mixed-integer linear programming mathematical model is developed and represented on OR-node network graph in the recent study proposed by Liu et al. [18]. A new hybrid evolutionary algorithm is designed on the basis of this MILP model, which combines a genetic algorithm with a simulated annealing algorithm. Moreover, the tournament selection with the varying tournament size was included to prevent the algorithm from premature convergence and increased randomness. Gao et al. [6] proposed the intelligent water drop algorithm in process planning optimization. A priority constraint matrix was used to generate feasible process plans. Falih and Shammari [5] proposed a hybrid constrained permutation algorithm and genetic algorithm approach. A constrained permutation algorithm is employed to generate feasible operation sequences and a genetic algorithm is then used to search for an optimal solution within the feasible search space. A mixed crossover operator in the GA is used to avoid premature convergence to local optima. In a study by Jiang et al. [19], authors proposed the novel fairness-based transaction packing algorithm for permissioned blockchain empowered industrial IoT systems. A heuristic algorithm and a min-heap-based optimal algorithm were employed to solve the fairness problem. Since the fairness problem is related to the sum of weighting time of the selected transaction, it is transformed into the subset sum problem and the extensive experiments were conducted to verify the performance of the fairness-based transaction packing algorithm.

In this paper, a novel improved metaheuristic approach called a genetic crow search algorithm (GCSA) is introduced for optimization of operation sequencing problem. The approach combines the classical CSA with the main components of the GA algorithm and through the joint action from local and global perspective of a population (flock) the evolution of crow individuals is achieved. The operation precedence graph (OPG) is utilized to represent the operation candidates, i.e., crow individuals with regard to corresponding precedence relationships among operations. Then, the initial population (flock) of crow individuals is constructed using vector-based representation to generate a group of feasible individuals. After the main steps of the CSA, a newly proposed nearest resource mechanism is applied to ensure the vectors of each crow individual are set to the nearest integer values. The evolution of the flock of crow individuals starts in the next stage, which consists of five strategies. Firstly, the tournament selection ensures that the most fit crow individuals from the flock are selected and the 3SX crossover operator is applied to generate new crow individuals. The shift mutation strategy is then used to exchange random genes from crow individuals. The feasibility of these crow individuals is questioned, and in that sense, the constraint repair mechanism is used to ensure that infeasible solutions are converted to the feasible domain. The evolution of crow individuals ends with resource mutation, which performs minor changes in genes of machine, tool and TAD vectors, respectively. Using adaptive crossover and mutation probabilities, the exploration ability of the GCSA is improved. The input parameters of the algorithm are optimized using the Taguchi design of experiments, in order to obtain the optimized parameter settings. In each iteration of the GCSA approach, the position of crow individuals is evaluated using the fitness function and the memory of crow individuals is updated. To verify the performance of the proposed algorithm, two case studies with different level of dimensionality are considered. The computational results show that the proposed GCSA performs well and can be considered an efficient method for solving the operation sequencing problem.

The rest of the paper is organized as follows. Section 2 describes the operation sequencing model by focusing on vector-based representation, data types and precedence constraints.

3. Operation Sequencing: Problem Formulation

Operation sequence in process planning consists of a number of selected operations required for feature machining, a sequence of these operations, machining resources such as machines, tools and tool approach directions (TADs), as well as setup plans. Since the capability and the availability of machining resources has an important role when generating operation sequences, several principles have to be considered according to Su et al. [17]:

(1) Setup plan includes a group operations that can be performed with the same TADs.
(2) These operations have to be machined using the same machine.
(3) Reduction of machine, tool and TAD changes is an objective.
(4) Only one operation can be performed on a single machine.
(5) Only one operation can be performed using a single cutting tool.
(6) Only one TAD can be used in a given time.

In this study, an operation sequence is represented using vectors with n bits of data. Those bits are integer values where each value contains relevant manufacturing information. Seeing that the optimization task considers the simultaneous sequencing of operations and selection of machining resources, four vectors are used to illustrate a single process plan. Figure 1 shows a simple example of a represented process plan with a total of 5 machining operations indexed from 1 to 5 (the first row). An operation sequence vector is marked with darker color and consists of bit numbers named *Operation[n]*. For example, *Operation[3]* considers the operation number 5. Under the same index, there are machine m3, tool t3 and TAD −z. These candidates are elements of the machine vector *Machines[n]*, the tool vector *Tools[n]* and the TADs vector *TADs[n]* respectively and are visualized in a light grey color.

Index of an operation sequence vector, *Operation[n]*

IDs of machining operations

IDs of randomly selected machines, tools and TADs that formulate vectors *Machines[n]*, *Tools[n]*, *TADs[n]*

1	2	3	4	5
3	2	5	1	4
m1	m1	m3	m2	m1
t6	t3	t7	t1	t4
+x	+x	−z	+z	+y

Figure 1. Representation of an operation sequence.

Since the MATLAB programming environment was selected for the implementation of the proposed GCSA approach, therefore, we will place emphasis on the types of data considered when dealing with the operation sequencing problem. All variables with their data types are shown in Table 1.

Table 1. Data types of the operation sequencing variables.

Data Types	Variables	Description
double	*Operation[n]*	The ID of the operation as a part of an operation sequence
double	*Machines[n]*	The ID of the machine as a part of a machine vector of n machines for performing n operations
double	*Tools[n]*	The ID of the tool as a part of a tool vector of n tools for performing n operations
double	*TADs[n]*	The ID of the cutting tool as a part of a TAD vector of n TADs for performing n operations
cell	*Machines{n}*	Array of machine candidates list for n cells with each cell containing alternative machines for the n-th operation
cell	*TADs{n}*	Array of tool candidates list for n cells where each cell contains alternative tools for the n-th operation
cell	*Tools{n}*	Array of TAD candidates list for n cells where each cell contains alternative TADs for the n-th operation

Precedence Constraints

The precedence constraints within operation sequencing are formulated according to relevant geometrical and manufacturing considerations. They indicate the precedence relationships between operations required to generate a feasible machining sequence. Hu et al. [16] classified constraint relationships into the precedence and the clustering constraints. The precedence constraints are obligatory conditions that must be met, while on the other side, the clustering constraints do not need to be met but have effect on machining efficiency and quality. The precedence constraints can be classified into eight types: datum relationships, primary and secondary relationships, face and hole relationships, processing stage relationships, feature priority relationships, thin-wall relationships, fixed order of operations relationships and material-removal relationships. Li et al. [20] made similar classification on hard and soft constraints, where hard constraints represent obligatory conditions, while soft constraints can be violated if necessary.

To visualize precedence relationships between operations, we adopted an operation precedence graph model (OPG). The OPG is a directed graph consisting of vertices and edges, OPG = (V, E), where vertices V represent the set of operations, while edges E denote the set of links between operation vertices. Each operation element within the operation vector is mapped to the corresponding vertex of the OPG. On the other hand, each precedence relationship is visualized using edges. Vertices that are not linked with edges have no precedence relationships between them. For easier manipulation of the data concerning precedence relationships, adjacency or precedence matrix has been adopted. Figure 2 shows the main elements of OPGs and their mapping to a matrix form. Using a simple example with five operations and four precedence constraints a precedence matrix PM is designed. The value *pc[i,j]* is a binary value which represents whether operation *i* precedes operation *j* or not. As shown, number 1 denotes that there is a precedence relationship between observed operations while 0 shows the lack of any precedences, i.e., no edge between operation *i* and operation *j*. By following the edges of an OPG, all the vertices should be traversed in order to obtain a feasible operation sequence.

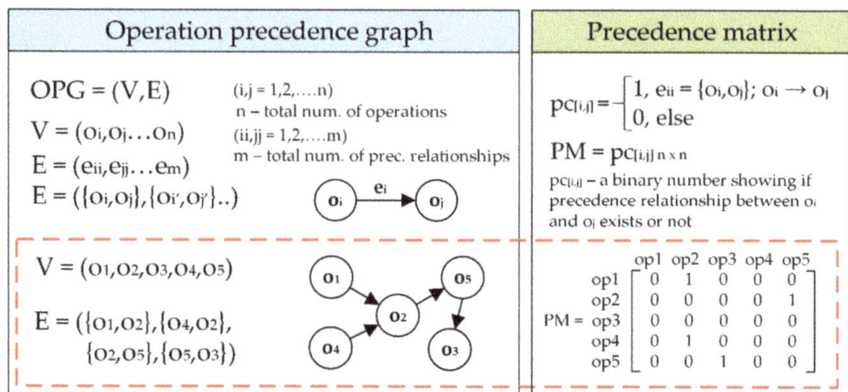

Figure 2. Operation precedence graph and precedence matrix.

4. A Genetic Crow Search Algorithm (GCSA) for Optimization of Operation Sequencing

4.1. Crow Search Algorithm

Crow search algorithm (CSA) was introduced in 2016 by Askarzadeh [21]. Its inspiration is derived from the crows' intelligent behavior exhibited in their ability to hide food and use memory to recall and restore the food when necessary.

Generally, crows are considered as one of the most intelligent living organisms on the planet, with many distinctive attributes such as self-awareness, sophisticated communication, face remembrance, food hiding, memorization, thievery, thievery avoidance and so on.

Their well-known greediness and cleverness is particularly demonstrated in situations when they observe other crow's movement with the purpose of finding better source of food and steal it once they are gone. Afterwards, crow thieves make sure they hide the stolen food and therefore avoid being victims of thievery as the previous owners. Another important element of cleverness is reflected in the fact that a crow can find out when the other crow is following it. Then, the crow that is being followed can fool the other crow by changing the direction of their flight.

From a computational point of view, the CSA algorithm is a population-based optimization algorithm which utilizes some of the exhibited crow's intelligence. A flock of crows may consider a population of search agents N where each crow i has its position x_i^{it} at iteration it. Every crow in a flock can memorize its hiding place whose position can be denoted as m_i^{it}. Similar to the personal best position (local best) of each particle in the PSO algorithm, the CSA considers the memory M_i^{it} the best food source (hiding place) the crow i found so far. Flying around the search space and moving towards another crow's hiding place helps crows to update their position and memory and possibly find a better food source. Two important states are possible when crow i tries to steal hidden food of crow j:

State 1: Crow j is not aware that is followed by crow i, resulting in the situation when crow i successfully commits thievery by moving towards the hiding place of crow j using the following expression:

$$X_i^{it+1} = X_i^{it} + rand_i \times fl \times \left(M_j^{it} - M_i^{it} \right) \tag{1}$$

where X_i^{it} and X_i^{it+1} are positions of the crow i in iterations it and $it + 1$ respectively, $rand_i$ stands for a random number between 0 and 1, fl is flight length of crow and M_i^{it} and M_i^{ij} are memorized positions of crow i and crow j respectively.

State 2: Crow j realizes that it is followed by crow i and fools crow i by choosing a new position randomly in order to save its food:

$$X_i^{it+1} = \begin{cases} X_i^{it+1} = X_i^{it} + rand_i \times fl \times \left(M_j^{it} - M_i^{it} \right) & \text{if } rand_j \geq AP \\ a \text{ random position} & \text{otherwise} \end{cases} \quad (2)$$

where $rand_j$ denotes a random number in range [0,1] and AP is awareness probability.

Two most important parameters of the flight length and the awareness probability. The role of the flight length parameter fl is to achieve equilibrium between local and the global search. Small values direct the search towards the local optimum (close to X_i^{it}) and large values prone towards the global optimum (away from X_i^{it}). Awareness probability AP also has the similar role in balancing intensification and diversification. Just as for the flight length, small values of AP increase local search capacity, and evidently, larger values emphasize global search capacity.

4.2. GCSA Methodology

The traditional CSA has been implemented mostly on continuous optimization problems with a small number of research studies done on their application in combinatorial optimization. Huang et al. [22] proposed a hybrid HCSA approach to solve the NP hard permutation flow shop scheduling problem (PFSP). They used several techniques to minimize the makespan of the PFSP, such as the smallest position value (SVP) rules to convert continuous to discrete values or job numbers, Nawaz-Enscore-Ham technique to generate a population of quality individuals and simulated annealing and variable neighborhood search algorithms to maintain the diversity of solution throughout the search process. Laabadi et al. [23] used the binary CSA to solve the two-dimensional bin packaging problem, which is also an NP-hard combinatorial problem. This approach utilizes a sigmoid function for mapping real-valued solution into binary ones and specific knowledge repair operator to improve diversity and restore feasibility of infeasible solutions.

The issues of the CSA are mostly expressed by low convergence rate and entrapment in the local optima, as reported in Shirke and Udayakumar [24]. Fine tuning of the algorithm and modifications performed by various authors significantly boosted the performances of the CSA. In our study, we attempt to improve the traditional CSA in order to solve the NP hard process planning optimization problem. Therefore, we introduce a hybridized approach based on the CSA and the main components of the GA. Moreover, adaptive probability coefficients for scaling crossover and mutation operators are included to achieve the trade-off between exploration and exploitation, Dou et al. [4]. The steps of the GCSA approach are depicted on the flowchart in Figure 3.

After initializing the starting parameters of both CSA and GA, the next step is to generate population of feasible individuals using vector-based representation that presents operation sequence, machine, tool and TAD vectors as components of a single process plan. Vectors representing positions of crows in a search space are then evaluated using fitness function described in Section 4.3. They are memorized as the best local solutions found so far by each crow, after which the best solution in the flock is stored. The main loop of the GCSA starts with the standard CSA steps described by Equations (1) and (2). After these steps, the nearest resource mechanism is employed to round the resulting vector elements to the nearest integers, according to the available machine, tool and TAD candidates. The process continues with the genetic strategies which are utilized to improve diversification of search process. Here, we adopted tournament selection, 3SX crossover operator and two mutation operators, shift and resource mutation. Repair mechanism is required after the initialization stage and shift mutation to ensure feasibility of crow individuals. Then, the GCSA steps continue with the evaluation of objective function and the update of crow's memory and the best solution (global). The search finishes after a predefined number of iterations. In the following sections, some steps are expressed in a more detailed manner.

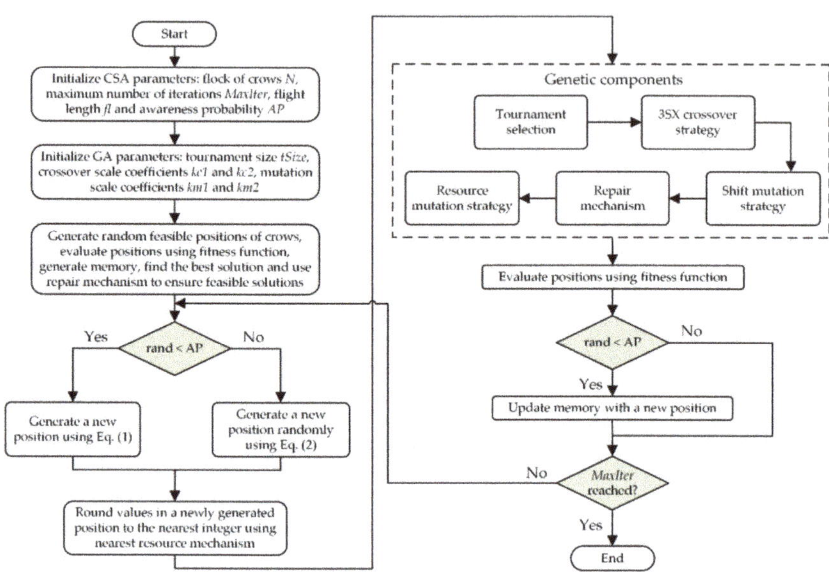

Figure 3. Flowchart of the genetic crow search algorithm (GCSA) for optimization of operation sequencing.

4.3. Evaluation Criterion of the Operation Sequencing Problem

Numerous studies conducted to this day have taken into account two most common evaluation criteria for operation sequencing, manufacturing (production) time and cost. According to [6,8], detailed information about tool paths and machining parameters is not available and therefore it is not possible to accurately determine the machining cost and machining time. Hence, the total cost of production is adopted as the optimization criterion in this research. The total production cost with weight coefficients consists of the five cost factors:

(1) **Total machine cost (TMC)** — the total sum of costs of used machines where machine cost indices denote the constant cost values for each machine.
(2) **Total tool cost (TTC)** — the total sum of costs of used cutting tools, where tool cost indices denote the constant cost values for each tool.
(3) **Total machine change cost (TMCC)**—the product of the number of machine changes and machine change cost indices, which denote the cost values of machine replacement.
(4) **Total tool change cost (TTCC)**—the product of the number of tool changes and tool change cost indices, which denote the constant cost values of tool replacement.
(5) **Total setup cost (TSC)**—the product of the number of setup changes and setup change cost indices which denote the constant cost values of a single machine setup.

The full description of all cost factors can be found in [4–8,11,14]. The total weighted production cost (TWPC) finally represents the sum of these five cost factors:

$$TWPC = w_1 \cdot TMC + w_2 \cdot TTC + w_3 \cdot TMCC + w_4 \cdot TTCC + w_5 \cdot TSC \tag{3}$$

where w_1–w_5 are the corresponding weight coefficients of cost factors.

4.4. Nearest Resource Mechanism

After the main CSA steps in the GCSA approach, the values in position vectors of crow individuals are not positive integers and the next step is to convert the real values obtained by the CSA to the closest number that matches an appropriate machine, tool or TAD candidate from the set. Here, we present the nearest resource mechanism to perform the round-off procedure and assure the newly generated positions are ready for the next stage. Besides the machine and the tool vectors, the TAD position vector is represented using integer values that match the appropriate directions which are denoted as letters with negative or positive signs. In Algorithm 1 below, we show the pseudocode of the most important steps of the nearest resource mechanism. The mechanism in this example is applied to machine vector, although the principle is the same for tool and TAD vectors.

Algorithm 1. The nearest machine mechanism

Machines{} – set of all machine candidates for each operation
currentMachine – current machine candidate
currentOperation – current operation in a process plan
diff = | Machines{currentOperation} – currentMachine |
for i = 1: number of operations
if currentMachine > Machines{ currentOperation }
for j = 1: diff
check if the set Machines{currentOperation} contains
| currentMachine – diff (j) |
Use | currentMachine – diff (j) | as a machine candidate;
else
continue loop;
end if
end for
elseif currentMachine < Machines{ currentOperation }
for j = 1: diff
check if the set Machines{currentOperation} contains
| currentMachine + diff (j) |
Use | currentMachine + diff (j) | as a machine candidate;
else
continue loop;
end if
end for
else
for j = 1: diff
check if the set Machines{currentOperation} contains
| currentMachine - diff (j) |
Use | currentMachine - diff (j) | as a machine candidate;
elseif the set Machines{currentOperation} contains
| currentMachine + diff (j) |
Use | currentMachine + diff (j) | as a machine candidate;
end if
end for
end if
end for

4.5. Genetic Components of the GCSA Approach

When producing a new generation of individuals, selection, crossover and mutation represent the crucial components of genetic algorithms. By mimicking the process of natural selection, they enable a population of solutions to evolve during a certain number of generations.

The first step in this process is a selection. A tournament selection operator has been adopted in this study. A few crow individuals from the flock are randomly selected to participate in the "tournament" and the crow individual with the highest fitness value wins the "tournament" and becomes a part of a new flock that consists of the best crow individuals that had previously won the competition.

The selection step is followed by the crossover operator whose purpose is to provide information exchange between two crow individuals from a flock. In this case, the three-string crossover (3SX) proposed by Falih and Shammari [5] was adopted. The 3SX operator affects only resource vectors, e.g., machine, tool and TAD vector respectively, leaving the operation sequence vector untouched and thereby assuring the feasibility of a process plan. The following steps characterize the 3SX crossover operator:

(1)   Two crows from a new flock are randomly selected as parent crows;
(2)   From random cutting point, two parent crows are divided into sections to produce two child crows;
(3)   Resources (machines, tools and TADs) from the right section of child crow 1 are replaced by resources of the same operations from parent crow 2;
(4)   Following the same pattern, resources in the right section of child crow 2 are replaced by resources of the same operations from parent crow 1.

Figure 4 illustrates the 3SX crossover operator applied on the 7-gene length crow individual, with a random cutting point between the 4th and 5th operation index.

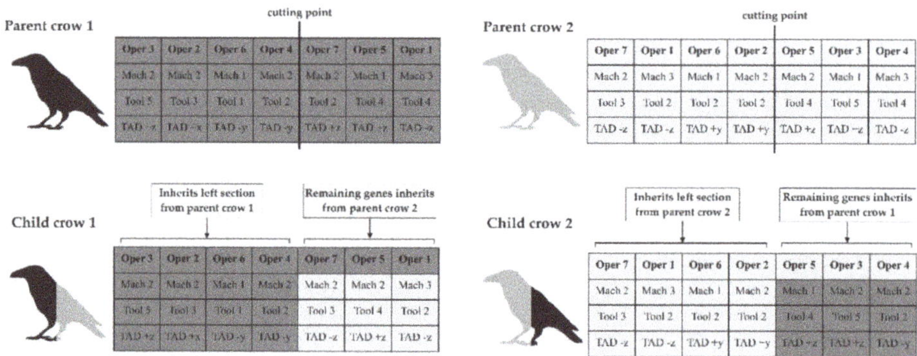

Figure 4. 3SX crossover operator.

Mutation strategies bring a finishing touch to the genetic process of producing new and fitter individuals. They aim to make further improvements in diversification and avoid premature convergence of the algorithm. The first mutation operator used within the GCSA approach is the shift mutation illustrated in Figure 5. A single crow individual is randomly selected from a newly formed flock of crows and two genes (operations with its resources) are randomly selected for exchange. After exchanging the two, some of the precedence constraints may be violated, which affects the operation sequence and therefore the feasibility of a process plan. In that case, the repair mechanism has to be employed to ensure all the crows represent feasible individuals. The same mechanism is applied after the initialization stage of the GCSA. We adopted an approach similar to the constraint handling heuristic proposed by Huang et al. [11].

The second mutation operator is the resource mutation which, similarly to the 3SX crossover operator, operates on resource vectors, machine, tool and TAD vector respectively. The feasibility of operation sequence is therefore not questioned. The following are the steps of the resource mutation operator illustrated in Figure 6:

(1)   Randomly select a crow individual;
(2)   Randomly select the mutation point, i.e., operation with its index;
(3)   Check available machines in the *Machines{}* for selected operation;
(4)   Randomly select a machine from the *Machines{}* as the current machine;
(5)   Identify other operations that have the same machine alternative in the *Machines{}*.
(6)   Assign the same machine alternative as the current machine for other operations.

(7) Repeat the same steps for tool and TAD vectors using the *Tool{}* and the *TADs{}* sets.

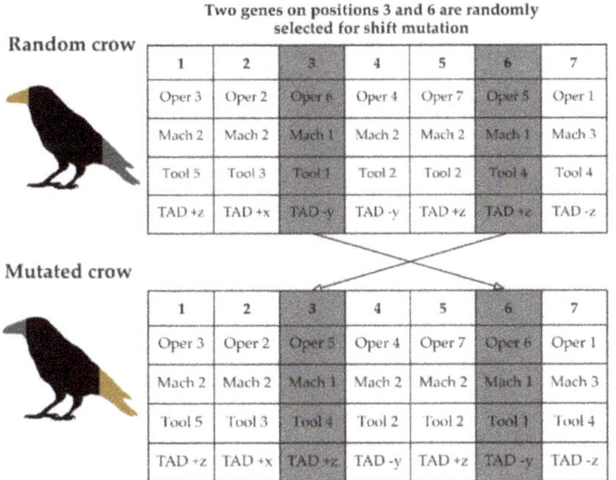

Figure 5. Shift mutation operator.

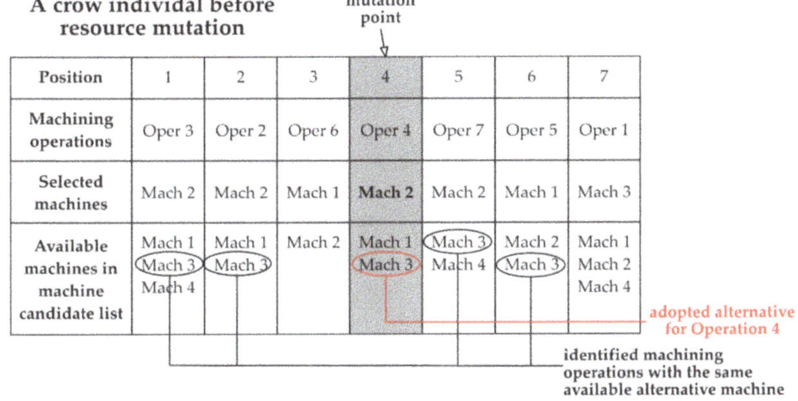

Figure 6. Resource mutation operator.

As reported by Dou et al. [4], fixed operators have shown to be less effective comparing to the adaptable operators, whose values vary in response to the fitness of individuals. To additionally improve search abilities of the GCSA, we included adaptive crossover and mutation probabilities:

$$p_c = k_{c1} \cdot \frac{f_{max} - f_i}{\cdot \left(f_{max} - f_{avg} \right)} + k_{c2} \tag{4}$$

$$p_m = k_{m1} \cdot \frac{f_{max} - f_i}{\cdot \left(f_{max} - f_{avg} \right)} + k_{m2} \tag{5}$$

where f_{max} and f_{avg} are the maximal and the average fitness value among all crow individuals in the flock, f_i is the fitness of the current crow individual and $k_{c1}, k_{c2}, k_{m1}, k_{m2}$ are specific probability scale factors for crossover and mutation operators, respectively.

5. Case Studies

In this paper, two experimental models are used to verify the feasibility and efficiency of the proposed GCSA optimization approach, which was coded using MATLAB environment and executed on a 1.99 GHz Intel i7 processor and 8 GB RAM computer. The first case study, which represents a common test example in process planning optimization, considers the prismatic part that was initially proposed by Guo et al. [25]. It contains 14 recognized machining features and 20 selected operations. The significant manufacturing information concerning features, operations, resources as well as precedence constraints is given in Table 2. The 3D solid model with the corresponding operation precedence graph (OPG) are shown in Figure 7.

Table 2. Features, operations, resource alternatives and predecessors of Part 1.

Feature Definition	Operations	Machines{}	Tools{}	TADs{}	Predecessors
F1—A planar surface	Milling (o1)	m2,m3	t6,t7,t8	+z	
F2—A planar surface	Milling (o2)	m2,m3	t6,t7,t8	-z	o1
F3—Two pockets arranged as a replicated feature	Milling (o3)	m2,m3	t6,t7,t8	+x	o1
F4—Four holes arranged as a replicated feature	Drilling (o4)	m1,m2,m3	t2	+z,-z	o1,o5,o18
F5—A step	Milling (o5)	m2,m3	t6,t7	+x,-z	o1
F6—A rib	Milling (o6)	m2,m3	t7,t8	+y,-z	o1
F7—A boss	Milling (o7)	m2,m3	t7,t8	-a	o1, o5
F8—A compound hole	Drilling (o8) Reaming (o9) Boring (o10)	m1,m2,m3 m1,m2,m3 m2,m3	t2,t3,t4 t9 t10	-a	o1, o7 o1, o7, o8 o1, o7, o8, o9
F9—A rib	Milling (o11)	m2,m3	t7,t8	-y,-z	o1
F10—A compound hole	Drilling (o12) Reaming (o13) Boring (o14)	m1,m2,m3 m1,m2,m3 m3,m4	t2,t3,t4 t9 t10	-z	o1, o2, o6, o11 o1, o2, o6, o11, o12 o1, o2, o6, o11, o12, o13
F11—Nine holes arranged as a replicated feature	Drilling (o15)Tapping (o16)	m1,m2,m3 m1,m2,m3	t1 t5	-z	o1, o2, o12, o13, o14 o1, o2, o12, o13, o14, o15
F12—A pocket	Milling (o17)	m2,m3	t7,t8	-x	o1, o18
F13—A step	Milling (o18)	m2,m3	t6, t7	-x,-z	o1
F14—A compound hole	Reaming (o19) Boring (o20)	m1,m2,m3 m3,m4	t9 t10	+z	o1, o12 o1, o12, o19

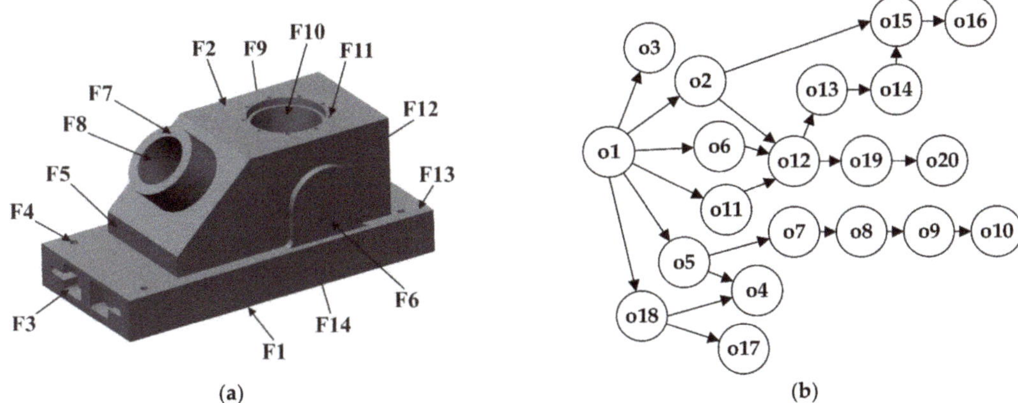

Figure 7. (**a**) Solid model and (**b**) the OPG of Part 1 with 20 operations.

The second case study considers a complex part taken from Huang et al. [8] that consists of 28 machining features and a total of 46 machining operations. Its main manufacturing information is given in details in Table 3. The solid model of this complex part with the corresponding OPG are represented in Figure 8. The cost indices for both Part 1 and Part 2 are given in Table 4.

Table 3. Features, operations, resource alternatives and predecessors of Part 2.

Feature Definition	Operations	Machines{}	Tools{}	TADs{}	Predecessors
F1—Flat surface	Rough turning (o1)	m1,m2	t1,t2,t3	+z	
	Half-finished turning (o2)	m2	t1,t2,t3	+z	o1,o22
	Finished turning (o3)	m2	t1,t2,t3	+z	o1,o22,o2
F2—Bearing hole	Rough boring (o4)	m7, m8	t4	-y,+y	o1,o22,o2,o3,o27
	Half-finished boring (o5)	m6,m7,m8	t5	-y,+y	o1,o22,o2,o3,o27, o4
	Finished boring (o6)	m6,m7,m8	t5	-y,+y	o1,o22,o2,o3,o27, o4,o5
F3—Angular surface	Rough milling (o7)	m5,m7,m8	t7,t8,t9	-a	o1,o22,o2,o3,o27, o4
	Finished milling (o8)	m3,m4,m6	t7,t8,t9	-a	o1,o22,o2,o3,o28, o31,o32,o33
F4—Four bosses	Rough milling (o9)	m3,m4,m5	t8,t9	+z	o1,o22
F5—Plane 1	Rough milling (o10)	m3,m4,m5	t8,t9	-z	o1
F6—Top boss	Rough milling (o11)	m3,m4,m5	t8,t9	-z	o1
	Finished milling (o12)	m4,m5	t8,t9	-z	o1,o22,o2,o3,o11
F7—Six bosses	Rough milling (o13)	m3,m4,m5	t8,t9	-z	o1
F8—Plane 2	Rough milling (o14)	m3,m4,m5	t8,t9	-z	o1
F9—Top window surface	Rough milling (o15)	m3,m4,m5	t7,t8	-z	o1,o22,o2,o3
	Finished milling (o16)	m3,m4,m5	t7,t8	-z	o1,o22,o2,o3,o15
F10—Inclined plane 1	Rough milling (o17)	m3,m4,m6,m7	t7,t8	-b	o1,o22,o2,o3
	Finished milling (o18)	m3,m4,m6,m7	t7,t8	-b	o1,o22,o2,o3,o17
F11—Inclined plane 2	Rough milling (o19)	m3,m4,m6,m7	t7,t8	-b	o1,o22,o2,o3
	Finished milling (o20)	m3,m4,m6,m7	t7,t8	-b	o1,o22,o2,o3,o19

Table 3. *Cont.*

Feature Definition	Operations	Machines{}	Tools{}	TADs{}	Predecessors
F12—The top holes	Drilling (o21)	m9,m10	t10	-z,+z	o1,o22,o2,o3,o12, o9
F13—Top plane	Rough milling (o22)	m4,m5	t7,t8,t9	-z	o1
F14—Counterbore hole	Spot facing (o23)	m9,m10	t20	-z	o1,o22,o2,o3,o27
F15—⌀18H7 hole	Drilling (o24)	m9,m10	t11	+z	o1,o22,o2,o3
	Reaming (o25)	m9,m10	t22	+z	o1,o22,o2,o3,o34
F16—2-⌀12.5 hole	Drilling (o26)	m9,m10	t12	+z	o1,o22,o2,o3
F17—12-⌀21 hole	Drilling (o27)	m9,m10	t13	+z	o1,o22,o2,o3
F18—18-⌀17 hole	Drilling (o28)	m9,m10	t14	+z	o1,o22,o2,o3
F19—Side hole	Rough boring (o29)	m7,m8	t6	-y	o1,o22,o2,o3
	Finish boring (o30)	m6,m7,m8	t6	-y	o1,o22,o2,o3,o29
F20—5-⌀20 hole	Drilling (o31)	m7,m8, m9,m10	t15	-a	o1,o22,o2,o3,o7, o28
F21—24-⌀8 hole	Drilling (o32)	m7,m8,m9,m10	t16	-a	o1,o22,o2,o3,o7,o28
	Tapping (o33)	m7,m8,m9,m10	t23	-a	o1,o22,o2,o3,o7, o28,o32
F22—Oil passage hole 1	Reaming (o34)	m6,m7,m8	t28	-c	o1,o22
	Tapping (o35)	m6,m7,m8	t24	-c	o1,o22,o34
F23—Oil passage hole 2	Drilling (o36)	m6,m7,m8	t17	-c	o1,o22
	Tapping (o37)	m6,m7,m8	t25	-c	o1,o22,o36
F24—⌀23 Counterbore hole	Spot facing (o38)	m3,m4,m9,m10	t21	-z	o1,o22,o2,o3,o24,o25
F25—Holes in inclined plane 1	Drilling (o39)	m9,m10	t18	-b	o1,o22,o2,o3,o17, o18,o27,o4
	Tapping (o40)	m9,m10	t26	-b	o1,o22,o2,o3,o17, o18,o27,o4,o39
F26—Holes in inclined plane 2	Drilling (o41)	m9,m10	t18	-b	o1,o22,o2,o3,o29, o19,o20
	Tapping (o42)	m9,m10	t26	-b	o1,o22,o2,o3,o29, o19,o20,o41
F27—Top holes	Drilling (o43)	m3,m4,m9,m10	t18	-z	o1,o22,o2,o3,o15, o16
	Tapping (o44)	m3,m4,m9,m10	t26	-z	o1,o22,o2,o3,o15,o16,o43
F28—Oil passage hole 3	Drilling (o45)	m6,m7,m8	t19	-x	o1,o22,o2,o3,o28
	Tapping (o46)	m6,m7,m8	t27	-x	o1,o22,o2,o3,o28, o45

5.1. Parameter Settings

In order to achieve the best possible parameter settings for the GCSA approach, we applied the Taguchi design of experiments to optimize the input parameters. The Taguchi design of experiments uses orthogonal arrays to obtain a certain number of experiments that can provide information of all factors that affect the performance of some process as expressed by Dou et al. [4]. Based on the input parameters of the GCSA and appropriate levels of variations, an experimental design matrix was formed using the standard L_{32} Taguchi orthogonal array with 32 experimental trials. In the mentioned case studies at the input of the experimental design, the following GCSA parameters are varied: maximal number of iterations, flock size, awareness probability, flight length, tournament size and scale factors k_{c1}, k_{c2}, k_{m1} and k_{m2}. Accuracy is increased by repeating every experiment 20 times. Therefore, the average result, the minimal result and the maximal result are used as the output parameters.

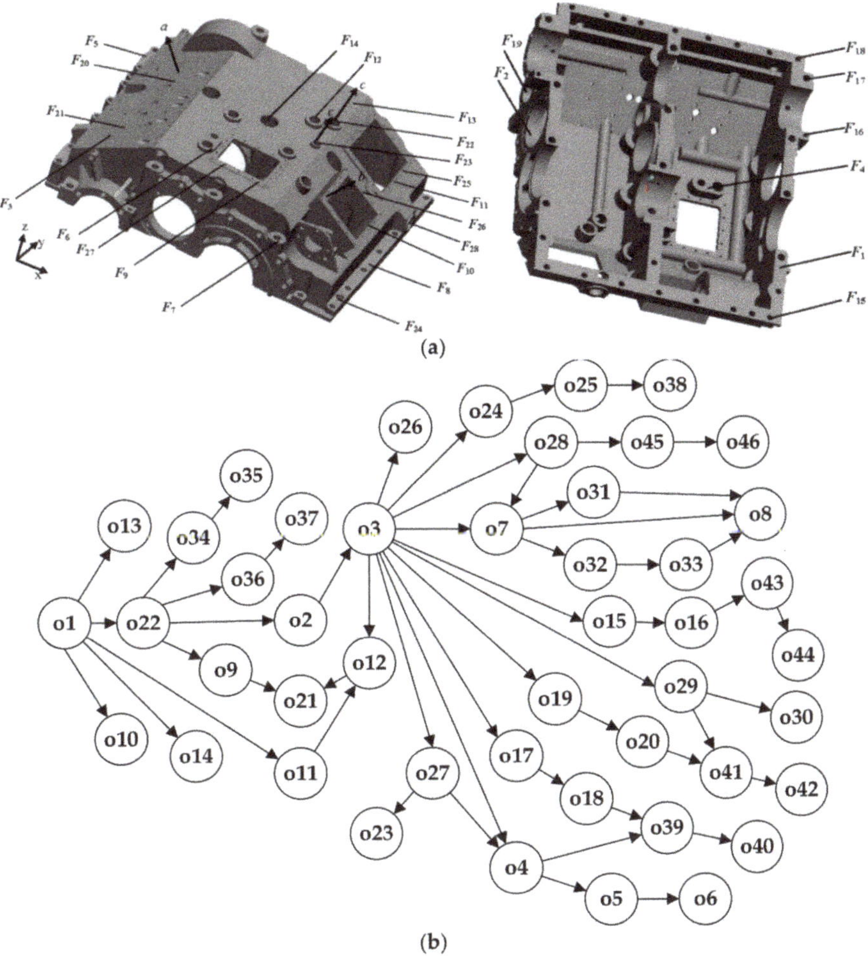

Figure 8. (**a**) Solid model and (**b**) the OPG of Part 1 with 46 operations.

Since the two cases significantly differ in dimensions, we formed two different experimental designs in terms of levels of variation to optimize the GCSA parameters for both. In that sense, the adopted levels of variation of Part 1 are: maximal number of iterations [400,1000], flock size [60,120], tournament size [2,5], k_{c1} [0.6, 0.9], k_{c2} [0.3, 0.6], k_{m1} [0.3, 0.6] and k_{m2} [0.1, 0.3]. In order to reduce the number of experiments, we applied L_{32} (2^9) orthogonal array, where 32 experiments are sufficient to make a conclusion. Full factorial design would require 8192 experiments to identify the effectiveness of parameters.

The adopted levels of variation of Part 2 are these: maximal number of iterations [1000,2000], flock size [80,100], awareness probability [0.2, 0.8], flight length [0.5, 2.5], tournament size [3, 5], k_{c1} [0.6, 0.9], k_{m1} [0.3, 0.7] and k_{m2} [0.1, 0.3]. In this design, the same L_{32} (2^9) orthogonal array with 32 experiments is used. Table 5 shows the optimal GCSA parameter settings for two case studies after the response optimization using Taguchi experimental design and Minitab statistical tool.

Table 4. The cost indices of Part 1 and Part 2.

Part 1				Part 2					
Cost per use									
M	MC	T	TC	M	MC	T	TC	T	TC
m1	10	t1	7	m1	25	t1	5	t15	4
m2	40	t2	5	m2	45	t2	6	t16	3
m3	100	t3	3	m3	50	t3	7	t17	4
m4	60	t4	8	m4	55	t4	12	t18	2
		t5	7	m5	20	t5	13	t19	2
		t6	10	m6	80	t6	9	t20	4
		t7	15	m7	45	t7	8	t21	3
		t8	30	m8	48	t8	9	t22	5
		t9	15	m9	16	t9	10	t23	3
		t10	20	m10	18	t10	4	t24	4
						t11	4	t25	4
						t12	3	t26	3
						t13	4	t27	4
						t14	3	t28	3
Cost per change									
MCCI = 160				MCCI = 120					
SCCI = 100				SCCI = 90					
TCCI = 20				TCCI = 15					

Table 5. The optimal parameter settings of the GCSA approach.

Case Study	MaxIter	FlockSize	AP	fl	TourSize	k_{c1}	k_{c2}	k_{m1}	k_{m2}
Part 1	1000	100	0.8	0.7	5	0.8	0.6	0.3	0.3
Part 2	2000	80	0.8	0.5	5	0.86	0.2	0.3	0.3

5.2. Computational Results and Analysis

To verify the flexibility and feasibility of the GCSA approach, we proposed two case studies with several conditions. The obtained results for each condition are discussed and the feasibility of this algorithm is verified by performing comparisons with traditional and modern algorithms.

For Part 1 proposed by Guo et al. [25], the optimized parameter settings shown in Table 5 is adopted. Using these parameter settings, Part 1 was considered for three different conditions:

(1) All manufacturing resources are available.
(2) Tool costs and tool cost changes are excluded.
(3) Same as the second with machine 2 and tool 8 not available.

The calculation results were obtained for each of the defined conditions and the best process schemes with minimized production cost are shown in Table 6. To present the experimental results more clearly, the comparison with results obtained by traditional and modern approaches from the literature is shown in Table 7 and includes the following comparative algorithms: IWD by Gao et al. [6], CPAGA by Falih and Shammari [5], HBMO by Wen et al. [14] and GA, SA and TS results also adopted from Wen et al. [14]. The GCSA was run 20 times to verify the consistency of the obtained results. For the first condition, the minimal production cost achieved by the GCSA is 2502, which appeared 9 times. The next minimal cost is 2507 which appeared in 7 runs of the GCSA. In terms of the minimal and the average TWPC (2507 and 2516,9), according to comparative results in Table 7, it can be concluded that the GCSA outperforms other approaches, traditional and modern ones. The worst result obtained by the GCSA slightly deviates from several algorithms in this comparison. As far as the second and third conditions are concerned, the GCSA showed similar performances. For the second condition, the minimal, the maximal and the average values of the TWPC are 2020, 2080 and 2047, respectively. In comparison with other approaches, the GCSA proved its superiority regarding all three statistical parameters. For the third condition, the TWPC of 2500 appeared 20 times in 20 runs, making it the most consistent result in this study. Apart from the CPAGA, other approaches did not show similar performances. All the results from Table 7 are graphically depicted in column charts presented in Figure 9. The x-axis shows the minimal, the maximal and the average result obtained by each algorithm in the comparative study, while the y-axis shows the values of the TWPC.

Table 6. The best process plans for Part 1 under three different conditions.

Condition 1																			
Operation	1	5	3	18	2	11	6	17	4	12	13	19	7	8	9	10	20	14	15	16
Machine	2	2	2	2	2	2	2	2	2	2	2	2	2	2	2	4	4	4	1	1
Tool	6	6	6	6	6	7	7	7	2	2	9	9	7	4	9	10	10	10	1	5
TAD	+z	+x	+x	-z	-z	-z	-z	-z	-z	-z	-z	+z	-a	-a	-a	-a	+z	-z	-z	-z

TMC = 800, TMCC = 320, NMC = 2, TTC = 247, TTCC = 180, NTC = 9, TSC = 900, NSC = 9, TWPC = 2447

TMCC—Total machine change cost; NMC—Number of machine changes; TTCC—Total tool change cost; NTC—Number of tool changes; NSC—Number of setup changes

Condition 2																			
Operation	1	6	18	17	5	2	11	12	4	13	7	8	9	3	19	20	10	14	15	16
Machine	2	2	2	2	2	2	2	2	2	2	2	2	2	2	2	4	4	4	1	1
Tool	7	7	7	7	6	8	8	4	2	9	8	4	9	6	9	10	10	10	1	5
TAD	+z	-z	-z	-z	-z	-z	-z	-z	-z	-z	-a	-a	-a	+x	+z	+z	-a	-z	-z	-z

TMC = 800, TMCC = 320, NMC = 2, TSC = 900, NSC = 9, TWPC = 2020

Condition 3																			
Operation	1	3	5	7	8	9	10	2	6	11	18	12	17	13	14	15	16	4	19	20
Machine	3	3	3	3	3	3	3	3	3	3	3	3	3	3	3	3	3	3	3	3
Tool	6	6	6	7	4	9	10	6	7	7	6	4	7	9	10	1	5	2	9	10
TAD	+z	+x	+x	-a	-a	-a	-a	-z	-z	-z	-z	-z	-z	-z	-z	-z	-z	-z	+z	+z

TMC = 2000, TMCC = 0, NMC = 0, TCTC = 250, TCTCC = 3200, NCTC = 16, TSC = 500, NSC = 5, TWPC = 2500

Table 7. The results of the GCSA compared to other algorithms for two case studies.

Part 1	Condition 1			Condition 2			Condition 3		
	Min	Max	Avg	Min	Max	Avg	Min	Max	Avg
GCSA*	2502	2700	2516,9	2020	2080	2047	2500	2500	2500
IWD, Gao et al. [6]	2527	2554	2553,5	2090	2380	2123	2590	2740	2615,3
CPAGA, Falih and Shammari [5]	2530	2535	2530,5	2090	2090	2090	2500	2500	2500
HBMO, Wen et al. [14]	2525	2557	2543,5	2090	2120	2098	2590	2600	2592,4
GA, Wen et al. [14]	2667	2885	2796	2220	2580	2370	2600	2840	2705
SA, Wen et al. [14]	2535	2829	2668,5	2120	2380	2287	2590	2740	2630
TS, Wen et al. [14]	2527	2690	2609,6	2120	2390	2208	2580	2740	2630

Part 2	Condition 1			Condition 2		
	Min	Max	Avg	Min	Max	Avg
GCSA*	4135	4496	4306,5	4338	4924	4544,7
CPAGA, Falih and Shammari [5]	4299	4315	4302	4503	4503	4503
GA-SA, Huang et al. [8]	4368	-	-	4450	-	-

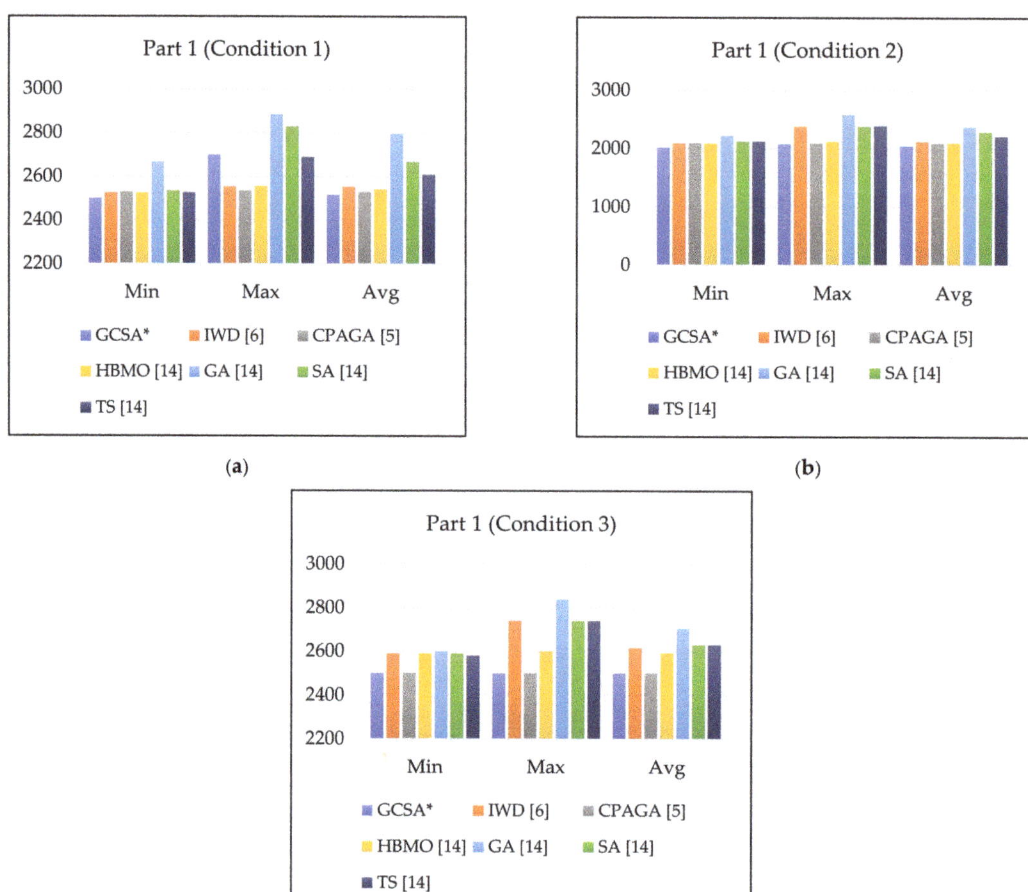

Figure 9. Comparison of the results under (**a**) condition 1, (**b**) condition 2 and (**c**) condition 3 for Part 1.

In the second study for Part 2 reported by Huang et al. [8], the optimized parameter settings also given in Table 5 are used. Since this part model has been recently proposed in scientific community, only two relevant approaches tested on this model can be adopted. The comparative results between the GCA with the CPAGA by Falih and Shammari [5] and GA-SA by Huang et al. [8] are shown in Table 7. Part 2 was carried out using two conditions:

(1)   All manufacturing resources are available.
(2)   Machines 3 and 7 and tool 8 are down.

For both conditions, the results are obtained after 20 runs. Under the first condition, the minimal TWPC is 4135, along with 4136 and 4137 which were the closest to the minimum. The average result is 4306.5, while the worst result is 4496. Only twice was the obtained TWPC above 4400, while all the other results were close to the average with four process plans reaching the TWPC below 4200. The GCSA achieved more superior result in terms of the minimal TWPC compared to the CPAGA and the GA-SA, while it showed slightly lower maximal and approximately the same average result compared to the CPAGA approach. The best solution for this condition is presented in Table 8.

Table 8. The best process plans for Part 2 (Condition 1).

Indices	1	2	3	4	5	6	7	8	9	10	11	12	13	14	15	16
Operation	1	22	2	3	10	11	15	16	12	14	13	17	18	19	20	9
Machine	1	5	2	2	5	5	5	5	5	5	5	5	5	5	5	5
Tool	1	7	1	1	8	8	8	8	8	8	8	8	8	8	8	8
TAD	+z	-z	+z	+z	-z	-z	-z	-z	-z	-z	-z	-b	-b	-b	-b	+z
Indices	17	18	19	20	21	22	23	24	25	26	27	28	29	30	31	32
Operation	27	24	28	26	25	45	46	36	37	34	35	7	32	33	31	4
Machine	9	9	9	9	9	7	7	7	7	7	7	7	7	7	7	7
Tool	13	11	14	12	22	19	27	17	25	28	24	8	16	23	15	4
TAD	+z	+z	+z	+z	+z	-x	-x	-c	-c	-c	-c	-a	-a	-a	-a	-y
Indices	33	34	35	36	37	38	39	40	41	42	43	44	45	46		
Operation	5	6	29	30	8	41	39	40	42	23	38	43	44	21		
Machine	7	7	7	7	3	9	9	9	9	9	9	9	9	9		
Tool	5	5	6	6	8	18	18	26	26	20	21	18	26	10		
TAD	-y	-y	-y	-y	-a	-b	-b	-b	-b	-z	-z	-z	-z	-z		

TMC = 1319, TMCC = 840, NMC = 7, TTC = 281, TTCC = 435, NTC = 29, TSC = 1260, NSC = 14, TWPC = 4135

Under the second condition, the best, the worst and the average TWPC using the GCSA are 4338, 4924 and 4544.7, respectively. The minimal result appeared three times in 20 runs, with more than half of the obtained TWPC being less than 4600. The best process plan for this condition is given in Table 9. Based on the comparative results in Table 7, a similar conclusion can be made as for the first condition. The GCSA overcome the results of the CPAGA and GA-SA in terms of the minimal TWPC. A smaller and larger deviation was shown in terms of the average and the maximal value, respectively. The graphical representation of the comparison with the two observed approaches is given in Figure 10.

Table 9. The best process plans for Part 2 (Condition 2).

Indices	1	2	3	4	5	6	7	8	9	10	11	12	13	14	15	16
Operation	1	22	2	3	15	16	11	14	13	10	12	9	24	28	27	25
Machine	1	5	2	2	5	5	5	5	5	5	5	5	9	9	9	9
Tool	1	7	1	1	7	7	9	9	9	9	9	9	11	14	13	22
TAD	+z	-z	+z	+z	-z	-z	-z	-z	-z	-z	-z	+z	+z	+z	+z	+z
Indices	17	18	19	20	21	22	23	24	25	26	27	28	29	30	31	32
Operation	26	7	31	32	33	4	5	6	29	30	45	46	34	35	36	37
Machine	9	8	8	8	8	8	8	8	8	8	8	8	8	8	8	8
Tool	12	7	15	16	23	4	5	5	6	6	19	27	28	24	17	25
TAD	+z	-a	-a	-a	-a	-y	-y	-y	-y	-y	-x	-x	-c	-c	-c	-c
Indices	33	34	35	36	37	38	39	40	41	42	43	44	45	46		
Operation	17	18	19	20	8	41	39	42	40	43	23	21	44	38		
Machine	4	4	4	4	4	9	9	9	9	9	9	9	9	9		
Tool	7	7	7	7	7	18	18	26	26	18	20	10	26	21		
TAD	-b	-b	-b	-b	-a	-b	-b	-b	-b	-z	-z	-z	-z	-z		

TMC = 1509, TMCC = 840, NMC = 7, TTC = 279, TTCC = 450, NTC = 30, TSC = 1260, NSC = 14, TWPC = 4338

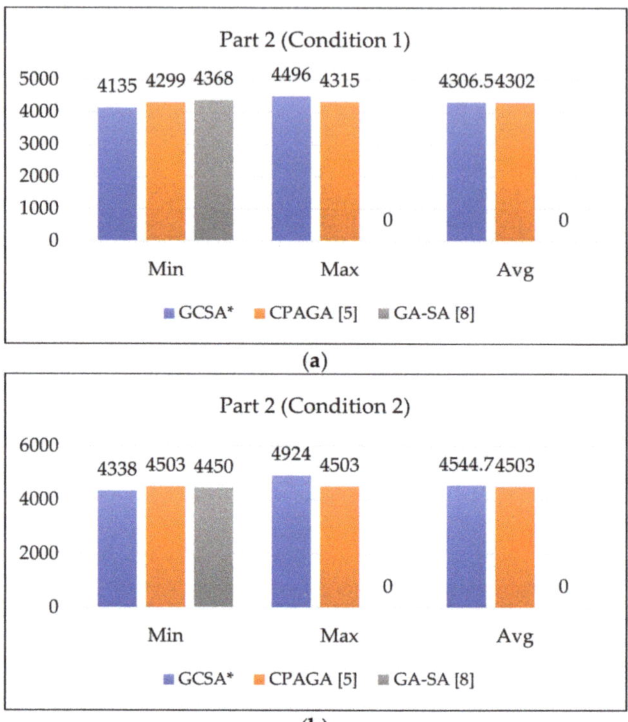

Figure 10. Comparison of the results under (**a**) condition 1 and (**b**) condition 2 for Part 2.

6. Conclusions

This paper proposed a genetic crow search algorithm (GCSA) approach to deal with the operation sequencing problem in computer-aided process planning. The traditional CSA algorithm was improved using genetic components of the GA, such as tournament selection, three-string crossover, shift mutation and resource mutation strategies which were employed to balance exploration and exploitation capabilities of the GCSA. Adaptive crossover and mutation probability coefficients were also added for that purpose. The OPG graphs were introduced to represent precedence relationships among operations within a sequence, and the vector representation was adopted to manipulate the data in Matlab programming environment. The nearest resource mechanism was proposed to apply the round-off procedure after the main CSA steps. In order to assure the feasibility of crow individuals, the repair mechanism was employed after the initialization stage and after the shift mutation. In the experimental stage, two dimensionally different case studies were adopted to test the performances and flexibility of the GCSA. The adopted evaluation criterion of the operation sequencing problem was the total production cost. For several conditions, the GCSA was tested and compared with the performances of other algorithms from the literature. The comparative results are presented in tables and graphs and the GCSA showed good performances for two case studies.

As far as future work is concerned, additional modifications of the proposed GCSA approach can be made to further increase global ability of the GCSA and improve the consistency of the obtained results. Moreover, ways of including relevant information about machining parameters into operation sequencing optimization should be considered. In addition, the focus can also be on further enriching the comparative analysis by taking into consideration various other metaheuristics, hybrid algorithms and optimization methods.

Finally, directions toward incorporating CSA with different heuristic mechanisms and metaheuristics other than genetic algorithms will be taken into account.

Author Contributions: Conceptualization, M.D. and M.M.; methodology, M.D., M.M. and R.C.; software D.L., B.P. and A.A.; validation, B.P. and M.D.; writing—original draft preparation, M.D. and M.M.; writing—review and editing, M.M. and D.L.; supervision R.C. and A.A. All authors have read and agreed to the published version of the manuscript.

Funding: This paper is part of a research on the projects: *"Application of smart manufacturing methods in Industry 4.0"*, No.142-451-3173/2020, supported by Provincial Secretariat for Higher Education and Scientific Research of the Autonomous Province of Vojvodina and *"Innovative scientific and artistic research from the FTS domain"*, No.451-03-68/2020-14/200156, supported by the Ministry of Education, Science and Technological Development of the Republic of Serbia.

Institutional Review Board Statement: Not applicable.

Informed Consent Statement: Not applicable.

Data Availability Statement: Data sharing is not applicable to this article.

Conflicts of Interest: The authors declare no conflict of interest.

References

1. Xu, X.; Wang, L.; Newman, S.T. Computer-aided process planning—A critical review of recent developments and future trends. *Int. J. Comput. Integr. Manuf.* **2011**, *24*, 1–31. [CrossRef]
2. Lukić, D.; Milošević, M.; Erić, M.; Đurđev, M.; Vukman, J.; Antić, A. Improving Manufacturing Process Planning Through the Optimization of Operation Sequencing. *Mach. Des.* **2017**, *9*, 123–132. [CrossRef]
3. Denkena, B.; Shpitalni, M.; Kowalski, P.; Molcho, G.; Zipori, Y. Knowledge Management in Process Planning. *CIRP Ann.* **2007**, *56*, 175–180. [CrossRef]
4. Dou, J.; Li, J.; Su, C. A discrete particle swarm optimisation for operation sequencing in CAPP. *Int. J. Prod. Res.* **2018**, *56*, 3795–3814. [CrossRef]
5. Falih, A.; Shammari, A.Z.M. Hybrid constrained permutation algorithm and genetic algorithm for process planning problem. *J. Intell. Manuf.* **2020**, *31*, 1079–1099. [CrossRef]
6. Gao, B.; Hu, X.; Peng, Z.; Song, Y. Application of intelligent water drop algorithm in process planning optimization. *Int. J. Adv. Manuf. Technol.* **2020**, *106*, 5199–5211. [CrossRef]
7. Petrović, M.; Mitić, M.; Vuković, N.; Miljković, Z. Chaotic particle swarm optimization algorithm for flexible process planning. *Int. J. Adv. Manuf. Technol.* **2016**, *85*, 2535–2555. [CrossRef]
8. Huang, W.; Lin, W.; Xu, S. Application of graph theory and hybrid GA-SA for operation sequencing in a dynamic workshop environment. *Comput. Aided Des. Appl.* **2017**, *14*, 148–159. [CrossRef]
9. Milošević, M.; Đurđev, M.; Lukić, D.; Antić, A.; Ungureanu, N. Intelligent Process Planning for Smart Factory and Smart Manufacturing. In *Proceedings of the 5th International Conference on the Industry 4.0 Model for Advanced Manufacturing*; Springer: Cham, Switzerland, 2020; pp. 205–214. [CrossRef]
10. Wang, Y.F.; Zhang, Y.F.; Fuh, J.Y.H. A hybrid particle swarm based method for process planning optimisation. *Int. J. Prod. Res.* **2012**, *50*, 277–292. [CrossRef]
11. Huang, W.; Hu, Y.; Cai, L. An effective hybrid graph and genetic algorithm approach to process planning optimization for prismatic parts. *Int. J. Adv. Manuf. Technol.* **2012**, *62*, 1219–1232. [CrossRef]
12. Lian, K.; Zhang, C.; Shao, X.; Gao, L. Optimization of process planning with various flexibilities using an imperialist competitive algorithm. *Int. J. Adv. Manuf. Technol.* **2012**, *59*, 815–828. [CrossRef]
13. Liu, X.-j.; Yi, H.; Ni, Z.-h. Application of ant colony optimization algorithm in process planning optimization. *J. Intell. Manuf.* **2013**, *24*, 1–13. [CrossRef]
14. Wen, X.-y.; Li, X.-y.; Gao, L.; Sang, H.-y. Honey bees mating optimization algorithm for process planning problem. *J. Intell. Manuf.* **2014**, *25*, 459–472. [CrossRef]
15. Wang, J.; Wu, X.; Fan, X. A two-stage ant colony optimization approach based on a directed graph for process planning. *Int. J. Adv. Manuf. Technol.* **2015**, *80*, 839–850. [CrossRef]
16. Hu, Q.; Qiao, L.; Peng, G. An ant colony approach to operation sequencing optimization in process planning. *Proc. Inst. Mech. Eng. Part B J. Eng. Manuf.* **2017**, *231*, 470–489. [CrossRef]
17. Su, Y.; Chu, X.; Chen, D.; Sun, X. A genetic algorithm for operation sequencing in CAPP using edge selection based encoding strategy. *J. Intell. Manuf.* **2018**, *29*, 313–332. [CrossRef]
18. Liu, Q.; Li, X.; Gao, L. Mathematical modeling and a hybrid evolutionary algorithm for process planning. *J. Intell. Manuf.* **2020**. [CrossRef]

19. Jiang, S.; Cao, J.; Wu, H.; Yang, Y. Fairness-based Packing of Industrial IoT Data in Permissioned Blockchains. *IEEE Trans. Ind. Inform.* **2020**, 1. [CrossRef]
20. Li, W.; Ong, S.; Nee, A.Y.C. *Integrated and Collaborative Product Development Environment: Technologies and Implementations*; World Scientific: Singapore, 2006; Volume 2. [CrossRef]
21. Askarzadeh, A. A novel metaheuristic method for solving constrained engineering optimization problems: Crow search algorithm. *Comput. Struct.* **2016**, *169*, 1–12. [CrossRef]
22. Huang, K.-W.; Girsang, A.S.; Wu, Z.-X.; Chuang, Y.-W. A Hybrid Crow Search Algorithm for Solving Permutation Flow Shop Scheduling Problems. *Appl. Sci.* **2019**, *9*, 1353. [CrossRef]
23. Laabadi, S.; Naimi, M.; Amri, H.E.; Achchab, B. A Binary Crow Search Algorithm for Solving Two-dimensional Bin Packing Problem with Fixed Orientation. *Procedia Comput. Sci.* **2020**, *167*, 809–818. [CrossRef]
24. Shirke, S.; Udayakumar, R. Evaluation of Crow Search Algorithm (CSA) for Optimization in Discrete Applications. In Proceedings of the 2019 3rd International Conference on Trends in Electronics and Informatics (ICOEI), Tirunelveli, India, 23–25 April 2019; pp. 584–589. [CrossRef]
25. Guo, Y.W.; Mileham, A.R.; Owen, G.W.; Li, W.D. Operation sequencing optimization using a particle swarm optimization approach. *Proc. Inst. Mech. Eng. Part B J. Eng. Manuf.* **2006**, *220*, 1945–1958. [CrossRef]

applied sciences

MDPI

Article

A Simulated Annealing Algorithm for Intermodal Transportation on Incomplete Networks

Mustapha Oudani

TICLab, ESIN, International Univesity of Rabat, Rabat 11103, Morocco; mustapha.oudani@uir.ac.ma

Abstract: Growing competition in the world enforces the need for an efficient design of transportation networks. Furthermore, a competitive transportation network should also be eco-friendly. As road transportation is responsible for the largest quantities of CO_2 emissions, Intermodal Transportation (IT) might be a potential alternative. From this perspective, intermodal terminals location is a cornerstone for building a sustainable transportation network. The purpose of this paper is to study and efficiently solve the Intermodal Terminal Location Problem on incomplete networks. We model this problem as a mixed integer linear program and develop a simulated annealing algorithm to tackle medium and large instances. The computational results show that the obtained solutions using simulated annealing are competitive and close to the exact solutions found by CPLEX solver for small and medium instances. The same developed algorithm outperforms the best found solutions from the literature using heuristics for larger instances.

Keywords: terminal location; intermodal transportation; simulated annealing; mixed integer program; incomplete networks

check for
updates

Citation: Oudani, M. A Simulated Annealing Algorithm for Intermodal Transportation on Incomplete Networks. *Appl. Sci.* **2021**, *11*, 4467. https://doi.org/10.3390/app11104467

Academic Editors: Farouk Yalaoui, Taha Arbaoui and Yassine Ouazene

Received: 22 February 2021
Accepted: 16 April 2021
Published: 14 May 2021

Publisher's Note: MDPI stays neutral with regard to jurisdictional claims in published maps and institutional affiliations.

1. Introduction

The growth in container-related activities over the last decades is having a remarkable impact on transportation demand. In fact, producers need more and more transportation services to move raw materials and manufactured products to final customers. Furthermore, worldwide economic growth has led to an increasing growth of domestic and international streams of goods and services. Accordingly, handling these flows of goods from production to transportation and storage induce major environmental concerns. Henceforth, in many countries, government, companies and Non-Government Organizations (NGO) are more and more inclined to purchase green products. They are also increasingly urged to incorporate green policy throughout their decision-making process. Therefore, service providers are motivated to provide solutions involving the integration of environmental aspects. It is well known that Intermodal Transportation (IT) is a cornerstone for achieving sustainable development through modal shift from road to eco-friendly modes. However, the IT competitiveness depends on several factors. Among the key elements for building a successful IT system, we cite: (i) optimal use of the most attribute of each transportation mode included in the system; (ii) overall network design; and (iii) optimal location of intermodal terminals. The European Conference of Ministers of Transport (ECMT) defined IT "as the carriage of goods by at least two different modes of transport in the same loading unit (an Intermodal Transport Unit or ITU) without stuffing or stripping operations when changing modes". This type of transport is the door-to-door chain from the initial shipper to the final client. A typical component of IT is the container-based system. In fact, container flows are strongly intermodal. Overall, an intermodal transportation chain is composed by three segments: (i) the first mile segment, which is, in most cases, a short segment (less than 50 km) that uses using road mode (or river); (ii) the main transportation segment, which represents the main segment and uses a consolidation mode such as rail, river, or maritime mode; and (iii) the last mile segment, which is similar to the first mile.

IT is a competitive alternative for door- to-door road transport when the unit cost of the consolidation mode (the second segment) is smaller than in road mode and compensates the additional fees of transshipment operations. IT is very successful and even crucial when we have the two following situations: (i) when the transport is not time critical; and (ii) when there is a large volume of goods. In fact, despite the great speed of the carriage (e.g., rail mode), the overall journey of an intermodal transport chain takes more than door-to-door road. Accordingly, intermodal transportation is not suitable for urgent deliveries. In addition, having a large amount of goods is necessary to benefit from the scale economy generated by the consolidation mode. Intermodal terminals are equipped centers for the transshipment of goods between different modes of transportation. The location of such terminals is a strategic decision that may have a strong and direct influence on the overall operational costs. This problem is addressed in the scientific literature by analogy to well-known Facility Location Problems (FLP) and to Hub Location Problems (HLP). The pioneering paper of [1] is the starting point of Intermodal Terminal Location Problem. The majority of later studies were based on this paper. The authors either proposed a modified version of the models or proposed solving approaches. The proposed works commonly assume that all located terminals are fully connected in a generated solution.

The main contributions of the current paper are the following:

- We propose a generalization of a well-known problem in the literature, namely the Intermodal Terminal Location Problem (ITLP) by introducing the incomplete version of the problem relaxing the completeness of the inter-terminal network.
- We build a mathematical MIP (Mixed Integer Program) model to formulate the problem and we discuss the meaning of each components.
- We solve small and medium instances to optimality using CPLEX solver.
- We develop an efficient simulated annealing is adopted to solve real life size instances.
- We present extensive numerical experiments and analyze the obtained results.

The remainder of the paper is structured as follows. We review related works in Section 2. The problem description and formulation are detailed in Section 3. We develop the simulated annealing method in Section 4. Numerical results are reported in Section 5. We provide results analysis and comments in Section 6. We give concluding remarks and perspectives in the last section.

2. Related Works

Although the research in the ITLP is a very young field, the number of papers related to this subject is steadily increasing. This reveals the importance of the Intermodal Terminal Location for both academics and industry. A review synthesis of works done in the field of IT can be found in the paper of [2]. Specifically, the Network Design Problem (NDP) is a well-studied problem in the framework of intermodal transportation. A stochastic approach for robust intermodal transportation plans in a real-world network was proposed by [3]. The relationship between IT and supply chain was discussed by [4]. They developed a scenario-based estimation to quantify the demand for the Italian IT Network. Owing to its safety, IT is increasingly considered for hazardous materials (Hazmat) distribution. For instance, Ref. [5] suggested using scheduled direct and faster trains to reduce the transport risk of hazmats. The proposed solution is as a bi-objective model solved by a tabu-search heuristic. Ref. [6] studied optimization challenges regarding rail processes at maritime terminals. They provided a survey of the related works in this important node of IT. Ref. [7] proposed a fuzzy version of the Intermodal terminal Location Problem. He studied the case in which the decision-maker may suggest locating a number of rail terminals around a fixed number "p". This scenario is modeled by adding a fuzzy constraint to the developed model. In [8], a compromise programming technique combined with fuzzy modeling is proposed to tackle the location problem of real-road terminals. The developed Multi-Criteria Decision Making Method (MCDM) is considered to model the conflicting objectives of different stakeholders involved in the intermodal transportation chain. Ref. [9] modeled the intermodal container terminal location problem in urban areas as a non-linear

Mixed Integer Program (NLMIP). They considered an entropy maximization approach and a Lagrangian relaxation technique to solve the problem in real-life case. A combined Delphi and Analytical Network Process (ANP) is proposed in [10] to solve the Intermodal Terminal location problem in accordance with sustainable requirements. Ref. [11] proposed a modeling approach for Intermodal Terminal Hub Location Problem with transfer modes and considering two-types of hubs. They proved that the problem is NP-hard and reported numerical results on real-world container distribution networks. In addition, multi-agent simulation tools are used in [12] to simulate the seaport-dry port system as intermodal terminal. The developed model is validated on the Ningbo-Zhouhan port. A combined SWOT analysis and Analytic hierarchy process (AHP) and Preference Ranking Organization Method for Enrichment of Evaluations (PROMETHEE) is proposed in [13] to study the location of intermodal terminals. Ref. [14] studied user-specified reliability levels using a model that provide the optimal route through road links, rail links and via intermodal terminals. A Markov Chain Monte Carlo method is proposed in [15] to solve the intermodal terminal location problem. The proposed approach is a two-layer approach that efficiently solve the problem with routing decisions. Computational results are reported. Ref. [16] proposed a Best-Worst Method (BWM) with a consensus model from the perspective of shipping line companies. The objective is to select the inland terminal locations from potential sites. A sensitivity analysis was performed by authors to assess the impact of the increasing volume of containers on the desirable locations. In [17], a quantitative method is proposed to locate inland rail terminals based on cost savings. The model was validated in real-world railway network from Croatia.

Table 1 summarizes the main research works related to the intermodal transportation. In addition to review papers that survey research works in this field, terminal location problems, route selection, network design, transport policy, and integrated location and routing problems are the main topics of related works. Terminal location problems are the strategic planning problem aiming to open a fixed number of rail terminals among a set of potential sites given by a decision-maker. The route selection problem is a tactical decision problem where the objective is to assign the best combination of transportation modes for each commodity. The network design is another example of strategic planning that aims to determine the initial network to use for the transshipment of demands. Some papers deal with transport policy issues where the general management rules of transport are addressed. Moreover, few research papers considered the integrated location and routing problem in the frame of the intermodal transportation. The particularity of the last problem lies in the fact that the two decisions of terminal location and the routing of demands are tackled simultaneously.

Table 1. Intermodal freight transportation: main themes.

Research Content	References
Review	[2,18–24]
Terminal location	[1,5,7–13,15–17,25–31]
Route selection	[14,32–37]
Network design	[33,38–41]
Transport policy	[42–45]
Integrated Location and Routing	[46,47]

We can clearly remark that terminal location problems are steadily attracting researchers attention. This proves the importance of locating rail terminal in the design of the intermodal transportation networks. In fact, while we can find different types of research papers in the field of intermodal transportation, various aspects of terminal location problem are studied. We can cite the strategic choice of potential sites, the study of uncertainty, the use of multi criteria decision making, etc.

Furthermore, we provide in Table 2 an overall classification of the models used for ITLP and a summary of works using either exact or heuristics methods. The first papers dealing with the studied problem use deterministic models to formulate the problem and exact methods to solve it. The use of such formulations enables applying the rich analytical deterministic tools to solve the problem. More recent papers use either stochastic or fuzzy modeling to capture the uncertainty, vagueness, and complexity of the problem. In such case, heuristics are often used as no exact method is able to solve non-linear or large size problems.

Table 2. Intermodal Terminal Location Problem: models and solving methods.

Research Type	References
Deterministic models	[1,5,9–11,13,14,17,25–27,32]
Stochastic/fuzzy models	[3,7,8,12,15,16,34,42,48,49]
Exact methods	[1,7,8,11,25]
heuristic methods	[10,12,15,27,32,50]

To our knowledge, we are the first to build a model for an intermodal transportation network considering the general case when the induced graph by the located rail terminals is an incomplete graph. The proposed model is a mixed linear program that provides optimal solutions to achieve a balanced intermodal network.

3. Problem Description and Formulation

3.1. The Complete and Incomplete Version of the ITLP

The ITLP aims to determine which of a set of potential sites locations to open as terminals and how to route the supply and demand of a set of customers (representing zones of supply and demand) through the network. The transportation of these demands may be done using a unimodal link (road mode) or a combination of two modes (rail-road). The objective is to minimize the total cost of the transportation as well as the installation cost of terminals. Up to now, the studied version of the ITLP introduced by [1,25], studied by [27,29], etc. assumes that the inter-terminals network is a complete graph. Figure 1 shows a small complete intermodal terminal network with three terminals T_1–T_3 and 5 customers (1–5).

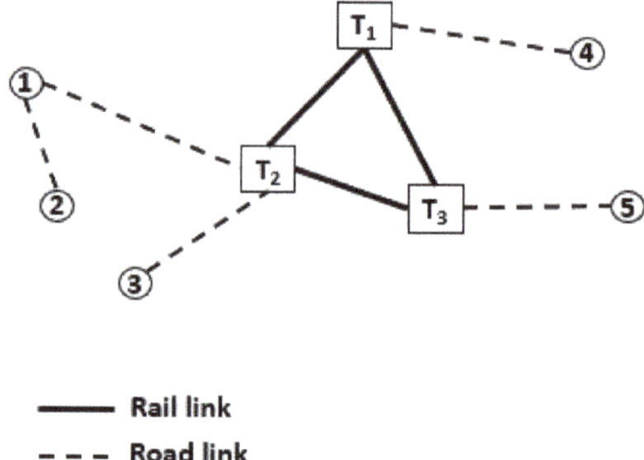

Figure 1. Small complete network.

As in hub problems, the assumption that the induced graph by the located terminals must be a fully-connected graph (complete graph) is a restrictive constraint that may increase the total investment cost. Moreover, this restriction is non-realistic since, in real life cases, a rail–road network is usually incomplete due to the heavy cost incurred for the construction of its infrastructure. The current work relax this assumption and proposes a general model in which the located rail-road terminals are partially connected. Figure 2 shows a small incomplete intermodal terminal network. The induced graph by terminals is not incomplete. For instance, there is no rail link between rail terminals T_1 and T_2.

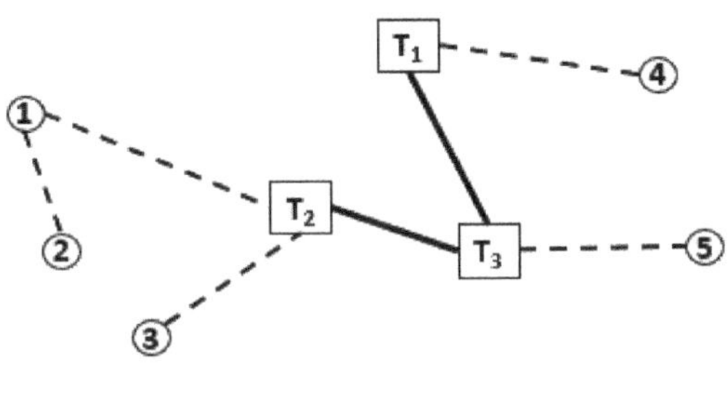

Figure 2. Small incomplete network.

3.2. Mathematical Model

Let us introduce the following parameters:

I : Set of senders/receivers.

K : Set of candidates sites for intermodal terminals.

l : Fixed number of links to be established.

q_{ij} : The amount of goods to be shipped from the customer i to customer j.

c_{ij}^{km} : The intermodal unit cost for transportation from sender i to receiver j using the two rail terminals k and m.

c_{ij} : The unimodal unit cost for transporting goods by truck from sender i to receiver j.

f_k : The investment cost for location of a terminal on site k.

C_k : Fixed capacity of a located rail terminal k.

Let us consider the following decision variables:

l_{km} : A binary variable equal 1 if the rail link between k and m is established, 0 otherwise. Accordingly, l_{kk} equal 1 if the node k is a terminal, 0 otherwise.

r_{ij} : The amount of goods routed from the sender i to receiver j.

s_{ij}^{km} : The amount of goods routed from the sender i to receiver j through the two rail terminals k and m.

The incomplete location problem of intermodal terminals may be formulated as follows:

$$\text{Min } C_1 = \sum_{i,j \in I} \sum_{k,m \in K} c_{ij}^{km} s_{ij}^{km} + \sum_{i,j \in I} c_{ij} r_{ij} + \sum_{k \in K} f_k l_{kk} \tag{1}$$

The linear objective function (1) minimizes the total cost for routing goods and the total cost for locating rail terminals. The first term expresses the cost for routing goods through the intermodal chain: from the initial sender i to the first rail terminal k using trucks, then from the terminal k to the second terminal m by rail mode, and later from the terminal m to the final customer j using road mode. The second term is the cost for routing goods directly using trucks from i to j while the third term is the overall investment cost for locating rail terminals.

$$\sum_{k,m \in K} s_{ij}^{km} + r_{ij} = q_{ij}, \forall i, j \in I \tag{2}$$

The constraint (2) states that the sum of goods transported via the intermodal chain through the terminals k and m and that routed directly from i to j equal exactly the total goods to be transported from i to j.

$$\sum_{i,j\in I}\sum_{m\in K} s_{ij}^{km} + \sum_{i,j\in I}\sum_{m\in K} s_{ij}^{mk} \leq C_k l_{kk}, \forall k \in K \tag{3}$$

The constraint (3) imposes to respect the terminals capacities.

$$l_{km} \leq l_{kk}, \forall k,m \in K \tag{4}$$

$$l_{km} \leq l_{mm}, \forall k,m \in K \tag{5}$$

The two inequalities (4) and (5) guarantee that a rail link is established if the two adjacent terminals are located. In fact, we have the following implications:

$$\text{if } l_{kk} = 0 \text{ or } l_{mm} = 0 \implies l_{km} = 0$$

$$l_{km} = l_{mk}, \forall k,m \in K \tag{6}$$

Equation (6) guarantees that a rail link is opened in the two directions.

$$\sum_{k,m\in K} l_{km} = l \tag{7}$$

Equation (7) guarantees the establishment of exactly the given number of rail links. The decision variable l_{km} controls the rail links to be opened. Thus, if due to economic, geographic or environmental constraints a railway link is impossible between two terminals, such scenario is forbidden by $l_{kl} = 0$.

$$s_{ij}^{km} \leq l_{km}, \forall k,m \in K, \forall i,j \in I \tag{8}$$

Equation (8) forbids the rail flows between terminals if they are not located. The developed model is a Mixed binary Integer Program (MIP). If we denote by p the number of potential sites (the cardinal of K) and by n the numbers of senders/receivers (the cardinal of I), then the program has $n^2 p^2 + 3p^2 + n^2 + p + 1$ constraints and $n^2 p^2 + n^2 + p^2$ variables.

Proposition 1. *If the triangular inequality holds for the transportation unit cost (assumed in the literature and holds for dataset used in the numerical experiments), then $s_{ij}^{kk} = 0, \forall i,j \in I, \forall k \in K$.*

This proves that the constraint $s_{ij}^{kk} = 0, \forall i,j \in I, \forall k \in K$, used in several mathematical models in recent papers in the literature, is an unnecessary constraint [27].

Proof. To express the fact that rail mode generate a scale economy (consolidation mode), the intermodal unit cost may be written as follows: $c_{ij}^{km} = c_{ik} + \alpha c_{km} + c_{mj}$ where c_{ik} is the cost between i and k, c_{km} the cost between the two rail terminals, and c_{mj} is the cost between the rail terminal m and the final receiver j with $\alpha < 1$. We denote by α the discount coefficient expressing the scale economy generated by the rail mode. This parameter is supposed to be equal to 0.5 in [27]. Since $c_{ij}^{kk} = c_{ik} + \alpha c_{kk} + c_{kj} = c_{ik} + 0 + c_{kj} = c_{ik} + c_{kj} \geq c_{ij}$ (the last inequality is the triangular inequality), and as we aim to minimize the total cost expressed by Equation (1), given the conservation flow constraint of Equation (2), it follows that all goods from i and destined to j will be routed directly using trucks, and then $r_{ij} = q_{ij}$, which implies that $\forall i,j \in I, s_{ij}^{kk} = 0$ □

Proposition 2. *Let l be the number of rail links to be located and $|K| = p$ the cardinal of the candidates sites set. The Incomplete ITLP modeled by the aforementioned MIP is unfeasible when $l > \frac{p(p-1)}{2}$.*

Proof. It is an immediate result of the handshaking lemma (a classical lemma in graph theory: an incomplete graph with p vertices has at most $\frac{p(p-1)}{2}$ edges). $\square$

3.3. Model Variants

The basic model introduced in the previous subsection may be enriched by adding other constraints to match other realistic situations.

For instance, one may model the situation where both the number of rail links l and the number of rail terminals t are given by the decision-maker by adding the constraint:

$$\sum_{k \in I} l_{kk} = t \tag{9}$$

In such case, as the number of rail terminals is known, the objective function becomes:

$$\text{Min } C_2 = \sum_{i,j \in I} \sum_{k,m \in K} c_{ij}^{km} s_{ij}^{km} + \sum_{i,j \in I} c_{ij} r_{ij} \tag{10}$$

Furthermore, the basic formulation may be extended to consider potential handling operations in the located rail terminals. In fact, let us denote by h_{km} the handling cost of a unit of goods transhipped from the rail terminal m to the rail terminal k (symmetry may be considered by assuming that $h_{km} = h_{mk}$). Therefore, the objective function considering asymmetric handling costs becomes:

$$\text{Min } C_3 = \sum_{i,j \in I} \sum_{k,m \in K} c_{ij}^{km} s_{ij}^{km} + \sum_{i,j \in I} c_{ij} r_{ij} + \sum_{k,m \in K} (h_{km} + h_{mk}) l_{km} \tag{11}$$

4. Simulated Annealing Algorithm

4.1. General Scheme

The Simulated Annealing (SA) algorithm is a local search metaheuristic used for finding an approximate global optimum. The SA is based on the Metropolis–Hastings criterion enabling to escape from being trapped in local minima. We used this single solution metaheuristic for the two reasons: (1) to solve the intractable instances for the CPLEX solver; and (2) because this method was successfully applied to similar location problems as in [51,52]. We adopted the classical general scheme of the SA, as shown in Algorithm 1.

Algorithm 1: Simulated annealing general scheme.

Data: $\epsilon = 10^{-4}, T_0 = 10^6$
Result: Best solution
initialization;
Generate initial solution S_0 ;
$T \leftarrow T_0$;
$S_c \leftarrow S_0$ $S_b \leftarrow S_c$;
while $(T > \epsilon)$ **do**
    Generate neighbor solution $S' \in N(S_c)$;
    Generate $r \in \text{random}(0,1)$;
    $\Delta E = f(S') - f(S)$;
    if $r < e^{\frac{-\Delta E}{T}}$ **then**
        $S \leftarrow S'$;
        if $f(S) < f(S_b)$ **then**
            $S_b \leftarrow S$
    else
  else
  $T \leftarrow T/2$

4.2. Initial Solution and Neighbors Generation

To generate the initial solutions, we use Algorithm 2.

Algorithm 2: Generation of initial solutions

Randomly generate rail links ;
if $l_{kk} = 1$ *and* $l_{mm} = 1$ **then**
    $C_{min} \leftarrow \min(C_k, C_m)$;
    $s_{ij}^{km} \leftarrow \frac{q_{ij}}{C_{min}}$;
    $r_{ij} \leftarrow q_{ij} - s_{ij}^{km}$
else
    $r_{ij} \leftarrow q_{ij}$

To generate neighbors of a solution, we define three types of moves:

- Permutation of rail links: We modify a given initial solution by interchanging two non-connected rail terminals with two connected ones. This move has the particularity to generate new feasible solution since the number of rail links remains the same.
- Permutation of terminal locations: This move interchanges a terminal node with a non-terminal one. As the generated solution may be infeasible, a correction phase is conducted to guarantee the respect of the model constraints. In fact, if the capacity of the new terminal is less than the old one, a new rail link is added. Otherwise, the flows are routed trough the new terminal.
- Changing the number of terminals: From a given solution with t number of terminals, we generate neighbor solutions by adding or deleting one or several rail terminals. The correction phase is made to respect links and capacities of terminals.

All this moves are used in a sequential way in the implemented algorithm. The best solution from the initial iteration to the last one is saved in a specific variable called *BestSol* and reported in the end of the SA execution.

4.3. Parameters Tuning of the SA Method

Since the simulated annealing algorithm is a parametric approach, the setting of values of these parameters has a significant effect on the objective function. For this, we ran a set of experiments to determine the most suitable values to reach good solutions. The controller parameters of the developed simulated annealing are the initial temperature T_0, the cooling rate δ, the neighbor moves rate γ and the minimal temperature T_{min}. The correspondent intervals of these parameters are presented in Table 3.

Table 3. SA parameters.

Parameter	Symbol	Interval
Initial temperature	T_0	$[100, 10^6]$
Cooling rate	δ	$[0.03, 0.9]$
Neighbor rate	γ	$[0.05, 0.9]$
Minimum temperature	T_{min}	$[0.1, 10]$

The Tagushi design of MINITAB software [53] was used for parameters tuning of the simulated annealing. Equation (12) defines the signal-to-noise ratio (S/N) of Tagushi method. This ratio expresses the effect of parameters on the objective function. A maximum value of this ratio minimizes the effect of uncontrollable factors. This ratio is given by:

$$S/N = -10\log(\frac{1}{z}\sum_{i=1}^{z} y_i^2) \tag{12}$$

The constant z represents the total number of experiments and y_i is the result of the experiment i. The results of parameters tuning using Tagushi design for a typical set of training instances are illustrated by charts of Figures 3 and 4.

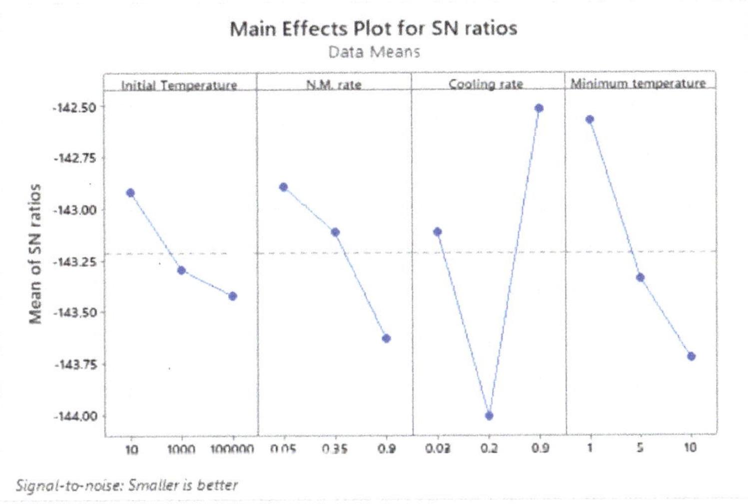

Figure 3. Main effects of parameters for SN ratios.

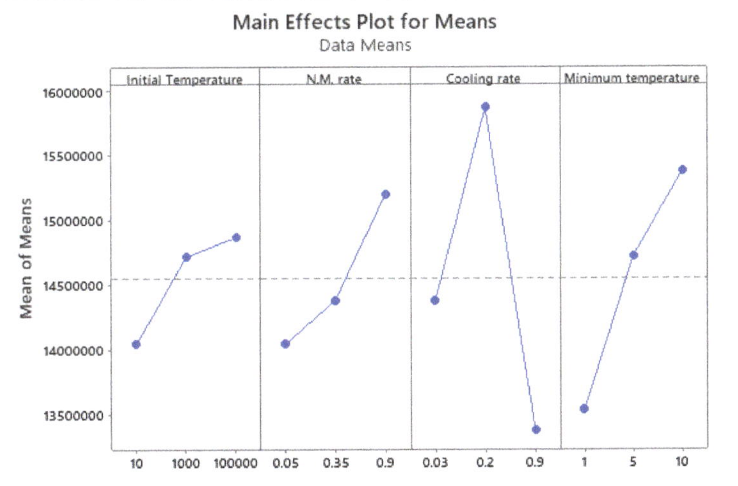

Figure 4. Main effects of parameters for Means.

5. Numerical Results

All experiments were done on a notebook Intel core i5, 2.53 GHz with 8 GB RAM memory under Windows 10. The SA algorithm is coded using Python 3.

5.1. SA Solutions vs. Optimal and Best Found Solutions

To assess the efficiency of the developed Simulated Annealing (SA) algorithm, we used the dataset introduced in [27]. In this dataset, customers and potentials sites coordinates are randomly generated between $(0,0)$ and $(10^4, 10^4)$. The goods demands are generated from the interval $[0, 500]$. The investment cost f_k in the interval $[0, 5.10^5]$ and potentials sites capacities are drawn from $[0, 10^4]$. After that, the direct cost c_{ij} is equal the Euclidean distance between the customers i and j, while the intermodal cost is computed using the following formula $c_{ij}^{km} = c_{ik} + \frac{1}{2}c_{km} + c_{mj}$. The objective function values of the studied model are detailed in [31]. For this, we lunched the SA method for 10 runs and we choose the best solutions. As shown in Figure 5, saving best solutions in an archive set enables to get a decreasing values of the objective function. Accordingly, the developed SA method is a descent strategy.

Table 4 reports the gaps of the SA solutions to CPLEX solutions. The second column gives the gap of the best obtained solution from the CPLEX solution, while the third column gives the average gap of the conducted 10 runs.

Note that we used the following formula to compute the gap:

$$gap = \frac{\text{Simulated Annealing Best solution} - \text{Optimal solution}}{\text{Simulated Annealing Best solution}}$$

The obtained results show that our method is efficient and may even reach the optimal solutions for the small size instances and reaches best solutions for medium instances in short times.

To further evaluate the performance of the developed SA, we compare our results with those of [27]. Note that, for a meaningful comparison, since our problem and that studied in [27] are not the same, we made a transformation of our problem to get the same structure of [27]. In fact, since our problem is a generalization of [27], we added the two following constraints to our formulation:

$$p = \sum_{k \in K} l_{kk}$$

$$\sum_{k,l \in K} l_{km} = \frac{p(p-1)}{2}$$

These two equations transform our incomplete (general case) to complete problem. The gaps of the last column in Table 5 show that our solutions by SA approach outperform best found solutions by [27] in 11 out of 15 instances.

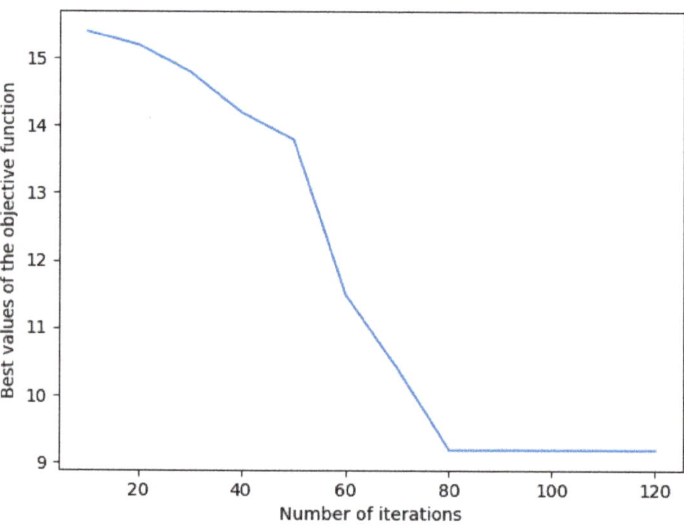

Figure 5. The best archived solution over the iterations for the instance 10C10L2TL.

Table 4. Simulated Annealing results.

Instance	Best Gap %	Average Gap %	Time (s)	# Terminals
10C10L2TL	0	0	0.32	6
10C10L4TL	0	0.03	0.89	6
10C10L6TL	0	0	1.00	7
10C10L8TL	0	0.1	0.56	8
10C10L10TL	0	0	1.6	6
10C10L12TL	0	0	1.23	9
20C10L2TL	0	0.4	3.45	3
20C10L4TL	0	0.23	3.65	5
20C10L6TL	0	0	4.6	5
20C10L8TL	0	0.3	4.6	5
20C10L10TL	1.3	1.8	3.87	5
20C10L12TL	2.8	3.8	2.7	6
40C10L2TL	1.6	1.9	6.7	7
40C10L4TL	1.2	2.1	5.4	8
40C10L6TL	2.5	3.8	4.3	8
40C10L8TL	1.6	4.1	5.7	8
40C10L10TL	3	3.5	5.7	8
40C10L12TL	1	2,1	7.8	8
80C10L2TL	1.1	1.9	8.7	3
80C10L4TL	3.6	5.8	10	4

Table 5. SA results versus Best found solutions.

Instance	Solution by [27] ($\times 10^6$)	Our Solution ($\times 10^6$)	Gap
20C40L	47.76	45.1	6%
20C50L	42.87	41.36	4%
20C60L	48.57	43.12	13%
20C70L	39.31	39.19	0%
20C80L	43.58	41.84	4%
20C90L	38.74	40.88	−5%
20C100L	48.54	46.54	4%
30C10L	117.72	105.84	11%
30C20L	113.5	116.51	−3%
30C30L	117.03	119.74	−2%
30C40L	91.31	88.14	4%
30C50L	105.48	102.94	2%
30C60L	103.24	114.64	−10%
30C70L	113.53	90.57	25%
30C80L	188.55	105.55	2%

5.2. Completeness Effect

To study the completeness effect on the solution cost, let

$$t = \sum_{k \in K} l_{kk}$$

which denotes the number of terminals in a solution. Then, we define the "Completeness rate" of inter-terminals induced graph as follows:

$$r = \frac{\sum\limits_{k,m \in K} l_{km}}{\frac{t(t-1)}{2}} \tag{13}$$

We modify the linear program introduced in the Section 3 by changing Equation (7) to become:

$$\sum_{k,m \in K} l_{km} \leq l \tag{14}$$

As shown in Figure 6, the completeness rate of the SA solutions is overall between 50% and 100%. This means two important things: (1) a sufficient number of rail links is necessary to reduce the total transportation cost as the rail mode enables to achieve scale economy by transporting huge volumes of flows; (2) a fully connected terminal network is unnecessary and increases the total transportation cost.

To further study the completeness rate effect on the generated solutions, we plot in Figure 7 the completeness rate over iterations of the archived best solutions for an instance with 40 nodes. This figure highlights that, while a sufficient number of rail links is necessary to reduce the total transportation cost (more than 50%), a fully connected rail network may not achieve an efficient transportation plan.

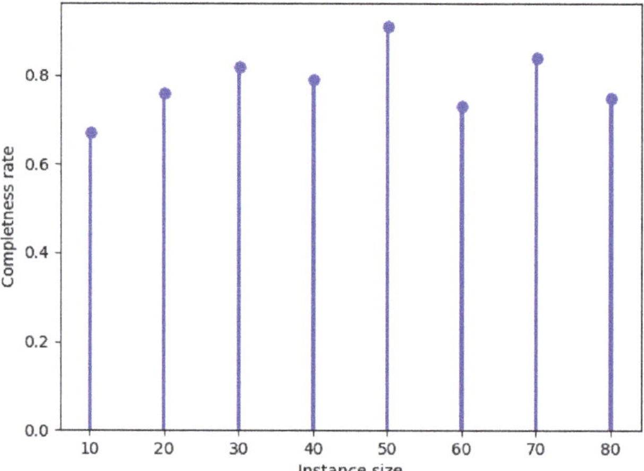

Figure 6. The completeness rate of the SA solutions.

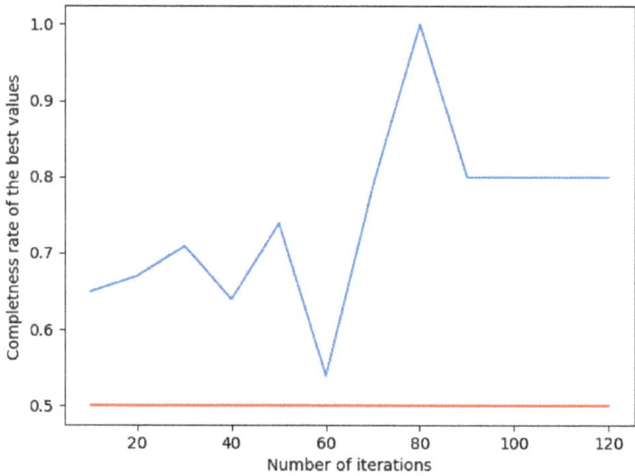

Figure 7. The completeness rate over iterations for the instance 40C10L2TL.

6. Concluding Remarks and Perspectives

Research in the freight terminal location problems is still in its beginning. However, this domain is attracting the attention of scholars and policy makers. The current paper contributes to this knowledge area by optimizing the location of intermodal terminals. Such terminals are the cornerstone in each intermodal freight system. Considering the incomplete inter-terminal network is relatively ignored in the literature, a mixed integer model for the intermodal terminal location problem is proposed in the current work. Due to its NP-hardiness, we develop an efficient Simulated Annealing Algorithm to tackle medium and large instances. Our solving approach allows us to retrieve the optimal solutions for small instances considering CPLEX 12.10. solver. In addition, our best found solutions are competitive in the medium and larger sized instances as the worst CPU time required appears to be less than 10 s. Specifically, on larger instances, our algorithm outperforms the best integer solutions returned by CPLEX, in both the objective cost and CPU time.

The numerical results show that considering the general case where all located terminals are not necessarily fully connected may substantially reduce the overall costs of location and transportation. Accordingly, this leads to the reduction of committed investments costs.

While we provide an efficient modeling and solving of the incomplete intermodal terminal location problem, an important extension of the current work is to integrate the routing decision in the mathematical formulation. In fact, the proposed formulation assumes that routing is efficiently managed once locations of terminals are fixed. However, routing and location are interdependent. Therefore, the current paper may be a first step to integrate routing decisions for the incomplete version of the ITL.

From methodological perspective, using other solving methods may be explored in a future study. Especially, exact methods such branch and bound or cutting planes methods may be adapted to the studied problem.

Another potential opportunity for future research may be the integration of the hub location formulation in the framework of the intermodal location considering the hubs as intermediate warehouse for the shipped commodities. Furthermore, it may be interesting to deal with a case study in the framework of intermodal transportation to consider more realistic situations.

Funding: This research was funded by TICLab laboratory of the International University of Rabat.

Institutional Review Board Statement: Not applicable.

Informed Consent Statement: Not applicable.

Data Availability Statement: Not applicable.

Acknowledgments: We thank the anonymous reviewers and the academic editor for their careful reading of our manuscript and their many insightful comments and suggestions.

References

1. Arnold, P.; Peeters, D.; Thomas, I.; Marchand, H. Pour une localisation optimale des centres de transbordement intermodaux entre réseaux de transport: formulation et extensions. *Can. Geogr./Géographe Can.* **2001**, *45*, 427–436. [CrossRef]
2. Bontekoning, Y.M.; Macharis, C.; Trip, J.J. Is a new applied transportation research field emerging?—A review of intermodal rail–truck freight transport literature. *Transp. Res. Part A Policy Pract.* **2004**, *38*, 1–34. [CrossRef]
3. Demir, E.; Burgholzer, W.; Hrušovský, M.; Arıkan, E.; Jammernegg, W.; Van Woensel, T. A green intermodal service network design problem with travel time uncertainty. *Transp. Res. Part B Methodol.* **2016**, *93*, 789–807. [CrossRef]
4. Colicchia, C.; Creazza, A.; Dallari, F. Lean and green supply chain management through intermodal transport: insights from the fast moving consumer goods industry. *Prod. Plan. Control* **2017**, *28*, 321–334. [CrossRef]
5. Verma, M.; Verter, V. A lead-time based approach for planning rail–truck intermodal transportation of dangerous goods. *Eur. J. Oper. Res.* **2010**, *202*, 696–706. [CrossRef]

6. Ambrosino, D.; Asta, V.; Crainic, T.G. Optimization challenges and literature overview in the intermodal rail-sea terminal. *Transp. Res. Procedia* **2021**, *52*, 163–170. [CrossRef]

7. Oudani, M. Intermodal Terminal Location Problem under fuzzy restrictions. In Proceedings of the 2020 IEEE 13th International Colloquium of Logistics and Supply Chain Management (LOGISTIQUA), Fez, Morocco, 2–4 December 2020; pp. 1–5.

8. Márton, P.; Milinković, S.; Belošević, I. Solving a container terminal location problem using decision support systems. *Transp. Res. Procedia* **2019**, *40*, 1459–1464. [CrossRef]

9. Teye, C.; Bell, M.G.; Bliemer, M.C. Entropy maximising facility location model for port city intermodal terminals. *Transp. Res. Part E Logist. Transp. Rev.* **2017**, *100*, 1–16. [CrossRef]

10. Tadić, S.; Krstić, M.; Roso, V.; Brnjac, N. Planning an intermodal terminal for the sustainable transport networks. *Sustainability* **2019**, *11*, 4102. [CrossRef]

11. Mokhtar, H.; Redi, A.P.; Krishnamoorthy, M.; Ernst, A.T. An intermodal hub location problem for container distribution in Indonesia. *Comput. Oper. Res.* **2019**, *104*, 415–432. [CrossRef]

12. Muravev, D.; Hu, H.; Rakhmangulov, A.; Mishkurov, P. Multi-agent optimization of the intermodal terminal main parameters by using AnyLogic simulation platform: Case study on the Ningbo-Zhoushan Port. *Int. J. Inf. Manag.* **2021**, *57*, 102133. [CrossRef]

13. Stoilova, S.D.; Martinov, S.V. Selecting a location for establishing a rail-road intermodal terminal by using a hybrid SWOT/MCDM model. *Iop Conf. Ser. Mater. Sci. Eng.* **2019**, *618*, 012060. [CrossRef]

14. Uddin, M.; Huynh, N. Reliable routing of road-rail intermodal freight under uncertainty. *Networks Spat. Econ.* **2019**, *19*, 929–952. [CrossRef]

15. Wu, X.; Cao, L. Using heuristic MCMC method for terminal location planning in intermodal transportation. *Int. J. Oper. Res.* **2018**, *32*, 421–442. [CrossRef]

16. Liang, F.; Verhoeven, K.; Brunelli, M.; Rezaei, J. Inland terminal location selection using the multi-stakeholder best-worst method. *Int. J. Logist. Res. Appl.* **2021**, 1–23. [CrossRef]

17. Rožić, T.; Rogić, K.; Ivanković, B. Modelling inland terminal locations based on transport cost optimisation. *Int. J. Shipp. Transp. Logist.* **2020**, *12*, 487–503. [CrossRef]

18. Crainic, T.G.; Kim, K.H. Intermodal transportation. *Handbooks Oper. Res. Manag. Sci.* **2007**, *14*, 467–537.

19. Dekker, R.; Bloemhof, J.; Mallidis, I. Operations Research for green logistics–An overview of aspects, issues, contributions and challenges. *Eur. J. Oper. Res.* **2012**, *219*, 671–679. [CrossRef]

20. Caris, A.; Macharis, C.; Janssens, G.K. Planning problems in intermodal freight transport: accomplishments and prospects. *Transp. Plan. Technol.* **2008**, *31*, 277–302. [CrossRef]

21. Ertem, M.A.; İşbilir, M.; Arslan, A.Ş. Review of intermodal freight transportation in humanitarian logistics. *Eur. Transp. Res. Rev.* **2017**, *9*, 10. [CrossRef]

22. Caris, A.; Macharis, C.; Janssens, G.K. Decision support in intermodal transport: A new research agenda. *Comput. Ind.* **2013**, *64*, 105–112. [CrossRef]

23. Sun, Y.; Lang, M.; Wang, D. Optimization models and solution algorithms for freight routing planning problem in the multi-modal transportation networks: a review of the state-of-the-art. *Open Civ. Eng. J.* **2015**, *9*, 714–723. [CrossRef]

24. Macharis, C.; Caris, A.; Jourquin, B.; Pekin, E. A decision support framework for intermodal transport policy. *Eur. Transp. Res. Rev.* **2011**, *3*, 167–178. [CrossRef]

25. Arnold, P.; Peeters, D.; Thomas, I. Modelling a rail/road intermodal transportation system. *Transp. Res. Part E Logist. Transp. Rev.* **2004**, *40*, 255–270. [CrossRef]

26. Verma, M.; Verter, V.; Zufferey, N. A bi-objective model for planning and managing rail-truck intermodal transportation of hazardous materials. *Transp. Res. Part E Logist. Transp. Rev.* **2012**, *48*, 132–149. [CrossRef]

27. Sörensen, K.; Vanovermeire, C.; Busschaert, S. Efficient metaheuristics to solve the intermodal terminal location problem. *Comput. Oper. Res.* **2012**, *39*, 2079–2090. [CrossRef]

28. Oudani, M.; El Hilali Alaoui, A.; Boukachour, J. An efficient genetic algorithm to solve the intermodal terminal location problem. *Int. J. Supply Oper. Manag.* **2014**, *1*, 279–296.

29. Lin, C.C.; Lin, S.W. Two-stage approach to the intermodal terminal location problem. *Comput. Oper. Res.* **2016**, *67*, 113–119. [CrossRef]

30. Lin, C.C.; Chiang, Y.I.; Lin, S.W. Efficient model and heuristic for the intermodal terminal location problem. *Comput. Oper. Res.* **2014**, *51*, 41–51. [CrossRef]

31. Oudani, M. Modelling the Incomplete Intermodal Terminal Location Problem. *IFAC-PapersOnLine* **2019**, *52*, 184–187. [CrossRef]

32. Huynh, N.; Fotuhi, F. A new planning model to support logistics service providers in selecting mode, route, and terminal location. *Pol. Marit. Res.* **2013**, *20*, 67–73. [CrossRef]

33. Bierwirth, C.; Kirschstein, T.; Meisel, F. On transport service selection in intermodal rail/road distribution networks. *Bus. Res.* **2012**, *5*, 198–219. [CrossRef]

34. Wang, Y.; Yeo, G.T. Intermodal route selection for cargo transportation from Korea to Central Asia by adopting Fuzzy Delphi and Fuzzy ELECTRE I methods. *Marit. Policy Manag.* **2018**, *45*, 3–18. [CrossRef]

35. Chang, T.S. Best routes selection in international intermodal networks. *Comput. Oper. Res.* **2008**, *35*, 2877–2891. [CrossRef]

36. Southworth, F.; Peterson, B.E. Intermodal and international freight network modeling. *Transp. Res. Part C Emerg. Technol.* **2000**, *8*, 147–166. [CrossRef]

37. Kengpol, A.; Tuammee, S.; Tuominen, M. The development of a framework for route selection in multimodal transportation. *Int. J. Logist. Manag.* **2014**, *25*, 581–610. [CrossRef]
38. Qu, Y.; Bektaş, T.; Bennell, J. Sustainability SI: multimode multicommodity network design model for intermodal freight transportation with transfer and emission costs. *Networks Spat. Econ.* **2016**, *16*, 303–329. [CrossRef]
39. Bouchery, Y.; Fransoo, J. Cost, carbon emissions and modal shift in intermodal network design decisions. *Int. J. Prod. Econ.* **2015**, *164*, 388–399. [CrossRef]
40. Meng, Q.; Wang, X. Intermodal hub-and-spoke network design: incorporating multiple stakeholders and multi-type containers. *Transp. Res. Part B Methodol.* **2011**, *45*, 724–742. [CrossRef]
41. Morlok, E.K.; Spasovic, L.N. Redesigning rail-truck intermodal drayage operations for enhanced service and cost performance. *J. Transp. Res. Forum* **1994**, *34*, 16–31.
42. Kayikci, Y. A conceptual model for intermodal freight logistics centre location decisions. *Procedia-Soc. Behav. Sci.* **2010**, *2*, 6297–6311. [CrossRef]
43. Tsamboulas, D.; Vrenken, H.; Lekka, A.M. Assessment of a transport policy potential for intermodal mode shift on a European scale. *Transp. Res. Part A Policy Pract.* **2007**, *41*, 715–733. [CrossRef]
44. Macharis, C.; Bontekoning, Y.M. Opportunities for OR in intermodal freight transport research: A review. *Eur. J. Oper. Res.* **2004**, *153*, 400–416. [CrossRef]
45. Spasovic, L.; Morlok, E.K. Using marginal costs to evaluate drayage rates in rail-truck intermodal service. *Transp. Res. Rec.* **1993**, 8–16.
46. Ghane-Ezabadi, M.; Vergara, H.A. Decomposition approach for integrated intermodal logistics network design. *Transp. Res. Part E Logist. Transp. Rev.* **2016**, *89*, 53–69. [CrossRef]
47. Braekers, K.; Caris, A.; Janssens, G.K. Integrated planning of loaded and empty container movements. *OR Spectr.* **2013**, *35*, 457–478. [CrossRef]
48. Hanaoka, S.; Kunadhamraks, P. Multiple criteria and fuzzy based evaluation of logistics performance for intermodal transportation. *J. Adv. Transp.* **2009**, *43*, 123–153. [CrossRef]
49. Ruan, J.; Wang, X.; Chan, F.; Shi, Y. Optimizing the intermodal transportation of emergency medical supplies using balanced fuzzy clustering. *Int. J. Prod. Res.* **2016**, *54*, 4368–4386. [CrossRef]
50. Bergqvist, R. Evaluating road–rail intermodal transport services–a heuristic approach. *Int. J. Logist. Res. Appl.* **2008**, *11*, 179–199. [CrossRef]
51. Murray, A.T.; Church, R.L. Applying simulated annealing to location-planning models. *J. Heuristics* **1996**, *2*, 31–53. [CrossRef]
52. Vincent, F.Y.; Lin, S.W.; Lee, W.; Ting, C.J. A simulated annealing heuristic for the capacitated location routing problem. *Comput. Ind. Eng.* **2010**, *58*, 288–299.
53. Mathews, P.G. *Design of Experiments with MINITAB*; ASQ Quality Press: Milwaukee, WI, USA, 2005.

applied sciences

Article

Performance Comparison between Particle Swarm Optimization and Differential Evolution Algorithms for Postman Delivery Routing Problem

Warisa Wisittipanich [1], Khamphe Phoungthong [2], Chanin Srisuwannapa [3], Adirek Baisukhan [3] and Nuttachat Wisittipanit [3,4,*

1 Center of Healthcare Engineering System, Department of Industrial Engineering, Faculty of Engineering, Chiang Mai University, Chiang Mai 50200, Thailand; warisa@eng.cmu.ac.th
2 Environmental Assessment and Technology for Hazardous Waste Management Research Center, Faculty of Environmental Management, Prince of Songkla University, Songkhla 90112, Thailand; khamphe.p@psu.ac.th
3 Department of Materials Engineering, School of Science, Mae Fah Luang University, Chiang Rai 57100, Thailand; chanin.sri@mfu.ac.th (C.S.); adirek.bai@mfu.ac.th (A.B.)
4 Center of Innovative Materials for Sustainability (iMatS), Mae Fah Luang University, Chiang Rai 57100, Thailand
* Correspondence: nuttachat.wis@mfu.ac.th

Abstract: Generally, transportation costs account for approximately half of the total operation expenses of a logistics firm. Therefore, any effort to optimize the planning of vehicle routing would be substantially beneficial to the company. This study focuses on a postman delivery routing problem of the Chiang Rai post office, located in the Chiang Rai province of Thailand. In this study, two metaheuristic methods—particle swarm optimization (PSO) and differential evolution (DE)—were applied with particular solution representation to find delivery routings with minimum travel distances. The performances of PSO and DE were compared along with those from current practices. The results showed that PSO and DE clearly outperformed the actual routing of the current practices in all the operational days examined. Moreover, DE performances were notably superior to those of PSO.

Keywords: postman delivery; vehicle routing problem; particle swarm optimization algorithm; differential evolution algorithm

check for updates

Citation: Wisittipanich, W.; Phoungthong, K.; Srisuwannapa, C.; Baisukhan, A.; Wisittipanit, N. Performance Comparison between Particle Swarm Optimization and Differential Evolution Algorithms for Postman Delivery Routing Problem. *Appl. Sci.* **2021**, *11*, 2703. https://doi.org/10.3390/app11062703

Academic Editor: Farouk Yalaoui

Received: 18 February 2021
Accepted: 14 March 2021
Published: 17 March 2021

Publisher's Note: MDPI stays neutral with regard to jurisdictional claims in published maps and institutional affiliations.

1. Introduction

The topic of global warming has been heavily addressed in recent decades due to increasing greenhouse gases [1]. Unsurprisingly, the transportation industry globally accounts for a large portion (mainly CO_2 emission), amounting to approximately 20% [2]. Since our economics and life wellness deeply depend on the industry, decent mitigation efforts are needed to enhance its operation efficiency. For logistics organizations, with transportation costs being their largest expenditure, an improved operation not only reduce the environmentally harmful gas, but also reduces the operational costs such as fuel, manpower and vehicle depreciation, significantly augmenting profits in the process [3]. As a result, a large logistics firm that handles thousands of transport vehicles per operational day should absolutely adopt an effective transportation management and scheduling system. The post office is one of the logistics establishments that has to deal with such management, involving area assignments for delivery vehicles and the type and number of vehicles, as well as delivery time windows. Some even engage in detailed routing of each vehicle, trying to determine the most optimized ways to deliver parcels to customers [4]. Such a challenge is essentially called the vehicle routing problem (VRP). However, most post offices do not feel the need to find solutions to the VRP and simply rely on the experience of their delivery drivers, which might be because of the high costs and complications involved in deploying such systems. Nevertheless, relying on skills and experience alone might not be suitable for a problem with exceptionally high complexities.

The VRP is one of the optimization problems that seeks optimized routing for vehicles that traverse either to deliver or pickup goods. An approach to finding a solution to the VRP might be to determine a set of routes having a minimized total distance traveled by all the vehicles involved [5]. Other VRP variations include enforcing the total delivery time to be within a certain time window [6], limiting vehicle capacity [7], having stochastic customer demands [8] or allowing customers to select their delivery options [9]. Generally, the problem sizes (such as number of vehicles and customer nodes) of the VRP that can be solved for exact optimized solutions by mathematical methods are limited [10], due to the inherent complexity of the VRP. However, actual VRP applications usually have large problem sizes, containing approximately hundreds or thousands of vehicles and customers (e.g., a beverage firm that has to deliver millions of bottles per day to a large number of depots and small businesses). Therefore, solving for optimized solutions by mathematical procedures alone is impractical for most applications. As a result, heuristic and metaheuristic optimization algorithms are usually selected as a pragmatic method to solve the VRP, since they can obtain feasible solutions relatively fast and the quality of those solutions is decent enough for real uses [11].

This study investigated the postman operations of the Chiang Rai post office, which is a logistics establishment that does not apply any VRP optimization techniques, and its delivery drivers have to rely on their own expertise and plans in order to traverse between customer locations. The routing of two delivery vehicles was tracked, in which they dispatched parcels to customers in their responsible areas for 50 operational days. The routing information consisted of traveled paths, distance covered and the actual geographic coordinates of customer locations which the vehicles visited each day. Then, we performed the routing optimization procedures based on those customer coordinates, using the metaheuristics algorithms for all the operational days of those delivery vehicles.

Since the delivery routes depended on the skills and experience in the area of their drivers, those routes might not be well optimized. Moreover, if new drivers were to replace the previous ones, they would have to learn anew and gain enough experience to better traverse the area. Thus, the efficiency of the delivery process was not consistent. As such, this research paper was focused on providing a smart routing decision, solving for VRP solutions such that the delivery performance could improve remarkably and be more consistent. Two metaheuristic algorithms were applied—particle swarm optimization (PSO) and differential evolution (DE)—and integrated with local search techniques. The algorithms sought to discover the most optimized routing those two delivery vehicles could take so that the total travel distance could be minimized. The solutions from the two algorithms, including those from current practices, had their performances compared in terms of the total distance covered.

The rest of the article is organized as follows. Section 2 provides literature reviews related to the VRP and the algorithms used in the study. The detailed dataset, mathematical model, algorithm frameworks and solution representation are given in Section 3. Finally, the experimental results and conclusion are provided in Sections 4 and 5, respectively.

2. Literature Review

The VRP, a nondeterministic polynomial time hardness (NP-hardness) problem [12], refers to an optimization problem related to routing management, first devised by Dantzig and Ramser [5] in 1959. It started as a simple problem of determining the optimal routes for a fleet of vehicles dispatching from a single terminal to multiple variants, including ones that limit vehicle capacity and delivery time, have multiple terminals or allow the customer to select a delivery option. Obtaining the exact optimal solution for the VRP is possible if the customer nodes are not too large for a mathematical solver. However, real VRP applications tend to have a substantial number of customer nodes. Therefore, heuristic methods are often chosen to solve the VRP [13].

There are several studies that focused on the VRP of post office deliveries with different strategies, objectives and constraints. Meira et al. [14] conducted an experiment to solve the multi-objective VRP with only one constraint—the route distance—using a large number of geographical delivery locations (~30,000 points) in Autur Nogueira, Brazil. Dumez et al. [9] tried to solve the VRP for a post office in which each customer could select a choice of vehicle routing problem delivery options (VRPDOs) with varying places and times, such as receiving parcels at home in the evening or at work during office hours. In addition, the VRP with a time window (VRPTW) was investigated by Niroomand et al. [15], who developed an optimization model for a network of post offices with aims to minimize the total operational cost within time constraints (i.e., office working hours).

Various algorithms were employed to solve the VRP and its variants. Most are metaheuristic, including differential evolution (DE) and particle swarm optimization (PSO). Other metaheuristic algorithms such as the large neighborhood search (LNS) [16] and hybrid intelligent algorithms [17] were also adopted for the VRP. DE is a population-based optimization method that is similar to other evolutionary algorithms in terms of conventional approaches (e.g., mutation and cross-over operations). However, it utilizes a less stochastic and greedier solving strategy [18]. DE has been successfully applied to many VRP studies. Kromer et al. [19] employed DE to find solutions to the stochastic VRP with simultaneous pickup and delivery by using actual bus service data in Anbessa, Ethiopia. Their objective was to minimize both the total travel distance and number of vehicles, and the algorithm yielded superior performance compared with that of a traditional savings algorithm. Sethanan [20] utilized DE (hybridized with a genetic operator) for a multi-trip VRP with backhauls and a heterogeneous fleet (MTVRPB). They devised an optimization method specifically for a company that needed to retrieve glass bottles (soft drinks) back from clients after delivery with a single objective (i.e., minimizing the total distance traveled). Erbao et al. [18] used both DE and genetic algorithms to find an optimal solution to the VRP with simultaneous delivery and pick-up and time windows (VRP-SDPTW). They built a mathematical model using mixed integer programming for their proposed VRP variant and yielded satisfying optimized results. Xing et al. [21] applied a hybrid discrete differential evolution (HDDE) to the split delivery VRP (SDVRP). The authors focused on vehicle routing management to deliver large supplies of hospital items split between multiple unmanned transport vehicles—a suitable solution in the pandemic era—with the objective of minimizing the total distance traveled, and the proposed HDDE algorithm produced superior performance compared with the forest-based tabu search (FBTS) and genetic algorithm (GA) in many problem instances. There are also studies that applied DE and local search to solve the VRP. For instance, Song and Don [22] proposed a DE algorithm integrated with local search to solve the capacitated VRP. They employed two strategies for the local search—two-random swap and two-opt testing of the combined method with the benchmark instances—and demonstrated that such a hybrid method performed better than other existing DE algorithms. Jun and Jian [23] modified DE further by using greedy subtour crossover in the mutation operation and modified ordered crossover in the crossover operation. In addition, local search with only the two-opt strategy was used in the decoding process.

In addition to DE, PSO was also used frequently in the VRP field. Moghaddam et al. [24] developed an advanced PSO to solve a VRP variant where the demands of customers were uncertain and demand distribution was undetermined, a special form of capacitated VRP (CVRP). They devised a new decoding method for the PSO and were able to obtain superior solutions compared with those from well-known studies using the same test instances. Another variant of the capacitated VRP, called the capacitated location-routing problem (CLRP), was also studied by Peng et al. [25], in which PSO was combined with clustering and local search techniques to solve the problem. Marinakis et al. [26] proposed the use of multi-adaptive PSO (MAPSO) with three different strategies to solve the VRPTW, where each strategy had a certain role in achieving optimal solutions faster, yielding promising results compared with those of other PSO variants. Chen and Shi [27] used hybrid PSO (HPSO) combined with simulated annealing (SA) to determine solutions for the multi-compartment VRP with a time window (MCVRPTW), where each vehicle possessed multiple compartments, each one containing goods with the same characteristics. They tested the algorithm with benchmark instances, and the results showed that HPSO had better performance than normal PSO (PSO without SA). Chen et al. [28] performed two types of PSO, namely discrete PSO (DPSO) and neural-like DPSO (NDPSO), on the periodic vehicle routing problem (PVRP). The PVRP deals with the tasks of finding an ideal customer service mode and determining optimal vehicle routing in accordance with the found mode. Sedighizadeh and Mazaheripour [29] combined PSO and the artificial bee colony (ABC) algorithm to solve the VRP, which converged faster than individual algorithms (PSO and ABC) and also yielded better solutions. Lastly, Yanwei et al. [30] employed PSO with self-adaptive inertia weight to solve the open VRP with time-dependent travel time (OVRPTD), a special VRP variant where vehicles do not have to return to the depot after finishing a delivery job and travel time is time-dependent, producing better solutions compared to those of PSO and the nearest-neighboring search (NNS) algorithm.

3. Materials and Methods

Details on the post office operation and how to obtain the necessary dataset and prepare the inputs for the study are presented in this section. Additionally, the mathematical model, algorithmic procedures of PSO and DE and process workflow for the postman delivery problem are included here.

3.1. Dataset

The dataset used in this study was obtained from the delivery operations belonging to the Chiang Rai post office in the Chiang Rai province of Thailand. The operations mainly involved postal delivery service to residential and commercial locations. The Chiang Rai post office holds delivery duties for customers in the area of the Viang and Rob-Viang districts, which are clustered into six delivery zones. Detailed data of the delivery operation were collected for 50 days, tracking two delivery vehicles equipped with Global Positioning System (GPS)-enabled android devices with location-tracking applications installed. The average number of customer locations to be delivered to per day by both vehicles was 97.22. Those two vehicles were motorcycles, each with large parcel pouches strapped on both sides, and both had duties to deliver parcels to customers in specific areas under the Chiang Rai post office's responsibility.

The GPS devices recorded routing data, including all customer locations and where the vehicles visited, in geographic coordinate form. Then, for each day of delivery operations, the total travel distance of both vehicles was extracted. Moreover, the stored latitude and longitude coordinates, which corresponded to the post office and customer locations, were used to construct a distance matrix of $n \times n$ dimensions, where n was the number of all location coordinates. Each cell of the distance matrix represented the smallest distance between a pair of location coordinates, and this distance value was determined by a python script using the googlemaps module from the Google Maps Application Programming Interface (API) [31]. The obtained matrix might not have been the symmetric type, since the shortest route from one coordinate to another may not have been the same the other way around, taking into account that one-way traffic could be enforced in some routes. In total, 50 matrices were constructed, each one representing one day of delivery operations. These matrices were essential inputs to the optimization algorithms, acting as a lookup distance table.

3.2. Mathematical Model

As mentioned earlier, this study addresses the delivery routing problem of the Chiang Rai post office. The problem is classified as a single-depot VRP, in which all vehicles that depart from the post office must return to the post office. For each operational day, a vehicle must decide on the delivery route for a set of different customers. All customers must be visited within an operational day with no specific time window. The traffic conditions were not considered in this case, due to the fact that there was little traffic in the investigated area. Thus, the delivery time and cost were primarily based on the total travel distance, and the objective of this problem was to find optimal routes with the lowest total travel distance. In addition, the capacity of the parcel pouches was assumed to be infinite, since the postman used two large pouches that could always contain all parcels, and there has been no case where the capacity was insufficient.

In this study, the postman delivery routing problem is represented as an integer programing (IP) model. The parameters and decision variables used in formulating the model are defined as follows:

Indices

i, j: Post office and customer locations ($i, j = 1$ for the post office location; $i, j = 2, 3, \ldots, n$ for customer locations).

k: Vehicle ($k = 1, 2, \ldots, m$).

Parameters

$d_{i,j}$: Distance from i to j
U_i, U_j: Constant number used to avoid a subtour.
N: Number of locations.
M: Number of vehicles.

Decision Variable

$x_{i,j,k}$: If vehicle k travels from i to j or 0 otherwise.

The minimization of the total travel distance is expressed as

$$Min\ Z = \sum_{i=1}^{n}\sum_{j=1}^{n}\sum_{k=1}^{m} d_{ij} * x_{ijk} \tag{1}$$

The constraints are as follows:

$$\sum_{i=1}^{n}\sum_{k=1}^{m} x_{ijk} = 1 \qquad \forall_j\ ;\ j = \{2,3,\ldots,n\} \tag{2}$$

$$\sum_{j=1}^{n}\sum_{k=1}^{m} x_{ijk} = 1 \qquad \forall_i\ ;\ i = \{2,3,\ldots,n\} \tag{3}$$

$$\sum_{i=1}^{n} x_{ipk} - \sum_{j=1}^{n} x_{pjk} = 0 \qquad \forall_p \forall_k\ ;\ p = \{1,2,\ldots,n\}, k = \{1,2,\ldots,m\} \tag{4}$$

$$\sum_{i=2}^{n} x_{i1k} \leq 1 \qquad \forall_k\ ;\ k = \{1,2,\ldots,m\} \tag{5}$$

$$\sum_{j=2}^{n} x_{1jk} \leq 1 \qquad \forall_k\ ;\ k = \{1,2,\ldots,m\} \tag{6}$$

$$U_i - U_j + N\sum_{k=1}^{m} x_{ijk} \leq N - 1\ \forall_i\ \forall_j\ ;\ i,j = \{2,3,\ldots,n\}\ and\ i \neq j \tag{7}$$

$$x_{ijk} \in \{0,1\} \qquad \begin{array}{l}\forall_i \forall_j \forall_k\ ;\ i,j = \{1,2,\ldots,n\}, \\ k = \{1,2,\ldots,m\}\end{array} \tag{8}$$

The objective of the model was to minimize the total travel distance as shown in Equation (1). Equations (2) and (3) ensured that each customer was visited once and by only one vehicle. The constraint in Equation (4) guaranteed the route continuity of all vehicles. Equations (5) and (6) ensured that each vehicle could take only one route, and it needed to depart from and return to the post office. Equation (7) was used to eliminate subtours. Equation (8) states that a decision variable is a binary number.

3.3. Application of PSO and DE to the Postman Delivery Routing Problem

As mentioned earlier, the postman delivery routing problem is NP-hard. Thus, solution methods based on metaheuristic approaches are more preferable than the traditional exact methods for finding solutions in real-world practices. This section presents the application of two effective metaheuristic approaches—particle swarm optimization (PSO) and differential evolution (DE)—to solve the postman delivery routing problem. A solution representation with an encoding and decoding scheme is proposed to transform a metaheuristic solution into a practical solution. The procedures of these algorithms are described below.

3.3.1. Particle Swarm Optimization (PSO)

Particle swarm optimization (PSO), proposed by Kennedy and Eberhart in 1995, is a population-based random search technique that imitates the behavior of fish schooling or birds flocking. In PSO, a solution is represented as a particle, and the population of solutions is called a swarm of NP particles. Each particle consists of two main attributes: The position and the velocity. The evolutionary concept of PSO is that each particle learns from the cognitive knowledge of its experiences (personal best, or *pbest*) and the social knowledge of the swarm (global best, or *gbest*) to guide itself to the better position. A particle moves to a new position using the updated velocity, which is adjusted according to the experiences of a particle. In particular, the updated velocity of a particle depends on three main control parameters: The inertia weight, personal best coefficient (cp) and global best coefficient (cg). Once a new position is obtained, the best position of each particle and the best position of the swarm are then updated as needed. These processes are repeated until a stopping criterion is met. Figure 1 shows the evolutionary procedure of the PSO.

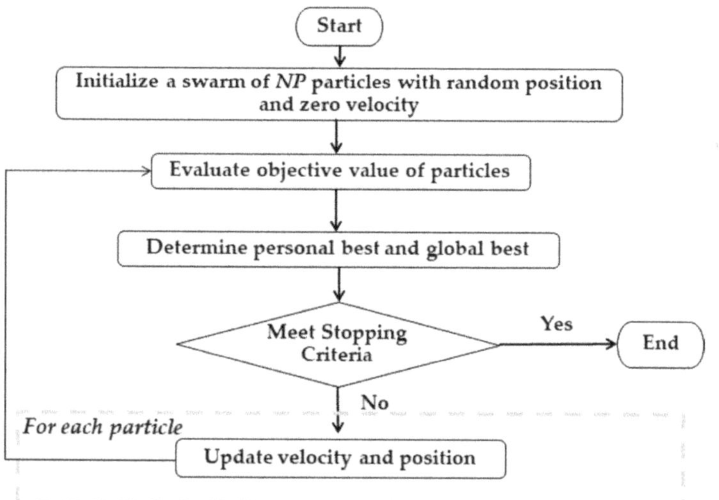

Figure 1. Evolutionary procedure of the particle swarm optimization (PSO) algorithm.

3.3.2. Differential Evolution (DE)

Differential evolution (DE), proposed by Storn and Price in 1995, is one of the evolutionary algorithms (EAs) used for global optimization over a continuous search space. Similar to other population-based random search techniques, DE starts by randomly generating an initial population of NV vectors. A solution in DE is represented by the D-dimensional vector, and each value in the D-dimensional space is represented as a real number. The crucial idea of DE is its mechanism for generating trial vectors by mutation and crossover operations with the use of a couple control variables: The scale factor (F)

and the crossover rate (*Cr*). In this study, the classic but efficient mutation operation of DE/rand/1 was applied to generate a mutant vector by adding a weighted difference (*F*) between two randomly selected vectors to the third randomly selected vector. DE is able to generate better diverse solutions, since the best solution in the population does not exert any influence on the other solutions in the population. In addition, the mutant vector is always a solution that is not from the original population; therefore, the crossover operation in DE is always between a solution from the population and a newly generated one [32]. Then, the selection operation or replacement of an individual vector occurs only if the trial vector outperforms its corresponding current vector. As shown in Figure 2, the evolution procedure of a DE population continues through repeated cycles of three main operations—mutation, crossover and selection—until a stopping criterion is met.

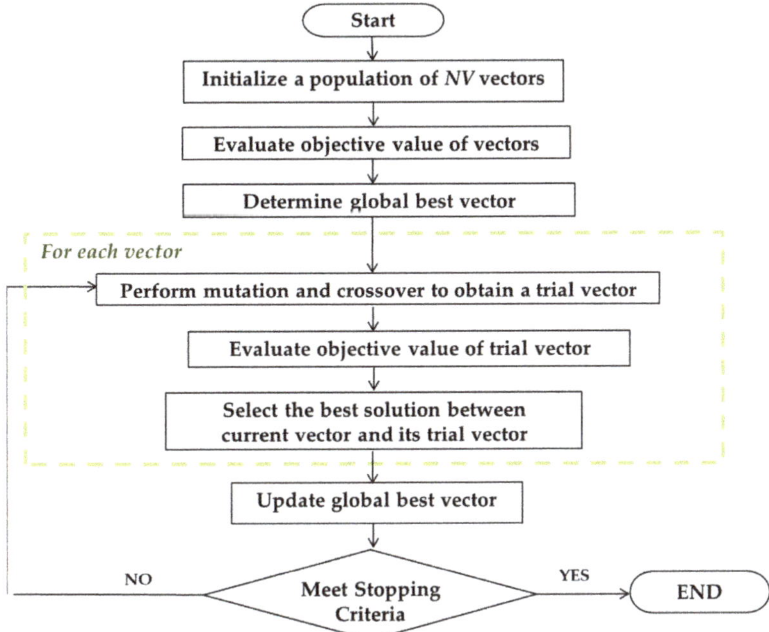

Figure 2. Evolutionary procedure of the differential evolution (DE) algorithm.

3.3.3. Solution Representation

In order to apply PSO and DE for solving a combinatorial problem like the postman delivery routing problem, solution representation or solution mapping was required to transform the real number of the *D*-dimensional space into practical solutions for vehicle routing. Figure 3 presents the procedure of solution representation with the encoding and decoding scheme.

The encoding process aimed to represent a solution as a string of dimensions. In this study, the postman delivery routing problem with $n - 1$ customers (noting that n is the number of locations, including the post office) and m vehicles was encoded as a string of D dimensions. The number of dimensions (D) was set to be equal to the total number of customers, and the dimension index stood for the customer index. Each value in a dimension was initially filled with a uniform random number in the range [0, 1]. Next, the process of decoding was divided into two stages: (1) the assignment of customers to vehicles and (2) the sequencing or routing for each vehicle. In the assignment stage, the range (max–min) of the dimension value was split by the number of vehicles as in Equation (9):

$$\text{Split} = (Max - Min)/m \tag{9}$$

where m is the number of vehicles, Min is the minimum value in the dimensions and Max is the maximum value in the dimensions.

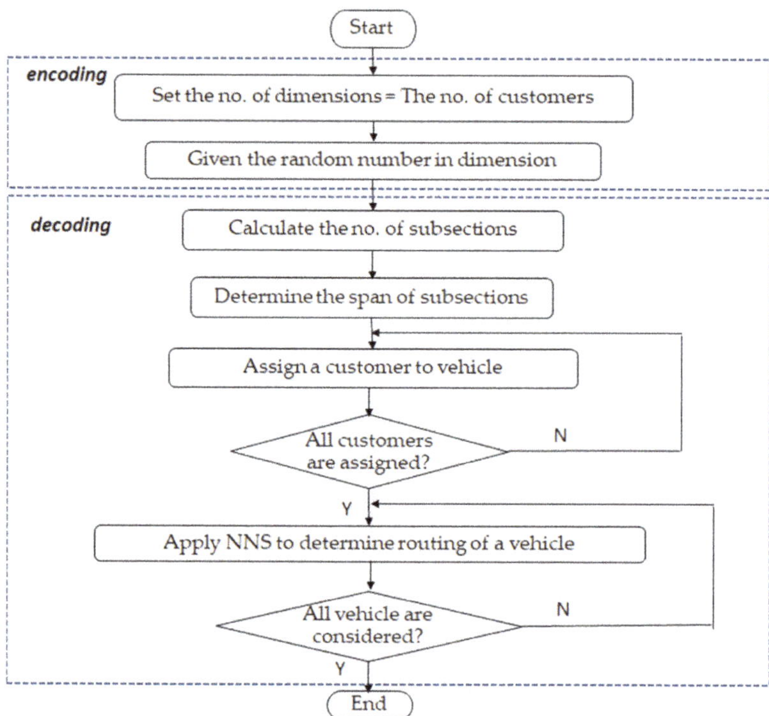

Figure 3. Solution representation procedures.

According to Equation (9), the range of the dimension value was equally partitioned into m subsections corresponding to m vehicles. The span of each subsection is determined in Figure 4. It is noted that a subsection index denoted a vehicle index.

Subsection 1 = [Min, Min + split]

Subsection 2 = [Min + split, Min + 2*split]

Subsection 3 = [Min + 2*split, Min + 3*split]

.

.

.

Subsection m = [Min + (m-1)*split, Max]

Figure 4. The span of m subsections.

Then, a customer was assigned to a vehicle, considering its customer dimension value with the corresponding subsection. For example, if the dimension value of customer 1 fell into Section 2, that customer was assigned to vehicle 2 correspondingly. This process continued until all customers were considered. Once the assignment of customers to vehicles was decided, a heuristic called the nearest-neighbor search (NNS) was applied to determine the delivery routing of each vehicle. Thus, for each vehicle, the next visited customer was the one with the nearest distance to the current location. The completed route of a vehicle was obtained when all customers assigned to that vehicle were considered. It is noted that all vehicles needed to depart from and return to the post office.

To illustrate the process of solution representation, let us consider an example of the postman delivery routing problem with six customers and two vehicles. A distance matrix between the post office and customers is shown in Figure 5. The sample matrix in this example is symmetric; however, a matrix constructed from the actual operation might be asymmetric, depending on the locations where the delivery process takes place.

Distance	Post office	1	2	3	4	5	6
Post office	-	193	71	79	165	43	54
1	193	-	198	44	168	173	112
2	71	198	-	66	80	67	123
3	79	44	66	-	176	21	78
4	165	168	80	176	-	112	52
5	43	173	67	21	112	-	79
6	54	112	123	78	52	79	-

Figure 5. A distance matrix for an example of the postman delivery routing problem with six customers and two vehicles.

In the encoding process, the number of dimensions was set to six, which was equal to the number of customers. Each value in a dimension was initially generated with a uniform random number between 0 and 1, as shown in Figure 6.

dimension	1	2	3	4	5	6
	0.51	0.25	0.72	0.43	0.87	0.34

Figure 6. Encoding example of the postman delivery routing problem with six customers and two vehicles.

Figure 7 illustrates the decoding procedure for obtaining the routing of a vehicle. First, according to Equation (9), the range of the dimension value is divided by the number of vehicles, which is $(0.87 - 0.25)/2 = 0.31$. In this example, two subsections are obtained according to the number of vehicles. Then, the condition of the customer assignment is set. The span of sections 1 and 2 are (0.25, 0.56) and (0.56, 0.87), respectively. When the customer dimension value falls into the first subsection, that customer is assigned to vehicle 1; otherwise, they are assigned to vehicle 2. Thus, customers 1, 2, 4 and 6 are assigned to vehicle 1, and customers 3 and 5 are assigned to vehicle 2. After that, the routing of each vehicle is determined using NNS method. According to the distance matrix in Figure 5, vehicle 1 starts from the post office, then goes to customer 6, customer 4, customer 2 and customer 1 before returning to the post office. Vehicle 2 starts from the post office, then goes to customer 5 and customer 3 before returning to the post office. The total distance of this particular example is 557 + 143 = 720 units.

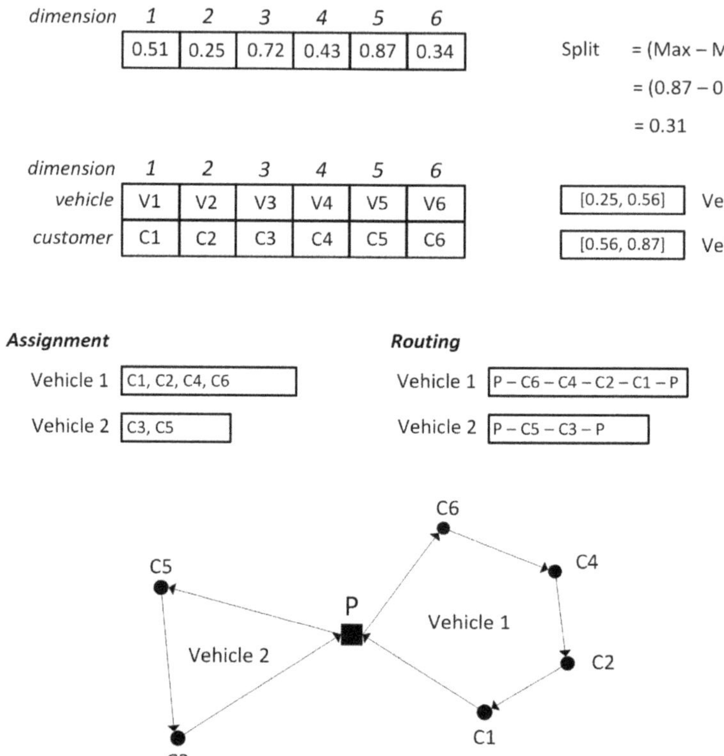

Figure 7. Decoding procedures to transform a *D*-dimensional solution into a vehicle routing solution.

4. Experimental Results

Information for the experiment, including the algorithm parameters for PSO and DE and the computational results, are shown in this section. The results emphasize the performance comparison among PSO, DE and current practices. In addition, box plots comparing the results of DE and PSO in some instances are provided.

4.1. Parameter Setting of Metaheuristics

Typically, the performance of metaheuristics is influenced by the searching and solution mapping mechanisms, while the setting of parameters decides how well the solutions converge. For a fair comparison between PSO and DE, the number of function evaluations for both algorithms was equally set to 2500, with population sizes of 50 and 50 iterations. Tables 1 and 2 show the parameter settings of PSO and DE, respectively. It is noted that these settings were empirically determined from preliminary experiments.

Table 1. The setting of PSO parameters.

Parameter	Parameter Value
Swarm Size	50
Iterations	50
Inertia Weight	0.9
Personal Best Coefficient (cp)	0.5
Global Best Coefficient (cg)	0.3

Table 2. The setting of DE parameters.

Parameter	Parameter Value
Population Size	50
Iterations	50
Scale Factor (F)	Random between 1.5 and 2.5
Crossover rate (Cr)	Linearly Increasing from 0.2 to 0.8
Crossover Type	Binomial Crossover

4.2. Computational Results

In this study, the metaheuristics were implemented using the Python programming language [33] under PyCharm IDE [34] version 2020.3.1 with Windows 10 as the operating system. The programs ran on an Intel® Core™ i5-4690 CPU 3.50 GHz platform with 32 GB of RAM. The performances of the algorithms were evaluated using the dataset derived from a real case study of postman delivery in association with the Chiang Rai post office for a period of 50 days. Table 3 shows the VRP solutions (total travel distance) obtained from the proposed metaheuristics, which include the best, average and standard deviation of the total distance from 10 replicated runs for each instance. The average time duration to run one optimization cycle for each instance is also reported.

Table 3. Experimental results of total distance.

Instance	Current Practice	PSO Best	PSO Average	PSO SD	Time (s)	DE Best	DE Average	DE SD	Time (s)
1	33.64	21.38	22.09	0.49	190.56	17.64	17.98	0.34	185.12
2	29.08	18.85	19.52	0.44	125.89	15.61	16.39	0.35	135.13
3	30.07	18.97	20.83	0.72	170.04	15.06	15.81	0.42	175.41
4	36.91	20.95	22	0.56	228.81	16.19	16.44	0.2	237.17
5	34.26	19.59	20.23	0.44	177.36	14.93	15.56	0.53	175.24
6	33.77	19.17	20.3	0.65	200.5	15.26	16.16	0.5	209.27
7	37.93	21.25	22.7	0.54	232.88	17.24	18.79	0.64	250.97
8	32.72	20.48	21.57	0.59	182.46	16.71	17.63	0.41	208.32
9	27.65	16.74	17.68	0.6	159.66	12.72	13.33	0.39	164.96
10	32.37	20.97	21.69	0.33	192.24	17.34	17.88	0.46	198.62
11	37.29	21.24	22.33	0.48	226.31	16.21	16.99	0.51	234
12	38.23	21.43	22.14	0.42	214.3	17.24	17.83	0.4	221.5
13	30.50	19.99	20.57	0.3	181.79	16.19	16.75	0.34	187.85
14	32.79	18.98	19.98	0.52	187.21	14.93	15.28	0.35	193.23
15	34.78	21.11	21.94	0.56	236.78	16.37	17.3	0.56	244.41
16	32.83	20.41	21.02	0.41	182.01	16.68	16.98	0.42	184.99
17	31.49	20.99	21.63	0.38	202.57	16.39	16.79	0.47	218.21
18	32.70	20.84	21.44	0.4	217.46	16.04	16.96	0.46	214.46
19	35.23	19.26	20.16	0.5	202.23	15.32	15.78	0.25	216.41
20	36.13	23.08	23.81	0.32	229.84	18.17	18.84	0.39	240.47
21	27.44	17.89	19.05	0.56	147.86	15.11	16.04	0.42	142.17
22	30.19	18.21	19.38	0.54	160.94	15.45	16.33	0.59	155.08
23	28.68	18.26	19.85	0.82	133.99	16.37	16.97	0.34	149.74
24	33.77	20.07	20.86	0.43	210.24	15.6	16.26	0.42	218.33
25	34.83	20.93	21.52	0.3	224.59	16.56	17.06	0.36	232.34
26	35.37	19.93	21.02	0.67	217.01	15.91	16.23	0.25	224.51
27	33.39	20.88	21.4	0.33	177.07	15.17	15.91	0.57	183.18
28	38.75	20.96	21.55	0.44	199.51	16.32	16.88	0.41	206.43
29	35.04	20.46	21.64	0.59	211.91	16.47	17.16	0.53	219.39
30	30.44	18.84	19.48	0.43	170.01	14.26	15.21	0.4	176.16
31	33.03	18.04	19.39	0.48	143.94	15.94	16.45	0.4	148.96
32	30.78	19.44	20.07	0.5	158.49	14.74	15.38	0.43	150.77
33	37.21	20.29	21.64	0.59	173.11	18.3	18.61	0.2	196.03
34	31.78	19.03	20.38	0.64	171.02	16.03	16.77	0.37	186.87

Table 3. *Cont.*

Instance	Current Practice	PSO			Time (s)	DE			Time (s)
		Best	Average	SD		Best	Average	SD	
35	35.12	20.7	22.09	0.7	166.5	17.49	17.89	0.37	174.06
36	34.74	20.72	21.66	0.55	180.39	16.98	17.91	0.38	185.7
37	30.19	19.34	20.57	0.48	164.51	14.54	16.33	0.95	164.86
38	31.82	18.49	19.36	0.46	155.19	15.56	16.55	0.55	162.95
39	31.39	17.82	18.84	0.48	152.59	14.74	15.28	0.4	155.57
40	29.36	18.68	19.59	0.41	148.9	14.19	14.91	0.47	147.4
41	31.88	18.53	19.21	0.4	157.6	14.53	15.13	0.39	156.06
42	32.20	19.32	20.33	0.44	200.58	15.15	16.23	0.53	203.92
43	30.42	18.77	19.46	0.29	132.47	15.16	15.86	0.4	124.59
44	33.57	18.64	19.31	0.43	163.96	15.72	15.95	0.16	172.84
45	30.40	17.99	18.65	0.42	129.07	15.66	16	0.3	137.2
46	38.38	21.45	22.05	0.41	220.58	16.74	17.48	0.53	216.86
47	28.97	17.31	19.16	0.67	157.94	15.34	15.64	0.29	153.27
48	30.18	19.46	20.37	0.46	190.09	15.69	16.34	0.4	202.63
49	33.70	20.81	21.44	0.27	177.48	16.57	16.95	0.44	201.89
50	31.68	18.19	18.85	0.38	172.02	14.32	14.85	0.52	163.4

To further analyze the effectiveness of the metaheuristics for solving the postman delivery routing problem, the percentage deviation (*PD*) between solutions obtained from the metaheuristics and those from the current practices were calculated using Equation (10). The results for the *PD* are shown in Table 4:

$$PD = ((Sol_{current\_practice} - Sol_{heu})/Sol_{current\_practice}) \times 100, \tag{10}$$

where $Sol_{current\_practice}$ is the solution obtained from the current practice and Sol_{heu} is the solution obtained from a metaheuristic.

Table 4. Experimental results of the percentage deviation (*PD*).

Instance	Current Practice	PSO		DE	
		Best	PD	Best	PD
1	33.64	21.38	36.44	17.64	47.56
2	29.08	18.85	35.18	15.61	46.32
3	30.07	18.97	36.91	15.06	49.92
4	36.91	20.95	43.24	16.19	56.14
5	34.26	19.59	42.82	14.93	56.42
6	33.77	19.17	43.23	15.26	54.81
7	37.93	21.25	43.98	17.24	54.55
8	32.72	20.48	37.41	16.71	48.93
9	27.65	16.74	39.46	12.72	54
10	32.37	20.97	35.22	17.34	46.43
11	37.29	21.24	43.04	16.21	56.53
12	38.23	21.43	43.94	17.24	54.9
13	30.5	19.99	34.46	16.19	46.92
14	32.79	18.98	42.12	14.93	54.47
15	34.78	21.11	39.3	16.37	52.93
16	32.83	20.41	37.83	16.68	49.19
17	31.49	20.99	33.34	16.39	47.95
18	32.70	20.84	36.27	16.04	50.95
19	35.23	19.26	45.33	15.32	56.51
20	36.13	23.08	36.12	18.17	49.71
21	27.44	17.89	34.8	15.11	44.93
22	30.19	18.21	39.68	15.45	48.82
23	28.68	18.26	36.33	16.37	42.92

Table 4. *Cont.*

Instance	Current Practice	PSO		DE	
		Best	*PD*	**Best**	*PD*
24	33.77	20.07	40.57	15.6	53.81
25	34.83	20.93	39.91	16.56	52.45
26	35.37	19.93	43.65	15.91	55.02
27	33.39	20.88	37.47	15.17	54.57
28	38.75	20.96	45.91	16.32	57.88
29	35.04	20.46	41.61	16.47	53
30	30.44	18.84	38.11	14.26	53.15
31	33.03	18.04	45.38	15.94	51.74
32	30.78	19.44	36.84	14.74	52.11
33	37.21	20.29	45.47	18.3	50.82
34	31.78	19.03	40.12	16.03	49.56
35	35.12	20.7	41.06	17.49	50.2
36	34.74	20.72	40.36	16.98	51.12
37	30.19	19.34	35.94	14.54	51.84
38	31.82	18.49	41.89	15.56	51.1
39	31.39	17.82	43.23	14.74	53.04
40	29.36	18.68	36.38	14.19	51.67
41	31.88	18.53	41.88	14.53	54.42
42	32.20	19.32	40	15.15	52.95
43	30.42	18.77	38.3	15.16	50.16
44	33.57	18.64	44.47	15.72	53.17
45	30.40	17.99	40.82	15.66	48.49
46	38.38	21.45	44.11	16.74	56.38
47	28.97	17.31	40.25	15.34	47.05
48	30.18	19.46	35.52	15.69	48.01
49	33.70	20.81	38.25	16.57	50.83
50	31.68	18.19	42.58	14.32	54.8
Average			39.93		51.62

According to the results in Table 3, it is obvious that both the PSO and DE algorithms performed very well in generating solutions with less distance, compared with those obtained from the current practices for all instances, with fast computing times. Both algorithms were able to provide robust solutions with small standard deviations. However, it was observed that DE clearly outperformed PSO, since DE was able to generate the better solution quality for the best and average travel distances in all instances. In Table 4, DE shows its superiority to PSO for all instances by providing higher *PD* values from the current practice solutions. The percentage deviation (*PD*) of the distance ranged from 33.34 to 45.91 with an average of 39.93 for the PSO algorithm and from 42.92 to 57.88 with an average of 51.62 for the DE algorithm.

Moreover, the performance differences between PSO and DE are visually demonstrated in Figure 8, showing box plots of the total travel distance from 10 replicated runs of the PSO and DE algorithms, which used solutions from instances 1, 20, 40 and 50. According to the results in Figure 8, it is clearly seen that the solutions obtained from DE were significantly better than those obtained by PSO.

Lastly, the behavior of the algorithms was investigated in terms of time and space complexity (i.e., how the algorithm runtime and space requirements scaled when the inputs grew). Specifically, the plots of the algorithm runtimes and memory usages versus the number of visited locations (customers plus the post office) were constructed as shown in Figures 9 and 10. The number of locations ranged from 10 to 100, with the distance matrix randomly generated specifically for the complexity examination of those algorithms.

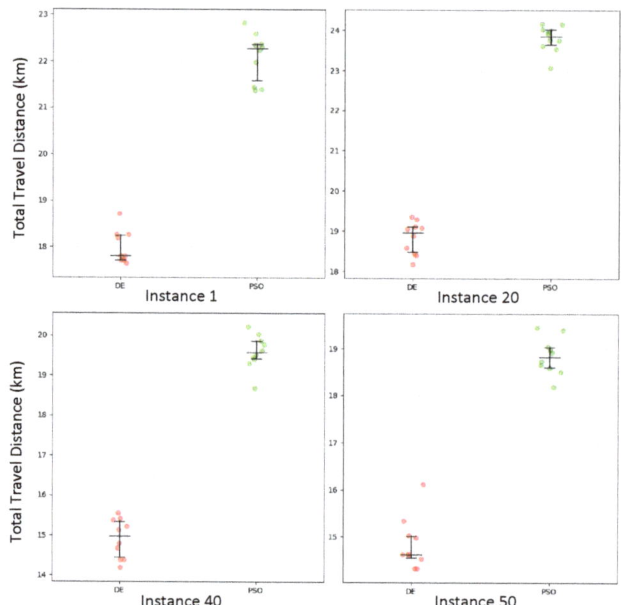

Figure 8. Box plot comparison of the total travel distance (km) from 10 replicated runs between the PSO and DE algorithms (instances 1, 20, 40 and 50).

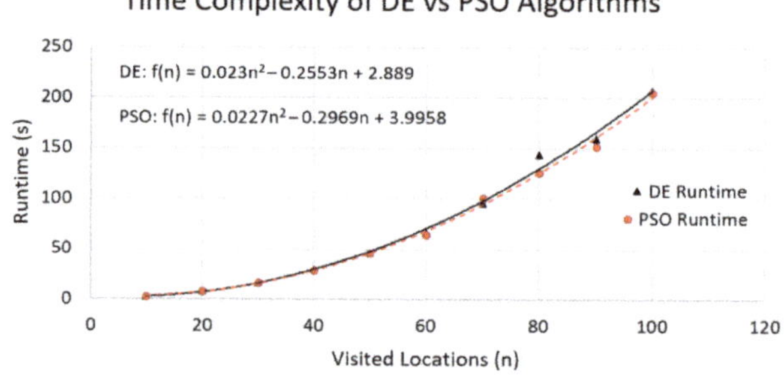

Figure 9. Plot between the algorithm runtimes (DE and PSO) and the visited locations. The quadratic regression fits and their functions are included.

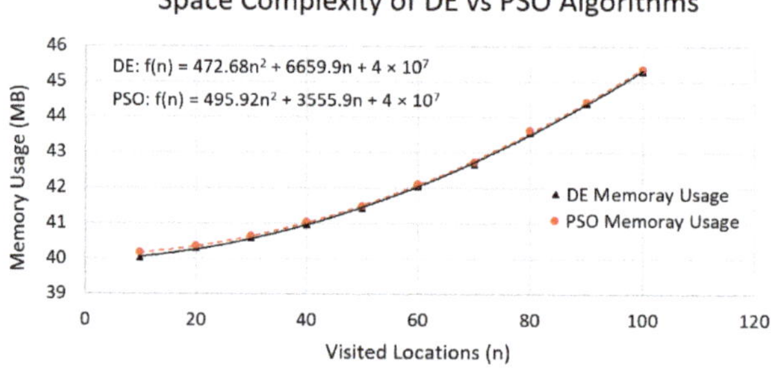

Figure 10. Plot between the algorithm memory usage (DE and PSO) and the visited locations. The quadratic regression fits and their functions are included.

According to Figures 9 and 10, the quadratic regression was used to fit both the time and space complexity plots for DE and PSO. The coefficients of determination (R^2) of the selected regression model in the time complexity plot were 99.43% for DE and 99.62% for PSO, and those in the space complexity plot were 99.96% for DE and 99.97% for PSO. Therefore, the plots clearly demonstrate that the time and space complexity of both algorithms were classified as having an order of n^2 complexity, or $O(n^2)$, since both the runtime and space requirements had quadratic relationships with the input sizes.

5. Discussion and Conclusions

This research employed metaheuristic algorithms, namely the PSO and DE algorithms, to solve the single-objective VRP for optimized routing solutions. The objective was to determine a solution with the lowest travel distance. The study used actual operational data from two delivery vehicles of the Chiang Rai post office over a period of 50 days. The data included geographic coordinates of the customer locations where those vehicles delivered parcels, which were then transformed into distance matrices for use by the optimization algorithms. For each delivery day, the algorithms tried to search and assign new routes to those two vehicles such that the total travel distance was minimized. The routing performances from the PSO and DE algorithms, as well as those from actual operations, were compared. The experimental results showed that both metaheuristic algorithms were far superior to the real delivery operations of the post office in all 50 days of the experiment. However, DE clearly outperformed PSO by providing outstanding solution quality for all instances. This result was mainly due to the different evolution mechanisms of DE and PSO. A PSO swarm has a high tendency to cluster rapidly and becomes stagnant, since the movement of a swarm is highly influenced by the global best solution. In addition, the diversification of DE is better than PSO, because the best solution in the population does not exert any influence on other solutions in the population. A new solution in DE would be selected only if it had better fitness, and thus the average fitness of the population would be equal or better from iteration to iteration. It is important to note that the results obtained from this study were based on the delivery constraints of the Chiang Rai post office. However, for the operations of other post offices, other constraints such as vehicle capacity, a customer's time window and the delivery time horizon could be taken into consideration. In such cases, the problem must be remodeled, and new solutions will be obtained under those circumstances.

Another essential factor was the duration for running one optimization cycle in order to obtain the optimized solution. For both the DE and PSO algorithms, the average duration was approximately six minutes. Combining the decent results and acceptable runtime duration, the methods used in this study have the potential to be employed in real post office operations. Additionally, even though there were two vehicles in this study, the proposed algorithms can also be applied to large cities in which the post office operates several delivery vehicles. In essence, the methods proposed in this study could drastically help reduce the delivery cost and time for postman delivery operations. Additionally, the vehicle routing optimization process (with about 100 customer locations) only takes minutes rather than hours, which is practical in the actual post office work environment where all parcels have to be processed and dispatched quickly.

There are some challenges to be considered if these methods are to be used in actual delivery operations. First, the post office operators need to determine a quick approach to convert customer addresses into geographic coordinates, which should be done systematically and automatically. Second, the duration to run an optimization cycle is proportional to the number of addresses and delivery vehicles. Therefore, the optimization parameters such as the population sizes and iterations should be well adjusted according to those numbers. Thus, the operators could utilize machine learning, artificial intelligence and image processing methods to quickly transform those customer addresses into definite GPS locations. They should also give enough lead time to account for the unexpected increase in processing time before all deliveries can start. In addition, the proposed PSO and DE use deterministic data, in which the customers and distance matrices are known in advance, and the routing is determined using these data. When some data are changed during delivery, the only way to adapt to such change is to rerun the algorithms. Thus, another important challenge is to develop an adaptive algorithm to respond to changes effectively.

In future studies, other metaheuristic algorithms, such as the adaptive large neighborhood search and genetic programming algorithms, are planned to be tested for their performance with datasets obtained from larger delivery operations, including not only postman delivery problems but also problems from other industries.

Author Contributions: Conceptualization, W.W. and N.W.; methodology, W.W.; software, A.B. and N.W.; validation, A.B.; formal analysis, W.W., K.P. and N.W.; investigation, W.W. and N.W.; resources, N.W.; data curation, C.S. and N.W.; writing—original draft preparation, W.W., K.P., C.S., A.B. and N.W.; writing—review and editing, W.W., K.P., C.S., A.B. and N.W.; visualization, W.W., C.S. and N.W.; supervision, N.W.; project administration, N.W.; funding acquisition, N.W. All authors have read and agreed to the published version of the manuscript.

Funding: This research was funded by Mae Fah Luang University.

Institutional Review Board Statement: Not applicable.

Informed Consent Statement: Not applicable.

Data Availability Statement: Data used in this study are available upon reasonable request to the corresponding author.

Acknowledgments: The authors are grateful to the Chiang Rai post office. The authors would like to acknowledge Mae Fah Luang University and Chiang Mai University for a partial financial sponsorship and their facility support in this research.

Conflicts of Interest: The authors declare no conflict of interest.

References

1. Case, M.J.; Johnson, G.G.; Bartowitz, K.J.; Hudiburg, T.W. Forests of the future: Climate change impacts and implications for carbon storage in the Pacific Northwest, USA. *For. Ecol. Manag.* **2021**, *482*, 118886. [CrossRef]
2. Our World in Data. Available online: http://ourworldindata.org/co2-emissions-from-transport (accessed on 21 January 2021).
3. Skerlic, S.; Muha, R. The importance of systems for controlling logistics costs in the supply chain: A case study from the Slovenian automotive industry. *Promet Traffic Traffico.* **2016**, *26*, 299–310. [CrossRef]
4. United States Postal Service, Office of Inspector General. Available online: http://uspsoig.gov/document/postal-service-dynamic-route-optimization-and-cost-savings-initiative (accessed on 21 January 2021).
5. Dantzig, G.B.; Ramser, J.H. The truck dispatching problem. *Manag. Sci.* **1959**, *6*, 80–91. [CrossRef]
6. Guzairov, M.B.; Yusupova, N.I.; Smetanina, O.N.; Rassadnikova, E.Y. Models and algorithms for the vehicle routing problem with time windos and other conditions. In Proceedings of the 2016 13th International Scientific-Technical Conference on Actual Problems of Electronics Instrument Engineering (APEIE), Novosibirsk, Russia, 3–6 October 2016; pp. 412–416.
7. Wang, Z.; Ren, X.; Ji, Z.; Huang, W.; Wu, T. A novel bio-heuristic computing algorithm to solve the capacitated vehicle routing problem based on Adleman-Lipton model. *BioSystems* **2019**, *184*, 103997. [CrossRef] [PubMed]
8. Niu, Y.; Zhang, Y.; Cao, Z.; Gao, K.; Xiao, J.; Song, W.; Zhang, F. MIMOA: A membrange-inspired multi-objective algorithm for green vehicle routing problem with stochastic demands. *Swarm Evol. Comput.* **2021**, *60*, 100767. [CrossRef]
9. Dumez, D.; Lehuede, F.; Peton, O. A large neighborhood search approach to the vehicle routing problem with delivery options. *Transp. Res. B* **2021**, *104*, 103–132. [CrossRef]
10. Baldacci, R.; Mingozzi, A. Lower bounds and an exact method for the capacitated vehicle routing problem. In Proceedings of the 2006 International Conference on Service System and Service Management, Troyes, France, 25–27 October 2006; pp. 1536–1540.
11. Ropke, S. Heurisic and Exact algorithms for Vehicle Routing Problems. Ph.D. Thesis, University of Copenhagen, København, Denmark, 2006.
12. Ladner, R.E. On the structure of polynomial time reducibility. *J. ACM* **1975**, *22*, 151–171. [CrossRef]
13. Tordecilla, R.D.; Copado-Mendez, P.J.; Panadero, J.; Quintero-Araujo, C.L.; Montoya-Torres, J.R.; Juan, A.A. Combining heuristics with simulation and fuzzy logic to solve a flexible-size location routing problem under uncertainty. *Algorithms* **2021**, *14*, 45. [CrossRef]
14. Meira, L.A.A.; Martins, P.S.; Menzori, M.; Zeni, G.A. Multi-objective vehicle routing problem applied to large scale office. *arXiv* **2018**, arXiv:1801.00712.

15. Niroomand, I.; Khataie, A.H.; Galankashi, M.R. Vehicle routing with time window for regional network services—Practical modelling approach. In Proceedings of the 2014 IEEE International Conference on Industrial Engineering and Engineering Management, Bandar Sunway, Malaysia, 9–12 December 2014. [CrossRef]
16. Nasri, M.; Hafidi, I.; Metrane, A. Multithreading parallel robust approach for the VRPTW with uncertain service and travel times. *Symmetry* **2021**, *13*, 36. [CrossRef]
17. Zhou, Z.; Ha, M.; Hu, H.; Ma, H. Half open multi-depot heterogeneous vehicle routing problem for hazardous materials transportation. *Sustainability* **2021**, *13*, 1262. [CrossRef]
18. Erbao, C.; Mingyong, L.; Kai, N. A Differential Evolution & Genetic Algorithm for Vehicle Routing Problem with Simultaneous Delivery and Pick-up and Time Windows. In Proceedings of the 17th World Congress the International Federation of Automatic Control, Seoul, Korea, 6–11 July 2008.
19. Kromer, P.; Abraham, A.; Snasel, V.; Berhan, E.; Kitaw, D. On the differential evolution for vehicle routing problem. In Proceedings of the 2013 International Conference on Soft Computing and Pattern Recognition (SoCPaR), Hanoi, Vietnam, 15–18 December 2013; pp. 384–389.
20. Sethanan, K.; Jamrus, T. Hybrid differential evolution algorithm and genetic operator for multi-trip vehicle routing problem with backhauls and heterogeneous fleet in the beverage logistics industry. *Comput. Ind. Eng.* **2020**, *146*. [CrossRef]
21. Xing, L.N.; Liu, Y.; Li, H.; Wu, C.; Lin, W.C.; Song, W. A hybrid discrete differential evolution algorithm to solve the split delivery vehicle routing problem. *IEEE Access* **2020**, *8*, 207962–207972. [CrossRef]
22. Song, L.; Dong, Y. An improved differential evolution algorithm with local search for capacitated vehicle routing problem. In Proceedings of the 2018 Tenth International Conference on Advanced Computational Intelligence (ICACI), Xiamen, China, 29–31 March 2008.
23. Lei, J.; Li, J. Solving capacitated vehicle routing problems by modified differential evolution. In Proceedings of the 2010 2nd International Asia Conference on Informatics in Control, Automation and Robotics (CAR 2010), Wuhan, China, 6–7 March 2010; pp. 513–516.
24. Moghaddam, B.; Ruiz, R.; Sadjadi, S.J. Vehicle routing problem with uncertain demands: An advanced particle swarm algorithm. *Comput. Ind. Eng.* **2012**, *62*, 306–317. [CrossRef]
25. Peng, Z.; Manier, H.; Manier, M.A. Particle swarm optimization for capacitated location-routing problem. *Comput. Ind. Eng.* **2020**, *146*. [CrossRef]
26. Marinakis, Y.; Marinaki, M.; Migdalas, A. A multi-adaptive particle swarm optimization for the vehicle routing problem with time windows. *Inf. Sci.* **2019**, *481*, 311–329. [CrossRef]
27. Chen, J.; Shi, J. A multi-compartment vehicle routing problem with time windows for urban distribution—A comparison study on particle swarm optimization algorithms. *Comput. Ind. Eng.* **2019**, *133*, 95–106. [CrossRef]
28. Chen, R.M.; Shen, Y.M.; Hong, W.Z. Neural-like encoding particle swarm optimization for periodic vehicle routing problems. *Expert Syst. Appl.* **2019**, *138*, 112833. [CrossRef]
29. Sedighizadeh, D.; Mazaheripour, H. Optimization of multi objective vehicle routing problem using a new hybrid algorithm based on particle swarm optimization and artificial bee colony algorithm considering Precedence constraints. *Alex. Eng. J.* **2018**, *57*, 2225–2239. [CrossRef]
30. Yanwei, Z.; Bin, W.; Wanliang, W.; Jingling, Z. Particle swarm optimization for open vehicle routing problem with time dependent travel time. In Proceedings of the 17th World Congress the International Federation of Automatic Control, Seoul, Korea, 6–11 July 2008.
31. Google Maps Platform, Geo-Location APIs. Available online: http://cloud.google.com/maps-platform (accessed on 15 May 2019).
32. Kachitvichyanukul, V. Comparison of three evolutionary algorithms: GA, PSO, and DE. *Ind. Eng. Manag. Syst.* **2012**, *11*, 215–223. [CrossRef]
33. Rossum, G.V. Python tutorial, Technical Report CS-R9526, Centrum voor Wiskunde en Informatica (CWI). Amsterdam, May 1995. Available online: https//www.python.org (accessed on 9 April 2019).
34. PyCharm, the Python IDE for Professional Developers. Available online: http://jetbrains.com/pycharm (accessed on 5 November 2020).

*applied
sciences*

Article

Multi-Criteria Optimization in Operations Scheduling Applying Selected Priority Rules

Zuzana Červeňanská, Pavel Važan, Martin Juhás * and Bohuslava Juhásová

Institute of Applied Informatics, Automation and Mechatronics, Faculty of Material Science and Technology in Trnava, Slovak University of Technology in Bratislava, 91724 Trnava, Slovakia; zuzana.cervenanska@stuba.sk (Z.Č.); pavel.vazan@stuba.sk (P.V.); bohuslava.juhasova@stuba.sk (B.J.)
* Correspondence: martin_juhas@stuba.sk

Abstract: The utilization of a specific priority rule in scheduling operations in flexible job shop systems strongly influences production goals. In a context of production control in real practice, production performance indicators are evaluated always en bloc. This paper addresses the multi-criteria evaluating five selected conflicting production objectives via scalar simulation-based optimization related to applied priority rule. It is connected to the discrete-event simulation model of a flexible job shop system with partially interchangeable workplaces, and it investigates the impact of three selected priority rules—FIFO (First In First Out), EDD (Earliest Due Date), and STR (Slack Time Remaining). In the definition of the multi-criteria objective function, two scalarization methods—Weighted Sum Method and Weighted Product Method—are employed in the optimization model. According to the observations, EDD and STR priority rules outperformed the FIFO rule regardless of the type of applied multi-criteria method for the investigated flexible job shop system. The results of the optimization experiments also indicate that the evaluation via applying multi-criteria optimization is relevant for identifying effective solutions in the design space when the specific priority rule is applied in the scheduling operations.

Keywords: multi-criteria optimization; simulation optimization; production control; multiple flexible job shop scheduling; priority rules

Citation: Červeňanská, Z.; Važan, P.; Juhás, M.; Juhásová, B. Multi-Criteria Optimization in Operations Scheduling Applying Selected Priority Rules. *Appl. Sci.* **2021**, *11*, 2783. https://doi.org/10.3390/app11062783

Academic Editor: Farouk Yalaoui

Received: 5 February 2021
Accepted: 17 March 2021
Published: 19 March 2021

Publisher's Note: MDPI stays neutral with regard to jurisdictional claims in published maps and institutional affiliations.

1. Introduction

In a competitive environment of manufacturing, production control, including the scheduling operations via modules associated with a variety of management tools, must reflect both customer's demands, and, on the other hand, the production capability, sustainability, and profit. The effective scheduling in a sense of achievement of a high production performance leads to generating benefits in the form of shortening processing time and reduction of costs. To ensure the changing requirements effectively and to generate suitable production schedules, implementation of heuristics, simulation, analytical models [1,2], artificial intelligence techniques [3], or dispatching rules, especially in case of dynamic scheduling [4,5], can be employed in this type of decision-making.

In the production sector, scheduling can be in a wider perspective defined as "a process of arranging, controlling, and optimizing work or workloads" [1]. With regard to finding the optimal schedule for a specific structure and production system conditions, the scheduling is considered as a complex combinatorial optimization problem [4], mostly proved of NP-hard type [1]. Exact optimization methods are mainly used only for the systems which have a specific topology where very strong simplifying assumptions must be used [6], so they are not too applicable in a real-world scenario for more complex systems. In that case, approximate optimization methods and metaheuristics [7] based on stochastic local search approach, machine learning techniques, especially artificial neural networks (ANN), fuzzy logic methods, and expert systems, are at the center of research interest to find optimal or near-optimal solutions [8,9] instead of exact mathematical optimization models.

145

In contrast to other methods, dispatching rules (we also use the term priority rules throughout the following text) represent the valuable practical and dominant approach of the shop floor control in the complex industry environment [10,11], such as, e.g., in semiconductor manufacturing for solving complex scheduling problems in real-time [12]. Priority rules are popular because they are characterized by the simplicity of implementation, satisfactory performance, and a substantially reduced computational requirement [2,11,13]. Nevertheless, the choice of appropriate dispatching rules is not a trivial task and depends on the relevant key performance indicators [11].

When considering the application of specific priority rule in production control, a question appears on how the selection of priority rule influences the production performance objectives. Years of research on the field of priority rules have shown that in general, no rule outperforms the others, whereas there are different system configurations, operating conditions, and production performance indicators [14,15]. This is understandable, considering that the rules are developed for a specific category of system configurations involving a specific set of performance objectives, therefore they generally do not work well in an environment where they were not intended [15].

The impacts generated by the selected priority rule are mostly difficult to explain by analytical methods [1], thus the simulation is used very often to evaluate the schedule efficiency in the complex scheduling problem [10,16]. As an illustration, in the simulation study, Vinod and Sridharan [13] evaluated the performance measures based on flow time and tardiness of jobs for the different combinations of due-date assignment methods and seven scheduling decision rules applied in a dynamic job shop system. Xanthopoulos et al. [5] compared seventeen dispatching rules in the study focused on stochastic dynamic scheduling problems with sequence-dependent setups. Performance measures were mean work-in-progress (WIP), mean cycle time, mean tardiness, and a fraction of tardy jobs.

In opposition to the simulation, the optimization approach offers to find the optimal or near-optimal scheduling concerning one or more desired objectives. Specifically, simulation-based optimization uses metaheuristics to determine nearly optimal parameter configurations for simulations, and it can solve even complex stochastic combinatorial optimization problems. Simulation-based optimization with genetic algorithm applyication has been successfully applied in the work of Vieira et al. [17]. Freitag and Hildebrandt [18] applied simulation-based optimization employing multi-objective genetic programming to solve the complex problem of automated design of scheduling rules for short-term scheduling in a semiconductor manufacturing environment. The authors investigated seven standard rules: FIFO (First In First Out), ERD (Earliest Release Date), EDD (Earliest Due Date), ODD (Operation Due Date), MOD (Modified Operation Due Date), SPT (Shortest Processing Time), and CR (Critical Ratio). Additionally, the ODD+ rule was used. Kuck et al. [11] proposed an adaptive simulation-based optimization approach to select suitable dispatching rules including FIFO, EDD, FASFS (First Arrival in the System First Served), CR, ODD, SPT, and MOD rules for each machine in the production system with different strategies to react to system changes. Solutions obtained via genetic algorithm were evaluated by applying a discrete-event simulation model.

In the related literature, research performed to investigate the impact of priority rules on production results in job shop systems has focused mostly on one or two optimization objectives, such as tardiness, makespan (completion time), mean flowtime, or proportion of tardy jobs, with one priority rule applied to all the machines on the shop floor. Zahmani et al. [16] presented a simulation model for makespan minimization using different priority rules for each machine on the shop floor. Obtained experimental results served for learning base construction, which was utilized for developing an inference model for the selection of the most suitable priority rule. In the work of Demir and Erden [19], Genetic Algorithm (GA) and Ant Colony Optimization (ACO) were used together to obtain the best combination of dispatching rule, due date assignment rule, and route of all jobs to minimize earliness, tardiness, and due dates of each job.

We can find a lot of literature that mainly tackles single and bi-criteria job shop scheduling problems [20], and more objectives are taken into account rather exceptionally. The reason is that when reflecting many aspects of production evaluation concurrently, multi-criteria optimization is needs to be considered to obtain satisfying results for production parameters setting. An identification of optimal solution is not trivial in this case because there is no single solution, but a set of possible optimal solutions, all equivalent in a mathematical sense. The individual solutions in this set form the so-called Pareto front (or Pareto frontier) in the objective space [21]. Methods for finding optimal solutions to the multi-criteria problem are still in progress nowadays, concerning an improvement in finding very accurate and highly diverse solutions for an approximation of true Pareto front to offer design-makers a diverse range of design options [22].

In this study, based on preliminary experiments with single-objective optimization performed for a flexible job shop production system with partially interchangeable work-stations, we focused on solving the multi-objective problem to evaluate and compare production results with respect to the priority rule applied for operations scheduling. The simulation-based optimization of five contradictory production objectives: Average flow-time, average machine utilization, average work in progress, average number of products, and an average costs per part unit is connected to the discrete-event simulation model of one-piece flow type when applying three specific well-known and often studied priority rules [5]. The rules are FIFO, which considers the order of incoming parts, EDD, which selects an operation of a part with the closest due date, and STR (Slack Time Remaining), where an operation of a part with the lowest time reserve is favored. We choose the FIFO priority rule as the most common one and simplest to implement. The reason for selection of another standard and very known EDD and STR scheduling principles was that they outperform other rules in a shortening flow time. It is important for the production of the type of make-to-order typical for flexible job shops and a lean production concept. Additionally, EDD and STR were involved in this investigation because they are widely used in the literature as benchmarks for metrics related to due date [5,23]. Two types of scalarization methods (Weighted Sum Method and Weighted Product Method) in a priori arrangement are applied for the definition of multi-objective function in the optimization model of the production system.

Compared with other related works, the main contribution of this study is considering more interesting conflicting performance production indicators of various types to offer the complex view to production results via multi-criteria evaluating many objectives simultaneously when employing a selected priority rule in scheduling operations. In addition, based on results of scalar multi-criteria simulation optimization, the global effective solutions can be recommended as support for decision-making in production control. Additionally, when solving a multi-criteria optimization problem (MOP), we have experimented with a modified approach in the transformation of individual objective functions in a scalar multi-criteria objective function comparing to works in the related literature. In this context, the most relevant publication to this work is the paper by Marler and Arora [24] dealing with practical aspects of multi-criteria methods usage and transformations for scalar multi-criteria objective function that inspired us to employ them in this work.

The main structure of the paper is as follows. Firstly, we briefly describe the problem of scheduling operations when priority rules are applied in the scheduling process. In the next part, a concept of Pareto-optimality in multi-criteria optimization and applied methods are mentioned. The part concerning the methodology of experiments introduces the simulation and optimization model and the design of optimization experiments. Finally, in the last section, we present and compare the obtained results focusing on the practical aspect of production control.

2. Scheduling Operations Based on Priority/Dispatching Rules

In job shop production systems, a shop floor control level makes decisions about all production and material handling resources including order release strategies, batch sizes, dispatching rules, and schemes related to preventive maintenance [4]. It ensures the generating production schedule that specifies the decision, at what time, and on what machine must start and end each job needed for manufacturing to optimize desired production objective/s.

Scheduling can be defined as the planning process that "deals with the allocation of resources to tasks over given periods" intending to optimize one or more performance measures [1]. Scheduling accomplishes two succeeding phases—loading and sequencing. The loading phase ensures an allocating of released jobs among workplaces. Sequencing specifies an order for jobs in the queue for the individual machine respecting the assigning in the loading phase. Grundstein et al. [25] differentiate sequencing into dispatching, which means an assignment of an order to a station, and queue processing, i.e., the sequence in which orders in a queue in front of a station are processed.

2.1. Scheduling in the Flexible Job Shop Systems

Flexible job shop (FJS) scheduling addresses problems of specific types of job shop production systems and faced to more complexity and flexibility comparing to job shop system scheduling. Generally, in job shop systems of each machine can perform only one operation for each job which is processed on machines with a given operation time. Flexible manufacturing job shop systems represent a specific more complex type of the job shop systems [20]. In FJS, the operation can be performed on any machine selected from a set of available identical machines with the flexible operation time depending on the processed product and the selected machine. To define deterministic FJS more precisely, we demand a set of N jobs to be processed on a set of M work centers. Within each job j, there is a set of operations that need to be processed in a specific order. Each operation can be processed by any identical machine of a given set. The processing time of the operation is explicitly known, and it may vary, depending on the machine on which it is running. The machine can only perform one operation at a time. The assignment of operation to the machine is given to the occupancy of the machine and the relevance of the machine to carry out the operation [7,26], so only some number of routings are possible for individual product.

Solving the flexible job shop scheduling problem means finding a schedule to ensure the utilization of sources optimally following production demands. Optimization techniques for generating a schedule performed in several main modes of scheduling—a forward scheduling mode, a backward scheduling mode, and a multi-pass mode [1]. The approach of the forward scheduling means that product processing starts in the first possible time, and jobs are assigned to machines towards the future [27]. Namely, this approach is built-in in algorithms of computer simulation methods that are employed to schedule production tasks. In this context, heuristic procedures, nature-inspired methods, neural networks, and utilizing priority dispatching rules are the most applied methods to solve the optimization problem in scheduling [1]. The most common metaheuristic procedures, such as Simulating Annealing [6], Tabu search [12], Random Search, or many types of nature-inspired optimization methods, such as evolutionary [7,28,29] and genetic algorithms [18], ant colony optimization [30], or particle swarm optimization [16,19,31], can provide acceptable solutions near the optimum. Although metaheuristics are advantageous in the sense of the solution quality and robustness, they do not guarantee solution optimality [1,27], they depend on the choice of method's parameters [11], and they are often too complex to implement computationally in a real-time system [2]. To enhance the efficiency of metaheuristic, a hybrid approach based on combining different methods has been often proposed by several authors in their works [9,19,32,33].

Objectives related to the scheduling problems can be categorized into three groups—objectives based on the completion time, due date, and on inventory and utilization costs [34]. According to an extensive survey [34], flowtime and makespan were the most frequent and significant objectives considered in practice because of the minimization of makespan increases the production system efficiency and minimization of flowtime ensures a stable resources utilization and minimization of work-in-progress. The optimization of these two objectives simultaneously reduces the production costs [34].

To include more production aspects, scheduling based on multi-criteria optimization in job shop systems has been at the center of interest for a long time [9,14]. According to related literature, the frequently studied performance objectives in this area were, e.g., mean flow time, the maximum makespan (completion time), the mean weighted completion time, the mean weighted tardiness, the maximum tardiness, the mean weighted earliness, the maximum earliness, maximum lateness, the ratio of tardy jobs, or mean WIP (Work In Progress) [20]. In real practice, the consideration of several performance goals is challenging from the control aspect because it demands either an a priori decision about the hierarchy of objectives or a responsive adjustment to varying conditions on the shop floor [23].

2.2. Priority/Dispatching Rules

In a flexible job shop scheduling area, dispatching rules are often applied as part of the operational control of a manufacturing system. Whenever a machine becomes status idle and some jobs are waiting and ready to process, the dispatching rule gives the priority to each job and determines the job with the highest priority to be started [18]. Results of optimization experiments in studies carried out in job shop systems have demonstrated that the schedule generation based on priority dispatching rules is a practicable method of finding good solutions, although without the optimality guarantee [27].

Pinedo [1] gives an overview of the most common and better-known elementary dispatching rules, together with the type of needed data and an environment where rules are suitable to apply. The main classification of the rules concerning their time dependence to static and dynamic and according to the system information to local and global is stated, too. Additionally, Baker and Trietsch [27] provide a comprehensive overview of dispatching rules and a comparison of their performance according to different performance measures. Similarly, for an excellent review of priority rules with an emphasis on the classification and identification of dominant rules we refer to [23]. Three selected rules FIFO, EDD, and STR which have been found very often in use [23], and they are relevant to this work are described below.

FIFO or FCFS (First Come First Served) rule is a simple planning method that is frequently favored because is easy to implement, and no specific information and calculation are needed. It is a random priority rule [35], the first job arriving at a work center is manufactured first and the order of the jobs determined by the pass through the first machine maintains throughout the system. After completion of the job at one machine, it leaves the queue at the next machine. Typically, all queues in the system are supposed to operate on the base rule, therefore, common is an issue with an occurrence of blocking, mainly in some types of assembly lines due to limited buffers. Although the FIFO rule is considered to not be very profitable for effective scheduling, it can reach slightly better results than other rules under some circumstances. In the investigation, Savasci and Herrmann [36] focused on automotive logistics in a real production environment of job shop production with replacing stations. They compared fifteen priority rules for delivery reliability and tardiness. They showed that seven of the rules provided no better solutions than the FIFO rule already used, whereas others improve the performance.

The EDD rule also belongs to the group of static rules which generate a schedule taking due date information into account. It can be applied in case, individual jobs have different due dates, and it belongs to one of the most used rules in practice [23,35]. It selects the operation associated with the job that has the earliest due date and uses the target due date as a priority key. The operation with the earliest target completion time receives the highest priority and is processed first. A list containing all available jobs ordered to non-decreasing due dates determines the sequence assigned to the machines for all available jobs [37]. Therefore, to use this rule, the information about the due dates of the jobs must be known in an ordinal sense, but we do not have to know job processing times. This rule and other similar based on due dates information are important in the customer-oriented complex job shop manufacturing environment. If production works based on "make-to-order", the due dates are agreed upon and must be kept. In this case, the due date adherence is the most significant production objective [25]. The EDD rule provides outstanding performance when there are sufficient production capacity, light load, and smaller shop configurations [23]. This coordinate mechanism is mainly useful when the goal is to reduce job lateness, minimize flow time and it also guarantees minimization of maximum and variance tardiness [23].

STR, MST (Minimum Slack Time) in [27], or MS (Minimum Slack) in [1] is a static rule concerning a specific operation, but a dynamic one with respect to a specific job [27], since the same job has assigned the different relative priorities to its individual operations. It chooses the waiting operation related to the job with minimum slack time. Slack time s_j of given job j with due date d_j and processing time p_j at time t is calculated by the expression (1):

$$s_j(t) = d_j - p_j - t \tag{1}$$

that represents the difference between the due date and the earliest possible finish time of the job [27]. The EDD rule or the STR rule ensures the optimal production performance when all due dates are sufficiently loose and spread out [1,23].

3. Multi-Criteria Optimization in Scheduling

Multiple conflicting objectives are a natural feature of most production scheduling environments. Therefore, at first, let us briefly introduce the multi-criteria optimization problem (MOP) and a concept of identification of MOP solutions based on the notion of Pareto-optimality and Pareto dominance.

3.1. Multi-Criteria Optimization Problem

In general, we consider a MOP (2) with n decision variables and k objectives ($k > 1$). Solving MOP is a procedure of simultaneous optimization of a set of k individual objective functions $Fi(x)$ subjected to constraints given by the system of inequalities and equalities. It can be represented in the form (2):

$$
\begin{aligned}
&\text{Minimize} \\
&x \in D \quad F(x) = [F_1(x),\ F_2(x),\ \ldots,\ F_k(x)]^T;\ k > 1 \\
&\text{subject to} \quad g_s(x) \leq 0, s = 1, 2, \ldots,\ m;\ m \geq 0, \\
&\qquad\qquad\quad h_l(x) = 0, l = 1, 2, \ldots,\ r;\ r \geq 0.
\end{aligned}
\tag{2}
$$

Vector x is the decision vector in the feasible region D in n-dimensional design (decision) space, constrained by inequalities $x_L \leq x \leq x_U$. In case of a combinatorial problem, the D region is discrete. Vector $F(x) \in E_k$ consists of k objective functions $F_i(x)$: $E_n \rightarrow E_1$ in the feasible objective (criterion) space Z, $Z \subset E_k$. As a result of solving MOP, if the objectives $F_i(x)$ are conflicting, one cannot be able to find one single solution as the best one, rather a set of trade-offs, all with a compromising nature. Due to the absence of an absolute ordering in the vector space, the way how to identify MOP solution is via the preference relationship of partial preorder based on Pareto relation. It is built on the following associated definitions:

Pareto dominance: A decision vector x ($x \in D$) (Pareto) dominates a decision vector y (denoted $x \prec y$) if and only if $F_i(x) \leq F_i(y)\ \forall\ i \in \{1, 2, \ldots, k\} \land F_i(x) < F_i(y)$ for at least one $i \in \{1, 2, \ldots, k\}$, $k \geq 2$.

Pareto optimality: A solution $x \in D$ is Pareto optimal if does not exists any other $x' \in D$ that dominates x [34].

The set of Pareto optimal solutions is called the Pareto optimal set. Its image in the objective space is referred to as the Pareto optimal front [38]. Pareto optima correspond to "best trade-off' solutions between different conflicting criteria" [39]. From a geometric perspective, the Pareto front represents a hypersurface in k-dimensional space of objective functions, where all non-dominated solutions are located.

The optimal solution of MOP can be also characterized by terms of efficient solution and non-dominated solution according to space where the MOP solutions are located. Typically, efficiency refers to a vector of design variables in the design space, whereas dominance refers to a vector of functions in objective space [24]. These basic terms are explained by the following definitions, adopted from Marler and Arora [24]:

A feasible solution x^* situated in design (decision) space D is efficient if and only if there does not exist any other $x \in D$ such that $F(x) \le F(x^*)$ with at least one $F_i(x) < F_i(x^*)$. With relation to the objective space, a corresponding vector $F(x^*)$ in the objective space is non-dominated if and only if there does not exist another vector $F(x) \in Z$ such that $F(x) \le F(x^*)$ with at least one $F_i(x) < F_i(x^*)$. That means all non-dominated solutions dominate the others, but do not dominate themselves.

Both introduced definitions represent a similar idea and "a recipe" for how to identify Pareto-optimal solutions via a vector´s coordinates comparison. We will use the term Pareto-optimal solution in the sense of non-dominated solution yielded by solving MOP throughout the following text. For an excellent survey of multicriteria scheduling, we refer to the book by T'Kindt and Billaut [39]. Additionally, Hoogeveen [40] provided a comprehensive introduction and survey of research results in multicriteria scheduling.

In the context of scheduling problems with multiple objectives, the concept of Pareto-optimality acts as commonly accepted. Pinedo [1] states that "a schedule is Pareto-optimal if it is impossible to improve on one of the objectives without making at least one other objective worse". The idea of Pareto-optimality in MOP solving implies the existence of the set of trade-off solutions instead of one single solution, therefore an expert in a position of decision maker (DM) is needed to support the identification of the most preferred solution of all available. If a priori method of MOP solving is used, DM expresses the preferences at the start of a solving process. In the case of the a posteriori method, the selection of MOP solution is generated after finding a set of optimal solutions. An interactive method is employing DM preferences directly into the running process of solving.

Within the classification of the a priori methods, several approaches are developed to deal with multi-objective scheduling problems in a priori arrangement. To mention some of them, in a hierarchical approach, the objectives are ranked in priority order and optimized in this order. In goal programming, all the objectives are considered as constraints expressing satisficing levels, and the aim is to obtain a solution that offers a value as close as possible to the desired goal for each objective. One objective can also be selected as the main one and then an optimized subject to other objectives [14]. The utility approach is based on an aggregation of multiple optimization objectives into one single utility or weighting function. The utility function very frequently employs a weighted linear combination of the objectives, and it represents the composite objective function in the most popular Weighted Sum Method (WSM). Marler and Arora [24] present other methods from the utility approach category, such as the more general weighted global criterion method, or weighted min-max method, weighted product method, and exponential weighted criterion method.

3.2. Applied Multi-Criteria Optimization Methods

In this study, the utility approach was employed to compare production system performance under different priority rules via WSM and Weighted Product Method (WPM).

WSM belongs to the most common scalarization procedure in the literature. It is a special type of global criterion method, expressed by the weighted sum of k individual objectives F_i (3):

$$U = \sum_{i=1}^{k} w_i.F_i \,; w_i > 0 \,; \sum_{i=1}^{k} w_i = 1 \tag{3}$$

where weights w_i as method parameters are used to model preferences. Assuming the vector w is fixed, then minimizing U function (3) is a necessary and sufficient condition for Pareto optimality under convex constraint conditions (Geoffrion's theorem [39]). This method suffers from a well-known disadvantage related to the geometric point of view. Since the scalar MOO function is generated as a linear combination of individual objective functions, the ability to yield the Pareto optimal solutions is restricted to the convex part of Pareto front only. Despite it, this method is rather popular in the application in a job shop scheduling area. To mention works that solved multicriteria problems applying WSM or via sum approach, e.g., Loukil et al. [14] published an extensive study of model types of production systems with seven objective functions studied alone or in different mutual combinations, and compared to the results in the literature. To perform multicriteria optimization, they used the multi-objective simulated annealing procedure for solving a given combinatorial optimization problem. Adibi et al. [9] studied dynamic job shop scheduling problems considering makespan and tardiness simultaneously when employing a variable neighborhood search method and ANN for finding optimal rescheduling. The work of Huang and Yu [41] addressed a multi-objective job-shop scheduling problem with equal-size lot splitting. It was solved via hybrid ant colony optimization-based algorithms. In the paper, the authors minimized scalar objective function generated in the form of the weighted sum of makespan, tardiness, and lot-splitting cost with an accent to issue related to lot-splitting of jobs and tradeoff between lot-splitting costs and makespan. To compare the effect of the first employed multi-objective method, the Weighted Product Method (WPM) was selected. The utility function U based on the multiplication of k objectives F_i was expressed in the form (4).

$$U = \prod_{i=1}^{k} F_i^{w_i} \, ; w_i > 0 \, ; \sum_{i=1}^{k} w_i = 1 \tag{4}$$

Weights w_i represent the relative importance of the individual objectives [24]. WPM is known mostly as a component of decision making methods in the Multiple Criteria Decision Making (MCDM) domain [42–44]. It supposed that the potential nonlinearities in the utility function and resulting computational issues can be the reason for lack of extensive application of this method [24]. According to Mateo [45], a major weakness of the method is that it overestimates the extremes leading to undesirable results, as it significantly favors/disfavors the final evaluation of any alternative that is far from the average in relation to one criterion.

4. Experimental Study

The effect of linking priority rules and multi-criteria optimization methods on optimization of production goals in the studied production system was explored via simulation-based optimization experiments performed in two phases. At first, single objective experiments were needed to find reference points for the transformation of production objectives. Secondly, multi-objective optimization according to general formulation (2) was conducted to determined MOP solutions for both WSM and WPM methods applied in definitions of multi-criteria objective functions in an optimization model (Equations (3) and (4)) when three different priority rules were consequently applied.

4.1. Simulation Model

The discrete-event simulation model in this study represents a hypothetical partially flexible job-shop production system adopted from [46]. It can be categorized as such a class of production systems that produce smaller numbers of different types of products according to several routings in a one-piece flow way, especially following the lean production concept. Regarding to its flexibility, some operations are only achievable on part of the available machines.

The model system has been built to maintain the main characteristics of FJS, which are mentioned in the Section 2.1. It meets the typical demands regarding the jobs and machines assumed in the deterministic flexible job shop scheduling problem [5,20,27]:

- All machines and all jobs are available at time t = 0.
- The number of jobs and machines are assumed to be finite;
- a job cannot be processed by two machines at the same time;
- each machine is always accessible for production, i.e., no machine breakdowns or order cancelations are supposed;
- the machine can execute only one operation at a time on any job;
- an operation of a job cannot be performed until its preceding operations are completed;
- once started operation must not be interrupted;
- operation is processed by any machine from a given set;
- the model does not include any assembly operations.

The studied model system consisted of six partially interchangeable machines with their input and output buffer [46]. A total of seven types of operations (A, B, C, D, E, F, G) were performed in the production system. The system manufactured four types of products, each with the specified determination of the sequence of four operations (that cannot be interchanged) defined in Table 1.

Table 1. Definition of order operations for coming parts.

Product Type	Operations Order
1	A, G, B, F
2	C, B, D, F
3	D, F, G, E
4	G, E, A, C

Transportation of parts between buffers was ensured by conveyors with sensors that allow identifying every coming part. If the appropriate attribute of a part was detected, then the part was accepted for the next processing. The interchangeability of workplaces according to the operations possible to perform at individual machines in the system is shown in Table 2.

Table 2. The interchangeability of workplaces.

Workplace Number	Possible Operation Performance
1	A, C, D
2	A, F, G
3	B, C
4	D, E, G
5	B, F, C
6	E, G

In this study, we implemented consequently selected priority rules FIFO, EDD, and STR to define the order in which the individual operations will be done. This order was determined by the calculated priority with respect to the rule performing. When implementing priority rule EDD into the simulation model, the priority key for the operation associated with a given job was updated with respect to the earliest due date information. In the case of the STR priority rule, Equation (1) was used for the calculation of the priority. Based on the determined priority rule, the machine always selected the operation that had the highest priority of all. If multiple operations had the same priority, these operations were grouped randomly. If the next operation can be done on different machines, the machine with minimal setup costs was selected to process it.

The simulation model was designed and built in such a way that it involved the possibility of choice of priority rule at the start of simulation. After selecting the priority rule by the user, the studied system was fully deterministic. Neither breakdowns due to order cancellations nor machine failures were allowed. All internal parameters, such like operation and setup times, and all costs items were keeping as constants during experiments, too. Most of operation times slightly differed due to the type of processing operation and type of product. Every machine was set up when the type of operation was changed. To simplify the investigated problem, the question of the impact of the different range of operating times was not included in the design of simulation experiments. The detailed information about operation times and costs for individual operation is given in Table 3.

Table 3. Values of internal system parameters.

Product Type	Routing	Operation Time	Costs for Operation
	A	4	5
1	G	6	7
	B	5	6
	F	3	4
	C	5	6
2	B	4	5
	D	3	4
	F	4	5
	D	3	4
3	F	3	4
	G	5	6
	E	3	4
	G	5	6
4	E	3	4
	A	4	5
	C	4	5

The model of the production system was created in simulation software Witness Horizon ver. 22, developed by Lanner. Simulation environment of Witness Horizon is designed for modelling, simulation, predictive analytics, and optimization of production and logistic processes, or services with the aim to verify consequences of different scenarios in the virtual world via simulation, or to find an effective set of system parameters. We used it for simulation and optimization experiments in the presented study. The discrete-event simulation model is depicted in Figure 1.

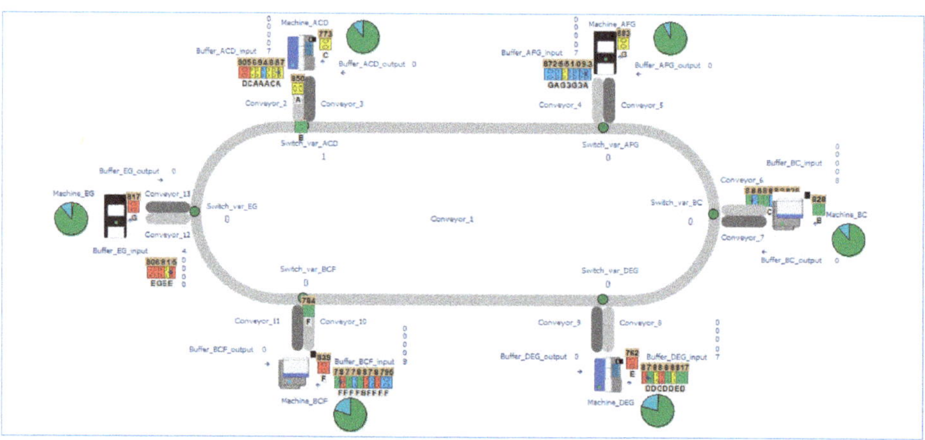

Figure 1. Simulation model of a flexible job-shop production system [46].

4.2. Preliminary Simulation Experiments

Conditions of optimization experiments were proposed on the base of preliminary simulation experiments performed on the presented simulation model for three priority rules consequently in Witness simulator. The experiments served for understanding the simulation model's behavior under different loadings and its limits associated with production results.

Different loadings were modelled by changing the time between arrivals (inter-arrival time) for the entry of each of four parts in the range 1 min to 20 min. Each of the parts entered in one piece lot size to the system.

Simulation runs lasted 1440 min, which represented standard period of 1 day covering three shifts. The warm-up period of 90 min was found out as the period which is needed to produce one product of each four types using simulation.

As system response, we observed values of five production indicators: Average flowtime (min.), average machine utilization (%), average work in progress (pcs), average number of products (pcs), and the average costs per part unit (€) as outputs from simulation runs. These experiments showed production limits according to the design variables and rules setting.

4.3. Optimization Model

When considering the avoidance of the extreme situations in production, which mean too few processed products and, on the other hand, the system overload, the space of design variables for optimization was determined as a four-dimensional domain defined by time between arrivals for the entry of each of four products in a reasonable range of 8 min to 16 min, based on the simulation results for this system. In this case, the modelled system showed acceptable values of production goals for all three priority rules.

Based on observed production results, the suitable values of constraints of production goals for optimization model were also suggested. As for the machine utilization constraints, from the practical point of view referring to [35], the decision was influenced by the fact that the machine utilization more than 80% generally corresponds to moderate and heavy shop load level in job shops. The constraints concerning production objectives applied in all optimization experiments are specified in Table 4.

Table 4. Constraints for optimization experiments.

Production Goal	Constraint
Average flow time (min)	<180
Average machine utilization (%)	>80
Average number of products (pcs)	>400

The optimization objectives represented five selected aforementioned key performance indicators: Average flowtime, average machine utilization, average work in progress, average number of products, and the average costs per part unit. For MOP solving, they were combined into two types of scalar multi-criteria objective functions. This scalarization procedure generating multi-criteria objective function based on WSM and WPM methods (Equations (3) and (4)) is described in detail in the following paragraphs.

4.3.1. Transformation (Normalization) of Production Objectives

When applying a weighted sum approach or other types of scalarization, it is reasonable to normalize individual objectives F_i entering the scalar multi-objective function to avoid the scaling effects and thus ensure their comparable impact in the objective function expression. We utilized the transformation which represented a robust min-max type of normalization. It can also be referred to as the upper-lower-bound transformation in the literature [47]. This transformation and many other potential ones are provided in a survey by Marler and Arora [24] in a comprehensive overview. The applied robust transformation was in this study implemented in a form (5) involving two so-called reference points as maximal and minimal values of the individual objective functions F_i.

$$F_i^{transform} = \frac{F_i - F_i^W}{F_i^U - F_i^W} \tag{5}$$

Here, the transformation (5) includes a component of an ideal point [38] vector F_i^U (maybe replaced the point, referred also as a utopia point [23]) and a component of the worst point vector F_i^W. They represent the unreachable solution for individual production objectives as the best (ideal), and as the worst one. The ideal point vector F^U is determined as the optimum of the single-criterion function according to the optimization goal regardless of other objectives. Vector F^W is composed of the worst values of production goals, obtained from single-optimization experiments. Value F_i in the expression (5) represents the individual objective value, resulted from the simulation run. The transformation scheme ensures the same range and magnitude for all objectives within the range between 0 and 1. Comparing to works [24,47] where the distance between solution F_i and utopia point is minimized, we modified the transformation scheme to (5) and maximized the distance between solution F_i and the worst point.

4.3.2. Single-Objective Optimizations

The presented approach demanded to find the minimum and maximum value for each of five selected production goals with respect to constraints presented in Table 4. We used a set of single-objective optimizations for this purpose instead of an estimation [24,38]. The objective functions for obtaining ideal and worst points were optimized according to the type of production objective. The objective functions and target of optimization for each of them are shown in Table 5.

Table 5. The target of the single-objective optimization.

Objective Function	Optimization Target When Finding Ideal Point	Optimization Target When Finding Worst Point
Flow time	minimize	maximize
Costs per part unit	minimize	maximize
Work in progress	minimize	maximize
Machine utilization	maximize	minimize
Number of products	maximize	minimize

Simulation environment Witness Horizon includes the optimization module Experimenter with its Advanced Mode for the performance of simulation-based optimization. Implemented optimization algorithms comprise Random solutions, Hill Climb, Min/Mid/Max, and All Combinations algorithms, and metaheuristic, such as Adaptive Thermostatistical Simulated Annealing (SA). The optimum of the individual objective functions representing production goals was obtained via simulation-based optimization in two steps. Firstly, employing a metaheuristic algorithm briefly explored the design space with respect to the local optima at the beginning of optimization process. None of the values reached by SA represented global optimum, only local ones, therefore we did not present them. Secondly, we used the Brute Force algorithm All Combinations, which runs all constrained combinations of design parameters. The optimization process via the All Combinations algorithm was time consuming, but with the guarantee that the optimal solution will be found. For all experiments, the length of the one simulation run was set up to 1440 min and the warm-up period was 90 min. We used one replication for one simulation run due to the deterministic nature of the model. The resulted values of the ideal and worst points are listed in Table 6.

Table 6. Results of single-objective optimization of production objectives. FIFO: First In First Out; EDD: Earliest Due Date; STR: Slack Time Remaining.

Priority Rule	Single-Objective Function	Coordinates of the Ideal Point in the Objective Space	Coordinates of the Worst Point in the Objective Space
FIFO	Flow time	47.534	179.884
	Costs per part unit	31.645	37.507
	Work in progress	3.189	21.229
	Machine utilization	87.245	84.902
	Number of products	467	421
EDD	Flow time	45.128	179.527
	Costs per part unit	31.410	37.297
	Work in progress	2.637	22.109
	Machine utilization	89.283	82.096
	Number of products	482	429
STR	Flow time	44.892	179.934
	Costs per part unit	31.250	37.455
	Work in progress	2.747	22.034
	Machine utilization	89.449	83.508
	Number of products	479	433

4.3.3. The Formation of Scalar Multi-Criteria Objective Functions and Conditions of Optimization

After the determination of all reference ideal and worst points necessary for the transformation of production goals, further experiments based on respecting all five production goals simultaneously are carried out under the same conditions. The scalar multi-criteria objective functions for each of the three priority rules were built on WSM and WPM methods.

Multi-objective U_{WSM} function (6) was formed on the base of the Weighted Sum Method (defined by Equation (3)) as the linear combination of the normalized individual objectives $F_i^{transform}$ expressed by (5) using the weight vector w.

$$U_{WSM} = \sum_{i=1}^{k} w_i F_i^{transform}, w_i > 0 \; \forall i, \sum_{i=1}^{k} w_i = 1 \tag{6}$$

All components w_i were equal to the value 0.2 to not give priority to any of five production goals. Due to the normalization of objectives given by (5), the structure of the proposed scalar objective function U_{WSM} demands the maximization of (6) to lead to the single preferred Pareto optimal solution finding.

When adopting the Weighted Product Method (defined by Equation (4)) with the same settings of weights, U_{WPM} function is defined by Equation (7), and, on the contrary, the minimization of (7) yields the Pareto optimal solution.

$$U_{WPM} = \prod_{i=1}^{k} \left(F_i^{transform} \right)^{w_i}, w_i > 0 \; \forall i, \sum_{i=1}^{k} w_i = 1 \tag{7}$$

The objectives that need to be maximized were transformed using a reciprocal of (5).

Box 1 and 2 contain an example of code which shows the transformation in WSM and WPM-based multi-criteria objective functions for EDD priority rule, written in the versatile WITNESS Action Language. (WITNESS also supports external code libraries written in common languages such as C++, C#, and VB.net).

Box 1. A code in a body of Weighted Sum Method (WSM)-based objective function using EDD priority rule in simulator.

```
IF Number_of_products >= 400 AND Flow_Time <= 180 AND Machine_utilization >= 80
RETURN 0.2 * ((Number_of_products − 429)/(482 − 429)) + 0.2 * ((Work_in_progress −
22.109)/(2.637 − 22.109)) + 0.2 * ((Flow_Time − 179.527)/(45.128 − 179.527)) + 0.2 * ((Ma-
chine_utilization − 82.096)/(89.283 − 82.096)) + 0.2 * ((Costs_per_part_unit − 37.297)/(31.410
− 37.297))
ELSE
RETURN 0.2 * ((Number_of_products − 429)/(482 − 429)) + 0.2 * ((Work_in_progress −
22.109)/(2.637 − 22.109)) + 0.2 * ((Flow_Time − 179.527)/(45.128 − 179.527)) + 0.2 * ((Ma-
chine_utilization − 82.096)/(89.283 − 82.096)) + 0.2 * ((Costs_per_part_unit − 37.297)/(31.410
− 37.297)) – 10
ENDIF
```

Box 2. A code in a body of Weighted Product Method (WPM)-based objective function using EDD priority rule in simulator.

```
IF Number_of_products >= 400 AND Flow_Time <= 180 AND Machine_utilization >= 80
RETURN ((482 − 429)/(Number_of_products − 429)) ** 0.2 * ((Work_in_progress − 22.109)/(2.637
− 22.109)) ** 0.2 * ((Flow_Time − 179.527)/(45.128 − 179.527)) ** 0.2 * ((89.283 − 82.096)/(Ma-
chine_utilization − 82.096)) ** 0.2 * ((Costs_per_part_unit − 37.297)/(31.41 − 37.297)) ** 0.2
ELSE
RETURN ((482 − 429)/(Number_of_products − 429)) ** 0.2 * ((Work_in_progress − 22.109)/(2.637
− 22.109)) ** 0.2 * ((Flow_Time − 179.527)/(45.128 − 179.527)) ** 0.2 * ((89.283 − 82.096)/(Ma-
chine_utilization − 82.096)) ** 0.2 * ((Costs_per_part_unit − 37.297)/(31.41 − 37.297)) ** 0.2 +
10
ENDIF
```

For other rules, the form of body function was created analogically, with corresponding ideal and worst points shown in Table 6. In all cases, the same weights of 0.2 were used as parameters expressed a priori articulated decision-maker´s preferences. Independently on the type of objective function, the key optimization idea was to ensure the maximization of a distance between the solution and the worst point. As can be seen in Box 1 and 2, all objective functions were constrained by the values presented in Table 4. For these optimization experiments, simulation-based optimization with All Combinations algorithm was applied with one replication for one simulation run lasting 1440 min, with the warm-up period 90 min to find global optimum.

5. Experimental Results

The performed optimization experiments were of two categories. Firstly, the single-objective optimizations exploring 4096 scenarios for each of the three priority rules were conducted to find the reference points involved in the robust transformation of production objectives (with the aims given in Table 5). All results were obtained under the same conditions, which are described in previous section with respect to design space, objective functions, constraints, and conditions of simulation runs.

The column graph in Figure 2 shows a comparison of ideal point values when applying different priority rules.

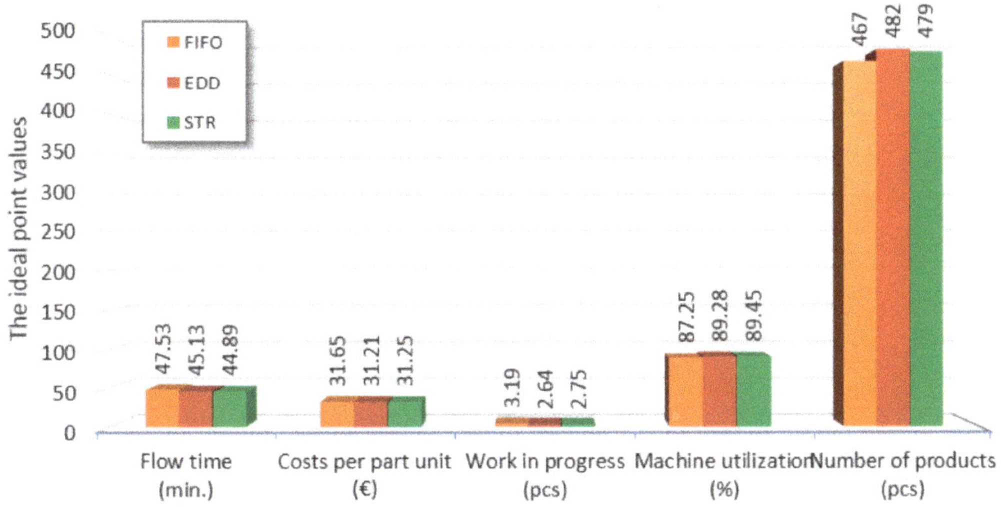

Figure 2. The comparison of the ideal point values with respect to different priority rules.

Analogically, Figure 3 depicts a comparison of obtained values of the worst point. All data were presented in Table 6 above.

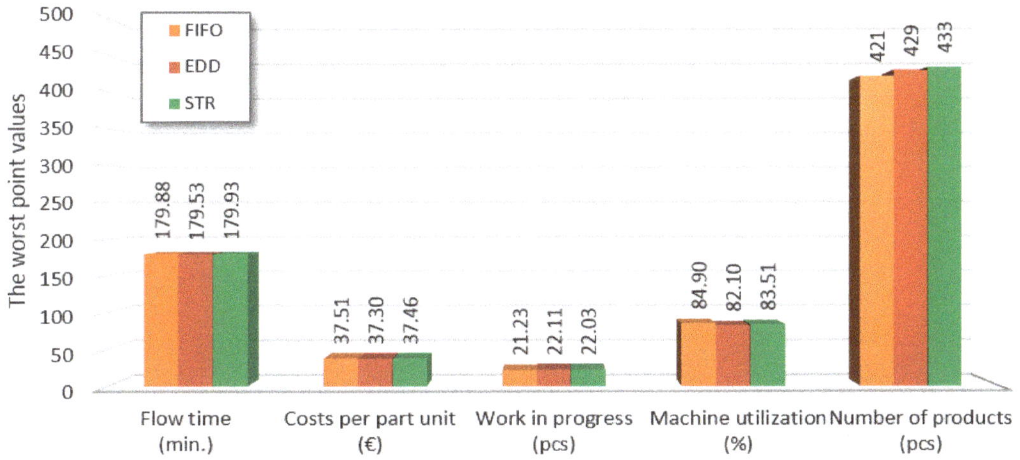

Figure 3. The comparison of the worst point values with respect to different priority rules.

Secondly, the experiments focused on multi-criteria optimization carried out to acquire the optimal daily production results according to given a priori preferences. When applying the WPM method, the minimization of objective function regarding the best and worst values of production performance, introduced in Box 2 and subjected to constraints, was performed. On the contrary, the objective function built via WSM scalarization method presented in Box 1 was maximized. In contrast to the approach in the works of Marler and Arora [24] or Xiang et al. [47], where the distance of the solution to the ideal point was minimized, we modified the structure of scalar objective function so that the maximization of the distance between solution and the worst point was needed.

The individual values of five production objectives for obtained optimal solutions in objective space applying WSM and WPM methods linked to three selected scheduling principles are shown in Table 7.

Table 7. Results of multi-criteria optimization based on scalarization for three priority rules.

Scalarization Method	Priority Rule	Optimum of the Objective Function	Number of Products (pcs)	Flow Time (min.)	Work in Progress (pcs)	Machine Utilization (%)	Costs Per Part Unit (€)
WSM	FIFO	0.826	460	69.173	5.544	86.569	31.940
	EDD	0.879	472	60.377	4.453	87.923	31.577
	STR	0.883	475	60.649	5.000	88.039	31.719
WPM	FIFO	1.042	447	47.946	3.826	82.756	31.766
	EDD	1.050	452	45.360	3.004	83.590	32.043
	STR	1.014	455	46.399	2.778	84.315	31.971

The graphical comparison of production goals obtained employing WSM and WPM methods for three different priority rules is presented in Figure 4a–e.

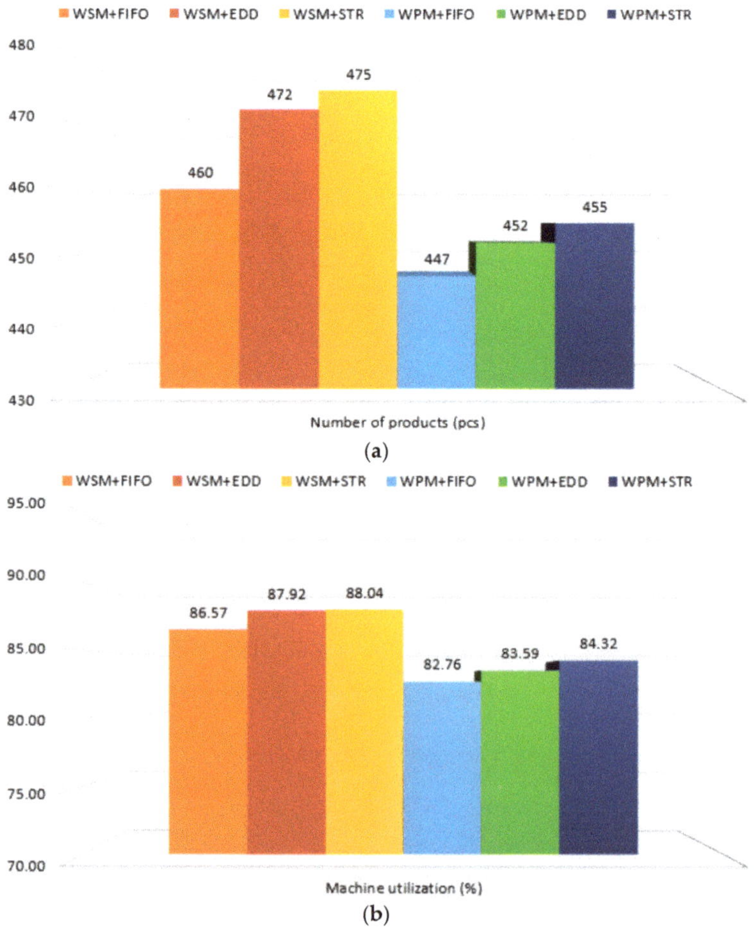

Figure 4. *Cont.*

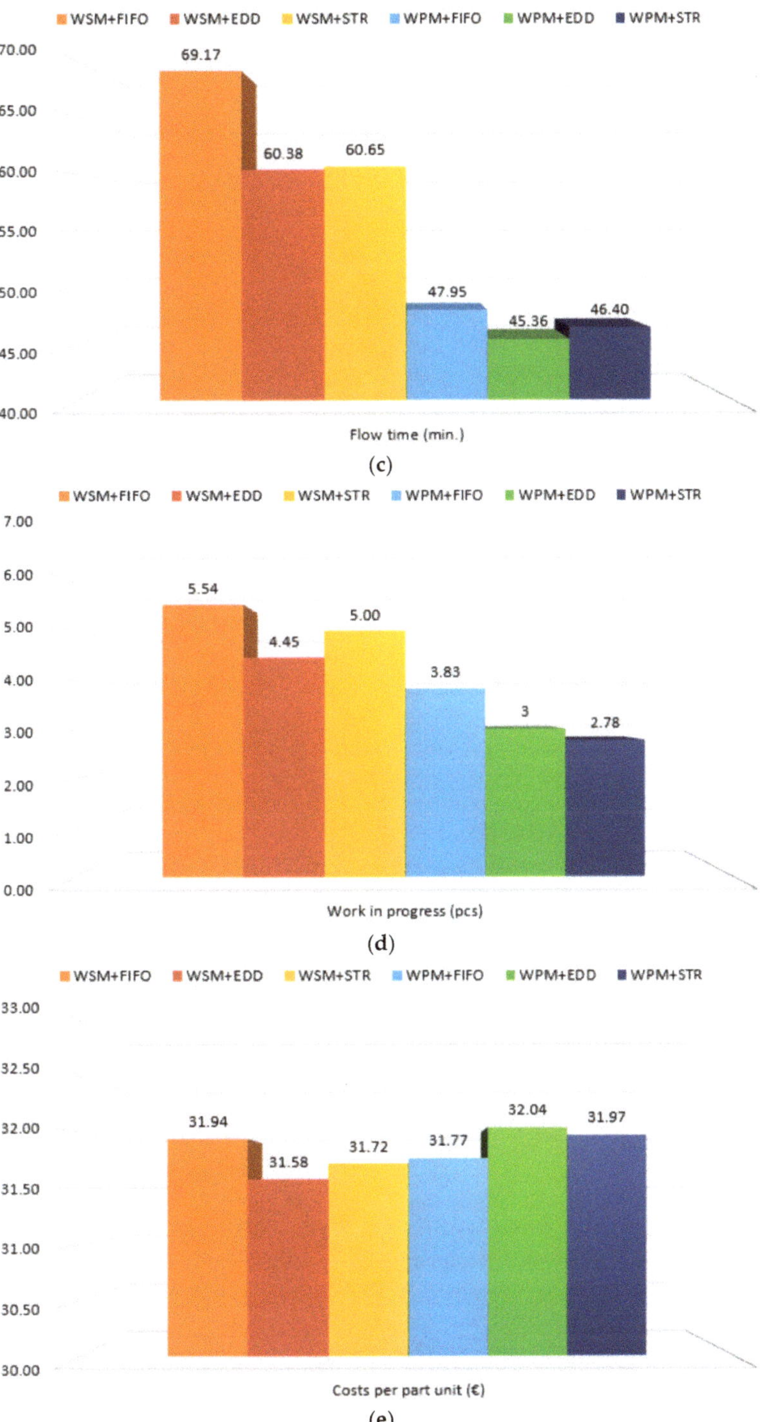

Figure 4. The comparison of production goals when applying two multi-criteria methods with different priority rules. (**a**) Number of products; (**b**) machine utilization; (**c**) flow time; (**d**) work in progress; (**e**) costs per part unit.

On the basis of obtained production goals as nondominated solutions in objective space presented in Table 7, the four design variables—the times between arrivals for entry of each of four parts P1, P2, P3, and P4 (the inter-arrival times (IATP1—IATP4)) were identified as effective solutions in the design space. They are presented for both methods and three priority rules in Table 8.

Table 8. Design input values of the inter-arrival times for parts P1—P4 in the design space corresponding to optimal production outputs (shown in Table 7) in objective space.

Scalarization Method	Priority Rule	IATP1 (min.)	IATP2 (min.)	IATP3 (min.)	IATP4 (min.)
	FIFO	12	15	15	9
WSM	EDD	12	15	14	9
	STR	14	13	13	9
	FIFO	15	15	15	9
WPM	EDD	14	14	14	10
	STR	15	12	15	10

In this work, we address not only solving multi-objective optimization problem in scheduling itself, but also the way, how to evaluate the impact of the applied priority rule to the production results via obtaining the trade-off solution, and comparing it to the best values of many production goals determined by limits of the studied system. To compare the obtained trade-off solutions, simple ordering based on obtained optimization results presented in Table 7 from the best one (1.) to the worst (5.) according to the optimization target for each of the production objectives was assigned. In addition, an average of the order values was calculated to evaluation the less is better, as it is presented in Table 9.

Table 9. The individual ordering production goals values according to the results of multi-criteria optimization.

Scalarization Method	Priority Rule	Number of Products	Flow Time	Work in Progress	Machine Utilization	Costs Per Part Unit	Average of the Order Values
	FIFO	3.	6.	6.	3.	4.	4.4
WSM	EDD	2.	4.	4.	2.	1.	2.6
	STR	1.	5.	5.	1.	2.	2.8
	FIFO	6.	3.	3.	6.	3.	4.2
WPM	EDD	5.	1.	2.	5.	6.	3.8
	STR	4.	2.	1.	4.	5.	3.2

On the base of average values relating to the ordering given in Table 9, we assigned the final ranking to each of the applied methods in the conjunction with the evaluated priority rule to determine the best performance result. A ranking "1" was given to the rule with the method that provided the best performance, i.e., to the summarized evaluation with the smallest average value. A ranking "2" to the next one, etc. The result of the final evaluation is introduced in Figure 5.

Finally, we compared the obtained compromise solution for the best evaluated scheduling to the coordinates of an unfeasible ideal point when the EDD priority rule was employed in production control. The differences due to the compromise character of the solution are presented in Figure 6.

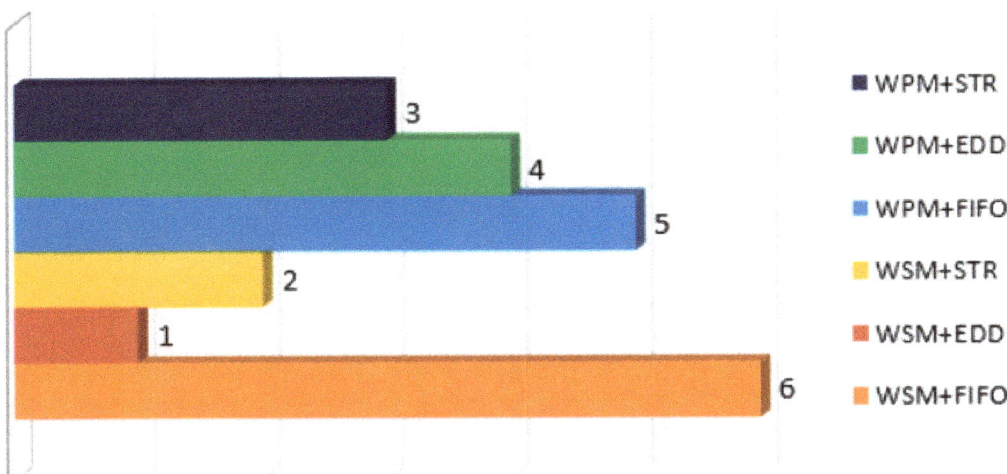

Figure 5. Ranking of applied methods of multi-criteria optimization in the conjunction with selected priority rule.

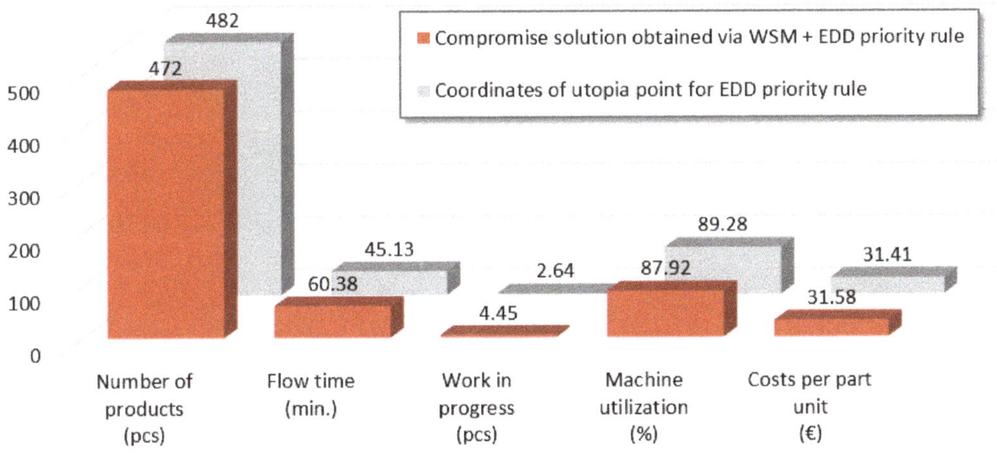

Figure 6. The comparison of coordinates of the ideal point for the EDD priority rule and the best evaluated compromise solution yielded by WSM + EDD.

6. Discussion

The obtained experimental outcomes reflect the dynamical structure of the system determined by applied priority rule. As the precise analytical expression of the transformation of input parameters (time between arrivals for the entry of the parts) to the outputs (production goals) is not known, all objective function values were determined by simulation in a simulation-based optimization process.

The results of experiments related to single-objective optimization of production objectives, which are presented in Table 6 and compared in Figures 2 and 3, show that priority rules EDD and STR provided the comparable ideal and worst values for almost all production objectives in the studied model system. That means the daily production limits of the investigated production system according to both rules were similar, therefore the reference points would have to influence the MOP solution only insignificantly. The ideal and the worst values were applied in the robust transformation involved in both employed multi-criteria optimization methods in the next step.

The following optimization experiments were conducted with the focus on comparing results of multi-criteria optimization with the relation to the evaluation of reached production goals, respecting the implemented specific priority rule. The results presented in Table 9 demonstrate that there does not exist a single dispatching rule that achieves the highest ranking in the connection to applied multicriteria methods according to all observed performance objectives. As can be seen in Figure 5, according to the type of scalar objective function used in the multi-criteria optimization, the best ranking overall was obtained for the WSM when the EDD priority rule was applying. Very similar although slightly worse results were achieved by the WSM using the STR rule. Both scalarization methods in the presence of the FIFO priority rule achieved the worst results with a relatively large loss. The result has confirmed what, e.g., Kemppainen [23] states, that the FIFO rule is not recommended for manufacturing operations since it is usually overcome by any other priority rule.

The presented findings point to that employing the EDD and STR rule brought similar production performance results in the context of the complex view to obtained compromise solutions in the studied production system. The relatively long inter-arrival times for EDD rule presented as effective solutions in design space in Table 8 are in consonance with the fact that due date based dispatching rules are expected to work well at low utilization levels [35], and outperformed other rules under light load and smaller job shop configurations [23].

Based on the findings of this investigation, a practical suggestion for production how to set up the input arrival time of each of four entering parts into the system under the specific rule can be proposed. It is determined by design variables presented in Table 8. When utilizing the EDD rule and WSM method, the effective solution is following the values of the design vector (12; 15; 14; 9), determined in minutes. Comparing to the unreachable best results determined by ideal point, the loss of production performance is 2.07% for an average number of manufactured products, 25.26% for an average flow time, 1.52% for average machine utilization, 40.78% for average work in progress, and 0.52% for average cost per part unit, respectively.

As for the recommendation of input arrival times when utilizing the STR rule with WSM, it is the design vector (14; 13; 13; 9), determined in minutes. The loss when comparing the compromise solution to the best values in case of setting relating to the WSM+STR rule, is 0.84% for an average number of manufactured products, 25.98% for an average flow time, 1.58% for average machine utilization, 45.06% for average work in progress, and 1.48% for average cost per part unit.

For the setup of design variables going up from the WPM + STR rule, the suggested design vector is (15; 12; 15; 10), determined in minutes. The corresponding loss is 5.01% for an average number of manufactured products, 3.25% for average flow time, 5.74% for average machine utilization, 1.12% for average work in progress, and 2.26% for average cost per part unit, respectively, all in the relation to the values of ideal point.

Despite the weights in scalar objective function were the same, we can notice that in contrast to WSM methods that prioritized some of the production objectives, the compromise solution yielded by WPM is in a relative evaluation more balanced. In this context, the results confirmed that the scalarization methods applying in a priori arrangement bring the problem with the precise reflection of the weights to desired Pareto optimal solutions [24]. The coordinates of the weight vector serve as the model parameters in a multi-objective function and influence the location of Pareto front points.

When exploring the compromise solutions found in the objective space from the perspective of the production goals control, we can see a noticeable impact of the applied scalarization method on the evaluation of production results. The comparison of the reached results in Figure 4a–e shows that all production goals that we wish to maximize, such as the number of products and machine utilization, were substantially higher in the case of the WSM method employed, regardless of the applied priority rule. On the contrary, the WPM method provided better results for production goal values that need to be minimized. As we can see in Figure 4, the production objectives, such as the average flow time and the average work in progress, achieved smaller values in comparison to results yielded by the WSM method. Finally, in the case of the average costs per part unit, we can notice very similar and comparable results for all applied methods.

Working with many contradictory objectives assumes that they are not completely independent, and a correlation can appear between some of them. As can be noticed in the results introduced in Figure 4, there exist two pairs of objectives which are correlated. We can observe the positive correlation between the number of products and machine utilization (Figure 4a,b), as well as between flow time and work-in-progress (Figure 4c,d). We supposed that correlations, applied transformation of production objectives, and different expressions defining the structure of objective function themselves could explain the obtained results significantly. To evaluate the direct influence of correlations on optimization results will need further experiments with a reduced set of objectives.

The rankings of priority rules strongly depend on selected performance measures involved in evaluation and the structure of the production system model itself. The interesting different approach, how to optimize multiple objectives in flexible job shop system, is introduced in a comparative study in the work of Sels et al. [48]. They conducted single-objective optimizations for five individual objectives in the presence of 30 priority rules consequently to evaluate the individual performance objective with respect to the scheduling principle. The observed objectives were makespan, mean flow time, proportion of tardy jobs, mean tardiness, and maximum tardiness. As for ranking of rules, each priority rule's performance was measured as the percentage deviation from the best performing priority rule. Based on the best results via ranking, they constructed the hybrid scheduling rule to ensure optimization of all five objectives simultaneously. EDD and Slack (called STR in our study) rules were the part of the hybrid priority rules in the top 5 of the priority rules that performed best on the five objective functions. Their experiments showed that an integrated optimization approach outperforms the hierarchical approach when solving the flexible job shop scheduling problem [48].

7. Conclusions

In this study, we presented the results of the wider investigation of common priority rules FIFO, EDD, and STR applied in the partially flexible job shop system in terms of several conflicting production performance indicators evaluated simultaneously. We employed scalar multi-criteria simulation-based optimization in a priori arrangement to identify the impact of applied priority rule to production results. From this perspective, the evaluation of the influence of the applied priority rule and system loading to daily production for five selected production objectives at the same time based on MOP solving offers the possibility to find effective settings of external input parameters with respect to the scheduling principle, and it can support decision-making in production control.

Combination of metaheuristic algorithm Simulated Annealing and full combinatorial analysis in the final phase of the optimization process yielded a global solution of defined MOPs for each of the selected priority rules. The experimental results demonstrated that the found compromise solutions strongly depended on a selected scalar multi-criteria method and applied priority rule. Although the weights in the scalar objective function were the same, based on the performed experiments, we can conclude that results for production goals which need to be minimized were better when the WPM method was applied, in contrast to WSM, which developed the values of other production objectives, regardless of the employed priority rule. EDD and STR priority rules were shown as more suitable than the FIFO rule, both with comparable production performance.

Despite these facts, from the complex point of view, all effective solutions in the design space resulted from scalar multi-criteria simulation-based optimization obtained independently from the applied rule and applied optimization method, and were situated in the area connected to the lower production system load. For the modelled production system, the recommended settings for inter-arrival time of entering part for four types of manufactured products were presented to obtain acceptable values of production goals.

Future research will be focused on other multi-criteria method applications, and the question of the effect of other priority rules is still open. The evaluation of stochastic system performance will be in the center of interest, too.

Author Contributions: Conceptualization, Z.Č. and P.V.; methodology, Z.Č. and P.V.; software, Z.Č. and P.V.; validation, Z.Č. and P.V.; investigation, Z.Č.; writing—original draft preparation, Z.Č. and P.V.; writing—review and editing, Z.Č., M.J. and B.J.; supervision, P.V.; project administration, Z.Č., P.V., M.J. and B.J.; funding acquisition, P.V. All authors have read and agreed to the published version of the manuscript.

Funding: This research was funded by VEGA agency, grant number 1/0232/18—"Using the methods of multi-objective optimization in production processes control".

Institutional Review Board Statement: Not applicable.

Informed Consent Statement: Not applicable.

Data Availability Statement: Not applicable.

Conflicts of Interest: The authors declare no conflict of interest. The funders had no role in the design of the study; in the collection, analyses, or interpretation of data; in the writing of the manuscript, or in the decision to publish the results.

References

1. Pinedo, M.L. *Scheduling—Theory, Algorithms, and Systems*, 5th ed.; Springer: Berlin, Germany, 2016.
2. Shahzad, A.; Mebarki, N. Learning dispatching rules for scheduling: A synergistic view comprising decision trees, Tabu search and simulation. *Computers* **2016**, *5*, 3. [CrossRef]
3. Priore, P.; Gómez, A.; Pino, R.; Rosillo, R. Dynamic scheduling of manufacturing systems using machine learning: An updated review. *Artif. Intell. Eng. Des. Anal. Manuf.* **2014**, *28*, 83–97. [CrossRef]
4. Vieira, G.E.; Herrmann, J.W.; Lin, E. Rescheduling Manufacturing Systems: A Framework of Strategies, Policies, and Methods. *J. Sched.* **2003**, *6*, 39–62. [CrossRef]
5. Xanthopoulos, A.S.; Koulouriotis, D.E.; Gasteratos, A.; Ioannidis, S. Efficient priority rules for dynamic sequencing with sequence-dependent setups. *Int. J. Ind. Eng. Comput.* **2016**, *7*, 367–384. [CrossRef]
6. Deroussi, L.; Gourgand, M.; Tchernev, N. Combining optimization methods and discrete event simulation: A case study in flexible manufacturing systems. In Proceedings of the 2006 International Conference on Service Systems and Service Management, Troyes, France, 25–27 October 2006; IEEE: Washington, DC, USA, 2006.
7. Ojstersek, R.; Brezocnik, M.; Buchmeister, B. Multi-objective optimization of production scheduling with evolutionary computation: A review. *Int. J. Ind. Eng. Comput.* **2020**, *11*, 359–376. [CrossRef]
8. Zarandi, F.M.H.; Sadat, A.A.A.; Sotudian, S.; Castillo, O. A state of the art review of intelligent scheduling. *Artif. Intell. Rev.* **2020**, *53*, 501–593. [CrossRef]
9. Adibi, M.A.; Zandieh, M.; Amiri, M. Multi-objective scheduling of dynamic job shop using variable neighborhood search. *Expert Syst. Appl.* **2010**, *37*, 282–287. [CrossRef]
10. Kaban, A.K.; Othman, Z.; Rohmah, D.S. Comparison of dispatching rules in job-shop scheduling problem using simulation: A case study. *Int. J. Simul. Model.* **2012**, *11*, 129–140. [CrossRef]
11. Kück, M.; Eike, B.; Freitag, M. Towards adaptive simulation-based optimization to select individual dispatching rules for production control. In Proceedings of the 2017 Winter Simulation Conference (WSC), Las Vegas, NV, USA, 3–6 December 2017; IEEE: Washington, DC, USA, 2017.
12. Gupta, A.; Sivakumar, A. Job shop scheduling techniques in semiconductor manufacturing. *Int. J. Adv. Manuf. Technol.* **2006**, *27*, 1163–1169. [CrossRef]
13. Vinod, V.; Sridharan, R. Simulation modeling and analysis of due-date assignment methods and scheduling decision rules in a dynamic job shop production system. *Int. J. Prod. Econ.* **2011**, *129*, 127–146. [CrossRef]
14. Loukil, T.; Teghem, J.; Tuyttens, D. Solving multi-objective production scheduling problems using metaheuristics. *Eur. J. Oper. Res.* **2005**, *161*, 42–61. [CrossRef]
15. Geiger, C.D.; Uzsoy, R.; Aytuǧ, H. Rapid modeling and discovery of priority dispatching rules: An autonomous learning approach. *J. Sched.* **2006**, *9*, 7–34. [CrossRef]

16. Zahmani, M.H.; Atmani, B.; Bekrar, A.; Aissani, N. Multiple priority dispatching rules for the job shop scheduling problem. In Proceedings of the 2015 3rd International Conference on Control, Engineering & Information Technology (CEIT), Tlemcen, Algeria, 25–27 May 2015; IEEE: Washington, DC, USA, 2015.
17. Vieira, G.E.; Kück, M.; Frazzon, E.; Freitag, M. Evaluating the Robustness of Production Schedules using Discrete-Event Simulation. *IFAC-PapersOnLine* **2017**, *50*, 7953–7958. [CrossRef]
18. Freitag, M.; Hildebrandt, T. Automatic design of scheduling rules for complex manufacturing systems by multi-objective simulation-based optimization. *CIRP Ann–Manuf. Technol.* **2016**, *65*, 433–436. [CrossRef]
19. Demir, H.I.; Erden, C. Dynamic integrated process planning, scheduling and due-date assignment using ant colony optimization. *Comput. Ind. Eng.* **2020**, *149*, 1–12. [CrossRef]
20. Xie, J.; Gao, L.; Peng, K.; Li, X.; Li, H. Review on flexible job shop scheduling. *IET Collab. Intell. Manuf.* **2019**, *1*, 67–77. [CrossRef]
21. Coello, C.A.; Brambila, S.G.; Gamboa, J.F.; Tapia, M.G.C.; Gómez, R.H. Survey and state of the art. Evolutionary multiobjective optimization: Open research areas and some challenges lying ahead. *Complex Intell. Syst.* **2020**, *6*, 221–236. [CrossRef]
22. Mirjalili, S.; Saremi, S.; Mirjalili, S.M.; Coelho, L.D.S. Multi-objective grey wolf optimizer: A novel algorithm for multi-criterion optimization. *Expert Syst. Appl.* **2016**, *47*, 106–119. [CrossRef]
23. Kemppainen, K. *Priority Scheduling Revisited–Dominant Rules, Open Protocols, and Integrated Order Management*, 1st ed.; Helsinki School of Economics–HSE Print: Helsinki, Finland, 2005.
24. Marler, R.T.; Arora, J.S. Survey of Multi-Objective Optimization Methods for Engineering. *Struct. Multidiscip. Optim.* **2004**, *26*, 369–395. [CrossRef]
25. Grundstein, S.; Freitag, M.; Scholz-Reiter, B. A new method for autonomous control of complex job shops–Integrating order release, sequencing and capacity control to meet due dates. *J. Manuf. Syst.* **2017**, *42*, 11–28. [CrossRef]
26. Xing, L.N.; Chen, Y.W.; Yang, K.W. An efficient search method for multi-objective flexible job shop scheduling problems. *J. Intell. Manuf.* **2009**, *20*, 283–293. [CrossRef]
27. Baker, K.R.; Trietsch, D. *Principles of Sequencing and Scheduling*; John Wiley & Sons, Inc.: Hoboken, NJ, USA, 2009.
28. Gao, K.; Cao, Z.; Zhang, L.; Chen, Z.; Han, Y.; Pan, Q. A review on swarm intelligence and evolutionary algorithms for solving flexible job shop scheduling problems. *J. Autom. Sin.* **2019**, *6*, 904–916. [CrossRef]
29. Chiang, T.C.; Lin, H.J. A simple and effective evolutionary algorithm for multiobjective flexible job shop scheduling. *Int. J. Prod. Econ.* **2013**, *141*, 87–98. [CrossRef]
30. Zhang, S.; Li, X.; Zhang, B.; Wang, S. Multi-objective optimisation in flexible assembly job shop scheduling using a distributed ant colony system. *Eur. J. Oper. Res.* **2020**, *283*, 441–460. [CrossRef]
31. Pezzella, F.; Morganti, G.; Ciaschetti, G. A genetic algorithm for the Flexible Job-shop Scheduling Problem. *Comput. Oper. Res.* **2008**, *35*, 3202–3212. [CrossRef]
32. Abualigah, L.; Diabat, A. A novel hybrid antlion optimization algorithm for multi-objective task scheduling problems in cloud computing environments. *Clust. Comput.* **2020**, *5*, 1–19. [CrossRef]
33. Cheng, H.C.; Chiang, T.C.; Fu, L.C. A two-stage hybrid memetic algorithm for multiobjective job shop scheduling. *Expert Syst. Appl.* **2011**, *38*, 10983–10998. [CrossRef]
34. Yenisey, M.M.; Yagmahan, B. Multi-objective permutation flow shop scheduling problem: Literature review, classification and current trends. *Omega* **2014**, *45*, 119–135. [CrossRef]
35. Baykasoğlu, A.; Göçken, M.; Unutmaz, Z.D. New approaches to due date assignment in job shops. *Eur. J. Oper. Res.* **2008**, *187*, 31–45. [CrossRef]
36. Savasci, F.; Herrmann, F. Application of priority rules in production planning and control using the example of automotive logistics. In Proceedings of the Applied Research Conference, Regensburg, Germany, 8 July 2018.
37. Schmidt, G. Performance guarantee of two simple priority rules for production scheduling. *Int. J. Prod. Econ.* **2000**, *68*, 151–159. [CrossRef]
38. Ruiz, A.B.; Saborido, R.; Luque, M. A preference-based evolutionary algorithm for multiobjective optimization: The weighting achievement scalarizing function genetic algorithm. *J. Glob. Optim.* **2015**, *62*, 101–129. [CrossRef]
39. T'Kindt, V.; Billaut, J.C. *Multicriteria Scheduling–Theory, Models and Algorithms*, 2nd ed.; Springer: Berlin, Germany, 2006.
40. Hoogeveen, H. Multicriteria scheduling. *Eur. J. Oper. Res.* **2005**, *167*, 592–623. [CrossRef]
41. Huang, R.H.; Yu, T.H. An effective ant colony optimization algorithm for multi-objective job-shop scheduling with equal-size lot-splitting. *Appl. Soft Comput. J.* **2017**, *57*, 642–656. [CrossRef]
42. Zavadskas, E.K.; Turskis, Z.; Antucheviciene, J.; Zakarevicius, A. Optimization of weighted aggregated sum product assessment. *Elektron. Ir Elektrotechnika* **2012**, *122*, 3–6. [CrossRef]
43. Fitriasari, N.S.; Fitriani, S.A.; Sukamto, R.A. Comparison of weighted product method and technique for order preference by similarity to ideal solution method: Complexity and accuracy. In Proceedings of the 2017 3rd International Conference on Science in Information Technology (ICSITech), Bandung, Indonesia, 25–26 October 2017; IEEE: Washington, DC, USA, 2017.
44. Wang, M.; Liu, S.; Wang, S.; Lai, K.K. A weighted product method for bidding strategies in multi-attribute auctions. *J. Syst. Sci. Complex* **2010**, *23*, 194–208. [CrossRef]

45. Mateo, J.R.S.C. Weighted Sum Method and Weighted Product Method. In *Multi Criteria Analysis in the Renewable Energy Industry*, 1st ed.; Mateo, J.R.S.C., Ed.; Springer: London, UK, 2012; pp. 19–22.
46. Vazan, P.; Cervenanska, Z.; Kotianova, J.; Krizanova, G. The impact of selected priority rules on production goals. In Proceedings of the 2019 20th International Carpathian Control Conference (ICCC), Krakow-Wieliczka, Poland, 26–29 May 2019; IEEE: Washington, DC, USA, 2019.
47. Xiang, Y.; Arora, J.S.; Rahmatalla, S.; Marler, T.; Bhatt, R.; Abdel-Malek, K. Human lifting simulation using a multi-objective optimization approach. *Multibody Syst. Dyn.* **2010**, *23*, 431–451. [CrossRef]
48. Sels, V.; Vanhoucke, M.; Gheysen, N. A comparison of priority rules for the job shop scheduling problem under different flow time- and tardiness-related objective functions. *Int. J. Prod. Res.* **2011**, *50*, 1–33. [CrossRef]

Article

A Proposed Extended Version of the Hadi-Vencheh Model to Improve Multiple-Criteria ABC Inventory Classification

Pei-Chun Lin [1,*] and Hung-Chieh Chang [2]

[1] Department of Transportation and Communication Management Science, National Cheng Kung University, Tainan 701, Taiwan
[2] Department of Mathematics, Southwestern Oklahoma State University, Weatherford, OK 73096, USA; hungchieh.chang@swosu.edu
* Correspondence: peichunl@ncku.edu.tw; Tel.: +886-62757575 (ext. 53222)

Received: 23 October 2020; Accepted: 17 November 2020; Published: 20 November 2020

Featured Application: Inventory Management.

Abstract: The ABC classification problem is approached as a ranking problem by the most current classification models; that is, a group of inventory items is expressed according to its overall weighted score of criteria in descending order. In this paper, we present an extended version of the Hadi-Vencheh model for multiple-criteria ABC inventory classification. The proposed model is one based on the nonlinear weighted product method (WPM), which determines a common set of weights for all items. Our proposed nonlinear WPM incorporates multiple criteria with different measured units without converting the performance of each inventory item, in terms of converting each criterion into a normalized attribute value, thereby providing an improvement over the model proposed by Hadi-Vencheh. Our study mainly includes various criteria for ABC classification and demonstrates an efficient algorithm for solving nonlinear programming problems, in which the feasible solution set does not have to be convex. The algorithm presented in this study substantially improves the solution efficiency of the canonical coordinates method (CCM) algorithm when applied to large-scale, nonlinear programming problems. The modified algorithm was tested to compare our proposed model results to the results derived using the Hadi-Vencheh model and demonstrate the algorithm's efficacy. The practical objectives of the study were to develop an efficient nonlinear optimization solver by optimizing the quality of existing solutions, thus improving time and space efficiency.

Keywords: non-linear programming; Hadi-Vencheh model; multiple criteria ABC inventory classification; nonlinear weighted product model

---

1. Introduction

To facilitate the successful management of a growing number of stock-keeping units (SKUs), inventory managers have found that inventory classification systems provide essential context for evaluating inventory management. ABC analysis is one of the most frequently used inventory classification techniques. Raw materials, subassemblies, intermediate products, parts, and end products can be divided into three classes: A (very important items), B (moderately important items), and C (relatively unimportant items). The ABC classification problem is presented as a ranking problem by the most current classification models [1–3]; that is, a group of inventory items is represented according to its overall weighted score of criteria in descending order. The idea of ABC analysis was applied to the inventory management at General Electric during the 1950s. This approach is based on Pareto's famous theory of inequality in the distribution of incomes. A conventional ABC study is conducted on the basis of one criterion: the annual dollar usage (value of an item times its annual usage) of SKUs.

Under Pareto's theory, all items are ranked based on a single criterion; within inventory management, dollar usage is the only criterion for managers to classify items into the A, B, and C categories. However, managers sometimes want to consider more attributes of an item when classifying goods. Many item characteristics could influence inventory control policy and must be considered. Flores [2] noted that other vital criteria can be adopted in addition to dollar usage, such as commonality, reparability, substitutability, lead time, and commonality. For instance, an enterprise must pursue efficient operations that can both minimize total costs and maximize satisfaction brought to their internal or external customers. If SKUs are only classified based on the single criterion of dollar usage, an item with a lower dollar usage, but a long lead time and high criticality, would be misclassified into the C category, resulting in serious damage to the company if the item were to run out of stock.

Detailed literary research has been carried out on multi-choice programming (MCP) theories and applications. MCP is a branch of multi-objective programming that stems from multiple-criteria decision-making (MCDM). MCDM tests several overlapping criteria of decision-making in various areas [4,5]. Multi-criteria inventory classification (MCIC) can be viewed as an application of multi-criteria decision analysis [6,7]. To solve the MCIC problem, the joint criteria matrix [8] is a simple and easy-to-understand tool, but it is not practical for more than two criteria and involves too much subjectivity. The analytic hierarchy process (AHP) is a popular methodology, but it involves subjectivity as well. Methods for solving the ABC inventory classification problem have been systematically and thoroughly reviewed and discussed in the relevant literature [9–12]. A number of methods were suggested in order to achieve multi-criteria classification of SKUs. These methods contribute much to the classification of items and help improve the efficiency and performance of a firm through better inventory management. However, these approaches contain some shortcomings, such as involving too much subjectivity or being overly complicated.

To facilitate better allocation of the priorities of items and further classification, it is worth developing a model that can accommodate multiple criteria to create guidelines for inventory control. This study builds a proper model for categorizing SKUs and demonstrates an efficient algorithm for solving the nonlinear programming model, in which the feasible solution set does not have to be convex. The rest of this paper is structured as follows. Section 2 provides the details of the model's development. The solution algorithm and its improvement are presented in Section 3. Section 4 details the results of the model constructed herein, with comparisons to previous studies using a benchmark data set. Conclusions and recommendations for future research are offered in the final section.

2. Literature Review on the HV Model and the WPM

Hadi-Vencheh [13] proposed a multi-criteria weighted nonlinear model for ABC inventory classification. The proposed model (hereafter the HV model) is an extension of the Ng model [1]. The Ng model transforms the inventory object to a scalar value. The grouping, according to the measured values, is then applied according to the ABC theory. Hadi-Vencheh also extended the Ng model to resolve the condition in which the score is independent of the weights from the model for each item. Despite the improvement in maintaining the influences of weights in the final score, one notable problem remains: the HV model calculates the scores assigned to each item using the weighted sum method (WSM) for criteria with different measurement units, which therefore requires converting the performance of individual inventory items into a normalized attribute value, expressed in terms of every criterion. Triantaphyllou [14] contended that if the problem involved criteria with different measurement units, the weighted product method (WPM) would be a more suitable tool to calculate the scores given to each item. To avoid an erroneous extreme value leading to inventory item misclassification, we propose the following nonlinear WPM to model the classification problem involving criteria with different measurement units. This study presents a broader version of the HV model, taking weight values into account in the ABC inventory classification for multiple criteria by using the WPM, which applies multiplication weights and forms a nonlinear optimization problem. To solve the nonlinear optimization problem efficiently, the canonical coordinates method (CCM) algorithm is used to calculate the weights of the criteria for each inventory item.

Suppose that I inventory items are present, and that the items must be graded as A, B, or C based on their results, according to J criteria. In particular, let the output of the ith inventory item be referred to as $y_{i,j}$ with respect to each criterion. For simplicity, all parameters are beneficial; in other words, they are positively connected with the degree of value of an item. The goal is to combine many performance scores in the subsequent ABC inventory classification, in regard to different parameters, into a single score. In both the Ng and HV models, a nonnegative weight $w_{i,j}$ is the weight of performance contribution of the ith item under the jth criteria to the score of the item. The parameters are supposed to be listed in descending order such that $w_{i,1} \geq w_{i,2} \geq \cdots \geq w_{i,J}$ for all items i. The proposed model by Hadi-Vencheh [13] is as follows:

$$
\begin{aligned}
\max S_i = \quad & \sum_{j=1}^{J} y_{i,j} w_{i,j} \\
\text{s.t.} \quad & \sum_{j=1}^{J} w_{i,j}^2 = 1 \\
& w_{i,j} - w_{i,j+1} \geq 0, \ j = 1, 2, \ldots, J-1 \\
& w_{i,j} \geq 0, \ j = 1, 2, \ldots, J
\end{aligned}
\tag{1}
$$

In the HV model, the performance in each criterion of the ith inventory item $y_{i,j}$ is further normalized to $s_{i,j}$, and the objective function of the nonlinear programming (NLP) model from Equation (1) is found to be

$$
\max \quad S_i = \sum_{j=1}^{J} s_{i,j} w_{i,j}
$$

Ng [1] indicated that normalization scaling involves extreme measurement values and would thus have an effect on all normalized measurements if the extremes changed. To avoid an invalid extreme value leading to inventory item misclassification, we propose the following nonlinear WPM to model the classification problem involving criteria with different measure units:

$$
\begin{aligned}
\max S_i = \quad & \prod_{j=1}^{J} y_{i,j}^{w_{i,j}} \\
\text{s.t.} \quad & \sum_{j=1}^{J} w_{i,j}^2 = 1 \\
& w_{i,j} - w_{i,j+1} \geq 0, \ j = 1, 2, \ldots, J-1 \\
& w_{i,j} \geq 0, \ j = 1, 2, \ldots, J
\end{aligned}
\tag{2}
$$

3. The Solution Algorithm

3.1. Nomenclature

3.1.1. Notation of the Weighted Product Method for ABC Classification

- I: set of inventory items;
- J: set of evaluation criteria;
- $y_{i,j}$: the ith inventory item in terms of the jth criteria;
- $w_{i,j}$: the weight of performance contribution of the ith item under the jth criteria;
- S_i: score of the item i.

3.1.2. Notation of the CCM Algorithm

- $\mathcal{R}$: the set of all real numbers;

- ξ: decision variables;
- ξ^0: feasible initial solution;
- $\phi_i(\xi)$: set of constraints, $i = 1, \ldots, m$;
- $f(\xi)$: the objective function;
- ξ^*: the optimal solution.

3.2. The CCM Algorithm

This section presents the canonical coordinates method (CCM) algorithm [15,16], which is applied to solve nonlinear programming problems in which the feasible solution set does not have to be convex. Convexity is a strong property that often replaces differentiability as a desirable property in most constrained optimization problems. However, the CCM efficiently addresses continuous search spaces and benefits from the low computational cost for solving constrained optimization. A set containing nonlinear constraints may or may not be convex. This study mainly demonstrates an efficient algorithm for solving nonlinear programming problems in which the feasible solution set does not have to be convex. The main difference between linear and nonlinear programming is that linear programming helps find the best solution from a set of parameters or requirements that have a linear relationship, whereas nonlinear programming helps find the best solution from a set of parameters or requirements that have a nonlinear relationship. The prerequisite for applying the CCM algorithm is that the implicit function theorem can be used in any feasible set. That is, at any point in the feasible set, one can find m variables, such as $z = (z_1, \ldots, z_m)$, in such a way that the Jacobian matrix of the constraint functions $\phi = (\phi_1, \ldots, \phi_m)$, with respect to z, are nonsingular. In the implicit function theorem, there exist functions g_j such that $z_j = g_j(x_1, \ldots, x_n)$, $j = 1, \ldots, m$. We describe the Algorithm 1 below:

Algorithm 1. CCM Algorithm

Input: The nonlinear program:

$$\max\{f(\xi)\,|\,\phi_i(\xi) = 0,\ i = 1, \ldots, m\}$$

with given differentiable functions f, $\phi_i : \mathfrak{R}^{m+n} \to \mathfrak{R}$, $i = 1, \ldots, m$, and a feasible point $\xi^0 \in \mathfrak{R}^{m+n}$ satisfying $\phi_i(\xi^0) = 0$, $i = 1, \ldots, m$.

Output: A critical point ξ^* of f satisfying $\phi(\xi^*) = 0$.

Steps:

1. ξ^0 is partitioned into $\xi^0 = (x^0, z^0)$, where $x^0 \in \mathfrak{R}^n$ and $z^0 \in \mathfrak{R}^m$ are such that $\det\left(J(\phi, z)\big|_{(x^0, z^0)}\right) \neq 0$.

2. For $i, j = 1, \ldots, m$ and $k = 1, \ldots, n$, we calculate the following partial derivatives at point $\xi^0 = (x^0, z^0)$: $\partial f / \partial x_k, \partial f / \partial z_j, \partial \phi_i / \partial x_k$, and $\partial \phi_i / \partial z_j$.

3. We then calculate the $m \times n$ matrix for the implicit function g (i.e., $(\partial g / \partial x) = -(\partial \phi / \partial z)^{-1}(\partial \phi / \partial x)$) and then find the direction $D^0 := (\partial f / \partial x) = (\partial f / \partial x_1, \ldots, \partial f / \partial x_n)$ and $D^0 = (\partial f / \partial x)^T + (\partial f / \partial z)(\partial g / \partial x)$.

4. We perform a line search along the ray through x^0 with the direction $D^0 = D(x^0)$; that is to say, we find a one-dimensional local optimal t^* of $\hat{F}(t) := f(x^0 + tD^0, z(t))$, $t \geq 0$. To do so, we need to solve for $z(t)$, which is done by solving the following system of ordinary differential equations:

$$\begin{cases} z(0) = z^0 \\ (\partial \phi / \partial x)(\partial F / \partial x)^T + (\partial \phi / \partial z)(dz / dt)^T = 0 \end{cases}$$

   where we set $x^* \leftarrow x^0 + t^* D^0$.

5. We compute $z_j^* = g_j(x^*)$, $j = 1, \ldots, m$ using Taylor polynomial approximation and then apply Newton's method to solve the system of ordinary differential equations at $t = t^*$ above.

6. If $\nabla f(x^*, z^*) \approx 0$, then we have found a local optimal point. Otherwise, we replace (x^0, z^0) with (x^*, z^*) and repeat the procedure.

The CCM algorithm helped us identify the local optimal points of NLP so that the feasible set fulfilled the requirements of the implicit function theorem. The problem could then be turned into an NLP problem on a subspace $\mathfrak{R}^n$ of the original space $\mathfrak{R}^{m+n}$.

3.3. Improvement of the Algorithm Using Efficient Selection of Bases

Step 1 in the CCM algorithm is to find a subset of m variables among the $(m+n)$ variables so that a resulting Jacobian matrix is non-singular [8]. This is equivalent to finding a subset of column vectors in the original $m \times (m+n)$ matrix that is linearly independent. Let $\phi_1, \cdots, \phi_m$ be differentiable functions in $(m+n)$ variables:

$$
A = \begin{bmatrix}
\partial\phi_1/\partial x_1 & \partial\phi_1/\partial x_2 & \cdots & \partial\phi_1/\partial x_{m+n} \\
\partial\phi_2/\partial x_1 & \partial\phi_2/\partial x_2 & \cdots & \partial\phi_2/\partial x_{m+n} \\
\vdots & \vdots & \vdots & \vdots \\
\partial\phi_m/\partial x_1 & \partial\phi_m/\partial x_2 & \cdots & \partial\phi_m/\partial x_{m+n}
\end{bmatrix}
$$

In order to find a subset of m columns of A that is linearly independent, the original method by Chang and Prabhu [15] was to choose any m subset of the $(m+n)$ columns to check if it qualified. There are two drawbacks to doing this. First, there are C_m^{m+n} choices of such subsets, and second, each choice will require a calculation of the determinant of an $m \times m$ matrix, which has the same complexity as a Gaussian elimination process. We will show that, by using Gaussian elimination on A to reach its reduced row echelon form, we can find one subset of columns of A that is linearly independent. The process of Gaussian elimination involves performing a sequence of row operations to a given matrix to reach its reduced row echelon form. Each type of row operation corresponds to a type of elementary matrix, all of which are nonsingular. Each time a row operation is performed, it is equivalent to multiplying the original matrix by an elementary matrix on the left. We can also see that in an $m \times (m+n)$ matrix with rank m, there is a subset of column vectors that is linearly independent. Now, let us state the proposition that yields the discovery of the desired linearly independent subset of columns of A.

Proposition 1. *Let m and n be positive integers, A be an $m \times (m+n)$ matrix with rank m, and $u_1, \ldots, u_{m+n}$ be the column vectors of A. Suppose B is the reduced row echelon form of A, and that $v_1, \ldots, v_{m+n}$ are the columns of B. Then, there exist integers $1 \le j_1 < j_2 < \cdots < j_m \le (m+n)$ such that $\left[v_{j_1}, \cdots, v_{j_m} \right] = I_{m \times m}$ forms the $m \times m$ identity matrix. Moreover, the corresponding subset of columns of A, $\left[u_{j_1}, \cdots, u_{j_m} \right]$, is non-singular.*

Proof. Matrix B must also have a rank of m because it is the reduced row echelon form of A, whose rank is m. Thus, there are m columns of B that form the $m \times m$ identity matrix. That is, there exist integers $1 \le j_1 < j_2 < \cdots < j_m \le (m+n)$ such that $\left[v_{j_1}, \cdots, v_{j_m} \right] = I_{m \times m}$. During the process of Gaussian elimination, performed to obtain the reduced row echelon form of A, we can find elementary matrices $E_1, E_2, \cdots, E_p$ such that

$$
B = E_p \cdots E_2 E_1 A
$$

Note that the kth column of B is also obtained from performing the same row operations on the kth column of A. Thus, we can say

$$
v_{j_k} = E_p \cdots E_2 E_1 u_{j_k}, \ k = 1, 2, \cdots, m
$$

Additionally, we can say

$$
E_p \cdots E_2 E_1 \left[u_{j_1}, \cdots, u_{j_m} \right] = \left[v_{j_1}, \cdots, v_{j_m} \right] = I
$$

Since all the elementary matrices $E_1, E_2, \cdots, E_p$ and the identity matrix I are nonsingular, then $\left[u_{j_1}, \cdots, u_{j_m}\right]$ must also be nonsingular. Now, we can simply apply Gaussian elimination to the matrix and find a linearly independent subset of column vectors that allows the implicit function theorem and the CCM algorithm to be applied. $\square$

3.4. Accuracy Improvement

A line search in the fourth step, whereby a system of nonlinear ordinary differential equations with initial values is resolved [15], must be carried out in the implementation of the CCM algorithm. The desired unidimensional direction can be approximated numerically from any line search, but its explicit functional expression cannot be calculated. This drawback impedes the output of the points found in any line search system. We present a modification of the CCM algorithm used by Chang and Prabhu [15], which adopted the gradient method to determine the next point without any line searching.

Suppose the feasible set $S = \left\{u \in \mathbf{R}^{m+n} \middle| \phi_i(u) = 0, \ i = 1, \cdots, m\right\}$ satisfies the condition of the implicit function theorem. That is, $\phi_i(u)$ is a holomorphic function in the $m + n$ variables, if one treats the $m + n$ variables as complex variables such that one of its Jacobians is nonsingular. Therefore, one can find an m subset of the $m + n$ variables, such as $z_1, \cdots, z_m$, so that the corresponding Jacobian matrix $(\partial\phi/\partial z)$ is nonsingular. Furthermore, there exist implicit functions g_j, $j = 1, \cdots, m$ in terms of the remaining n variables, $x_1, \cdots, x_n$ for example, such that $z_j = g_j(x_1, \ldots, x_n)$, $j = 1, \cdots, m$. The original NLP can now be viewed as the following induced NLP:

$$\text{Maximize} \quad F(x)$$
$$\text{Subject to} \quad x \in U$$

where $F(x) = f(x, g_1(x), \cdots, g_m(x))$ and U is a neighborhood of the point $x^0 \in \mathbf{R}^n$ that the implicit function theorem holds. Because U contains an open subset of $\mathbf{R}^n$, the induced NLP can be viewed as a locally non-constrained NLP model. One important benefit is that moving along the induced gradient direction $D = (\partial F/\partial x)$ will stay in U if the distance is small enough.

One common issue with solving an NLP model using the gradient method is that it is likely to leave S by traveling along the gradient direction of a feasible point. This causes a big problem in keeping the NLP model's feasibility. Applying the CCM algorithm does not present such a problem, as every iteration remains within the feasible region. This is because the gradient of the inducted objective function F, with regard to the selected value $x_1, \cdots, x_n$, will locally move inside the feasible set U if selected carefully. When using the CCM algorithm in a small-scale NLP model, one can conduct a line search along the gradient direction of the induced objective function. The relation between the induced line search and the movement along the original feasible set is illustrated in Figure 1.

Because there is one-to-one mapping between U and a neighborhood of $\left(x^0, z^0\right)$ in S, there is a one-dimensional curve C in S, corresponding to the line $L = \{x(t) | t \geq 0\}$ in U such that

$$C = \left\{x(t), z(t) \middle| x(t) \in L, z(t) = g(x(t))\right\}$$

The problem can be viewed as lifting a straight line in $\mathbf{R}^n$ to a curve in $\mathbf{R}^{m+n}$. In performing a line search, one has to find a one-dimensional, local, optimal point on such a curve with only the knowledge of the projection of the curve, while the other coordinates are unknown. Fortunately, we also know the explicit objective and constraint functions, so we can approximate the change of the unknown coordinates Δz with the derivatives dz/dt. That is, we can approximate Δz by

$$\Delta z \sim \left(\frac{dz}{dt}\right)\Delta t$$

Let $D^0 = D(x^0) = (d_1, \cdots, d_n)$, in which case

$$\frac{dz_j}{dt} = \frac{\partial g_j}{\partial x_1}\frac{dx_1}{dt} + \cdots + \frac{\partial g_j}{\partial x_n}\frac{dx_n}{dt}$$

In addition,

$$\left(\frac{dz}{dt}\right) = \left(\frac{\partial g}{\partial x}\right)\begin{pmatrix} d_1 \\ \vdots \\ d_n \end{pmatrix} = -\left(\frac{\partial \phi}{\partial z}\right)^{-1}\left(\frac{\partial \phi}{\partial x}\right)\begin{pmatrix} d_1 \\ \vdots \\ d_n \end{pmatrix}$$

We can now move the previous point along its gradient direction $(\Delta x, \Delta z)$ where $\Delta x = D^0 \Delta t$, in which D^0 is the exact induced gradient of the induced objective function in the projection space and Δz is the change in z by the above approximation. We can choose Δt carefully to avoid any line searching. There are two reasons to avoid a line search.

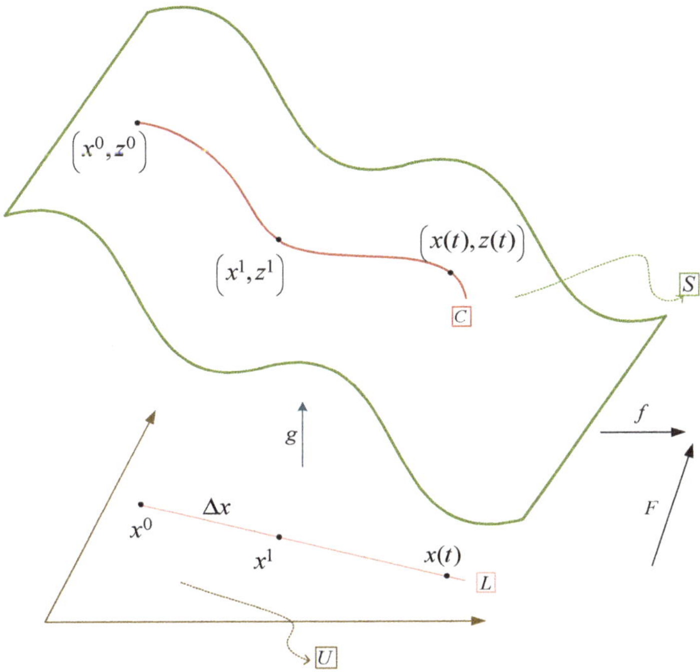

Figure 1. The relationship between induced line search and movement along the original feasible set.

First, it requires the knowledge of the lifted curve for any $z > 0$. When an NLP model is a small-scale problem and the corresponding system of ordinary differential equations is possible to solve, the CCM algorithm can locate or approximate the exact lifted curve C at any $t > 0$. To do so, we may need to apply some Ordinary Differential Equation methods, such as the Euler and Runge–Kutta methods. As t gets larger, the feasibility of the point $(x(t), z(t))$ is likely disappearing.

Second, we use numerical data to approximate the partial derivatives at all the points involved. The ODE problem in the fourth step is a point-by-point case without an explicit global expression for the coefficients $(\partial \phi/\partial x)(\partial F/\partial x)^T$ and $\partial \phi/\partial z$ in it. Thus, it might be impossible to solve for it in practice. We have chosen to avoid any line searching. Instead, as mentioned above, we move a point to the next one in its gradient direction $(\Delta x, \Delta z)$ with a chosen Δt. This way, we have control over staying as close to the feasible set as we need. Not only is the feasibility better kept, but the calculation is also reduced, since we are not solving for the system of ordinary differential equations globally.

4. Illustrative Example

We applied the WPM to the same problem of the multi-criteria inventory classification problem as reported in the referenced literature [1,6,7,13,17]. Following Ng [1] and Hadi-Vencheh [13], we considered three criteria for inventory classification: annual dollar usage (ADU), average unit cost (AUC), and lead time (LT). We also assumed the importance of the criteria to be ADU, AUC, and LT, in descending order. All of the criteria held positive inventory item scores.

4.1. Quality of Solutions

The 47 inventory items' optimal scores and weights are shown in Table 1. As demonstrated, the optimal scores derived by using the CCM algorithm to solve the WPM of multicriteria ABC classification were as good as those derived using LINGO [18], the off-the-shelf optimization software. If we look carefully at the weights derived using LINGO, it is obvious that most item weights for ADU and AUC were identical, and that some item weights (4, 25, 27, and 30) for LT were zero. This is because the primary underlying technique used by LINGO's nonlinear solver is to reach a feasible solution for nonlinear models quickly. The weight values derived by using the CCM better fit the assumption that the criteria were graded in descending order, such that $w_{i,1} \geq w_{i,2} \geq \cdots \geq w_{i,J}$ for all items i. Therefore, the CCM is superior to LINGO in terms of solution quality in this illustrative example.

Next, the maximal overall scores were sorted in descending order, and inventory classification was conducted based on the WPM (shown in Table 2). For comparison purposes, we maintained the same distribution of items in the A, B, and C classes as in studies within the cited literature [1,13,17]; that is, there were 10 class A items, 14 class B items, and 23 class C items. The ABC analysis using the Ng [1], Hadi-Vencheh [13], and Zhou and Fan (ZF) models [17] are also shown in Table 2. There were ten items (8, 29, 15, 16, 27, 33, 39, 40, 34, and 45) that did not have the same classifications in the WPM model as in the Ng, HV, and ZF models. The difference in classification was due to the difference in score computation schemes. Of the 10 class A items identified in the WPM, only item 8 was recognized as a class B item in the Ng, ZF, and HV models. Moreover, in these models, item 29 was classified as a class A item, while the WPM reclassified it as a class B item. A comparison of items 8 and 29 revealed that item 8 was superior to item 29 in terms of ADU value ($y_{8,1} = 2640 > y_{29,1} = 268.68$). Although item 29 outperformed item 8 in AUC ($y_{8,2} = 55 < y_{29,2} = 134.34$) and LT ($y_{8,3} = 4 < y_{29,3} = 7$), the differences were not significant. Therefore, based on the most important consideration, the value of the annual consumption of inventory items (ADU) in a year, the WPM provided a more reasonable classification.

Regarding the 14 class B items in the HV model, eight items (6, 7, 23, 18, 19, 28, 12, and 31) were retained in class B when the WPM was adopted, five of the class B items (33, 39, 40, 34, and 45) were reclassified as C, and the remaining one (item 8) was moved up to class A. Out of the 23 class C items, 18 items were retained as such, whereas the remaining five (15, 16, 22, 20, and 27) were moved up to class B. Items 33, 39, 40, 34, and 45, classified as class B items in the Ng, HV, and ZF models, were reclassified as class C items when using the WPM (see Table 3) and had relatively higher LT measurements (more than 4), but lower performance in terms of AUC and ADU, which were the two more important criteria. However, the maximum ADU value (197.92) of these five items was much less than the minimum ADU value (336.12) of items 15, 16 and 27, which were reclassified as class C items by the WPM, while their AUC values were about even. Therefore, the WPM provided a more reasonable ranking of items.

Table 1. Measures of inventory items, including their optimal scores and weights.

	Item Parameter			LINGO				CCM			
				Objective Value	Decision Variable			Objective Value	Decision Variable		
	ADU	AUC	LT	Score	ADU Weight	AUC Weight	LT Weight	Score	ADU Weight	AUC Weight	LT Weight
1	5840.64	49.92	2	8.92	7.050×10^{-1}	7.050×10^{-1}	7.770×10^{-2}	8.92	7.050×10^{-1}	7.049×10^{-1}	7.790×10^{-2}
2	5670	210	5	10.02	6.979×10^{-1}	6.979×10^{-1}	1.606×10^{-1}	10.02	6.980×10^{-1}	6.978×10^{-1}	1.608×10^{-1}
3	5037.12	23.76	4	8.38	6.974×10^{-1}	6.974×10^{-1}	1.654×10^{-1}	8.38	6.974×10^{-1}	6.973×10^{-1}	1.658×10^{-1}
4	4769.56	27.73	1	8.34	7.071×10^{-1}	7.071×10^{-1}	0.000×10^{0}	8.34	7.072×10^{-1}	7.071×10^{-1}	2.360×10^{-4}
5	3478.8	57.98	3	8.71	7.015×10^{-1}	7.015×10^{-1}	1.262×10^{-1}	8.71	7.015×10^{-1}	7.014×10^{-1}	1.264×10^{-1}
6	2936.67	31.24	3	8.15	7.007×10^{-1}	7.007×10^{-1}	1.347×10^{-1}	8.15	7.007×10^{-1}	7.006×10^{-1}	1.350×10^{-1}
7	2820	28.2	3	8.05	7.005×10^{-1}	7.005×10^{-1}	1.364×10^{-1}	8.05	7.005×10^{-1}	7.004×10^{-1}	1.367×10^{-1}
8	2640	55	4	8.52	6.977×10^{-1}	6.977×10^{-1}	1.627×10^{-1}	8.52	6.977×10^{-1}	6.976×10^{-1}	1.630×10^{-1}
9	2423.52	73.44	6	8.73	6.921×10^{-1}	6.921×10^{-1}	2.051×10^{-1}	8.73	6.921×10^{-1}	6.920×10^{-1}	2.053×10^{-1}
10	2407.5	160.5	4	9.20	6.990×10^{-1}	6.990×10^{-1}	1.507×10^{-1}	9.20	6.991×10^{-1}	6.989×10^{-1}	1.509×10^{-1}
11	1057.2	5.12	2	6.12	7.026×10^{-1}	7.026×10^{-1}	1.133×10^{-1}	6.12	7.025×10^{-1}	7.024×10^{-1}	1.145×10^{-1}
12	1043.5	20.87	5	7.24	6.894×10^{-1}	6.894×10^{-1}	2.222×10^{-1}	7.24	6.894×10^{-1}	6.893×10^{-1}	2.226×10^{-1}
13	1038	86.5	7	8.30	6.874×10^{-1}	6.874×10^{-1}	2.346×10^{-1}	8.30	6.875×10^{-1}	6.873×10^{-1}	2.347×10^{-1}
14	883.2	110.4	5	8.28	6.936×10^{-1}	6.936×10^{-1}	1.944×10^{-1}	8.28	6.938×10^{-1}	6.934×10^{-1}	1.945×10^{-1}
15	854.4	71.2	3	7.87	7.002×10^{-1}	7.002×10^{-1}	1.397×10^{-1}	7.87	7.003×10^{-1}	7.000×10^{-1}	1.399×10^{-1}
16	810	45	3	7.51	6.995×10^{-1}	6.995×10^{-1}	1.463×10^{-1}	7.51	6.996×10^{-1}	6.994×10^{-1}	1.465×10^{-1}
17	703.68	14.66	4	6.68	6.917×10^{-1}	6.917×10^{-1}	2.075×10^{-1}	6.68	6.917×10^{-1}	6.916×10^{-1}	2.081×10^{-1}
18	594	49.5	6	7.49	6.866×10^{-1}	6.866×10^{-1}	2.391×10^{-1}	7.49	6.867×10^{-1}	6.865×10^{-1}	2.393×10^{-1}
19	570	47.5	5	7.39	6.902×10^{-1}	6.902×10^{-1}	2.177×10^{-1}	7.39	6.902×10^{-1}	6.900×10^{-1}	2.178×10^{-1}
20	467.6	58.45	4	7.36	6.944×10^{-1}	6.944×10^{-1}	1.885×10^{-1}	7.36	6.946×10^{-1}	6.942×10^{-1}	1.887×10^{-1}
21	463.6	24.4	4	6.74	6.920×10^{-1}	6.920×10^{-1}	2.056×10^{-1}	6.74	6.920×10^{-1}	6.919×10^{-1}	2.058×10^{-1}
22	455	65	4	7.41	6.946×10^{-1}	6.946×10^{-1}	1.871×10^{-1}	7.41	6.948×10^{-1}	6.944×10^{-1}	1.873×10^{-1}
23	432.5	86.5	4	7.57	6.952×10^{-1}	6.952×10^{-1}	1.830×10^{-1}	7.57	6.954×10^{-1}	6.949×10^{-1}	1.832×10^{-1}
24	398.4	33.2	3	6.80	6.978×10^{-1}	6.978×10^{-1}	1.616×10^{-1}	6.80	6.979×10^{-1}	6.977×10^{-1}	1.618×10^{-1}
25	370.5	37.05	1	6.74	7.071×10^{-1}	7.071×10^{-1}	0.000×10^{-0}	6.74	7.071×10^{-1}	7.071×10^{-1}	2.570×10^{-4}
26	338.4	33.84	3	6.70	6.975×10^{-1}	6.975×10^{-1}	1.640×10^{-1}	6.70	6.975×10^{-1}	6.975×10^{-1}	1.642×10^{-1}
27	336.12	84.03	1	7.25	7.071×10^{-1}	7.071×10^{-1}	0.000×10^{0}	7.25	7.071×10^{-1}	7.071×10^{-1}	2.860×10^{-4}
28	313.6	78.4	6	7.37	6.859×10^{-1}	6.859×10^{-1}	2.431×10^{-1}	7.37	6.859×10^{-1}	6.859×10^{-1}	2.433×10^{-1}
29	268.68	134.34	7	7.67	6.840×10^{-1}	6.840×10^{-1}	2.537×10^{-1}	7.67	6.840×10^{-1}	6.840×10^{-1}	2.539×10^{-1}
30	224	56	1	6.67	7.071×10^{-1}	7.071×10^{-1}	0.000×10^{0}	6.67	7.071×10^{-1}	7.071×10^{-1}	2.880×10^{-4}
31	216	72	5	7.01	6.882×10^{-1}	6.882×10^{-1}	2.295×10^{-1}	7.01	6.882×10^{-1}	6.882×10^{-1}	2.297×10^{-1}
32	212.08	53.02	2	6.63	7.032×10^{-1}	7.032×10^{-1}	1.045×10^{-1}	6.63	7.032×10^{-1}	7.032×10^{-1}	1.048×10^{-1}
33	197.92	49.48	5	6.69	6.864×10^{-1}	6.864×10^{-1}	2.404×10^{-1}	6.69	6.864×10^{-1}	6.864×10^{-1}	2.406×10^{-1}
34	190.89	7.07	7	5.46	6.606×10^{-1}	6.606×10^{-1}	3.567×10^{-1}	5.46	6.606×10^{-1}	6.606×10^{-1}	3.595×10^{-1}
35	181.8	60.6	3	6.67	6.975×10^{-1}	6.975×10^{-1}	1.647×10^{-1}	6.67	6.975×10^{-1}	6.975×10^{-1}	1.649×10^{-1}
36	163.28	40.82	3	6.32	6.963×10^{-1}	6.963×10^{-1}	1.738×10^{-1}	6.32	6.963×10^{-1}	6.963×10^{-1}	1.740×10^{-1}

Table 1. *Cont.*

| Item Parameter | | | LINGO | | | | CCM | | | |
| | | | Objective Value | Decision Variable | | | Objective Value | Decision Variable | | |
ADU	AUC	LT	Score	ADU Weight	AUC Weight	LT Weight	Score	ADU Weight	AUC Weight	LT Weight
37	150	30	5	6.16	6.826×10^{-1}	6.826×10^{-1}	2.612×10^{-1}	6.16	6.826×10^{-1}	6.826×10^{-1}	2.613×10^{-1}
38	134.8	67.4	3	6.54	6.971×10^{-1}	6.971×10^{-1}	1.680×10^{-1}	6.54	6.971×10^{-1}	6.971×10^{-1}	1.683×10^{-1}
39	119.2	59.6	5	6.47	6.849×10^{-1}	6.849×10^{-1}	2.486×10^{-1}	6.47	6.849×10^{-1}	6.849×10^{-1}	2.488×10^{-1}
40	103.36	51.68	6	6.33	6.782×10^{-1}	6.782×10^{-1}	2.831×10^{-1}	6.33	6.782×10^{-1}	6.782×10^{-1}	2.833×10^{-1}
41	79.2	19.8	2	5.25	7.009×10^{-1}	7.009×10^{-1}	1.321×10^{-1}	5.25	7.009×10^{-1}	7.009×10^{-1}	1.323×10^{-1}
42	75.4	37.7	2	5.67	7.018×10^{-1}	7.018×10^{-1}	1.223×10^{-1}	5.66	7.018×10^{-1}	7.018×10^{-1}	1.227×10^{-1}
43	59.78	29.89	5	5.53	6.765×10^{-1}	6.765×10^{-1}	2.908×10^{-1}	5.53	6.765×10^{-1}	6.765×10^{-1}	2.910×10^{-1}
44	48.3	48.3	3	5.59	6.933×10^{-1}	6.933×10^{-1}	1.964×10^{-1}	5.59	6.933×10^{-1}	6.933×10^{-1}	2.000×10^{-1}
45	34.4	34.4	7	5.37	6.590×10^{-1}	6.590×10^{-1}	3.625×10^{-1}	5.37	6.590×10^{-1}	6.590×10^{-1}	3.639×10^{-1}
46	28.8	28.8	3	4.88	6.889×10^{-1}	6.889×10^{-1}	2.252×10^{-1}	4.88	6.889×10^{-1}	6.889×10^{-1}	2.291×10^{-1}
47	25.38	8.46	5	4.12	6.510×10^{-1}	6.510×10^{-1}	3.903×10^{-1}	4.12	6.510×10^{-1}	6.510×10^{-1}	3.931×10^{-1}

Table 2. Comparison of ABC classifications using the optimal weighted product model (WPM), Zhou and Fan (ZF) model, Ng model, and Hadi-Vencheh (HV) model inventory scores.

Item	Optimal Score (CCM)	ADU	AUC	LT	WPM Model (CCM)	WPM Model (LINGO)	HV model	Ng Model	ZF Model
2	10.0222	5670	210	5	A	A	A	A	A
10	9.20134	2407.5	160.5	4	A	A	A	A	A
1	8.92424	5840.64	49.92	2	A	A	A	A	A
9	8.73406	2423.52	73.44	6	A	A	A	A	A
5	8.70633	3478.8	57.98	3	A	A	A	A	B
8	8.51791	2640	55	4	A	A	B	B	B
3	8.38316	5037.12	23.76	4	A	A	A	A	A
4	8.33829	4769.56	27.73	1	A	A	A	A	C
13	8.29587	1038	86.5	7	A	A	A	A	A
14	8.28055	883.2	110.4	5	A	A	A	B	A
6	8.15406	2936.67	31.24	3	B	B	B	A	C
7	8.05394	2820	28.2	3	B	B	B	B	C
15	7.86615	854.4	71.2	3	B	B	C	C	C
29	7.67051	268.68	134.34	7	B	B	A	A	A
23	7.57315	432.5	86.5	4	B	B	B	B	B
16	7.50776	810	45	3	B	B	C	C	C

Table 2. *Cont.*

Item	Optimal Score (CCM)	ADU	AUC	LT	WPM Model (CCM)	WPM Model (LINGO)	HV model	Ng Model	ZF Model
18	7.49247	594	49.5	6	B	B	B	B	A
22	7.40991	455	65	4	B	B	C	C	B
19	7.39402	570	47.5	5	B	B	B	B	B
28	7.36954	313.6	78.4	6	B	B	B	B	A
20	7.35514	467.6	58.45	4	B	B	C	C	B
27	7.24598	336.12	84.03	1	B	B	C	C	C
12	7.24393	1043.5	20.87	5	B	B	B	B	B
31	7.01167	216	72	5	B	B	B	B	B
24	6.79949	398.4	33.2	3	C	C	C	C	C
21	6.74367	463.6	24.4	4	C	C	C	C	C
25	6.73618	370.5	37.05	1	C	C	C	C	C
26	6.69892	338.4	33.84	3	C	C	C	C	C
33	6.69389	197.92	49.48	5	C	C	B	B	B
17	6.67996	703.68	14.66	4	C	C	C	C	C
30	6.67213	224	56	1	C	C	C	C	C
35	6.67164	181.8	60.6	3	C	C	C	C	C
32	6.63135	212.08	53.02	2	C	C	C	C	C
38	6.53696	134.8	67.4	3	C	C	C	C	C
39	6.47356	119.2	59.6	5	C	C	B	B	B
40	6.32774	103.36	51.68	6	C	C	B	B	B
36	6.32156	163.28	40.82	3	C	C	C	C	C
37	6.16169	150	30	5	C	C	C	C	B
11	6.11791	1057.2	5.12	2	C	C	C	C	C
42	5.66488	75.4	37.7	2	C	C	C	C	C
44	5.59151	48.3	48.3	3	C	C	C	C	C
43	5.5337	59.78	29.89	5	C	C	C	C	C
34	5.45517	190.89	7.07	7	C	C	B	B	B
45	5.36813	34.4	34.4	7	C	C	B	B	B
41	5.24818	79.2	19.8	2	C	C	C	C	C
46	4.87682	28.8	28.8	3	C	C	C	C	C
47	4.12265	25.38	8.46	5	C	C	C	C	C

Table 3. The 10 items reclassified using the WPM.

Item	ADU	AUC	LT	WPM	HV Model	Ng Model	ZF Model
8	2640	55	4	A	B	B	B
29	268.68	134.34	7	B	A	A	A
15	854.4	71.2	3	B	C	C	C
16	810	45	3	B	C	C	C
27	336.12	84.03	1	B	C	C	C
33	197.92	49.48	5	C	B	B	B
39	119.2	59.6	5	C	B	B	B
40	103.36	51.68	6	C	B	B	B
34	190.89	7.07	7	C	B	B	B
45	34.4	34.4	7	C	B	B	B

4.2. Elapsed Runtime and Iterations

The efficiency of solving a problem is also an important criterion for algorithm comparisons; a good algorithm should solve problems within an acceptable time. This section compares the number of iterations of implementations of the CCM and LINGO to solve the WPM of multi-criteria ABC classification. The main difference between implementing the CCM algorithm and the LINGO solver was that we did not need to specify a starting point or moving step size. When the solution region is a polyhedron, determining the first basic solution (the starting point) would be vital, as the local optimal solution is usually located near the basic feasible solution (BFS); the quality of the solution is highly related to the location of the BFS. An unsuitable starting point could lead to a worse local optimal solution. The search region of an algorithm is related to the step size of the line search, and the search region decides whether a feasible solution can be found. The step size also determines the quality of the final solution. A large step size within a line search could jump over the optimal solution, whereas smaller search distances could trap in and require a significant amount of time to reach the local optimal solution. A good starting point and step size for a search can help researchers find acceptable solutions in less time. Table 4 illustrates the process of tuning the step size in order to reach a feasible solution. The CCM algorithm provides more flexibility than other solvers from commercial package software, which presents a higher probability of finding better solutions. In this study, using CCM to solve the problem required more time to achieve the local optimal solution than using LINGO because the step size of the line search determined whether the CCM could find feasible solutions. In this study, LINGO took mere seconds to find a feasible solution, whereas CCM took more time—sometimes nearly a minute—to find a solution.

Table 4. Tuning the step size to reach a feasible solution.

	Step size	$\sum_{j=1}^{J} w_j^2$		Step size	$\sum_{j=1}^{J} w_j^2$
Item 4	0.0003	1.000009	Item 5	0.00045	1.000062
	0.000295	0.999993		0.000448	1.000059
	0.000298	1.000003		0.00044	1.000046
	0.000297	1		0.00041	0.999997
				0.000413	1.000002
				0.000412	1

When looking at the number of iterations each algorithm needs, CCM requires more iterations to solve the problem, even with a larger step size. As presented in Table 5, LINGO can solve most problems in only 60 iterations, whereas CCM might need approximately one thousand iterations. In conclusion, LINGO is more efficient than CCM in this study, which contradicts the results of the previous study, which stated that CCM can solve nonlinear problems more quickly than other software packages. The result of the current study might stem from the fact that the problem in this study is too simple to demonstrate the power of CCM.

Table 5. The number of iteration for LINGO and the canonical coordinates method (CCM) to find the local optimal solution.

Item	LINGO	CCM	Item	LINGO	CCM	Item	LINGO	CCM	Item	LINGO	CCM
1	60	1720	13	60	1505	25	35	2482	37	60	1367
2	60	1866	14	55	1852	26	60	1646	38	50	2369
3	60	1207	15	60	1911	27	35	3213	39	50	1888
4	35	1867	16	59	1673	28	55	1741	40	50	1657
5	60	1627	17	60	1090	29	50	2024	41	58	1903
6	60	1419	18	59	1377	30	35	3045	42	50	2452
7	60	1388	19	57	1456	31	55	1900	43	55	1496
8	60	1495	20	60	1729	32	55	2397	44	75	555
9	58	1428	21	60	1287	33	55	1645	45	55	215
10	57	1995	22	55	1802	34	60	1158	46	75	492
11	60	1086	23	55	2009	35	50	2223	47	60	2712
12	60	1105	24	60	1612	36	55	1932			

5. Conclusions

In this paper, we presented an extended version of the HV model to improve multi-criteria ABC inventory classification. Our proposed nonlinear weighted product model (WPM) incorporates multiple criteria with different measurement units without converting the performance of each inventory item in terms of each criterion into a normalized attribute value. This represents an improvement over the model proposed by Hadi-Vencheh. The WPM could also be viewed as providing a more reasonable classification for inventory items from the illustrated example, presented and used to compare our model with the HV model. In this paper, we also presented the improved CCM algorithm for solving the WPM, in which nonconvex nonlinearity was present in both the objective function and the constraints. The strategy presented here involved greatly reducing the steps in choosing m variables among $(m + n)$ variables, such that the corresponding $m \times m$ Jacobian matrix was nonsingular. Using the improved algorithm, we applied Gaussian elimination to the original matrix to determine which m variables to choose. Our second improvement was to remove solving the nonlinear differential equations system, which occurs in the line search method of the CCM algorithm. This paper demonstrates an efficient algorithm for solving nonlinear programming problems, in which the feasible solution set does not have to be convex. The practical implication of this study is to further improve the efficient nonlinear optimization solver based on the CCM by optimizing the quality of existing solutions, thus improving time and space efficiency.

Future research should continue to investigate the feasibility of implementing this proposed CCM algorithm in discrete domain issues for engineering applications in order to decide if the algorithm could be superior to off-the-shelf software. Future studies could apply the CCM to other nonlinear programs that arise in practice. For instance, autonomous vehicles represent one of many developments that will influence future mobility needs and planning needs. Traffic assignment models seek the same objective as route guidance strategies and provide turning points with information for implementing route guidance control strategies. Faster algorithms developed specifically for traffic assignment can be adapted and used in vehicle route guidance systems. The minimization of total travel time is a common goal, both globally and from a traffic administration perspective. The current road network manages more traffic by achieving system optimization. Some researchers have focused their efforts on dynamic traffic assignment because of the unrealistic assumptions of static traffic assignment. Difficulties encountered by the dynamic model result from route calculation being related to travel time on an arc, which is also dependent on the traffic along the route. It is difficult to solve such relationships analytically under a dynamic circumstance. In response to the difficulties of dynamic traffic modeling, Jahn et al. [19] therefore developed a model in which flow represents the traffic patterns in a steady state, and the results are the boundary for the total travel time. However, the algorithm by Jahn et al. [19] only solves problems with convex, nonlinear objective functions and linear constraints. To avoid this restriction, future studies could adopt the CCM to solve nonlinear optimization models and provide strategies for route guidance.

Author Contributions: Conceptualization, H.-C.C. and P.-C.L.; methodology, H.-C.C.; software, H.-C.C. and P.-C.L.; validation, H.-C.C.; data curation, P.-C.L.; writing—original draft preparation, P.-C.L.; writing—review and editing, P.-C.L.; visualization, P.-C.L.; funding acquisition, P.-C.L. All authors have read and agreed to the published version of the manuscript.

Funding: This research was funded by the Ministry of Science and Technology, R.O.C., grant number 101-2221-E-006-215-MY2.

Conflicts of Interest: The authors declare no conflict of interest.

References

1. Ng, W.L. A simple classifier for multiple criteria abc analysis. *Eur. J. Oper. Res.* **2007**, *177*, 344–353. [CrossRef]
2. Flores, B.E.; Whybark, D.C. Multiple criteria ABC analysis. *Int. J. Oper. Prod. Man.* **1986**, *6*, 38–46. [CrossRef]
3. Teunter, R.H.; Babai, M.Z.; Syntetos, A.A. ABC classification: Service levels and inventory costs. *Prod. Oper. Manag.* **2010**, *19*, 343–352. [CrossRef]
4. Harirchian, E.; Jadhav, K.; Mohammad, K.; Aghakouchaki Hosseini, S.E.; Lahmer, T. A comparative study of MCDM methods integrated with rapid visual seismic vulnerability assessment of existing RC structures. *Appl. Sci.* **2020**, *10*, 6411. [CrossRef]
5. Carneiro, J.; Martinho, D.; Alves, P.; Conceição, L.; Marreiros, G.; Novais, P. A multiple criteria decision analysis framework for dispersed group decision-making contexts. *Appl. Sci.* **2020**, *10*, 4614. [CrossRef]
6. Flores, B.E.; Olson, D.L.; Dorai, V. Management of multicriteria inventory classification. *Math. Comput. Model.* **1992**, *16*, 71–82. [CrossRef]
7. Ramanathan, R. ABC inventory classification with multiple-criteria using weighted linear optimization. *Comput. Oper. Res.* **2006**, *33*, 695–700. [CrossRef]
8. Flores, B.E.; Whybark, D.C. Implementing multiple criteria ABC analysis. *J. Oper. Manag.* **1987**, *7*, 79–85. [CrossRef]
9. Adida, E.; Perakis, G. A nonlinear continuous time optimal control model of dynamic pricing and inventory control with no backorders. *Nav. Res. Log.* **2007**, *54*, 767–795. [CrossRef]
10. Aggarwal, V. A closed-from approach for multi-item inventory grouping. *Nav. Res. Log.* **1983**, *30*, 471–485. [CrossRef]
11. Leap, N.J.; Bauer, K.W. A confidence paradigm for classification systems. *Nav. Res. Log.* **2011**, *58*, 236–254. [CrossRef]
12. Liang, X.; Yan, H. Inventory control and replenishment with flexible delivery-time upgrade. *Nav. Res. Log.* **2014**, *61*, 418–426. [CrossRef]
13. Hadi-Vencheh, A. An improvement to multiple criteria ABC inventory classification. *Eur. J. Oper. Res.* **2010**, *201*, 962–965. [CrossRef]
14. Triantaphyllou, E. *Multi-Criteria Decision Making Methods: A Comparative Study*; Boston, M.A., Ed.; Springer: Berlin/Heidelberg, Germany, 2000; Chapter 2; pp. 5–21.
15. Chang, H.-C.; Prabhu, N. Canonical coordinates method for equality-constrained nonlinear optimization. *Appl. Math. Comput.* **2003**, *140*, 135–158. [CrossRef]

16. Chang, H.-C.; Lin, P.-C. A demonstration of the improved efficiency of the canonical coordinates method using nonlinear combined heat and power economic dispatch problems. *Eng. Optimiz.* **2014**, *46*, 261–269. [CrossRef]
17. Zhou, P.; Fan, L. A note on multi-criteria ABC inventory classification using weighted linear optimization. *Eur. J. Oper. Res.* **2007**, *182*, 1488–1491. [CrossRef]
18. Lindo System Inc. *Extended Lingo (Release 8.0) [Computer Software]*; Lindo System Inc.: Chicago, IL, USA, 2003.
19. Jahn, O.; Möhring, R.; Schulz, A.; Stier-Moses, N. System-optimal routing of traffic flows with user constraints in networks with congestion. *Oper. Res.* **2005**, *53*, 600–616. [CrossRef]

Publisher's Note: MDPI stays neutral with regard to jurisdictional claims in published maps and institutional affiliations.

applied sciences

Article

Financial Optimization of the Resource-Constrained Project Scheduling Problem with Milestones Payments

Marcin Klimek

Department of Computer Science, State School of Higher Education in Biala Podlaska, 21-500 Biala Podlaska, Poland; m.klimek@dydaktyka.pswbp.pl

Featured Application: The proposed model of stage settlements with milestones payments may be useful in practical projects and be beneficial both for the client and for the contractor. Its application may reduce delays in the completion of project stages, which is a significant problem in information technology (IT) or industry projects.

Abstract: This article presents the resource-constrained project scheduling problem with the discounted cash flow maximization criterion from the perspective of a contractor. Cash flows are considered as the contractor's expenses related to the execution of activities and client's payments (revenue to the contractor) after the completion of contractual stages. To solve the problem, dedicated techniques to generate solutions and a simulated annealing algorithm are proposed. Finally, the proposed procedures are examined using the test library, Project Scheduling Library (PSPLIB). An experimental analysis identified the efficient moves and techniques for creating solutions, that is backward scheduling with optimization of completion times of project stages and triple justification.

Keywords: resource-constrained project scheduling problem; discounted cash flow maximization; milestones payments; simulated annealing algorithm

Citation: Klimek, M. Financial Optimization of the Resource-Constrained Project Scheduling Problem with Milestones Payments. *Appl. Sci.* **2021**, *11*, 661. https://doi.org/10.3390/app11020661

Received: 18 December 2020
Accepted: 11 January 2021
Published: 12 January 2021

Publisher's Note: MDPI stays neutral with regard to jurisdictional claims in published maps and institutional affiliations.

1. Introduction

The resource-constrained project scheduling problem (RCPSP) is one of the most frequently undertaken issues within operational research. For RCPSP, optimizing models are often analyzed which only include time criteria, that is makespan minimization etc. [1]. In practice, however, an analysis of the financial effects of planned projects, in particular large-scale projects executed over a long period of time, plays an important role. For long-term projects it is recommended to take into account the change in the value of money over time, which occurs in Max-NPV (net present value maximization) models [2]. The optimization criterion of NPV maximization is the most common criterion that takes into account the financial aspects of project planning [3]. When calculating NPV, the sum of all discounted cash flows associated with the project is calculated. RCPSP models with discounted cash flows (RCPSP-DC) are considered. An overview of NPV maximization research can be found in the literature [4,5].

Research works analyze payment project scheduling (PPS) models from the perspective of the contractor and/or the client. Payment schedules determine the frequency, amounts and delivery dates of client payments to the contractor. There are different methods for making financial settlements by compromise between the client and the contractor, whose interests in respect of the amount, timing and location of progress payments are divergent. The contractor aims to obtain the highest possible client payments as soon as possible. The client prefers to make a one-off payment after work. According to Szmerekovsky [6] both parties (client and contractor) are willing to maximize their profit instead of looking for a plan that satisfies them both. PPS problems are analyzed, where the solution from the perspective of client [7], the contractor [4,8,9] or both is sought [10–13].

This work discusses the problem of financial optimization from the contractor's perspective. Contractor expenditures (cash outflows) are generally known or may be estimated by them prior to the project, while client payments are determined by negotiation. Different approaches are considered in which total costs for executing activities may be incurred at their beginning [6,9] or completion [14], or may be proportionally allocated to the beginning and end of their execution [15,16], or distributed depending on their progress [11]. Fixed costs related to the activities are most often incurred at the start time and other costs, such as resource consumption, may arise during their execution.

As contractors are forced to spend money on project activities, it is advisable for them to acquire as much money as possible from the client prior to completing the project, such as, funds corresponding to the progress of the project. An improperly designed project settlement plan between the contractor and the client can cause financial problems for the contractor. Payments can be made in many ways [6,9,14,17]. One-off lump-sum payments after completion of a lump sum payment (LSP) project are one of the forms of settlement between client and contractor, most preferred by the client. For long-term projects, the LSP model may be unacceptable to a contractor who has to spend a long time financing the activities. In practice, progress payments are often determined throughout the duration of the project. Earlier client payments are typically associated with certain events and with milestones (completion of specific activities, stages). In general, payments can be made according to a fixed schedule, such as once a month (Equal Time Intervals—ETI model, Progress Payments–PP model) or associated with specific events (Payments at Event Occurrences–PEO model), i.e., completion of activity (Payments at Activities' Completion times–PAC model) etc. [9,17,18].

PPS models serve as a decision support tool for negotiations between the client and the contractor and can be used when planning the contractor's budget. Elmaghraby [19] suggests that the contractor can use information about the project and its costs to construct an appropriate client-contractor payment model where the delivery dates for these payments are assigned to the various milestones and the amounts are estimated based on the cost of performance of activities. All progress payments are arranged during the scheduling at milestones specified while planning or payment models are created that allow contractors to identify milestones and related cash flows.

For the project settlement between the client and the contractor, the bonus-penalty system [8,20,21] can be used to encourage the contractor to execute project (stages) as soon as possible. Penalties are determined for exceeding the contractual completion time of the project (stage), bonuses are given for early completion of the project (stage), etc. Contractual deadlines are specified for the project (stage) or time windows are defined where project (stage) completion is not awarded or penalized.

For the RCPSP also the capital constrained project scheduling (CCPSP) models are analyzed [22,23], where capital is one of the limited, non-renewable resources that is considered in the scheduling. At any time during the project, cash inflows and outflows must balance. Activities can be executed only if funds are available, such as those obtained by completing previous activities or stages of the project.

In this work we propose the RCPSP model with the criterion of maximizing the discounted cash flows associated with the project from the contractor's perspective, with stage settlements in which client agrees with the contractor the work stages (milestones) with deadlines and the amounts of payment for completion of stage work. In the event of late completion of the contractual stages, penalties reducing stage client payments are applied. The proposed optimization model can be useful in practice where milestones are an important part of work planning, for example, to be used to determine the progress of a project. In the RCPSP studies, the stage settlement of project work, apart from the author's work, was not considered in this form. Milestone payments are considered in studies but optimization models differ from those discussed in this work [16,24]. Milestone events are often assigned to events such as completion of individual activity [4,19]. In this work, payments are made after the completion of the activity group as a stage of the project [8,21,25,26].

For the proposed financial model used to optimize a multi-stage project, it is recommended to schedule the work so that activities are started as late as possible at the earliest possible completion of contractual stages. The aim of this work is to check the efficiency of the simulated annealing algorithm using the techniques of generating solutions, taking into account the specific stage settlements, that is: backward scheduling with optimization of project completion times, justification and right-shifting of activities for schedule with fixed resource allocation.

The remainder of this paper is organized as follows. Section 2 contains a formulation of the problem. Section 3 presents the definition of an exemplary project illustrating the analyzed problem. Section 4 provides a simulated annealing algorithm and techniques for generating solutions that are presented for the example project. Section 5 describes the experimental analysis performed for the Project Scheduling Library (PSPLIB) test instances [27] with defined additional project financial settlements. At the end of the work, a summary is presented and directions for further study are provided.

2. Problem Formulation

A project is a unique set of activities executed using available resources (employees, machines, materials). This work discusses the classic non-preemptive single-mode RCPSP. The project is shown as a directed graph $G(V, E)$, in representation of activity-on-node (AON), where V is a set of nodes corresponding to the activities and E is a set of arcs describing the order relations between the activities of finish-start zero-lag precedence. Activities are executed using the limited, renewable resources whose number is constant over time. The schedule (start times for all activities) is determined with the optimization criterion (makespan minimization, DCF maximization etc.) when the order constraints (1) and resource constraints (2) are fulfilled.

$$ST_i + d_i \leq ST_j, \qquad \forall(i,j) \in E \tag{1}$$

$$\sum_{i \in J(t)} r_{ik} \leq a_k, \qquad \forall t : t = 1, ST_{N_A+1}, \forall k : k = 1, \ldots, K \tag{2}$$

where:

- i is the activity index, $i = 1, \ldots, N_A$,
- N_A is the number of activities in a project,
- ST_i is the start time of the activity i,
- d_i is the duration of activity i,
- $J(t)$ is the set of activities executed in time interval $[t-1, t]$,
- a_k is the number of available resources of type k; at any time t, the number of resources used must not exceed the number of available resources a_k for $k = 1, \ldots, K$,
- r_{ik} is the demand of activity i for the resource of type k,
- K is the number of types of renewable resources.

This work offers an optimization model from the perspective of the contractor, which applies stage and financial settlements between the client and the contractor. Arrangements regarding joint settlements may be conducted for the adopted baseline schedule, which is the basis for defining inter alia, the number of contractual stages of the project, the dates of their completion, the activities to be taken at these stages and the method of payment determination. The proposed optimization model maximizing the discounted cash flow from the contractor's perspective with the costs incurred for the completion of the activities and the revenue obtained for the completed stages is described by the Equations (3)–(5):

$$F = \sum_{i=1}^{N_A} \left(CFA_i \cdot e^{-\alpha \cdot ST_i} \right) + \sum_{m=1}^{N_M} \left(CFM_m \cdot e^{-\alpha \cdot MT_m} \right) \tag{3}$$

$$MT_m = \max_{i \in MA_m} (FT_i), \qquad \forall m : m = 1, \ldots, N_M \tag{4}$$

$$CFM_m = MP_m - MC_m \cdot \max(MT_m - MD_m, 0), \qquad \forall m : m = 1, \ldots, N_M \tag{5}$$

where:

- F is the sum of discounted cash flows, objective function,
- CFA_i is the contractor's expenses related to the execution of activity i,
- m is the index of project stage (milestone), $m = 1, \ldots, N_M$ (N_M-number of stages),
- FT_i is the finish time of activity i ($FT_i = ST_i + d_i$),
- CFM_m is the client payment for completion of m—project stage established for the current schedule;
- α is the discount rate;
- MT_m is the completion time of m—project stage established for the current schedule;
- MA_m is the set of activities to be executed in m—project stage;
- MP_m is the agreed client payment to the contractor for completion of m—project stage;
- MD_m is the contractual deadline for m—project stage;
- MC_m is the contractual unit cost for exceeding MD_m deadline.

Client payments CFM_m (cash inflows from contractor's perspective) are made right after completion of contractual stages within MT_m completion time within defined amount MP_m reduced by any contractual penalties that include the delays unit cost of MC_m. It is assumed that all contractor's expenditures (cash outflows from contractor's perspective) are related to the execution of the activities and are incurred for each activity exactly as scheduled (in start time). The problem of possible delays in cash transfers is omitted in this model—for delayed payments, a system of non-periodic compensation may be used.

For contractors, the client payments for the project stages are financial means that can be used for the current business such as the purchase of materials necessary for the execution of subsequent activities, the salaries of the employees, etc. For the client, the need to incur earlier expenses is compensated for by the contractual penalties for failure to meet deadlines of the project stages and the possibility of controlling the course of work while the project is being executed.

3. Illustrative Example

To illustrate the analyzed optimization model we consider the sample project shown in Figure 1. The project consists of 8 activities and dummy activities 0 and 9 representing the starting and ending node in the graph G (V, E) respectively. Activities are executed using one type of limited, renewable resources of availability of 10. Discount rate used in calculations is $\alpha = 0.01$.

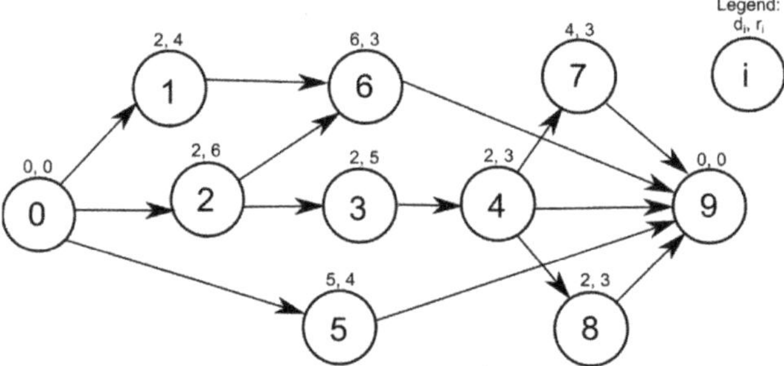

Figure 1. Example of project in activity-on-node (AON) representation.

Assume that the baseline schedule, which is the basis for determining the stage settlement of the project between the client and the contractor, is shown in Figure 2. It can be generated, for example, using the schedule generation scheme (SGS) serial, for the activity list {1, 5, 2, 3, 6, 4, 7, 8}.

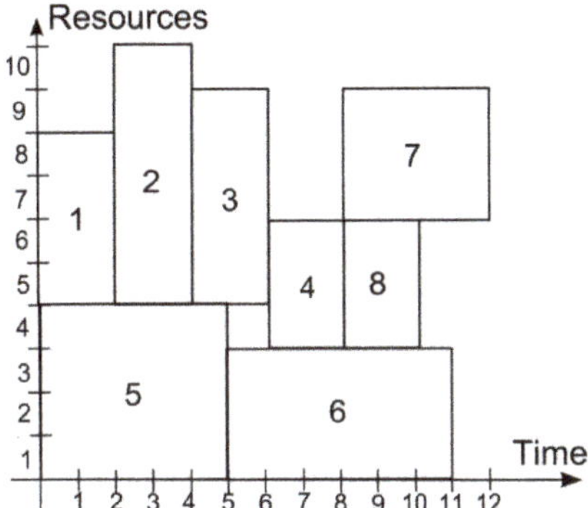

Figure 2. Baseline schedule.

Assume that based on the baseline schedule of Figure 2 three contractual stages of work that are distinguished that have the following completion times: $MD_1 = 4$, $MD_2 = 8$, $MD_3 = 12$. Activity sets to be completed in further stages are as follows $MA_1 = \{1, 2\}$, $MA_2 = \{3, 4, 5\}$, $MA_3 = \{6, 7, 8\}$. The cash flow related to CFA_i activities and the data for stage project settlements such as: contractual completion times MD_m, client contractual payment amounts MP_m and delay costs MC_m, are shown in the AON representation in Figure 3. Milestone data is denoted by the "square" symbol. The same colours show the nodes representing activities from the same stage of the project. The stage settlements should be arranged to encourage the contractor to execute the project as quickly as possible and not to exceed MD_m completion times.

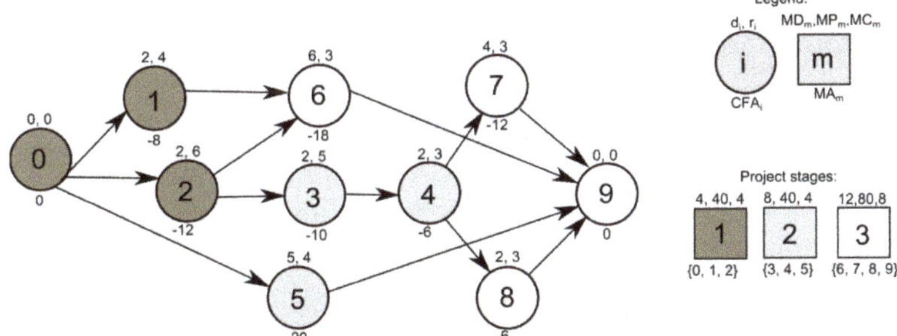

Figure 3. Project in AON representation with defined stages.

Once the stage settlement has been completed, the contractor will prepare a schedule which maximizes the sum of the discounted cash flows function F. It is advisable to delay the expenditure (negative cash flow) incurred at the start of activities (lower discounted value) and accelerate client payments for project stages (positive cash flows) acquired at the time of their completion (higher discounted value). The baseline schedule shown in Figure 2 can be improved, for example, by the deferred start of activities 6 and 8, which will not delay the completion of stage 3. It is also possible to complete the work stages earlier. The proposed techniques of generating solutions taking into account the specificity of the discussed problem, are described further in this work.

4. Simulated Annealing Algorithm

The resource-constrained project scheduling problem as a job shop generalization is a strongly NP-hard problem [28]. Due to the complexity of the problem for RCPSP, it is reasonable to use heuristic algorithms with polynomial computational complexity that find solutions within acceptable time also for larger projects. A wide overview and comparison of the efficiency of algorithms for the problem of project scheduling can be found in [5,29].

In this work, when searching for schedules, a simulated annealing (SA) algorithm is used, which is effective and has repeatedly been used to solve the RCPSP problem [29]. The SA is flexible and has the ability (when it is fine-tuned) to approach global optimality; it explores the solutions space and prefers better neighbours but also accepts worse neighbours in order to avoid becoming stuck in local optima. The SA improves a single candidate's solutions. Population-based metaheuristic algorithms (scatter search—SS, genetic algorithm—GA [30], particle swarm optimization–PSO [31,32]) can be better for global searches due to global exploration and local exploitation ability and will be examined for proposed problems in future work.

The SA algorithm developed in this work is similar to the basic variant of this meta-heuristics [33]. The solutions are represented as a list of activities–permutation of activity numbers taking into account the precedence relations. The activity list is converted to an executable schedule using the SGS [34]. In this work, the serial SGS is used, which in the forward (backward) scheduling strategy, determines the starting time for the first (last) unscheduled activity from the activity list at the earliest (latest) time, taking into account the precedence and resource constraints.

The SA algorithm starts the search from the random list of activities for which the serial SGS generates a schedule. Then iterations of the algorithm are repeated until the stop condition is satisfied: viewing the specified maximum number of schedules. Each iteration of the SA algorithm performs a move generating a new activity list from the neighborhood of the current solution. Moves that will be used are:

- Insert—an activity is randomly selected, which is then moved to a random position between the last predecessor in the activity list and the first successor;
- Swap—an activity is randomly selected, which is then exchanged with a randomly selected activity taking into account precedence relations;
- Adjacent swap—a random activity is selected, which is then exchanged with the next activity in the activity list, unless it violates the precedence restrictions;
- Insert all—an activity is randomly selected, which is then moved to the position between the last predecessor in the activity list and the first successor; all possible insert positions are checked and the position for which the schedule with the highest value of objective function F is generated;
- Swap all—an activity is randomly selected, which is then exchanged with every activity occupying the last position between the last predecessor in the list and the first successor; this exchange operation is performed for which the schedule with the highest value of objective function F is generated.

For the problem of maximizing discounted negative and positive cash flows in the PAC model, the schedule established by the SGS procedures can be improved by starting activities (stages) related to the positive cash flow(s) as early as possible and by starting activities (stages) related to negative cash flow as late as possible. For this problem, what is analyzed are, among other things, bi-directional SGS [29,30], shift operation algorithms that move activities with negative cash flows to the right and/or activities with positive cash flows to the left [4], and recursive forward/backward improvement procedures [35,36].

For a schedule that is appropriate for the problem analyzed in the article, with objective function F, the positive cash flow, that is the client payment for completing the contractual stages of work, should be obtained as soon as possible and the negative cash flow, that is the contractor's expenses related to the activities started, incurred according to the as late as possible (ALAP) principle. When planning activities, it is advisable to carry out a "balance sheet" of losses from subsequent payments for the completed project stages and the benefits of deferred expenses for starting activities. Increase in NPV always defers these activities in time, the delayed start of which does not change the completion time of the project stages.

The development of effective procedures for generating schedules dedicated to the problem of financial multi-stage optimization is the subject of previous research [8,26,37]. The following subsections describe and illustrate the scheduling techniques used by the analyzed SA algorithm:

- backward scheduling with optimization of completion times of contractual project stages;
- forward scheduling with activity right-shift procedure;
- justification.

4.1. Backward Scheduling with Optimization of Completion Times of Stages

A baseline schedule is created for the processed activity list assuming scheduled completion times of MT_m equal to the contractual MD_m deadlines. In the baseline schedule, the MT_m stage completion times can be larger than the contractual MD_m, if it is not possible to generate the executable schedule for the processed activity list from $MT_m = MD_m$ for each $m = 1, \ldots, N_M$. In the backward scheduling procedure with optimization of the completion times of the contractual stages of the project, the iterative, unit left-shift of planned MT_m are performed. Shifting operations are performed for the following stages, starting from the first and ending with the last. Unit shifts of a given stage are executed as long as this operation increases the value of the objective function F.

For an example project, the baseline schedule in Figure 2 is generated using the forward serial SGS. For backward serial SGS it is recommended to use a different activity list than for forward scheduling. It is suggested to set up a list of activities based on the baseline schedule—activities are sorted in the order of increasing activity finish times FT_i (similarly to the experimental studies for the initial solution). The activity list determined in this way is as follows: {1, 2, 5, 3, 4, 8, 6, 7}.

The backward schedule with the value of the objective function $F = 58.50$, determined for the activity list {1, 2, 5, 3, 4, 8, 6, 7} for completion times of stages such as: $MT_1 = MD_1 = 4$, $MT_2 = MD_2 = 8$, $MT_3 = MD_3 = 12$, is shown in Figure 4a. The procedure of left shifts of project stages starts from shifting the completion time of the first stage:

- for $MT_1 = 3$ a schedule with higher objective function is generated $F = 58.77$,
- for $MT_1 = 2$ a schedule with higher objective function is generated $F = 58.96$,
- for $MT_1 = 1$ it is not possible to create a feasible schedule and the procedure moves to the optimization of second stage and MT_2 completion time.

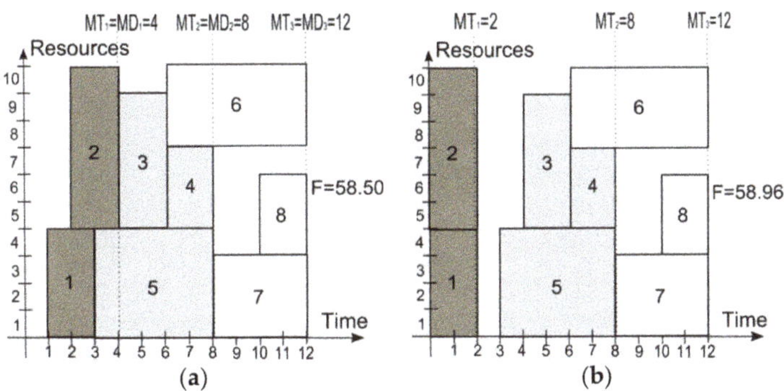

Figure 4. (a) Backward schedule for completion times of project stages $MT_1 = 4$, $MT_2 = 8$, $MT_3 = 12$, (b) schedule determined as a result of optimization of the completion time of first stage with $MT_1 = 2$, $MT_2 = 8$, $MT_3 = 12$.

The schedule established as a result of optimizing the completion time of the first stage with completion times of stages $MT_1 = 2$, $MT_2 = 8$, $MT_3 = 12$ is shown in Figure 4b. The optimization of the completion time of the second stage is as follows:

- for $MT_2 = 7$ the schedule with higher objective function $F = 58.98$ is generated;

- for $MT_2 = 6$ it is not possible to create a feasible schedule and the procedure moves to optimizing the completion time of the third stage MT_3.

The schedule determined by optimizing the completion time of the first and second stage with $MT_1 = 2$, $MT_2 = 7$, $MT_3 = 12$ is shown in Figure 5a. The optimization of the completion time of the third stage is as follows:

- for $MT_3 = 11$ the schedule with higher objective function $F = 59.36$ is generated;
- for $MT_3 = 10$ the schedule with higher objective function $F = 59.59$ is generated;
- for $MT_3 = 9$ it is not possible to create a feasible schedule and the procedure ends the first step of the procedure for optimizing the completion times of project stages.

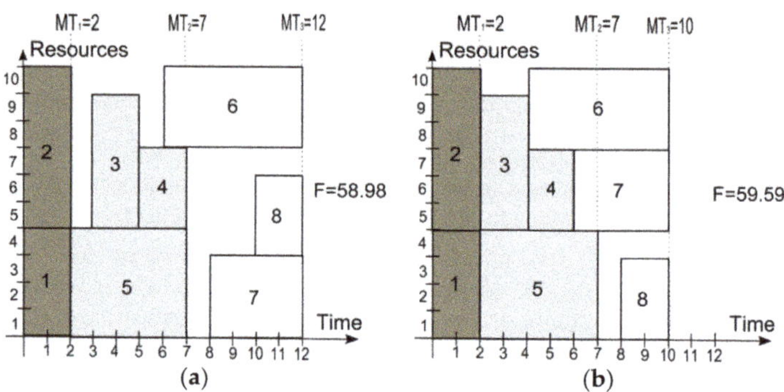

Figure 5. (**a**) Schedule determined as a result of optimization of the completion time of 2nd stage with $MT_1 = 2$, $MT_2 = 7$, $MT_3 = 12$, (**b**) schedule as a result of optimization of completion times of all stages: $MT_1 = 2$, $MT_2 = 7$, $MT_3 = 10$.

In the second step of the procedure, no changes are made (in optimizing of completion times of MT_1, MT_2, MT_3) to the schedule and the algorithm ends.

The final schedule generated during optimizing the completion times of all project stages with objective function $F = 59.59$ is shown in Figure 5b.

4.2. Forward Scheduling with Activity Right-Shift Procedure

Shifts of activities are performed taking into account the resource and order constraints. Shifting an activity to the right (or left) can cause various changes to the schedule. This problem does not occur if resource allocation is found for the schedule. Multiple resource allocations can be generated for a given schedule, characterized by various properties that affect the operations of right shifts and the quality of the resulting solutions. Assignment of resources to activities is analyzed during a proactive, robust scheduling [26,38–40]. An overview of resource allocation procedures is presented by Deblaere et al. [41]. For the analyzed problem, the resource allocation with as few additional order relations (additional arcs in $G(V,E)$) as possible are preferred as in robust scheduling. The new precedence restrictions resulting from the resource allocation reduce the possibility of shifting activities to the right to increase the cumulative cash flow F. To minimize the number of extra arcs, resource allocation algorithms allocate activities between which are order or unavoidable dependencies to be performed by the same resources, and maximize the sum of flows between the activities etc. [41]. For a schedule with a fixed resource allocation, it is easy to carry out the right-shift operations (resource constraints are taken into account)—unambiguously and with short calculation time which changes in the schedule will lead to starting activity shifted to the right can be determined.

In this work the resource allocation is generated using the ISH-UA (iterative sample heuristic with unavoidable arcs) procedure [7] whose effect on the schedule in Figure 2 is presented in Figure 6a.

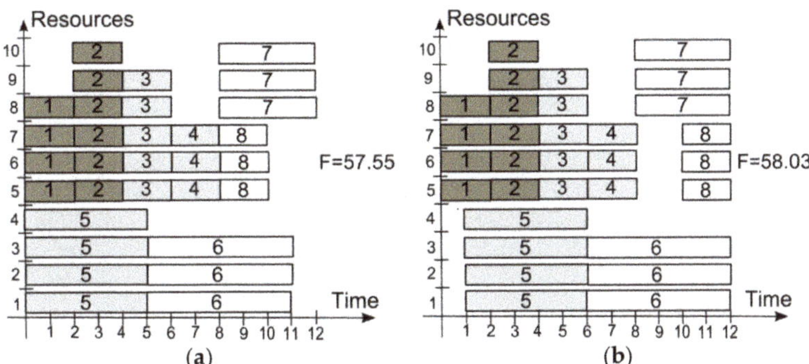

Figure 6. (**a**) Schedule with resource allocation set by the ISH-UA (iterative sample heuristic with unavoidable arcs) procedure, (**b**) schedule set using the activity right-shift procedure.

The right-shift procedure is iterative. In subsequent iterations, right-shifting of activities considered in the sequence of decreasing time of their completion in the schedule for which the allocation of resources is determined is analyzed. For the analyzed activity, unit shifts are performed until the operation does not increase the value of F. For the resource allocation in Figure 6a, the right-shift of the subsequent activities is analyzed:

- activity 7 shift will not increase F,
- one unit shift of activity 6 will increase F from 57.55 to 57.72,
- activity 8 shifts by two time units will increase F from 57.72 to 57.83,
- activity 4 and 3 shifts will not increase F,
- one unit shift of activity 5 will increase F from 57.83 to 58.03,
- activity 2 and 1 shifts will not increase F.
- Schedule with right shifts with $F = 58.03$ is shown in Figure 6b.

The activity right-shift procedure does not make any changes to the completion of the project stages. It can be effective in improving schedules with the best possible MT_m. For the analyzed example, the optimal solution with $F = 59.59$ (Figure 7b) is found using the SGS serial for the activity list {1, 2, 3, 4, 5, 6, 7, 8} with the resource allocation found using the ISH-UA procedure (Figure 7a).

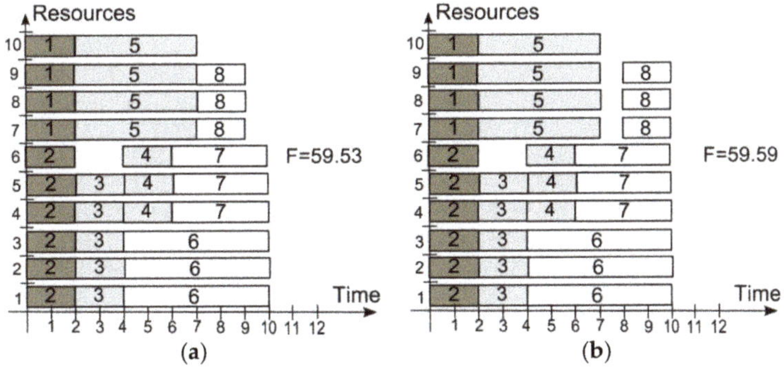

Figure 7. (**a**) Schedule generated for the activity list {1, 2, 3, 4, 5, 6, 7, 8} with the allocation of resources determined by the ISH-UA procedure. (**b**) Schedule improved by the right-shift procedure.

4.3. Justification

The right justification (RJ) and left justification (LJ) techniques transform the schedule generated by SGS procedures. Their use improves the quality of generated schedules [42]. They are used, among others. for RCPSP with a makespan minimization criterion or for

RCPSP with due dates. The justification of an activity to the right (to the left) consists in setting the latest (earliest) start time if possible taking into account the order and resource constraints. When generating solutions, RJ and LJ are combined: double justification is used, RJ + LJ or LJ + RJ respectively. Activities for RJ (LJ) can be selected:

- by any means from available activities whose all successors (predecessors) have already been subjected to justification when analyzing all activities from the beginning (end) of the schedule subjected to the justification (justification by eligible);
- in the order of decreasing finish times (increasing start times) in the schedule subjected to the justification (justification by extremes).

Most of the work, like this one, uses the effective justification by extremes technique, which is a special case of justification by eligibles.

Research conducted for the analyzed problem indicated that the triple justification technique (RJ + LJ + RJ) and the use of the modified RJ technique is effective for the forward schedule, and that the RJ technique is used to move an activity to the right (delaying its start), taking into account the current completion times of the project stages.

We will now show the RJ + LJ + RJ technique for the baseline schedule of Figure 2. First, RJ is used for activities with maximum finish time (justification by extremes)-activities in order 7, 6, 8, 4, 3, 5, 2, 1 (at the equal finish time, the activity with the higher number is analyzed earlier). For each activity, the latest start time is determined taking into account the precedence and resource relations and the actual completion time of the project stage to which this activity is assigned ($MT_1 = MD_1 = 4$, $MT_2 = MD_2 = 8$, $MT_3 = MD_3 = 12$). As a result of RJ, a schedule is presented in Figure 8a with the objective function $F = 58.50$.

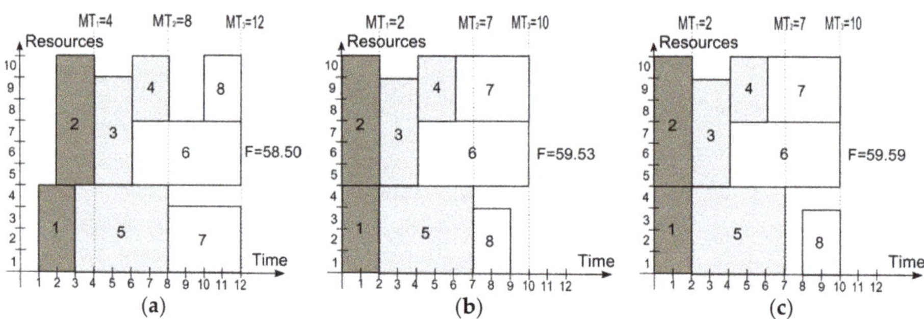

Figure 8. (**a**) Baseline schedule after right justification (RJ), (**b**) baseline schedule after RJ + left justification (LJ), (**c**) baseline schedule after RJ + LJ + RJ.

Technique LJ is performed after RJ. The activities with the minimum start time for the schedule in Figure 8a (justification by extremes) are selected for LJ. Activities in the order 1, 2, 3, 5, 4, 6, 7, 8 are justified (at equal start time, the activity with lower number is analyzed first). For each activity, the earliest start time is determined, taking into account the precedence and resource relations. As a result of LJ, the schedule shown in Figure 8b with the higher value of the objective function $F = 59.53$ than the schedule of Figure 8a is generated, thanks to the earlier completion of stages 1, 2 and 3 ($MT_1 = 2$, $MT_2 = 7$, $MT_3 = 10$).

The schedule in Figure 8b can be improved by delaying the start of the activity while meeting the completion times of the contractual stages of the project. The activities are shifted by re-applying the modified RJ technique. For RJ, the activities are chosen with the maximum time of completion in the schedule from Figure 8b-activities in the order 7, 6, 8, 5, 4, 3, 2, 1. During the activity justification, current end times of project stages are considered ($MT_1 = 2$, $MT_2 = 7$, $MT_3 = 10$). As a result of RJ, the schedule presented in Figure 8c with a higher value of the objective function $F = 59.59$ than the schedule of Figure 8b is generated by the delayed start of activity 8.

5. Experimental Analysis

The experiments were conducted for 480 test instances from a J30 (30-activity projects) set and 480 test instances from J90 set (90-activity projects) from the PSPLIB [27]. For each PSPLIB project, three contractual stages are defined based on the baseline schedule S created using the SGS serial for the activity list $\{1, 2, \ldots, 30\}$ or $\{1, 2, \ldots, 90\}$. The contractual deadlines of stages are determined as $MD_1 = T/3$, $MD_2 = 2T/3$, $MD_3 = T$, where T is the makespan of the project in the baseline schedule S. The sets of activities to be executed in each stage are created as follows:

- MA_1 contains all activities whose end time in schedule S is less than or equal to MD_1,
- MA_2 contains all activities whose end time in schedule S is less than or equal to MD_2 and greater than MD_1,
- MA_3 includes other activities that are not included in MA_1 or MA_2.

The data for determining cash flows for each test instance: contractual client payment amounts are $MP_1 = 60$, $MP_2 = 60$, $MP_3 = 120$, contractual penalties are $MC_1 = 1.5$, $MC_2 = 1.5$, $MC_3 = 3$, while the costs of executing CFA_i activity are calculated proportionally to the total demand for resources and time for a given activity, while the sum of CFA_i for all the activities is equal to 100. Discount rate used in experiments is $\alpha = 0.01$.

The experiments were conducted on an Intel Core i7-4770, 3.4 GHz CPU, 8 GB RAM using C# application implemented in Visual Studio 2017. The main goal was to evaluate the simulated annealing algorithm performance, moves (Swap, Insert, Swap all, Insert all, Adjacent swap), and techniques of generating schedules (backward scheduling, activity right-shift procedure, justification). Parameters for the SA algorithm such as: the geometric cooling scheme and the initial temperature of 0.1 were established in preliminary experiments.

In all analyzed configurations of the SA algorithm, 5000 activity lists are analyzed. Because of the stochastic nature of the calculations, experiments are conducted twice. The results of the computational experiments are presented in Tables 1 and 2. In these tables: FS—forward scheduling using serial SGS without improvement, BS—backward scheduling procedure with optimization of completion times of project stages, SH—forward scheduling with improvement using activity right-shift procedure, RJ—forward scheduling with improvement using right justification technique, TJ—forward scheduling with improvement using triple justification technique (RJ + LJ + RJ).

Table 1. Results of the experiments—J30 projects.

	Average Value of the Function F					Number of the Best Solutions					CPU
	Swap	Ins.	Swap All	Ins. All	Adj. Swap	Swap	Ins.	Swap All	Ins. All	Adj. Swap	Time [s]
FS, 1st run	76.19	76.13	76.17	76.02	75.91	0	0	0	0	0	0.05
FS, 2nd run	76.18	76.11	76.18	76.01	75.98	0	0	0	0	0	0.05
BS, 1st run	77.38	77.33	77.38	77.32	77.17	328	319	324	302	275	0.80
BS, 2nd run	77.38	77.32	77.37	77.30	77.16	329	306	322	299	270	0.81
SH, 1st run	77.25	77.16	77.21	77.08	76.99	55	48	51	46	39	4.46
SH, 2nd run	77.23	77.15	77.22	77.07	77.01	50	49	53	44	40	4.45
RJ, 1st run	77.50	77.43	77.49	77.36	77.29	275	255	271	234	223	0.17
RJ, 2nd run	77.48	77.44	77.46	77.37	77.29	266	255	263	243	218	0.17
TJ, 1st run	77.58	77.55	77.58	77.52	77.46	315	299	301	277	258	0.45
TJ, 2nd run	77.57	77.55	77.58	77.49	77.48	309	293	305	273	259	0.45

The average values of the objective function F for the baseline schedules are 74.04 for projects from J30 and 48.09 for projects from J90. The average values of the objective function F for the best found schedules are 80.63 for projects from J30 and 57.22 for projects from J90. The use of developed scheduling techniques improves the quality of generated solutions. The highest average value of the objective function F for both J30 and J90 project sets is reached for RS that uses the triple justification technique (RJ + LJ + RJ). An effective technique for generating solutions is also the backward scheduling procedure with the optimization of completion times of the project stages. It constructs solutions with a lower mean value of F, but often generates schedules with the highest value of F. The least efficient and most costly in terms of calculation is the activity right-shift procedure.

Table 2. Results of the experiments—J90 projects.

	Average Value of the Function F					Number of the Best Solutions					CPU
	Swap	Ins.	Swap All	Ins. All	Adj. Swap	Swap	Ins.	Swap All	Ins. All	Adj. Swap	Time [s]
FS, 1st run	50.23	50.44	49.99	49.79	49.54	0	0	0	0	0	0.23
FS, 2nd run	50.23	50.44	50.03	49.82	49.57	0	0	0	0	0	0.23
BS, 1st run	52.88	52.96	52.85	52.67	52.53	133	209	89	65	21	4.44
BS, 2nd run	52.87	52.97	52.83	52.66	52.54	111	208	85	79	36	4.45
SH, 1st run	51.49	51.67	51.24	51.07	50.81	0	0	0	0	0	44.00
SH, 2nd run	51.46	51.67	51.27	51.07	50.85	0	0	0	0	0	44.01
RJ, 1st run	52.60	52.71	52.36	52.17	52.04	20	38	15	11	10	1.18
RJ, 2nd run	52.59	52.70	52.38	52.21	52.06	18	23	13	13	10	1.18
TJ, 1st run	52.94	53.03	52.82	52.78	52.63	54	82	27	25	10	3.21
TJ, 2nd run	52.95	53.00	52.86	52.81	52.65	59	75	37	32	13	3.22

For projects from J30 set, the best solutions are generated using Swap or Swap all moves, while for J90, the most effective is Insert move. The worst results are achieved for the Adjacent swap move.

6. Summary

This work examines the problem of maximizing the discounted cash flow for a multi-stage project from the perspective of its contractor with client payments made after the contractual stages of work, penalties for late completion and expenses incurred by the contractor when performing the activities. The problem is illustrated and techniques for generating appropriate solutions are described. A simulated algorithm based on dedicated scheduling methods is provided. An experimental analysis was performed, which identified the efficient moves (Swap, Swap all for J30, Insert for J90), and techniques for creating solutions, that is backward scheduling with optimization of completion times of project stages and triple justification.

The problem discussed is actual. The proposed model of project stage settlements can be useful in practice and beneficial for both the client and the contractor. It may lead to the reduced untimely execution of practical projects. The data used for financial settlements, such as deadlines, contractual payments and penalties can be negotiated between the client and the contractor to determine a solution that satisfies both parties.

Further works focus on comparing the effectiveness of various metaheuristics for the proposed model, especially population-based algorithms such as genetic algorithms and particle swarm optimization. Another direction of the work will be the development of the capital-constrained model (CCPSP) with staged project settlement for which the use of the described techniques of generating solutions will be tested.

Funding: This research received no external funding.

Conflicts of Interest: The authors declare no conflict of interest.

References

1. Hartmann, S.; Briskorn, D. A Survey of Variants and Extensions of the Resource-Constrained Project Scheduling Problem. *Eur. J. Oper. Res.* **2010**, *207*, 1–14. [CrossRef]
2. Russell, A.H. Cash Flows in Networks. *Manag. Sci.* **1970**, *16*, 357–373. [CrossRef]
3. Rosłon, J.; Książek-Nowak, M.; Nowak, P. Schedules Optimization with the Use of Value Engineering and NPV Maximization. *Sustainability* **2020**, *12*, 7454. [CrossRef]
4. Leyman, P.; Vanhoucke, M. A New Scheduling Technique for the Resource–Constrained Project Scheduling Problem with Discounted Cash Flows. *Int. J. Oper. Prod. Res.* **2015**, *53*, 2771–2786. [CrossRef]
5. Hartmann, S.; Kolisch, R. Experimental Evaluation of State-of-the-Art Heuristics for the Resource-Constrained Project Scheduling Problem. *Eur. J. Oper. Res.* **2000**, *127*, 394–407. [CrossRef]

6. Szmerekovsky, J.G. The Impact of Contractor Behavior on the Client's Payment-Scheduling Problem. *Manag. Sci.* **2005**, *51*, 629–640. [CrossRef]
7. Dayanand, N.; Padman, R. On Modelling Payments in Projects. *J. Oper. Res. Soc.* **1997**, *48*, 906–918. [CrossRef]
8. Klimek, M.; Łebkowski, P. Heuristics for Project Scheduling with Discounted Cash Flows Optimisation. *Bull. Pol. Acad. Tech.* **2015**, *63*, 613–622. [CrossRef]
9. Ulusoy, G.; Sivrikaya-Şerifoğlu, F.; Şahin, Ş. Four Payment Models for the Multi-Mode Resource Constrained Project Scheduling Problem with Discounted Cash Flows. *Ann. Oper. Res.* **2001**, *102*, 237–261. [CrossRef]
10. He, Z.; Wang, N.; Li, P. Simulated Annealing for Financing Cost Distribution Based Project Payment Scheduling from a Joint Perspective. *Ann. Oper. Res.* **2014**, *213*, 203–220. [CrossRef]
11. Kavlak, N.; Ulusoy, G.; Şerifoğlu, F.S.; Birbil, Ş.İ. Client-Contractor Bargaining on Net Present Value in Project Scheduling with Limited Resources. *Nav. Res. Log.* **2009**, *56*, 93–112. [CrossRef]
12. Ulusoy, G.; Cebelli, S. An Equitable Approach to the Payment Scheduling Problem in Project Management. *Eur. J. Oper. Res.* **2000**, *127*, 262–278. [CrossRef]
13. Bahrami, F.; Moslehi, G. Study of Payment Scheduling Problem to Achieve Client–Contractor Agreement. *J. Adv. Manuf. Technol.* **2013**, *64*, 497–511. [CrossRef]
14. Vanhoucke, M.; Demeulemeester, E.; Herroelen, W. Progress Payments in Project Scheduling Problems. *Eur. J. Oper. Res.* **2003**, *148*, 604–620. [CrossRef]
15. Dayanand, N.; Padman, R. Project Contracts and Payment Schedules: The Client's Problem. *Manag. Sci.* **2001**, *47*, 1654–1667. [CrossRef]
16. He, Z.; Wang, N.; Jia, T.; Xu, Y. Simulated Annealing and Tabu Search for Multi-Mode Project Payment Scheduling. *Eur. J. Oper. Res.* **2009**, *198*, 688–696. [CrossRef]
17. Leyman, P.; Vanhoucke, M. Payment Models and Net Present Value Optimization for Resource-Constrained Project Scheduling. *Comput. Ind. Eng.* **2016**, *91*, 139–153. [CrossRef]
18. Mika, M.; Waligóra, G.; Węglarz, J. Simulated Annealing and Tabu Search for Multi-Mode Resource-Constrained Project Scheduling with Positive Discounted Cash Flows and Different Payment Models. *Eur. J. Oper. Res.* **2005**, *164*, 639–668. [CrossRef]
19. Elmaghraby, S.E. Activity Nets: A Guided Tour through Some Recent Developments. *Eur. J. Oper. Res.* **1995**, *82*, 383–408. [CrossRef]
20. He, Z.; Xu, Y. Multi-Mode Project Payment Scheduling Problems with Bonus–Penalty Structure. *Eur. J. Oper. Res.* **2008**, *189*, 1191–1207. [CrossRef]
21. Klimek, M.; Łebkowski, P. Financial Optimisation of the Scheduling for the Multi-Stage Project. *Bull. Pol. Acad. Tech.* **2017**, *65*, 899–908. [CrossRef]
22. Smith-Daniels, D.E.; Padman, R.; Smith-Daniels, V.L. Heuristic Scheduling of Capital Constrained Projects. *J. Oper. Manag.* **1996**, *14*, 241–254. [CrossRef]
23. Leyman, P.; Vanhoucke, M. Capital- and Resource-Constrained Project Scheduling with Net Present Value Optimization. *Eur. J. Oper. Res.* **2017**, *256*, 757–776. [CrossRef]
24. He, Z.; Liu, R.; Jia, T. Metaheuristics for Multi-Mode Capital-Constrained Project Payment Scheduling. *Eur. J. Oper. Res.* **2012**, *223*, 605–613. [CrossRef]
25. Klimek, M. Insertion Algorithms to Solve the Resource-Constrained Multi-Stage Project Scheduling Problem with Discounted Cash Flow Maximization. *Manag. Prod. Eng. Rev.* **2018**, *9*, 109–116. [CrossRef]
26. Klimek, M.; Łebkowski, P. Robustness of Schedules for Project Scheduling Problem with Cash Flow Optimisation. *Bull. Pol. Acad. Tech.* **2013**, *61*, 1005–1015. [CrossRef]
27. Kolisch, R.; Sprecher, A. PSPLIB—A Project Scheduling Problem Library: OR Software—ORSEP Operations Research Software Exchange Program. *Eur. J. Oper. Res.* **1997**, *96*, 205–216. [CrossRef]
28. Blazewicz, J.; Lenstra, J.K.; Kan, A.H.G.R. Scheduling Subject to Resource Constraints: Classification and Complexity. *Discret. Appl. Math.* **1983**, *5*, 11–24. [CrossRef]
29. Kolisch, R.; Hartmann, S. Experimental Investigation of Heuristics for Resource-Constrained Project Scheduling: An Update. *Eur. J. Oper. Res.* **2006**, *174*, 23–37. [CrossRef]
30. Sari, V.; Cakir, S.; Kilic, E. Ece Evaluation of Scatter Search and Genetic Algorithm at Resource Constrained Project Scheduling Problems. In Proceedings of the 2011 15th IEEE International Conference on Intelligent Engineering Systems, Poprad, Slovakia, 23 June 2011; pp. 127–130.
31. Yu, Y.; Li, Y.; Li, J. Parameter Identification of a Novel Strain Stiffening Model for Magnetorheological Elastomer Base Isolator Utilizing Enhanced Particle Swarm Optimization. *J. Intell. Mater. Syst. Struct.* **2014**, *26*, 2446–2462. [CrossRef]
32. Chen, R.-M. Particle Swarm Optimization with Justification and Designed Mechanisms for Resource-Constrained Project Scheduling Problem. *Expert Syst. Appl.* **2011**, *38*, 7102–7111. [CrossRef]
33. Kirkpatrick, S.; Gelatt, C.D.; Vecchi, M.P. Optimization by Simulated Annealing. *Science* **1983**, *220*, 671–680. [CrossRef] [PubMed]
34. Kolisch, R. Serial and Parallel Resource-Constrained Project Scheduling Methods Revisited: Theory and Computation. *Eur. J. Oper. Res.* **1996**, *90*, 320–333. [CrossRef]
35. Vanhoucke, M.; Demeulemeester, E.; Herroelen, W. Maximizing the Net Present Value of a Project with Linear Time-Dependent Cash Flows. *Int. J. Oper. Prod. Res.* **2001**, *39*, 3159–3181. [CrossRef]
36. Vanhoucke, M. A Scatter Search Heuristic for Maximising the Net Present Value of a Resource-Constrained Project with Fixed Activity Cash Flows. *Int. J. Oper. Prod. Res.* **2010**, *48*, 1983–2001. [CrossRef]

37. Klimek, M. Techniques of Generating Schedules for the Problem of Financial Optimization of Multi-Stage Project. *Appl. Comp. Sci.* **2019**, *15*, 18–35. [CrossRef]
38. Leus, R.; Herroelen, W. Stability and Resource Allocation in Project Planning. *IIE Trans.* **2004**, *36*, 667–682. [CrossRef]
39. Klimek, M.; Łebkowski, P. Resource Allocation for Robust Project Scheduling. *Bull. Pol. Acad. Tech.* **2011**, *59*, 51–55. [CrossRef]
40. Pang, N.; Su, H.; Shi, Y. Project Robust Scheduling Based on the Scattered Buffer Technology. *Appl. Sci.* **2018**, *8*, 541. [CrossRef]
41. Deblaere, F.; Demeulemeester, E.; Herroelen, W.; Van de Vonder, S. Robust Resource Allocation Decisions in Resource-Constrained Projects *. *Decis. Sci.* **2007**, *38*, 5–37. [CrossRef]
42. Valls, V.; Ballestín, F.; Quintanilla, S. Justification and RCPSP: A Technique That Pays. *Eur. J. Oper. Res.* **2005**, *165*, 375–386. [CrossRef]

applied sciences

Article

Slotting Optimization Model for a Warehouse with Divisible First-Level Accommodation Locations

Pablo Viveros [1,*], Katalina González [1], Rodrigo Mena [1], Fredy Kristjanpoller [1] and Javier Robledo [2]

[1] Department of Industrial Engineering, Technical University Federico Santa María, Av. España 1680, Valparaíso, Chile; katalina.gonzalez.12@sansano.usm.cl (K.G.); rodrigo.mena@usm.cl (R.M.); fredy.kristjanpoller@usm.cl (F.K.)

[2] Department of Computer Engineering, Technical University Federico Santa María, Av. Santa María 6400, Santiago, Chile; javier.robledo@usm.cl

* Correspondence: pablo.viveros@usm.cl

Abstract: Efficiency in supply chains is critically affected by the performance of operations within warehouses. For this reason, the activities related to the disposition and management of inventories are crucial. This work addresses the multi-level storage locations assignment problem for SKU pallets, considering divisible locations in the first level to improve the picking operation and reduce the travel times associated with the routes of the cranes. A mathematical programming model is developed considering the objective of minimizing the total travel distance, and in the background, maximizing the use of storage capacity. To solve this complex problem, we consider its decomposition into four subproblems, which are solved sequentially. To evaluate the performance of the model, two analysis scenarios based on different storage strategies are proposed to evaluate both the entry and exit distance of pallets, as well as the cost associated with the movements.

Keywords: slotting; order picking; storage strategies; stackability; SKU; product family

check for **updates**

Citation: Viveros, P.; González, K.; Mena, R.; Kristjanpoller, F.; Robledo, J. Slotting Optimization Model for a Warehouse with Divisible First-Level Accommodation Locations. *Appl. Sci.* **2021**, *11*, 936. https://doi.org/ 10.3390/app11030936

Received: 9 December 2020
Accepted: 13 January 2021
Published: 20 January 2021

Publisher's Note: MDPI stays neutral with regard to jurisdictional claims in published maps and institutional affiliations.

1. Introduction

Nowadays, companies pay more attention to the management of their warehouses, seeking to optimize the location of their products in the facilities based on the remaining time before being dispatched. Their supply chains tend to be dynamic, responding to the high response times demanded by customers. This requires agile warehouse management given a higher product turnover. However, there are no two businesses that work in the same way due to the multiple factors that can affect the dynamics of a supply chain, where the operations of a warehouse play a fundamental role in determining operating costs and customer time response throughout the distribution chain [1].

In this sense, the order picking process figure as one of the fundamental processes within warehouse management. This can be defined as the process by which items in storage are handled to fulfill a specific customer order. This operation has been evaluated as the most critical in the supply chain: ref. [2] establish that this process involves approximately 55% of the total operating costs related to the warehouse activities. This process normally considers tasks such as searching, handling, and product transferring, tasks that represent, respectively, 20%, 50%, and 30% of the total time of order preparation. Travelling is considered an unproductive activity, which does not add value to the operation, and considering that this item represents the largest component of order times, it reveals an opportunity to improve the performance of storage centers. In recent years, companies have sought to reduce travel times through the use of technologies, such as advanced storage management software, flow racks, voice picking, radio frequency identification, among others, spending around 10% to 30% more per year for this purpose. This is consistent with the information found in the literature, from which it is estimated that less than 15% of the SKUs (Stock-Keeping Units) are correctly located [3]. Better solutions can result from

199

a deeper analysis of the correct positioning of products within a warehouse based on the demand for each product.

In order to address the problems described in the order picking process, the concept of slotting has recently been studied and analyzed in industry and specialized literature. In this sense, the slotting process seeks to address order picking by searching for an intelligent arrangement of products within the warehouse. Specifically, it is the correct assignment of SKUs to available storage locations, establishing the storage mode, the volume of space to be assigned, and the exact location for product storage [4]. Slotting aims to improve efficiency in order preparation and reduce operating costs [5]. For the Slotting process, it is necessary to answer two basic questions: (1) How to classify SKUs? and (2) How to assign classified SKUs to locations? [6].

According to [7], it is possible to use a variety of strategies to optimize the operational process when carrying out the location assignment. However, it is possible to distinguish three main strategies, which are classified as dedicated, class-based, and random storage. Dedicated storage consists of assigning a fixed slot for a certain item within the warehouse [4]; this strategy is aimed at quick access in manual systems, thus reducing search time and improving the handling and transfer of products by operators. In class-based storage, items are classified, grouped, and stored in certain slots according to their turnover rate, which enables higher handling or sorting throughput as a result of the connection between products, i.e., locations are assigned adjacently for those items that are usually ordered together. Random storage, in turn, results in higher utilization of warehouse capacity due to location assigning characteristics, resulting in a uniformly distributed allocation. However, the latter strategy can be cumbersome and sometimes expensive, which is why it is commonly used under the closest position rule [4]. As mentioned by [8], if the storage process is random or inappropriate, this will affect the cost of movement, waiting time, and transfer. The study also concludes that the efficiency of the storage strategy depends on a wide variety of factors, but as a general rule, ABC analysis yields better results.

The present work is divided into two stages. The first stage addresses a storage location assignment problem focused on divisible horizontal locations, generating more storage positions of lower height (see Figure 1). This problem will be called Storage Location Assignment Problem with Variable Height or SLAP-VH.

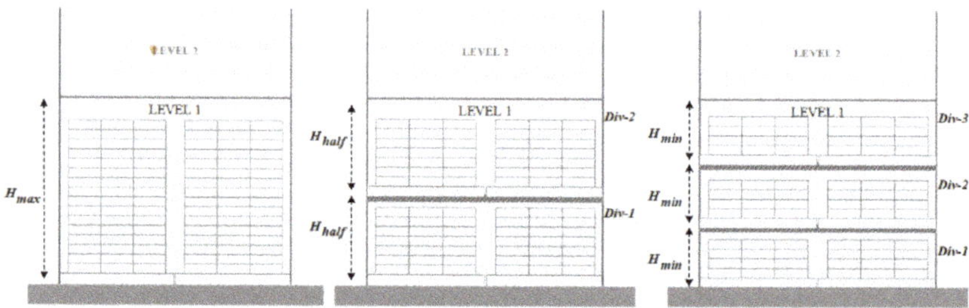

Figure 1. Representation of a divisible location of the first level of variable height accommodation. Where H_{max}, H_{half} and H_{min} correspond to the maximum, half the maximum height, and minimum height respectively.

In this way, SLAP-VH considers selecting locations to store products based on their demands, positioning customer's most requested items near the exit, maintaining the priority of stackability (a factor that determines ergonomic efficiency), and arrange them in a way that minimizes cranes travel during operations.

In this work, SLAP-VH has been approached through mathematical programming and solved from its decomposition into four subproblems, which are solved sequentially. The subproblems correspond to assignment models based on an estimate of the collection route, i.e., the result determines the best position for each SKU considering the minimum travel distance in a linear way for the accommodation route, starting from the main route with side trips within aisles, and considering serpentine selection performed by an operator to assemble an order.

The second stage of the work consists of the economic evaluation of the crane path, based on the solution found by the proposed model, for two different analysis scenarios: one based on a random class-based storage strategy, and the other scenario is based on a storage strategy purely by class, where two decision criteria are defined to select the best storage route.

The rest of this article is organized as follows: a literature review is presented in Section 2; Section 3 describes the study warehouse and the approach to the problem that motivated the proposed methodology; Section 4 presents the global model of SLAP-VH; Section 5 presents the decomposition of the SLAP-VH problem and the models for each subproblem; Section 6 defines the proposed solution methodology; the assessment of the scenarios is described in Section 7; Sections 8 and 9 presents the results obtained and the comparison of scenarios; Section 10 presents the discussion of the results; Section 11 provides final comments and conclusions.

2. Previous Literature

Slotting optimization research can be divided into two areas. The first is the modeling of the slotting process to improve efficiency in order preparation, load movement, handling of heavy over lightweight items, among others. And the second area is related to model resolution, using for example, exact, heuristic, and/or metaheuristic algorithms, among others [9]. Several researchers have developed both mathematical and heuristic models to address Slotting related to the Storage Location Assignment Problem (SLAP). Some of these studies can be consulted in Table 1, where the models developed tend to be very complex due to the SLAP NP-Hard complexity. In this sense, ref. [10] affirm that most of the storage and recovery problems fall into this category, which motivates the use of heuristics for their resolution. Besides, certain investigations propose decompositions of the problem, where initially the number of locations is addressed and the assignment is considered afterward. Using this strategy, ref. [11] addressed the SLAP for outbound containers in a maritime terminal, which was divided into two stages. They first determined the number of locations from a mixed-integer linear programming model and then defined the exact storage location based on a hybrid sequence stacking algorithm. This experimental study demonstrated effectiveness in comparing and analyzing performance in total travel distance between the yard and the bunk, the imbalance of the workload between different blocks of containers, and the percentage of repetitions in loading operations. In this way, the hybrid sequence stacking algorithm reduces the unnecessary movement of containers in cargo operations to 18%, compared to 30% of the average rehandle percentage in the case study analyzed.

It is interesting to note that the vast majority of the studies carried out consider as an objective function the minimization of time, cost, or distance. However, authors such as [12] developed a mathematical model that minimizes storage time by incorporating a two-objective optimization approach: maximization of space availability and minimization of production due date measurements. Ref. [1] developed a mathematical model that considers two simultaneous goals: minimize travel distance and maximize average storage usage. The SLAP-VH also incorporates this feature, considering the minimization of cranes route as the main objective function, and the maximization of storage use as a secondary function.

One of the main parameters of the SLAP modeling is the planning horizon. Research in the field reveals that evaluation periods adjusted to the problem achieve greater effectiveness in the correct arrangement of SKUs. For example, in [13] was defined the time horizon in periods of discrete units (days) to consider the real situation of the research, where the storage warehouse releases outgoing objects and stores incoming objects in each unit period. For the case of SLAP-VH, the period (t) is based on the normal distribution of the demand for each SKU (see Figure 2), this allows the Slotting to be defined as active since the correct accommodation adapts to the temporality of the products according to the defined periods.

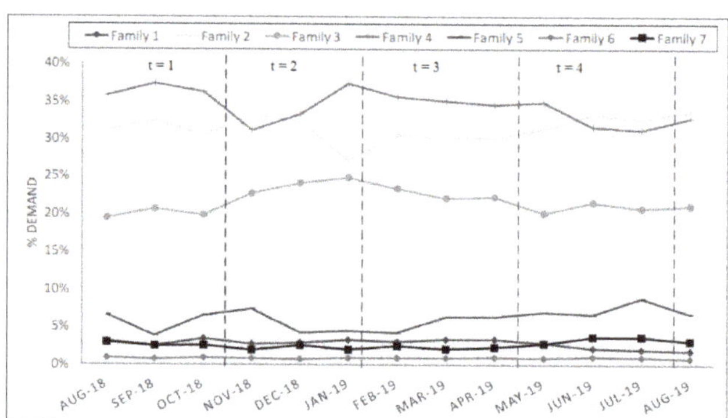

Figure 2. Definition of periods for SLAP-VH based on the monthly demand of each SKU grouped by families.

The models proposed by the researchers are commonly solved by heuristics or metaheuristics, considering the complexity of the SLAP. However, these heuristic and metaheuristic algorithms are not able to guarantee the optimality of the solutions, although, in some cases, they offer nearly optimal solutions depending on the execution times and the parameter setting. Researchers have evaluated the effectiveness of different algorithms for the SLAP. For example, ref. [14] made a comparison between an optimization method, a Tabu Search metaheuristic, and an empirical rule. The authors concluded that the size of the accommodate list has a significant effect on its structure and the optimality of the results. Additionally, the optimization method reduced labor costs compared to the solutions provided by metaheuristics. Reducing costs has a strong impact on the entire order picking process since all activities are connected. Ref. [15] state that the optimization of positioning not only reduces costs and energy in the operation but also improves the efficiency of the storage of goods. Therefore, optimizing the Slotting process helps to complete orders at a faster pace, benefiting the consolidation work (i.e., a process that groups the demand for orders from different areas, into complete orders), and the service level delivered to end consumers. Managers want to reduce the time spent on movements within the warehouse, not only to decrease labor cost and production downtime but also to shorten customer response time. Ref. [16] developed a study of the time of the activities carried out in the preparation of orders in Maines Paper and Food Services (Inc in Conklin, NY), concluding that 40% of the time is spent on travel. Therefore, reducing this percentage could significantly save a warehouse operating expense. Reducing travel time implies reducing the distance of cargo transfer and therefore the number of operators required to place an order. Ref. [17] state that the distance largely depends on the design of the warehouse. It is logical, in the case of slotted storage, to place the merchandise that has a high turnover in privileged locations close to the dispatch point. In contrast, when placing an order where items need to be selected in a single tour, it makes more sense to assign SKUs to locations that minimize travel time for routes.

Table 1. Main characteristics of the presented storage location assignment models.

Reference	Characteristic of the Model	Storage Strategy	Function to Optimize	Solution Method
[16]	Warehouse with Crossover-Aisles with one direction of movement.	Weight order.	Minimize total distance traveled.	Mathematical model programming language C. Simulation Microsoft Visual Basic 6.0.
[13]	Linear movement of objects.	Single storage.	Minimize the number of obstructive objects.	Mathematical programming model.Genetic algorithm (GA).
[15]	Based on the principle of lower energy and shorter time.	Random.	Minimize total energy consumption. Minimize storage time.	Discrete particle swarm optimization (PSO).
[11]	Storage of containers in a maritime terminal.	Category of containers.	Minimize travel distance and time required load transfer.	Mixed-integer programming model. Hybrid stacking algorithm sequence procedure.
[18]	Travel distance considering three axes.	Symmetric storage.	Minimize total travel distance.	Mathematical model evaluated in LINGO Software.
[19]	Stored cycle travel times and double-depth retrieval.	Random. Always positioned first to the second lane.	Minimize travel time.	Analytical model. Simulation.
[20]	Warehouse symmetrical storage locations for a single product. Considers two horizontal axes and one vertical.	Symmetric storage.	Minimize total travel distance.	Mixed-integer programming model. Metaheuristics: Differential evolution (DE), Global optimization of the local and near neighbor particle swarm (GLNPSO).
[21]	Placing of accommodation lots with a fleet of k homogeneous teams.	Priority rule FCFS (First Come First Served).	Minimize the total time of lot formation.	Statistical design of divided parcels. Metaheuristic of intelligent neighborhood search (INS).
[1]	Bi-objective. Storage of blocks of pallets with storage locations for a single product.	Random.	Minimization of the total travel distance. Maximization of the average use of storage.	Mathematical model and a constructive heuristic algorithm.
[12]	Storage and recovery. Storage location for a single product.	Random.	Minimize travel time. Secondary function: maximize storage use.	Decomposition of 3 subproblems. Metaheuristic methods: Random, priority rule, and Genetic Algorithm.
[14]	Storage of homogeneous characteristics and WMS integration.	FIFO rule (First In First Out).	Minimize total costs.	Optimization model. Taboo search metaheuristic. Empirical rule.
[22]	Application of new storage strategy.	The selection area is divided into two sections called free seats and start locations.	Minimize the time required for order preparation.	Simulation (FlexSim). Matlab.
[23]	The model is based on correlated data products parameters, clients' orders and warehouse layout.	Correlated data from the analysis of the ABC, XYZ, EIQ, AHP and COI indices.	Ensure the shortest time to access products of the highest frequency from the gates	software Pickup Simulo codified in PHP language and MySQL relational databases. Application of artificial neural networks

Table 1. *Cont.*

Reference	Characteristic of the Model	Storage Strategy	Function to Optimize	Solution Method
[24]	Machine Learning techniques for Product Allocation Problem	Product classification strategy comparing with reference picking data	Minimization of statistical dispersion indicators and maximization of clustering quality using Calinski-Harabasz criterion	Simulation using feed-forward artificial neural network and automatic clustering

On the other hand, travel time depends entirely on the location of the SKUs in the warehouse and, in turn, on the storage strategy. The latter is an important factor in warehouse performance, since optimizing the slotting process consists of optimizing storage locations based on the characteristics of the racks and also the requirements of the processes within the warehouse.

Given the above, this work proposes an optimization model for the Slotting process (SLAP-VH), where its distinctive feature is the consideration of a warehouse whose locations corresponding to the first accommodation level are divisible. To the best of our knowledge, this feature has not been previously considered within SLAP literature. This work is divided into two stages: determining the optimal location for each SKU in a planning horizon with four defined periods, and determine the best of two allocation scenarios to reduce travel distance and costs. SLAP-VH is solved through a mixed-integer linear programming model (MILP), considering a representative sale day belonging to the most critical period (highest number of orders). The choice of the solution method addressed for SLAP-VH was determined from a comparative table between three metaheuristics and an optimization method (namely: PSO, GA, Tabu Search, and optimization), where the high execution times of the metaheuristics motivated the construction of these evaluation scenarios, solving them from a MILP optimization model with the Gurobi solver through an exact procedure. On the other hand, the combination of storage locations (racks and floor locations) is a unique feature compared to the optimization models investigated, in addition to the incorporation of four outputs from the warehouse under study. The main characteristics of the study problem compared to the research carried out are presented in Table 2.

Table 2. Attributes of SLAP-VH.

Model Characteristics	Storage Strategy	Function to Optimize	Solution Method
Warehouse with four outputs. Storage in Racks and storage a floor. Locations in first accommodation level with multi-level.	Two storage scenarios: • Storing by random classes • Storage purely by classes	Minimize travel cranes. Maximize storage use.	Mathematical programming model MILP. Python programming language, Pyomo Gurobi solver.

The literature reviewed in this research shows how the storage location assignment problem has been addressed in different warehouses. None of the models proposed by the reviewed authors solves the problem with different types of storage locations and divisible locations developed in this paper; however, they provide relevant factors to consider in storage strategies and problem modeling for locations storage both in-floor locations ("Perchas"), described by [1], and also in racks, developed by [12]. Besides, this latter considers the subdivision of the assignment problem, influencing the development of this research. On the other hand, the objective proposed in this work is equivalent to most of the models addressed in the state-of-the-art literature, that is, to reduce the travel time of cranes or to minimize the operating cost of the warehouse picking operation, functions that are strongly related. Therefore, combining and adapting the models presented in the literature to the proposed problem allows to obtain nearly optimal solutions, minimizing time, cost, and also increasing the warehouse space utilization.

3. Warehouse Description and Methodology

The mathematical model is built and evaluated based on real data from a distribution center (DC) of a beverage company that supplies Chile and neighboring countries, provided with around 1200 SKUs on inventory in its dependencies. The DC picking area is modeled to develop the SLAP-VH, where the SKUs are represented as loaded pallets of homogeneous dimensions. As can be seen in Figure 3, the warehouse is composed of columns of racks with equal height and the same number of slots. The width and depth of the storage positions of racks are the same for the entire warehouse with dimensions $1 \times 1.2 \text{ m}^2$. The height varies between 0.7 m to 1.2 m according to the division of the location and SKU format dimensions. Thus, there are three operations (o) where only one of them can be applied for each location, that is: the location cannot be divided, can be divided into two parts, or can be divided into three parts.

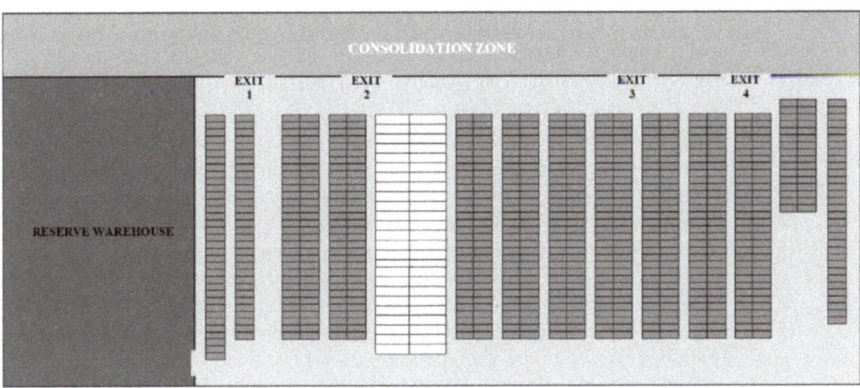

Figure 3. Layout of the study distribution center. The lighter slots represent Perchas, while the darker slots are racks.

Warehouse aisles correspond to the space between each rack, which are listed from left to right. The design of the warehouse is represented as a rectangular section, where the entrance of the cranes to the locations are on both sides of the aisle, except for the first and last side. Additionally, the warehouse has four exits to the consolidation and dispatch area.

Into the warehouse, only trilateral cranes are considered with one operator onboard and an average speed of 5 km/h, where minimizing the distance of crane movements allows determining the number of trips during an 8-h shift, concluding in this way the number of operators needed to satisfy the requirements. This fact, however, can be addressed with greater depth in future researches.

In the warehouse under study, there are three storage considerations: (a) locations should be assigned to the pallets according to affinity; (b) The pallets with the highest demand should ideally be assigned in locations without division, and; (c) two columns of locations, which do not include racks, cannot be divided. These columns are called "Perchas". These are worked separately, treating as a fourth subproblem for SLAP-VH.

Perchas are locations stacking pallets that are stored on the floor at the top, arranged in stacks. This area is managed according to the dedicated storage strategy, since, as shown in Figure 4, each location has capacity for twelve pallets, that is, five pallets can be assigned in base and depth, and then start to be stacked while maintaining ergonomics and operator protection. It is important to store only one type of SKU for each Percha location, since a great storage depth generates inconveniences such as, for example, not finding a pallet of SKU "A" since it is under a stack of SKU pallets "B". In other cases, this situation can cause the operator may forget that they stored a pallet under other SKUs, passing the expiration date. Regarding time, stacking pallets of different SKU types demands the operator to spend more time searching and moving the necessary pallets until reaching the desired one.

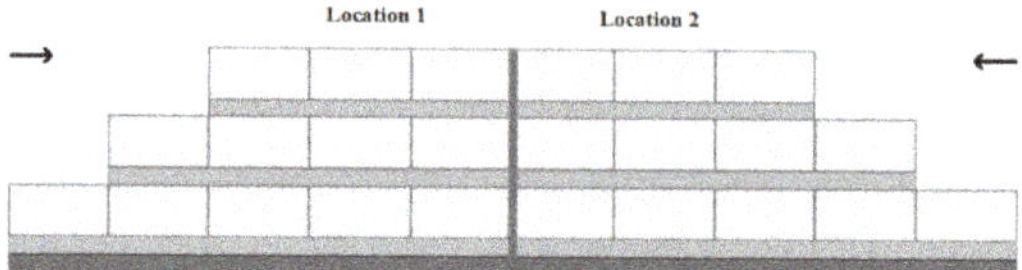

Figure 4. Representation of Perchas locations with a two-sided view, where each rectangle represents an SKU pallet. Bold arrows indicate pallet entry.

The decision to divide locations is based on the demand for the products: for example, if an SKU is less requested in a period, this will be assigned to locations of smaller capacity and further from the warehouse. The opposite happens with a highly demanded SKU, which will be assigned to a high capacity location and close to the exit. Finally, the Perchas are locations defined by the company for a certain type of SKU where the decision consists of positioning the pallet according to its proximity to the exit point since these locations are not divisible.

SLAP-VH only considers the first-level stacking locations, that is, the first location from the floor of each rack. Storage is governed by order priority rules, specifically by EDD (Earliest Due Date), except for the picking slots, corresponding to the first level of accommodation. For this reason, this research focuses exclusively on the assignment of these locations, which can be divided into three positions, in two, or they can be maintained without performing any division. In this sense, three, two, or one pallet can be stored respectively. Each first-level storage slot has capacity for two pallets of maximum height (Figure 5(a.1,a.2)), while a divided slot (Figure 5(b.1,b.2)) has capacity for four medium height pallets, or six pallets of minimum height (Figure 5(c.1,c.2)). Therefore, the term "location" will be used to denote the storage space of a rack with a capacity of one pallet of maximum height, while the storage space of a divided slot of less height will be called "position".

Within the warehouse, SKUs are classified based on the Pareto principle according to the class or product family to which they belong, where 80% of the total quantity demanded corresponds to type "A" SKUs, 15% to the "B" and 5% to the "C". The ABC classification is essential to know the need for accommodation of each SKU based on its demand. Each of the families has a different number of SKUs and each family has a regrouping called a subfamily, which is based on the weight of the products. In this way, the grouping by families only allows determining the product category, while the regrouping by subfamilies allows classifying with more information about each SKU based on the size of the product. This creates a greater connection between stackability priority and affinity between the products. In this sense, it is necessary to clarify that grouping by product families and subfamilies is a strategy defined by the company, which is enhanced by the ABC classification by periods. Initially, there were 12 families, however, in order to group those with a smaller quantity without affecting stackability, 7 families were finally obtained.

As [3] indicates, the general assignment problem is classified as an NP-hard complexity problem. From this statement, SLAP-VH can also be considered as an NP-hard problem, since it applies a stricter constraint set for the general storage assignment problem. The case study considered 726 SKUs that must be distributed in 1170 locations, a value that may increase considering the division of locations. Therefore, evaluating all the possible combinations becomes an impractical strategy given the order compilation time $O(n!)$ for exact algorithms. However, based on the assumptions used and presented in Table 3, the underlying complexity tends to decrease, reducing the number of decisions and the combination of cases that arise when trying to model reality. In this way, the presented model results in a mixed-integer linear programming problem (MILP).

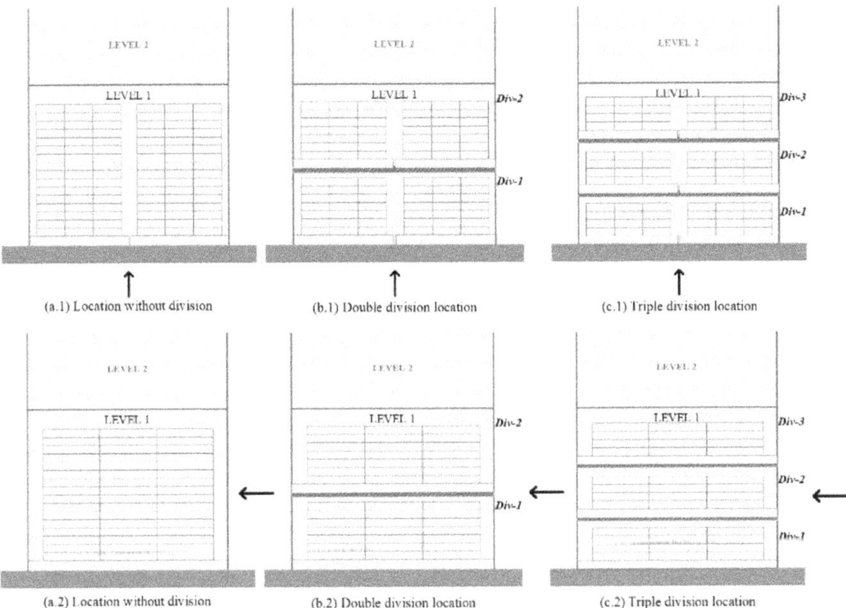

Figure 5. Representation of the first accommodation level locations with simple, double, and triple division, where (**a.1**), (**b.1**), and (**c.1**) correspond to the front view of a storage location in a rack, while (**a.2**), (**b.2**), and (**c.2**) correspond to the side view. Bold arrows indicate pallet entry.

Table 3. Main assumptions considered in SLAP-VH.

- The Warehouse Starts Empty to Accommodate SKUs Packing
- All accommodation operations are performed by homogeneous cranes
- Stored pallets are homogeneous and loaded by a single SKU
- All SKUs must be stored in at least one location or position
- All locations used in the problem corresponds to the first accommodation level
- All locations can be divided, except for *Perchas*
- The distance is measured with respect to the warehouse exits and the end of the aisle in a linear way
- The distance is equal for both divided and without division locations
- A not divided location is assigned when an SKU demands at least one pallet bin
- A two-position location is assigned when an SKU demands between 0.5 and 1 pallet bin
- A three-position location is assigned when an SKU demands less than half of a pallet bin

It is important to consider that for modeling purposes, only the use of full storage pallets will be considered. This assumption applies since the cranes are loaded with either a full pallet or with a load conformed for different clients until creating a complete pallet, given the volume of demand of the company. It should be noted that the full pallet denomination refers to a pallet loaded in its full storage capacity. In this sense, the operators move full pallets to a location or position and then return without cargo to repeat the packing operation, before the order picking process. In order picking, there are three different activities to move SKU pallets:

1. Transfer of a complete pallet: the operator only transfers a pallet loaded with a single SKU to the nearest exit. This activity is normally done when an SKU is highly demanded.
2. Transfer of a replenishment pallet: the operator moves a complete pallet to a location or position when it has become empty during the picking operation.
3. Transfer of a conformed pallet: the operator makes collection trips until a complete pallet is formed and goes to the closest exit.

These activities are considered in the evaluation of scenarios through the picking operation once the product layout has been obtained.

Appendix A lists the notations, abbreviations, and symbols used in the document.

4. Global Solution Model

In this section, the global model of SLAP-VH is presented, which will be denominated P_1, as can be seen in Figure 6. The problem begins by solving the first subproblem, which determines the number of locations through the demand for the SKUs, to continue with the second subproblem, and start the construction of the two different Slotting scenarios. Once the results obtained are known, we proceed to the third subproblem, defining which locations should be divided. In parallel, the fourth subproblem is solved to later introduce the results obtained in the global model of P_1. The specifications of each subproblem are defined in the next section. It is necessary to clarify that the sequentiality in the development of the problem is determined from the key decisions considered in Section 5, since the simultaneous resolution of the analyzed subproblems does not allow obtaining feasible scenarios. Finally, the SLAP-VH solution is reflected in two slotting layout scenarios, which are evaluated to determine the optimal solution in terms of distance and costs. These scenarios show the three types of storage strategies that have been applied in the warehouse under study. Since locations without racks already have a defined strategy, the two scenarios defined in Section 7 are generated.

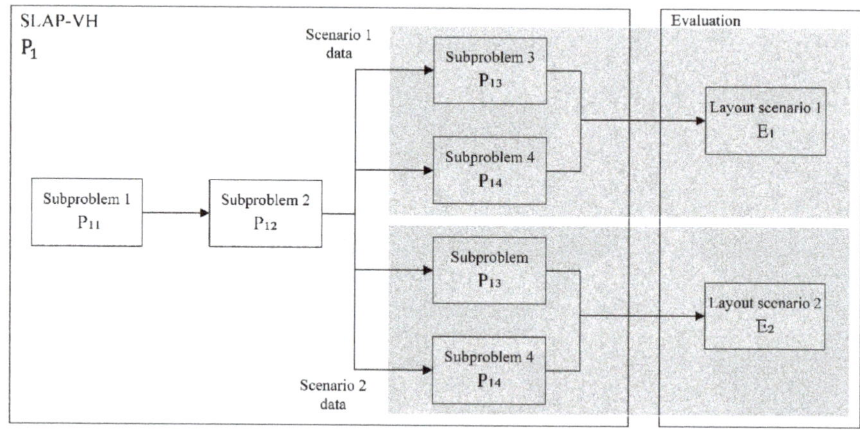

Figure 6. Solution framework for the study problem.

The mathematical model proposed for P_1 aims to determine which is the best position for each SKU within the warehouse that minimizes the travel distance of cranes, considering the ABC classification and the stackability priority. The relevant parameters and decision variables are presented in Appendix A.

The formulation of the model P_1 is presented below:

$$Min \; \sum_I \sum_J \sum_K \left(Y_{ji} \cdot D_{jk} \cdot value\_sku_i \right) \tag{1}$$

Subject to

$$\sum_S \sum_I \left(Y_{ji} \cdot sku\_subfamiliy_{si} \cdot position\_subfamily_{js} \right) = 1, \; \forall \, j \in J \tag{2}$$

$$\sum_J \left(Y_{ji} \cdot type_{jo} \right) \geq sku\_type_{io}, \; \forall \, i \in I, \, o \in O \tag{3}$$

The objective of the proposed problem is to secure to each SKU at least one location, considering the minimum distance from the warehouse exit. Constraint set (2) establishes that each location/position must have a single SKU, and in set (3) it is determined that an SKU can have more than one location/position. Both restrictions consider the type of operation that position has, and the family to which it corresponds.

5. Decomposition of the Problem

The problem P_1 with its corresponding objective function F_1 is divided into four subproblems, called $P_{1.1}$, $P_{1.2}$, $P_{1.3}$, and $P_{1.4}$, where their corresponding objective functions are called $F_{1.1}$, $F_{1.2}$, $F_{1.3}$, and $F_{1.4}$ respectively. These are solved sequentially, where the result of each one is used as input to the immediately following subproblem. Each subproblem determines the following key decisions: (1) Number of locations to divide, (2) assign product family per aisle, (3) sort according to stackability priority within the aisle and define the location to be divided, and (4) define the number of locations for each family that go to Perchas sorting by subfamily. As stated above, it is common to solve these subproblems separately considering their complexity, although they are strongly interconnected [25]. Table 4 presents a list of the problems studied in this research.

Table 4. List of addressed problems, and objective functions considered in each problem.

Abbreviation	Description	
SLAP-VH or P_1	Storage location assignment problem with variable height	
$P_{1.1}$	Subproblem of P_1 that determines how many locations to divide	
$P_{1.2}$	Subproblem of P_1 that assigns families to aisles	
$P_{1.3}$	Subproblem of P_1 that determines which locations to divide	
$P_{1.4}$	Subproblem of P_1 that determines how many locations correspond to each family that goes to *Perchas*	

Abbreviation	Description	Related Problem
F_1	Minimize the distance from an SKU to the warehouse exit	P_1
$F_{1.1}$	Minimum number of divided locations	$P_{1.1}, P_{1.2}$
$F_{1.2}$	Affinity analysis	$P_{1.2}, P_{1.3}$
$F_{1.3}$	Minimize the distance, from a location to warehouse exit	$P_{1.3}, P_1$
$F_{1.4}$	Minimize the number of locations for "B" classification SKUs	$P_{1.4}, P_1$

In P_1, locations are assigned to each SKU from the results obtained from $F_{1.1}$, $F_{1.2}$, $F_{1.3}$, and $F_{1.4}$, where each subproblem considers different search spaces to determine key decisions. Appendix A shows the main parameters and decision variables used in the models for each subproblem.

5.1. Subproblem $P_{1.1}$: Locations to Divide

The model for $P_{1.1}$ is presented below, in which the decision variables define the quantity of simple, double, and triple locations that each SKU must occupy.

$$Min \sum_I \left(\frac{uB2_i}{2} + \frac{uB3_i}{3} + \frac{uC2_i}{2} + \frac{uC3_i}{3} \right) \tag{4}$$

Subject to:

$$total\_UA \geq \sum_I uA_i \tag{5}$$

$$total\_UB \geq \sum_I uB_i \tag{6}$$

$$total\_UB2 \geq \sum_I \frac{uB2_i}{2} \tag{7}$$

$$total\_UB3 \geq \sum_I \frac{uB3_i}{3} \tag{8}$$

$$total\_UC \geq \sum_I uC_i \tag{9}$$

$$total\_UC2 \geq \sum_I \frac{uC2_i}{2} \tag{10}$$

$$total\_UC3 \geq \sum_I \frac{uC3_i}{3} \tag{11}$$

$$total\_locations \geq total\_UA + total\_UB + total\_UB2 + total\_UB3 + total\_UC \\ + totaL\_UC2 + total\_UC3 \tag{12}$$

$$uA_i \geq \alpha * dda\_locations_i \\ \forall i \in I \tag{13}$$

$$uA_i = 1 \\ \forall i \in I: dda\_locations_i = 0 \wedge SKU\_A_i = 1 \tag{14}$$

$$uB_i \geq \beta * dda\_locations_i \\ \forall i \in I: SKU\_B_i = 1 \tag{15}$$

$$u2B_i \geq dda\_decimal_i \\ \forall i \in I: dda\_decimal_i \geq \gamma \wedge SKU\_B_i = 1 \tag{16}$$

$$u3B_i \geq dda\_decimal_i \\ \forall i \in I: dda\_decimal_i \leq \gamma \wedge SKU\_B_i = 1 \tag{17}$$

$$uC_i \geq \beta * dda\_locations_i \\ \forall i \in I: SKU\_C_i = 1 \tag{18}$$

$$u2C_i \geq dda\_decimal_i \\ \forall i \in I: dda\_decimal_i \geq \gamma \wedge SKU\_C_i = 1 \tag{19}$$

$$u3C_i \geq dda\_decimal_i \\ \forall i \in I :: dda\_decimal_i \leq \gamma \wedge SKU\_C_i = 1 \tag{20}$$

$$total\_locations \leq loc\_total \tag{21}$$

$$total\_sum2_s \geq \frac{\sum_I (uB2_i + uC2_i) \cdot sku\_subfamily_{si}}{4} \\ \forall s \in S \tag{22}$$

$$\sum_{I}(uB2_i + uC2_i)\cdot sku\_subfamily_{si} - total\_sum2_s\cdot 4 = 0$$
$$\forall s \in S \tag{23}$$

$$total\_sum3_s \geq \frac{\sum_I(uB3_i + uC3_i)\cdot sku\_subfamiliy_{si}}{6}$$
$$\forall s \in S \tag{24}$$

$$\sum_{I}(uB3_i + uC3_i)\cdot sku\_subfamiliy_{si} - total\_sum3_s * 6 = 0$$
$$\forall s \in S \tag{25}$$

The objective is to minimize the number of total double and triple divisions in the warehouse. Constraints sets (5) to (11) indicate the total number of simple, double, and triple locations. In (12), the total amount of locations obtained is grouped, which are limited by equation (21). Constraints sets (13) to (20) guarantee the coverage of locations according to the ABC classification of the SKUs, reflected in the parameters α, β, γ. Expressions (22) and (23) consider that the number of double locations for each subfamily is divisible by 4. The previous condition is also considered in (24) and (25), where the number of triple locations for each subfamily is divisible by 6.

5.2. Subproblem $P_{1.2}$: Allocation of Families to Aisles

This subproblem has been approached from the experience with the affinity between the studied families, which aims to define those families that should be assigned close to one another. It is possible to highlight three main classifications to determine affinity: (1) SKUs that belong to different families and are commonly requested together by the customers; (2) SKUs that are not frequently requested, and therefore have no affinity with other SKUs of other families, and; (3) given that Perchas have greater storage capacity per location, those SKUs that are less demanded by customers and belong to the families in Perchas are considered to be positioned near these locations. Table 5 describes the analysis carried out from the families under study.

Table 5. Affinity or compatibility between families.

Compatibility		Closeness to Perchas	Without Compatibility
Family 1	Family 6	Family 2	Family 3
Family 2	Family 4	Family 4	Family 1
Family 7	Family 5		Family 6

The affinity between families allows determining the physical storage arrangement within the aisles. Assigning families to aisles, that is, constructing accommodation lines, facilitates the work of operators when traveling along the aisle selecting only what is necessary, without going back or travel long distances when making the collection route.

5.3. Subproblem $P_{1.3}$: Determine Which Location to Divide

To determine which locations should be divided, the following linear programming problem is proposed:

$$Min \sum_{S}\sum_{J}\sum_{K} X0_{js}D_{jk}\left(\frac{1}{weight_s}\right) \tag{26}$$

Subject to

$$\sum_{J} X0_{js} \leq ub0_s$$
$$\forall s \in S \tag{27}$$

$$\sum_{J} X2_{js} \leq ub2_s$$
$$\forall s \in S \tag{28}$$

$$\sum_J X3_{js} \leq ub3_s$$

$$\forall s \in S \tag{29}$$

$$\sum_S \left(X0_{js} + X2_{js} + X3_{js} \right) = 1$$

$$\forall j \in J \tag{30}$$

The objective function of the subproblem $P_{1.3}$ is to minimize the distance from simple locations to the exit, for each family according to the priority order. Constraints sets (27), (28), and (29) determine the number of locations that will not be divided, or otherwise divided into two and three, respectively. Constraint set (30) establishes that each location can only perform one operation.

5.4. Subproblem $P_{1.4}$: Perchas

The mathematical model for $P_{1.4}$, has the objective to determine the necessary number of locations (see Figure 7) for each SKU of families that must be positioned on *Perchas*.

$$Min \sum_I u_i sku\_b_i \tag{31}$$

Location to floor
(*Percha*)

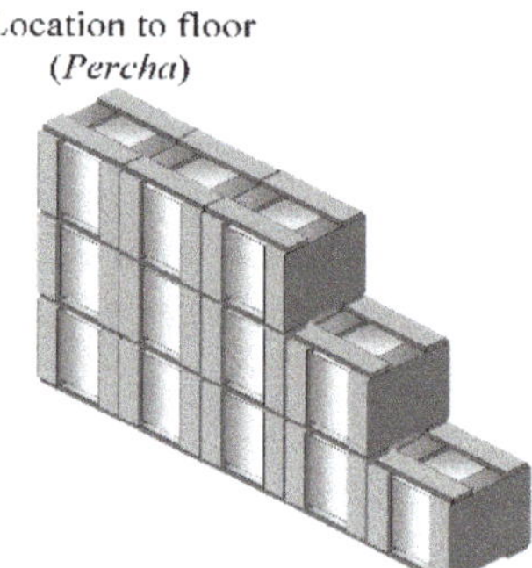

Figure 7. Representation of *Percha* location, ordered with five base pallets and three levels.

Subject to

$$total\_per \geq \sum_I u_i \tag{32}$$

$$u_i \geq 1$$
$$\forall\, i \in I \text{ with } dda\_locations \leq \theta \tag{33}$$

$$1 \leq u_i \leq \mu * dda\_locations_i$$
$$\forall\, i \in I \text{ con } dda\_locations > \theta \tag{34}$$

The objective function in (31) corresponds to the minimization of the number of locations for those SKUs that are rated as "B". Constraint (32) establishes that the total number of locations must not exceed the number of available locations. Constraints sets (33) and (34) determine the number of locations according to the ABC classification of the SKUs, reflected in the parameters θ and μ.

5.5. General Algorithm

A flow diagram is built to solve the global model, which establishes the inputs required for each subproblem graphed in Figure 8, where (A), (B), (C), and (D) correspond to key decisions that solve the subproblems $P_{1.1}$, $P_{1.2}$, $P_{1.3}$, and $P_{1.4}$ respectively, ending with the assignment of locations in P_1. The details of each step are described in the next section.

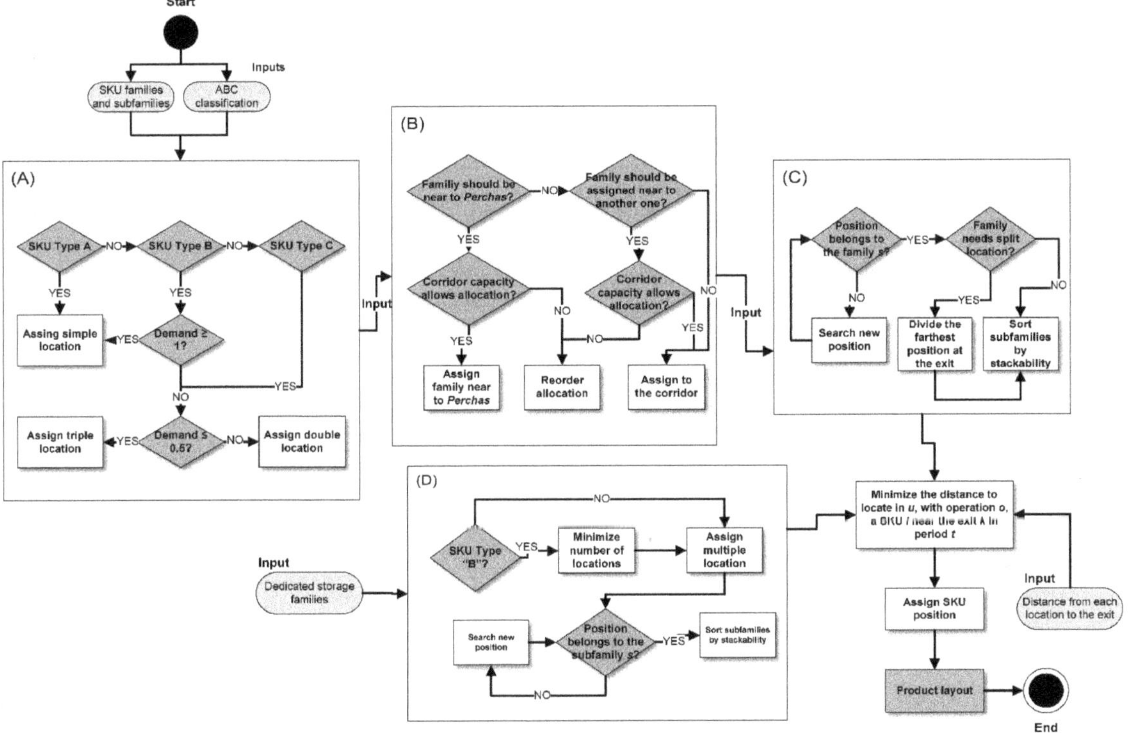

Figure 8. General flow diagram for resolution model of the SLAP-VH study problem.

6. Proposed Methodology

This section defines the proposed solution models to address each subproblem.

6.1. Solution Approach Proposed for $P_{1.1}$

The problem $P_{1.1}$ can be considered a variant of the integer linear programming demand satisfaction problem. The mathematical model presented defines demand as the number of locations that each SKU needs, which is satisfied with simple, double, or triple locations. The total demand for locations corresponds to a non-integer value, therefore, to obtain the best combination of possible operations, the maximum coverage is determined for SKUs, where those articles that demand more or equal capacity than a location can offer, requires the assignment of simple positions. If demand is less than one location, double locations must be assigned if it is between one or half pallet, and triple if it is less than half of a pallet.

The structure of the rack in storage space symbolizes two locations, therefore, it is not possible to divide a simple location, since if one of these is divided, the paired location must do so. Therefore, the number of double locations assigned by subfamilies must consider a multiple of four, and a multiple of six for triple locations.

The problem model $P_{1.1}$ is implemented using Python programming language and the Pyomo library. This mixed-integer linear optimization model (MILP) is solved using Gurobi optimizer. The problem involves $4,574$ variables, $4,573$ of an integer nature, and one continuous variable, in addition to $3,454$ constraints and one objective minimization function.

6.2. Solution Approach Proposed for $P_{1.2}$

Family affinity and stackability priority require full attention from managers and operators. On the one hand, the incoming and outgoing products must be considered to accommodate them correctly and, on the other hand, the ergonomics of the pallet must be studied securing stability in its dimensions and weight, avoiding unsafe actions for the operators. In this way, the distribution of families and subfamilies is carried out based on affinity and stackability, merging the analyzed classification to the study warehouse, and finally generating the assignment order of subfamilies.

From Figure 9, a logical order emerges that addresses the priorities of assigning families to aisles. It is important to emphasize that the warehouse is inflexible, therefore, there is only one possible arrangement to meet the priorities considering the solution obtained in $P_{1.1}$, this is represented in Figure 10. It should be considered that the assignment of families to aisles must maintain the continuity of the family, i.e., not to position an "X" family in the first aisle and continue its assignment in the last. This helps operators to find products in aisles in a faster way. In this sense, it can be observed that those families that must go near the "Perchas" location are the closest, without interrupting the continuity of the following families. The same observation applies to those who have compatibility. Finally, the one that does not have compatibility, being the most numerous, is positioned in the aisles furthest from "Perchas" avoiding losing continuity.

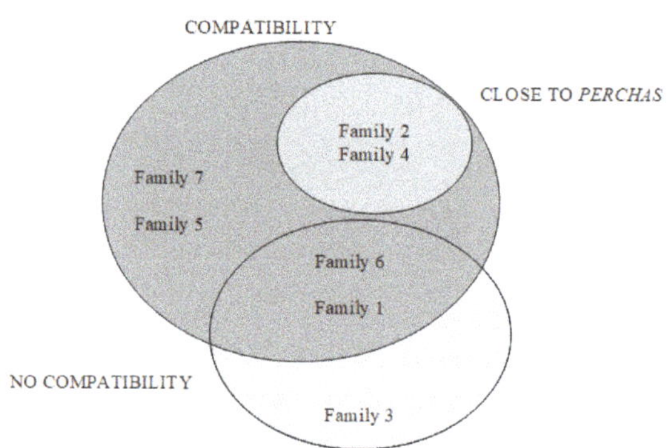

Figure 9. Classification diagram by affinity.

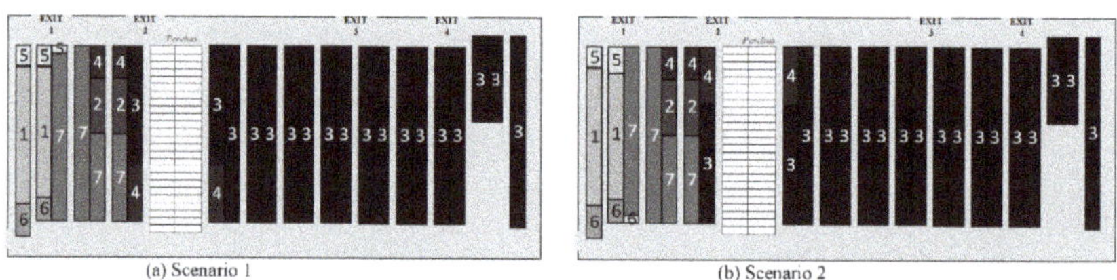

Figure 10. The fusion between the affinity classification diagram and the store. The numbers indicate the families positioned by racks in each aisle.

6.3. Solution Approach Proposed for $P_{1.3}$

Once the families are assigned to each aisle and the division for locations are known, the objective of $P_{1.3}$ is to define which locations will be divided according to the number of locations assigned to each family. The assignment operation is done for those families with a large volume of SKUs since those with less than 10% of the total SKUs can be assigned manually, and the optimal result is predictable with respect to the priorities.

The model of this subproblem is coded, like the subproblem $P_{1.1}$, in Python language from the Pyomo software packages. The model is implemented with the mixed-integer linear optimization solver Gurobi, considering $13,717$ variables used, being $13,698$ of them of binary nature, $13,716$ integer (including binary), and one continuous variable. Besides, 798 constraints and one objective minimization function were considered. These values decrease as the number of SKUs per family decreases.

Once each position by families is defined, we proceed to sort by ABC classification, since the model's solution sends all double and triple locations at the end of the aisles. Despite being the optimal global solution, it affects the grouping of subfamilies. On the other hand, the stackability priority is defined by the model, which remains fixed. The ABC classification allows defining the prioritization of each SKU based on its demand and thus establishing its proximity to the point of departure, so that it takes less time for the crane to execute the movements and supply the dispatch area.

6.4. Solution Approach Proposed for $P_{1.4}$

In the case of "Perchas" locations, the number of SKUs that must be assigned to this section of the warehouse is 170, a value that exceeds the capacity in 76 locations, since each SKU must have at least a position or location based on its classification. To balance the number of SKUs and locations, those SKUs of type "C" are positioned in racks.

Thus, the decision of how many locations assign to each SKU depends, as in $P_{1.1}$, of the demand and its classification. However, in $P_{1.4}$, it is sought to assign the least amount of locations to SKUs of type "B", since the demand for each SKU of type "A" of these families is very high compared to those assigned to the racks.

6.5. Global Proposed Solution Approach

Finally, each solution obtained from the defined subproblems is specified as parameters in P_1, which turns out to be a one-to-one assignment problem. At this point, how many locations each SKU needs, in which aisle it should go, and where it can be positioned according to the type of location are known data, thus making it possible to know in which position or location each SKU should go.

Figure 11 illustrates the entry of each SKU family to the bottleneck, which corresponds to the solution of the model $P_{1.1}$. Then the route that the SKU takes is decided according to the subfamily to which it belongs, which is a solution of the model $P_{1.2}$. Then, the next decision is defined according to the SKU type, since this classification defines in what type of position it can be located, and the final optimization defines the destination of the SKU by the minimum distance to the exit of the warehouse.

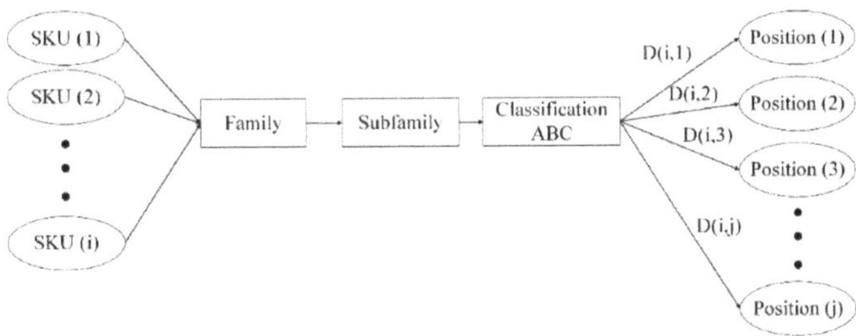

Figure 11. Assignment problem diagram P_1.

7. Evaluation Scenarios

In this section, the second stage of the research is presented. Two scenarios are carried out once the solution of the subproblem $P_{1.2}$ is known, which result from a combination of storage strategies, where both have dedicated storage due to "Perchas", but differ in the racks, being by random classes or purely by classes. These are based on the main strategies used by the company.

As can be seen in Figure 12, the first scenario, called E_1, is defined as a dedicated assignment in "Perchas" locations and by randomized classes in racks. Said randomness is determined by aisles, that is, a family that requires more than one aisle can randomly assign subfamilies among these, without losing the priority of the lightest at the beginning of the aisle and the heaviest at the end. In this way, the Subfamily Layout is built, which begins at the end of each aisle, where every two locations are filled on one side and pass to the other side of the aisle until all are completed. The second scenario, called E_2, corresponds to a dedicated allocation in "Perchas" locations and purely by classes in racks. The allocation by product family determines sequentiality in the aisles, that is, for a family that requires more than one aisle, it will not assign a subfamily S_1 until the subfamily S_2 has been assigned, always considering the stackability priority. Thus, the Subfamily Layout is built from the end of an aisle and the beginning of the next, where, as in E_2, the corridors are filled every 2 locations on one side and pass to the other side of the aisle, until all are completed.

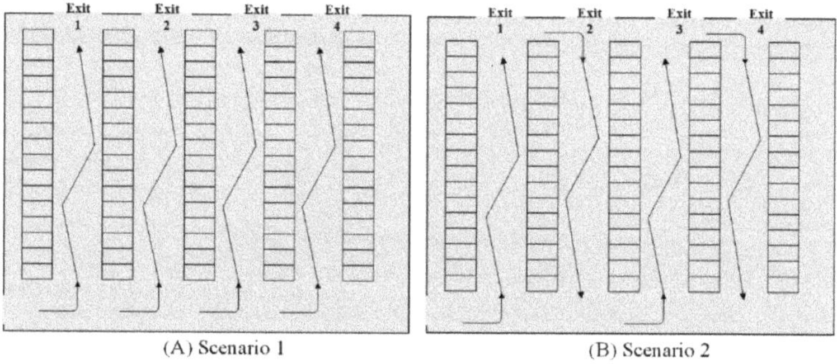

Figure 12. Representation of the filling sequence.

The evaluation of each scenario is carried out considering the following criteria, in the Table 6, based on a high sales day, called V_{high}, corresponding to period 2. V_{high} has a set of trucks where each of them dispatches between one to four loads.

Table 6. Slotting scenario evaluation rubric, where the minimum score corresponds to 1 and the maximum score is 4.

Criteria	[%]	4	3	2	1
Total Distance • Full pallet output distance • Conformed pallet output distance • Replenishment entry distance	70	Reduced over 10% versus the base scenario.	Reduced between 5–10% versus the base scenario.	Reduced between 1–5% versus the base scenario	Does not reduce the distance versus the base scenario
Movement costs • Use of cranes for a complete pallet • Use of cranes for conformed pallets • Use of cranes for replenishments	30	Reduced over 10% versus base scenario	Reduced between 5–10% versus the base scenario	Reduced between 1–5% versus the base scenario	Does not reduce the costs versus base scenario

The number of full pallets that are dispatched is calculated from the number of total boxes requested in V_{high} over the number of boxes per pallet of each SKU, then the distance traveled from the SKU location to the exit, multiplied by the number of movements, corresponds to the exit distance of complete pallets. Thus, everything that does not leave the warehouse as a complete pallet is collected per unit of boxes, which represents the exit distance of the conformed pallet. Replenishment comes from movements that deplete the availability of a pallet of SKUs in the warehouse. This calculation is determined by the number of times a position is empty when removing full pallets and/or boxes. Therefore, the route from entering the warehouse to a depleted position multiplied by the number of times it needs to be filled corresponds to the input distance for replenishment. Finally, the total distance arises from the aggregation of these three items. The above is summarized in Equation (35):

$$Total\ Distance = \sum_{c}\sum_{ca}[P_i(d_S) + PP_i(d_S)] + R_i(d_E) \tag{35}$$

On the other hand, the cost criterion assesses the use of cranes based on the distances traveled for each defined operation, namely: cranes for a complete pallet, collection cranes, and replenishment cranes, as well as required operators. Average speed is defined in the Picking operation for cranes, considering high congestion, of 5 km/h, where these have 6.5 h of average use per operator and work shift. Each crane consumes approximately 11.13 Kwh, which is equivalent to 0.11 USD/Kwh (the data used for the accounting of operators and cranes are provided by the company). Besides, an average salary of 708.78 USD per operator is considered [26]. Thus, the calculation of the number of operators necessary for the picking process by salary, added to the number of cranes by the value of electricity consumption and monthly rent, allows obtaining the total monthly cost, as indicated in Equation (36):

$$Movement\ costs = (W_{ps} + W_{pc} + W_{pe})\cdot Salary + (G_{ps} + G_{pc} + G_{pe})A\omega_1\omega_2 \tag{36}$$

8. Results

8.1. Computational Results

The algorithms were coded in Python, and the results were carried out on an Intel Core i5-8250U CUP 1.60 GHz 1.80 GHz core with 4 GB RAM.

Table 7 shows the execution time and results for subproblems $P_{1.1}$ and $P_{1.4}$, while Table 8 provides duration time and results for each scenario evaluated. Complementary, the Execution Times with Gurobi's Python Interface are presented in Appendix B.

Table 7. Computational results for the general subproblems of P_1.

Model	Solver Application [sg]	Results Time [sg]
$P_{1.1}$	28.43	0.29
$P_{1.4}$	0.54	0.15

Table 8. Computational results of the P_1 model for each scenario.

Model	Scenario	Solver Application [sg]	Results Time [sg]
P_1	E_1	158.06	15.68
P_1	E_2	137.28	16.75

Table 9 shows the execution time and results for subproblems $P_{1.3}$, considering scenarios and family.

Table 9. Computational results of the $P_{1.3}$ model for each scenario according to the evaluation family.

Family	Scenario	Solver Application [sg]	Results Time [sg]
Family 1	E_1	0.05	0.19
Family 3	E_1	0.53	1.47
Family 7	E_1	0.09	1.80
Family 1	E_2	0.05	0.15
Family 3	E_2	2.59	0.60
Family 7	E_2	0.78	0.25

8.2. Results by Subproblems

For subproblem $P_{1.1}$, the minimum number of locations to be divided corresponds to 16.17% of the total locations, a value that includes 24.14% locations with double position and 75.86% with a triple position.

Table 10 shows the number of locations for each family and subfamily, in which the latter is described by the set S with index ε, where ε indicates the number of regrouping of a family.

Table 10. Corresponding percentages of the total locations available in the defined warehouse by family and subfamily.

Family	Total [%]	Subfamily	Simple Locations [%]	Double Locations [%]	Triple Locations [%]	Total [%]
Family 1	6.23	s_1	0.00	0.19	0.56	0.74
		s_2	2.97	0.00	1.12	4.09
		s_3	0.00	0.00	0.19	0.19
		s_4	0.00	0.00	0.19	0.19
		s_5	0.28	0.00	0.19	0.46
		s_6	0.37	0.00	0.19	0.56
Family 2	2.79	s_1	0.09	0.19	0.37	0.65
		s_2	0.46	0.37	0.56	1.39
		s_3	0.19	0.00.	0.56	0.74
Fily 3	70.72	s_1	1.86	0.00	0.19	2.04
		s_2	30.67	0.56	1.12	32.34
		s_3	5.86	0.56	0.56	6.97
		s_4	0.74	0.00	0.74	1.49
		s_5	12.55	0.00	0.37	12.92
		s_6	13.29	0.74	0.93	14.96
Family 4	3.72	s_1	0.00	0.19	0.19	0.37
		s_2	0.09	0.19	0.19	0.46
		s_3	0.56	0.00	0.00	0.56
		s_4	0.56	0.00	0.19	0.74
		s_5	1.21	0.19	0.19	1.58
Family 5	0.74	s_1	0.37	0.19	0.19	0.74
Family 6	2.23	s_1	1.30	0.19	0.74	2.23
Family 7	13.57	s_1	0.37	0.19	0.19	0.74
		s_2	8.55	0.19	2.23	10.97
		s_3	1.49	0.00	0.37	1.86

From Figure 13, it is possible to notice that the largest number of locations belongs to family 3, which has the highest percentage of simple locations. This fact reveals that this family is the most requested during this period. On the contrary, family 2 represents the highest percentage of triple locations, from which it can be concluded that this family contains mostly SKUs of type "B" and "C".

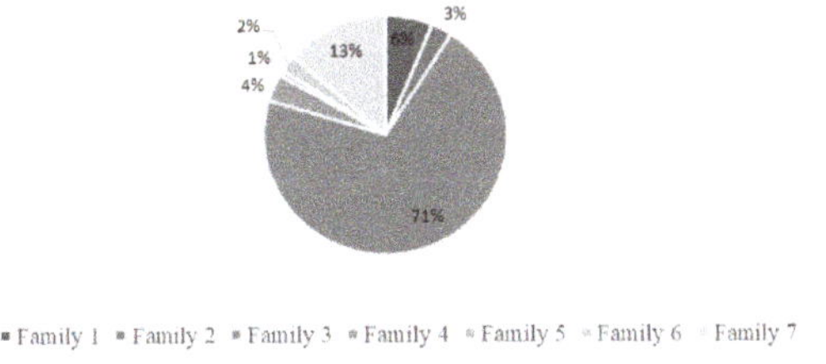

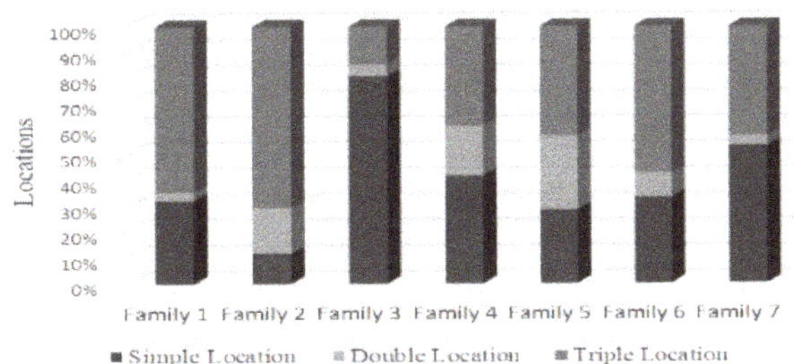

(**a**) Number of locations for each family

(**b**) Assignment of single double and triple location for each family

Figure 13. Percentage representation for the results obtained in $P_{1.1}$.

For the subproblem $P_{1.2}$, the design of the distribution of families per aisle follows the order described in Figures 14 and 15, where they are assigned into the warehouse from left to right. Furthermore, the described configuration considers the results obtained in $P_{1.1}$ and the relationship generated in Figure 9.

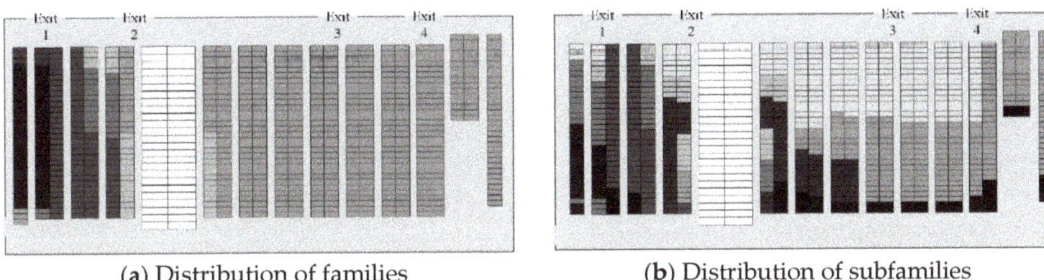

(**a**) Distribution of families

(**b**) Distribution of subfamilies

Figure 14. Representation of the family layout in the study warehouse for E_1. The gray variation corresponds in (**a**) for each family and in (**b**) for each subfamily.

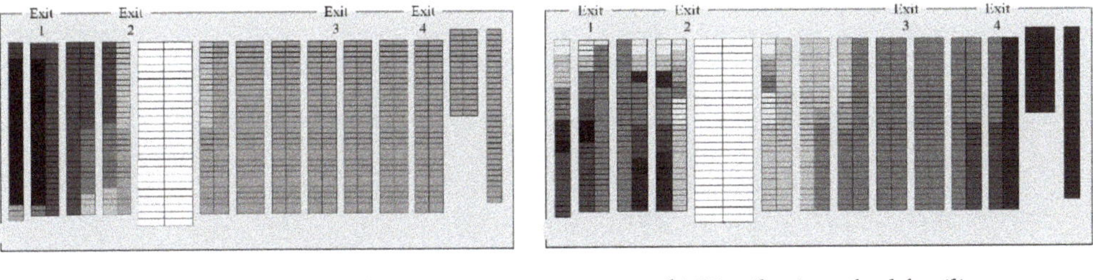

(**a**) Distribution of families

(**b**) Distribution of subfamilies

Figure 15. Representation of the family layout in the study warehouse for E_2. The gray variation corresponds in (**a**) for each family and in (**b**) for each subfamily.

The difference between both scenarios is highly observable in the distribution by subfamily, specifically in the right section of the warehouse, since it corresponds to the assignment of the largest family in SKUs covering the largest number of aisles, as can be seen in Figure 15. The gradient color (from light to dark) represents from lower to higher weight by subfamilies.

On the other hand, the division of locations is represented in Figures 16a and 17a, where the execution of the model $P_{1.3}$, reveals that the divided locations should be applied to the end of each aisle, while the simple ones should be considered at the beginning.

However, the stackability priority motivates a reorder within each aisle, without changing the family layout (see Figures 16b and 17b).

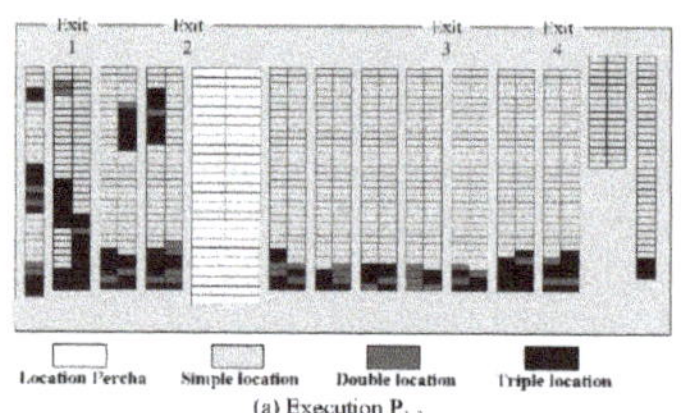

(a) Execution $P_{1.3}$

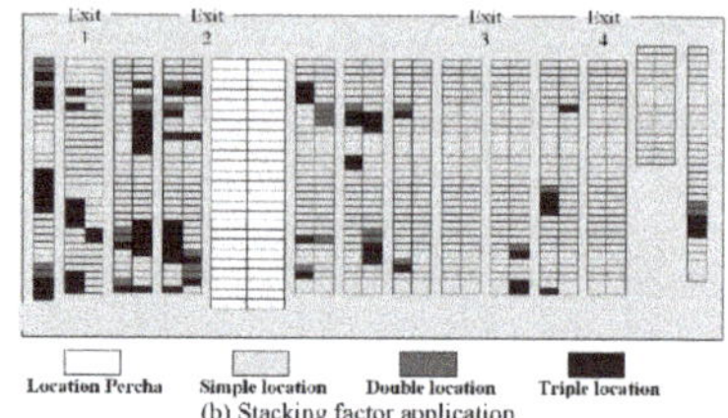

(b) Stacking factor application

Figure 16. Layout of types of location in the warehouse, where (**a**) corresponds to the execution of $P_{1.3}$ for E_1 and (**b**) to the reorder by stackability.

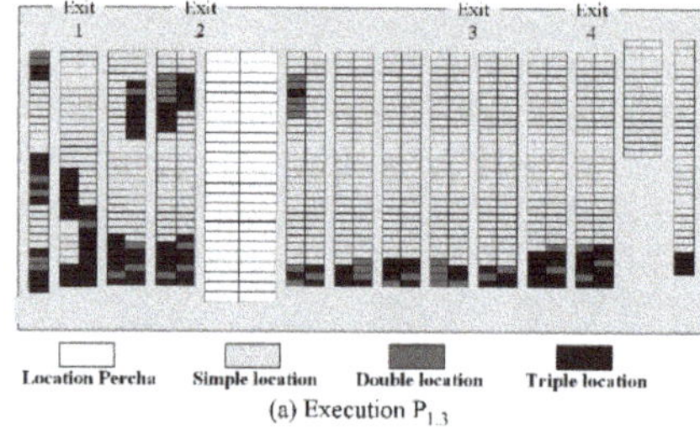

(a) Execution $P_{1.3}$

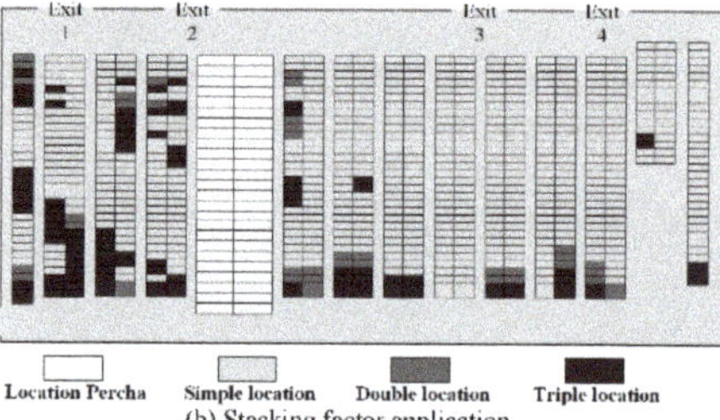

(b) Stacking factor application

Figure 17. Layout of types of location in the warehouse, where (**a**) corresponds to the execution of $P_{1.3}$ for E_2 and (**b**) to the reorder by stackability.

Simple locations correspond mainly to type "A" SKUs, representing 86.6% of the total of these locations, while double and triple locations mainly represent type "C" with SKUs 76.19% and 76.52%, respectively. The rest are assigned for SKUs of type "B".

Finally, the solution obtained in $P_{1.4}$ is observed in Table 11, as in the subproblem $P_{1.1}$, the number of locations for each family and subfamily is shown, where the latter is described by the set S with indices ε, where ε indicates the number of regroups in a family.

Table 11. Corresponding percentages of the total *Perchas* locations available in the warehouse defined by family and subfamilies.

Family	Total [%]	Subfamily	Total [%]
Family 2	45.74	S_1	24.47
		S_2	21.28
Family 4	54.26	S_1	5.32
		S_2	6.38
		S_3	9.57
		S_4	14.89
		S_5	18.09

From these results, the distribution of families can be seen in Figure 18 for each scenario.

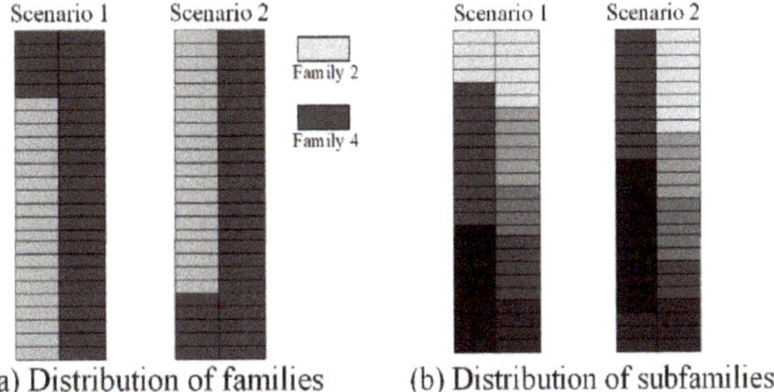

(a) Distribution of families   (b) Distribution of subfamilies

Figure 18. Families layout in *Perchas* location corresponding to sides with non-divisible locations.

Finally, each previous result allows obtaining the SLAP-VH solution, that is, assigning at least one storage location to each SKU. According to this, the warehouse is increased due to the divided locations by 18.17%.

9. Results by Scenarios

The number of SKUs under the case study corresponds to 56.55% of the SKUs V_{high}, where the rest belongs to the values warehouse, storage of merchandise of high monetary value that is normally sold for the minimum unit. Thus, based on the evaluation criteria described, E_1 scores 2, while E_2 scores 1. As can be seen in the Table 12.

Table 12. Evaluation of Slotting scenarios based on the criteria defined in the rubric.

Criteria	E_1	E_2
	[km]	[km]
Total distance	1792.38	1875.48
• Full pallet output distance	36.63	39.78
• Conformed pallet output distance	1706.41	1788.84
• Replenishment entry distance	49.34	46.86
Total	3535.42	3750.96
	[$USD]	[$USD]
Movement costs	73,582.70	76,559.99
• Use of cranes for complete pallets	12,754.34	12,760.00
• Use of cranes for conformed pallets	51,998.44	54,966.15
• Use of cranes for replenishments	8829.92	8833.85
Total	147,168.4	153,159.99

Regarding the minimum total distance for both scenarios, it is possible to observe, from Table 9, that the number of complete pallets is closer to the exit than those pallets that are conformed on the collection. However, the total exit distance is less in E_1. The distance to the entrance, represented by the replenishment pallet path, is closer for E_2, this indicates that those locations that run out of availability are closer to the reserve warehouse than in E_1. Thus, the total distance (input and output) is reduced by 4.64% in E_1 with respect to E_2, as seen in Figure 19. Therefore, it can be stated that the most demanded SKUs are those closest to the exit on stage E_1. The preparation time is reduced in the same proportion, approximately 5%, since the time and the distance are directly proportional, but both are inversely proportional to the order in the warehouse, that is, the higher the order obtained, the lower is the distance traveled by the cranes and thus less time is spent on this item, clarifying that the calculated time does not consider the specific time between the loading or leaving a pallet at a location, this is purely the perspective of horizontal movement from one point to another.

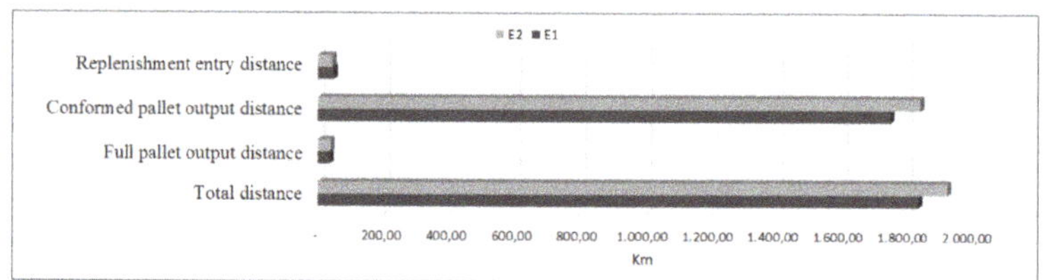

Figure 19. Variation between both scenarios regarding the total distance criterion.

Finally, E_1 saves 4.05% compared to E_2 in movement costs. Although the number of operators and cranes for full pallet picking and replenishment turn out to be similar since the distances traveled are similar, the greatest difference is observed in conformed pallets, where the number of operators, and therefore cranes, increases for E_2, since the travel time for this scenario corresponds to 5.19% more than E_1, considering only the outbound route (pallet loading and unloading times are not included).

From the above it is possible to conclude that the dedicated and randomized storage strategy by classes allows reducing the route for cranes in the Picking process, achieving a cost reduction evaluated at 4% per month for this period of the study cycle as presented in Figure 20. Although the preparation of the best scenario requires greater precision, just by defining a better location for each SKU, this fact directly impacts the trip that the operator makes during Picking. The quantity required for each scenario with respect to the exit of the complete pallet and entry of the replenishment pallet turns out to be the same, given the movements that must be carried out and the defined route. However, the quantity in the unit collection varies significantly, being favorable for E_1, where not only the distance is reduced, but also the crane congestion is reduced since it requires a smaller number of operators than in E_2.

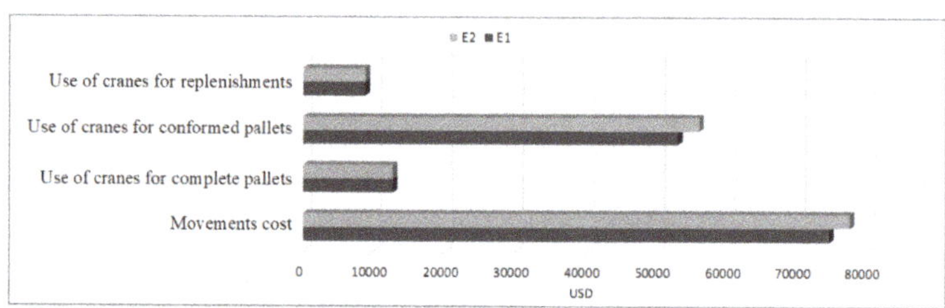

Figure 20. Variation between both scenarios with respect to the cost of movements criterion.

On the other hand, in both scenarios, the operation reports benefits due to the maximum use of the warehouse (considering the first level storage locations). However, the first scenario generates more advantages, such as better handling, avoiding damage to the products thanks to the priority of stackability. Most in-demand SKUs are closer to the exit, which means faster dispatching, facilitating the delivery process. Finally, distributing the products intelligently in the warehouse clears the aisles and facilitates the transfer of pallets both for the cranes that leave and those that enter to restock.

10. Discussion

Considering those authors who propose different ways of approaching the Slotting optimization problem, the vast majority of these focus on solving it through metaheuristics based on search and accommodation placement. However, this study is carried out under a decomposition of the problem that determines each key decision. While decomposition is one of the optimal methods for tackling problems with broad key decisions, separating the problem allows greater handling of data manipulation and a dedicated focus on each decision to be resolved. The computational results reveal the impact on the flow time of the executions for each subproblem and at the same time for each scenario, where each model provides the necessary data to address the problem in a sequential method and additional information that allows generating an effective analysis of each subproblem which helps to create a faster and better result in the SLAP-VH solution procedure. Ref. [1] analyzed the efficiency between solving the assignment problem metaheuristically and mathematically, where the latter does not turn out to be very fast in terms of the average execution time ($26,640$ s, while the heuristic algorithm took 155 s). However, it provides a solution that is very close to optimal, where the mean gap of the solutions is 4.3% after excluding an outlier.

This study groups each factor of the Slotting design to improve the Picking operation, such as distance, cost of operation, crane movements, use of human resources, and storage strategies based on implementation alternatives Although the studies carried out by other authors evaluate the application of the methodology, the results obtained before the application are similar when using both decomposition and a metaheuristic. Comparing the result obtained by [14] which obtains a 4% reduction in the cost of human resources considering the departure of a specific batch of accommodation, the same percentage was obtained in the reduction of costs for accommodation in first level locations corresponding to the SLAP-VH problem. In addition, Slotting has been applied in the same way as in each of the compared investigations, where the methodology used is the same, but the impact depends completely on the storage space since it is not the same case to apply a Slotting optimization in a maritime terminal [11] versus a beverage distribution center. As for the infrastructure, the handling of the type of material, and, most importantly, the specific objective of performing the Slotting. For example, the results obtained by [12] are related to the objective of programming and assigning cranes, according to their specifications, to carry out operations in the locations where they can work. While for this study, the results obtained are related to improving the picking process and reducing the distance of cranes.

Finally, the solution to the problem will depend on four major factors: the storage location type, the storage strategy, the number of SKUs to be stored, and the warehouse's size. All these factors are strongly related when it comes to minimizing travel time, operational cost, and storage usage. The model used in this research connects these factors considering divisible locations in racks and "Percha" locations, the two storage strategies evaluated in the different scenarios, families, subfamilies and SKUs, and the warehouse entry/exit distance. Comparing the different scenarios, E_1 exceeded E_2 in terms of the quality of the solution and the computational time optimally, significantly improving the distance traveled by the cranes and, simultaneously, the warehouse space usage as well as the operating cost.

11. Conclusions and Summary

The proposed research developed the SLAP-VH problem, which was decomposed into four subproblems of mathematical modeling solved sequentially in a storage environment of variable height and floor locations, where each one covered a key decision to achieve the proposed general objective. The results of the evaluation of the Slotting scenarios are alternatives that seek to approach the main problem in the Picking operation: the unproductive travel time during the pallet transfers by the cranes in order preparation. The application of this model allows defining the best location for each SKU, considering that the appropriate allocation strategy for the size of the study warehouse corresponds to E_1. According to the results of the case study, there is a decrease in the distance traveled of 5% of E_1 with respect to E_2.

On the other hand, the benefits granted by the best scenario are reflected in the maximum utilization of the warehouse, since the products are assigned efficiently, avoiding their deterioration due to the priority of stackability. In addition, this strategy improves the use of human resources and vehicles in the operation, the workload distribution, and also reduces the operating costs. The results obtained from the research provide valuable information for managers, where they can define, according to their warehouse capacity and the ABC classification of their SKUs, the intelligent allocation of each product within the warehouse. In parallel, minimizing the number of trips made in order to reduce the distance of travel of cranes by optimizing Slotting, is equivalent to reducing the number of operators, a favorable strategy for operating costs.

The solution design can be adapted and applied to other similar storage systems, modifying four main input parameters: warehouse capacity, demand for SKUs, incorporation of families, and the proper distances to evaluate, where the objective function used could be further enriched by including the picking routes, creating a combination of studies between the best Picking route and the slotting optimization.

Author Contributions: P.V.: Management of the research project, design of the proposal methodology, research problem modeling, article writing, analysis and discussion of the mathematical model and coordination of reviews; K.G.: Theoretical and empirical research, article writing, computational model development, data processing and results analysis; R.M.: Design of the proposal methodology, modeling of the problem, computational development, article writing, results analysis and discussion; F.K.: Theoretical research, data processing and numerical analysis results; J.R.: Computational model design and optimized result processing. All authors have read and agreed to the published version of the manuscript.

Funding: This research received no external funding.

Institutional Review Board Statement: Not applicable.

Informed Consent Statement: Not applicable.

Data Availability Statement: The data presented in this study are available on request from the corresponding author. The data are not publicly available due to confidentiality of the company.

Acknowledgments: The authors wish to thank the anonymous reviewers and the editorial board of the journal.

Conflicts of Interest: The authors declare no conflict of interest.

Appendix A. Notations, Abbreviations, and Symbols

Abbreviations and Concepts			
SKU	Stock Keeping Unit	Position	Divided storage area of a capacity rack for half or a third of the height of a pallet
Family	Set of SKUs that are part of the same classification	Simple	A location that is not divided, only stores a pallet of SKU
Subfamily	Set of SKUs that are part of the same classification within a family	Double	A location that is divided into two sections, allowing to position two pallets of medium-height SKUs
Perchas	Location of the warehouse where the pallets are stacked on the floor	Triple	A location that is divided into three sections, allowing the positioning of three SKU pallets of a third of the height
Location	Storage area of a rack with a capacity for a single pallet	Rack	A metal frame that, according to its characteristics, allows storage goods.

Parameters	
i	SKU index
j	Warehouse locations racks index
k	Warehouse exit index
s	Subfamilies index
o	Operations index
D_{jk}	Distance from location j to exit k
$position\_subfamily_{js}$	1, if position j belongs to subfamily s; 0, if not
$sku\_subfamiliy_{si}$	1, if SKU i belongs to subfamily s; 0 if not
sku_type_{io}	Amount of positions of an SKU i with operation o
$type_{jo}$	1, if position j has operation o; 0 if not
$value\_sku_i$	Value of the i SKU
$dda\_locations$	The integer part of the number of locations demanded by the SKU i
$dda\_decimal_i$	Difference between the quantity of locations demanded and its Entire Part of the SKU i
loc_total	Total number of simple locations (without considering *Perchas*)
SKU_A_i	1, if SKU i is of type "A"; 0, if not
SKU_B_i	1, if SKU i is of type "B"; 0, if not
SKU_C_i	1, if SKU i is of type "C"; 0, if not
α	Minimum percentage of demand for SKUs locations of type "A"
β	Minimum percentage of demand for SKUs locations of type "B" and "C"
$value\_sku_i$	Value of the i SKU
γ	Cut-off value to define locations with double or triple position
$ub0_s$	Amount of locations without division
$ub2_s$	Amount of locations with double position
$ub3_s$	Amount of locations with triple position
D_{jk}	Distance from location j to exit k
$weight_s$	Standard liters value for the SKUs of a subfamily s
per_total	*Perchas* location available in the warehouse
θ	Cut-off value to consider a number of locations greater than zero
μ	Minimum percentage of demand for locations of SKUs

Decision Variables	
Y_{ji}	1, if SKU i is located at j; 0, if not
uA_i	Number of Simple Type "A" SKU i locations
uB_i	Number of Simple Type "B" SKU i locations
$uB2_i$	Number of Double Type "B" SKU i locations
$uB3_i$	Number of Triple Type "B" SKU i locations
uC_i	Number of Simple Type "C" SKU i locations
$uC2_i$	Number of Double Type "C" SKU i locations
$uC3_i$	Number of Triple Type "C" SKU i locations
$total\_UA$	Total number of Simple Type "A" locations
$total\_UB$	Total number of Simple Type "B" locations
$total\_UB2$	Total number of Double Type "B" locations
$total\_UB3$	Total number of Triple Type "B" locations
$total\_UC$	Total number of Simple Type "C" locations
$total\_UC2$	Total number of Double Type "C" locations
$total\_UC3$	Total number of Triple Type "C" locations
$total\_locations$	Total number of locations
$total\_sum2_s$	Number of double locations in subfamily s
$total\_sum3_s$	Number of triple locations in subfamily s
$X0_{js}$	1, if subfamily s does not divide location j; 0, if not
$X2_{js}$	1, if subfamily s divides location j in two; 0, if not
$X3_{js}$	1, if subfamily s divides location j in three; 0, if not
u_i	Number of locations *"perchas"* of the SKU i
$total\_per$	Total number of *Perchas* locations

Parameters of Evaluation	
P_i	Number of complete pallets of SKU i
d_S	Distance to the Warehouse exit from a position to a port
PP_i	Number of SKU pickups i
R	SKU replenishment quantity i
d_E	Distance to the Warehouse entrance from the end of an aisle to the position
c	Porting Trucks Set
W_{ps}	Number of operators for full pallet output
W_{pc}	Number of operators for the output of conformed picking pallets
W_{pe}	Number of operators for replenishment pallet input
$Salary$	Monthly salary of an operator
G_{ps}	Number of cranes for the output of complete pallets
G_{pc}	Number of cranes for the output of conformed picking pallets
G_{pe}	Number of or cranes for the entry of replenishment pallet
A	Rent price of a monthly crane
ω_1	Hours of monthly use of a crane
ω_2	Monthly consumption price of a crane

Appendix B. Execution Times with Gurobi's Python Interface

Model		Variable		Constraint	Objective Function	Time Solver Application [sg]
		Integers	Continuous			
$P_{1.1}$		4.573	1	3.454	1	28.43
$P_{1.3}$		13.716	1	798	1	3.613
$P_{1.4}$		82	1	84	1	0.54
P_1	E_1	891.390	1	3.318	1	158.06
P_1	E_2	891.390	1	3.318	1	137.28

References

1. Öztürkoğlu, Ö. A bi-objective mathematical model for product allocation in block stacking warehouses. *Int. Trans. Oper. Res.* **2017**, *27*, 2184–2210. [CrossRef]
2. Bartholdi, J.; Hackman, S. *Warehouse and Distribution Science*; 2014; Available online: http://www2.isye.gatech.edu/~{}jjb/wh/book/editions/wh-sci-0.96.pdf (accessed on 1 December 2020).
3. Frazelle, E. Chapter 9: Logistics and Supply Chain Information Systems. In *Supply Chain Strategy: The Logistics of Supply Chain Management*; McGraw-Hill: New York, NY, USA, 2002; p. 287.
4. Manzini, R. (Ed.) *Warehousing in the Global Supply Chain. Advanced Models, Tools and Applications for Storage Systems*; Springer: London, UK, 2012. [CrossRef]
5. Yang, L.; Liu, W. The Study of Warehousing Slotting Optimization Based on the Improved Adaptive Genetic Algorithm. In Proceedings of the 6th International Conference on Advanced Materials and Computer Science (ICAMCS 2017), Zhengzhou, China, 29–30 April 2017. [CrossRef]
6. Petersen, C.G.; Siu, C.; Heiser, D.R. Improving order picking performance utilizing slotting and golden zone storage. *Int. J. Oper. Prod. Manag.* **2005**, *25*, 997–1012. [CrossRef]
7. Hompel, M.; Schmidt, T. Chapter 1: Introduction. In *Warehouse Management: Automation and Organisation of Warehouse and Order Picking Systems*; Springer: Heidelberg/Berlin, Germany, 2007; pp. 1–12. [CrossRef]
8. Lorenc, A.; Lerher, T. Effectiveness of product storage policy according to classification criteria and warehouse size. *FME Trans.* **2019**, *47*, 142–150. [CrossRef]
9. Ai-Min, D.; Jia, C. Research on slotting optimization in automated warehouse of pharmaceutical logistics center. *Int. Conf. Manag. Sci. Eng. Annu. Conf. Proc.* **2011**, 135–139. [CrossRef]
10. Van den Berg, J.; Gademann, A.J.R.M. Optimal routing in an automated storage/retrieval system with dedicated storage. *IIE Trans. (Inst. Ind. Eng.)* **1999**, *31*, 407–415. [CrossRef]
11. Chen, L.; Lu, Z. The storage location assignment problem for outbound containers in a maritime terminal. *Int. J. Prod. Econ.* **2012**, *135*, 73–80. [CrossRef]
12. Ballestín, F.; Pérez, Á.; Quintanilla, S. A multistage heuristic for storage and retrieval problems in a warehouse with random storage. *Int. Trans. Oper. Res.* **2017**, *27*, 1699–1728. [CrossRef]
13. Park, C.; Seo, J. Mathematical modeling and solving procedure of the planar storage location assignment problem. *Comput. Ind. Eng.* **2009**, *57*, 1062–1071. [CrossRef]
14. Gómez, R.; Cano, J.; Campo, E. Gestión de la asignación de posiciones (Slotting) eficiente en centros de distribución agroindustriales. *Espacios* **2018**, *39*, 23.
15. Li, H.P.; Fang, Z.F.; Ji, S.Y. Research on the slotting optimization of automated stereoscopic warehouse based on discrete particle swarm optimization. In Proceedings of the 2010 IEEE 17th International Conference on Industrial Engineering and Engineering Management, IE and EM2010, Xiamen, China, 29–31 October 2010; pp. 1404–1407. [CrossRef]
16. Damodaran, P.; Koli, S.; Srihari, K.; Kimler, B. Optimal product slotting. In Proceedings of the IIE Annual Conference, Norcross, GA, USA, 14–18 May 2005; pp. 1–6.
17. Mantel, R.J.; Schuur, P.C.; Heragu, S.S. Order oriented slotting: A new assignment strategy for warehouses. *Eur. J. Ind. Eng.* **2007**, *1*, 301–316. [CrossRef]
18. Kasemset, C.; Meesuk, P. Storage Location Assignment Considering Three-Axis Traveling Distance: A Mathematical Model. In *Logistics Operations, Supply Chain Management and Sustainability*; Golinska, P., Ed.; Springer: Cham, Switzerland, 2014; pp. 15–30. [CrossRef]
19. Lerher, T. Travel time model for double-deep shuttle-based storage and retrieval systems. *Int. J. Prod. Res.* **2015**, *54*, 2519–2540. [CrossRef]
20. Wisittipanich, W.; Kasemset, C. Metaheuristics for warehouse storage location assignment problems. *Chiang Mai Univ. J. Nat. Sci.* **2015**, *14*, 361–377. [CrossRef]
21. Gómez, R.A.; Giraldo, O.G.; Campo, E.A. Conformación de Lotes Mínimo Tiempo en la Operación de Acomodo Considerando k Equipos Homogéneos usando Metaheurísticos. *Inf. Tecnol.* **2016**, *27*, 53–62. [CrossRef]

22. Bahrami, B.; Aghezzaf, E.H.; Limère, V. Enhancing the order picking process through a new storage assignment strategy in forward-reserve area. *Int. J. Prod. Res.* **2019**, *57*, 6593–6614. [CrossRef]
23. Lorenc, A.; Lerhet, T. PickupSimulo–Prototype of Intelligent Software to Support Warehouse Managers Decisions for Product Allocation Problem. *Appl. Sci.* **2020**, *10*, 8683. [CrossRef]
24. Lorenc, A.; Kuznar, M.; Lerher, T. Solving Product Allocation Problem (PAP) by Using ANN and Clustering. *FME Trans.* **2021**, *49*, 206–213. [CrossRef]
25. Zäpfel, G.; Wasner, M. Warehouse sequencing in the steel supply chain as a generalized job shop model. *Int. J. Prod. Econ.* **2006**, *104*, 482–501. [CrossRef]
26. BNE. Bolsa Nacional de Empleo. 2020. Available online: https://www.bne.cl/ (accessed on 1 December 2020).

Article

Wave Planning for Cart Picking in a Randomized Storage Warehouse

Jiun-Yan Shiau * and Jie-An Huang

Department of Logistics Management, National Kaohsiung University of Science and Technology, Kaohsiung City 80778, Taiwan; eg218520@gmail.com
* Correspondence: sho@nkust.edu.tw

Received: 16 October 2020; Accepted: 11 November 2020; Published: 13 November 2020

Abstract: Randomized storage strategy is known as a best practice for storing books of an online bookstore, it simplifies the order picking strategy as to retrieve books in purchase orders from closest locations of the warehouse. However, to be more responsive to customers, many distribution centers have adopted a just-in-time strategy leading to various value-added activities such as kitting, labelling, product or order assembly, customized packaging, or palletization, all of which must be scheduled and integrated in the order-picking process, and this is known as wave planning. In this study, we present a wave planning mathematical model by simultaneously consider: (1) time window from master of schedule (MOS), (2) random storage stock-keeping units (SKUs), and (3) picker-to-order. A conceptual simulation, along with a simplified example for the proposed wave planning algorithm, has been examined to demonstrate the merits of the idea. The result shows the wave planning module can improve the waiting time for truck loading of packages significantly and can reduce the time that packages are heaping in buffer area. The main contribution of this research is to develop a mixed integer programming model that helps the bookseller to generate optimal wave picking lists for a given time window.

Keywords: order picking; wave planning; warehouses; distribution centers; mixed integer programming

---

1. Introduction

1.1. The Purpose of Randomized Storage Warehouse

Randomized storage is a storage assignment strategy which determines how an item can be stored in any empty location in a warehouse. One stock keeping unit (SKU) could therefore be stored in several different locations, and the advantage of this strategy is to exploit spaces more efficiently. Randomized storage strategy is especially suitable for storing books of an online bookstore. Customers like to make multiple purchases in a single order to eliminate shipping cost. By using a randomized storage warehouse for an online bookstore, the order picking strategy can be simplified as to retrieve books in purchase orders from closest locations of the warehouse.

1.2. The Importance of Order Picking for a Warehouse and the Type of Order Picking

Warehouse order picking is a simple concept, but in practice, the picking process can be complicated. Warehouse order picking refers to the necessary labor and parts involved in pulling items out of inventory to fulfill customer orders [1]. This process may sound like the simplest aspect of a business, but when one deals with numbers and realizes that in any given distribution center, order picking counts as the most time-consuming operation and an average of 55% of operating costs. A well-thought-out warehouse will soon find out these ineffective systems may prevent them from further innovation.

For a business to develop a warehouse order picking strategy to meet the demand shifts of the future, the emphasizes on speediness, accuracy, and organization should always be bear in mind.

There are many types of order picking. Some of the order pickings are classified by the customer order handling policies, some are classified by equipment used by pickers, some are classified by warehousing areas, and the others are classified by materials handling equipment. For example, discrete order picking and batch picking are those classified by the customer order handling policies; cart picking, radio frequency (RF) picking, voice picking, tote picking, and put to store are those classified by equipment used by pickers; pick and pass or zone order picking are classified by warehousing areas; and full pallet picking and materials handling equipment assisted picking are those classified by materials handling equipment used. Discrete picking is when a picker goes through one order at a time, grabbing an item line by line before moving onto the next order. It is the simplest of all order picking methods and can also be one of the most accurate. Discrete order picking with a RF scanner or voice assistant device helps to streamline the process and increase accuracy. Cases where discrete order picking can be ideal include (1) when a warehouse has a small staff, (2) when orders are for multiple items, and (3) when the items ordered are very large. Batch picking is the process of collecting inventory for identical orders at once, rather than picking items one order at a time or picking orders with different SKUs and quantities consecutively. Batch picking focused on less walking and faster order processing. It is particularly good for warehouses that stock small items. In pick and pack operations, instead of batching the items into an order tote, you can pick orders directly into a shipping tote. Put-to-store is often used with picking carts to facilitate batch picking of multiple orders in one pass. It is ideal for retail store replenishment. It also optimizes cross-docking operations where a percentage of full case quantities are broken down to store level cartons. Zone order picking is a picking technique in which warehouses have been sectioned into several zones, and pickers have been assigned to each zone. Pickers pick all of the items from a customer order that are located in their zone and then passing the order on to the next zone, so that zone order picking is also called pick and pass order picking. Zone picking is particularly useful in large warehouses that deal with a high number of SKUs. Because each picker is assigned to a designated area, it allows pickers to gain deep familiarity with their assigned zones and the SKUs stored there, which can increase picking speed and reduce errors.

In this paper, we suggest looking at an order picking problem from forms and workflow of a warehouse management system (WMS) perspective. This would help one to tackle (i.e., finding out the considering factors/attributes) an order picking planning problem.

From information technology perspective (i.e., data and workflow), the inputs of a WMS are customer orders and pre-packaged stock keeping units [2]. For control and management purposes, the most important attributes in customer orders are what SKU (or item), how many quantities, and where (location) to deliver for each customer order (illustrated as (I Q L) in Figure 1), and in pre-packaged SKUs what are the SKUs and how many quantities are for pre-packaging (illustrated as (I Q) in Figure 1). A master order schedule (MOS) is another form recording what SKU and how many quantities are going to process for delivering. MOS is more like a filter that filters what customer orders and/or pre-packaged SKU are going to process based on similarity of delivery routing [3] or similarity of in-coming customer orders' time windows [3,4]. The important attributes in MOS are what SKU, how many quantities, and when to load into what routing trucks (illustrated as (I Q R) in Figure 1).

By considering SKU storage locations, customer orders in MOS are further batched or split [5,6] based on similarity of their SKUs' storage locations. The major concerns of batching or splitting are what SKU, how many quantities, and where storage (illustrated as (I Q S) in Figure 1). Batched or split order picking calls for managers to assess current orders for popularity and place them into batches by their SKU. From there, pickers are in the best location of the warehouse to get orders fulfilled as quickly as possible. Batch picking works especially well for e-commerce businesses with somewhat predictable ordering trends, like those specializing in food and apparel. There are lots of order picking policies proposed at this research field, discrete picking, zone picking, and wave picking [5,7]. Discrete picking policy is when a picker picks items of multiple orders by batching during a single trip. Zone picking policy divides the picking area into a few subzones and the dedicated picker will pick items by splitting in that subzone only. The main research topic is the number of subzones and pickers. Wave picking is used if orders by splitting and/or batching for a common destination or common truck loading time are released simultaneously for picking in multiple warehouse areas.

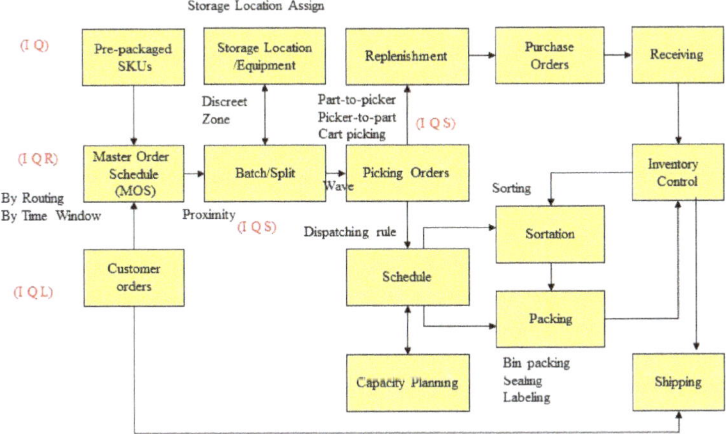

Figure 1. Forms and workflow of a warehouse management system (WMS).

We summarized advantages, disadvantages, and applications of those different types of order picking policies or strategies as Table 1 below.

Table 1. Summaries of types of order picking policies or strategies.

Types	Advantages	Disadvantages	Applications
Discrete picking [5,8]	• Simplest for implement • One of the most accurate • Minimizes the number of touches to ship an order	• May take more time in the pick area	• Warehouse has a small staff • Orders are for multiple items • Items ordered are large
Batch picking [1,2,4–7][9–13]	• Less walking • Faster order processing • Reduce walk times for individual pickers	• Having additional sorts • The risk of congestion is high • Necessity of preparation area • Lots of errors can occurred • Necessity of accumulating some of customer orders till creating batches	• Retail store replenishment • E-commerce businesses with predictable ordering trends

Table 1. *Cont.*

Types	Advantages	Disadvantages	Applications
Zone picking (Pick and pass) [2,5,7,14,15]	• Pickers gain deep familiarity with their assigned zones • Decreases congestion	• Multi-part orders • Less interaction • Inconsistent demand per zone	• Warehouses that deal with a high number of SKUs
Wave picking [5,7,16,17]	• Maximize shipping and picking operations • Orders are prioritized by time and importance	• More complicate planning and computing are required • Difficult to process more than one wave simultaneously • Idle time while some workers wait for others • Not easy to add similar work or additional tasks to a wave that is already in progress	• E-commerce businesses
Put to store [4,5,8,18,19]	• Eliminating a formal packing operation	• Select the shipping carton prior to the order picking process • Difficult to get this type of logic to work effectively	• Bulk pick and/or case pick • Distribution-to-store order carton • Convey complete cartons to shipping and start a new carton
Cart picking [5,6,8,18–20]	• Easy to deal with many customer order lines	• Leave a lot of room for human error • Slower picking speeds	• E-commerce companies
Radio frequency (RF) picking [5,8,21]	• Increasing accuracy level	• Slightly more expensive than a magnetic strip • Tags must be placed in hangtags at the source • Labor must be expended to tag items • Tags sometimes cannot be read	• Case pick • Pallet load • Put away • Order checking operation

Table 1. *Cont.*

Types	Advantages	Disadvantages	Applications
Voice picking [16,18]	• Very effective in both productivity and accuracy	• Slower than RF picking • Too much noise in the warehouse will make it difficult for pickers to hear the commands	• Case pick • Pallet load • Put away
Tote picking [14,15]	• Easy to deal with many customer order lines	• Need piece-picking technology and equipment	• Chain drug stores • Mail order catalog companies • Repair parts distributors • Pharmacies • Office products • Tobacco retail outlets
Full pallet picking [5,14,20]	• Much simpler than either tote pick or case pick	• Need storage equipment and lift trucks	• Paper towels company

Upon order picking equipment, picking orders (also called picking lists) are finally generated. Normally, order picking equipment (or systems) are classified as part-to-picker, picker-to-part, or put systems [14]. The important attributes in picking orders are what SKU, how many quantities, and where stored (illustrated as (I Q S) in Figure 1). Picker-to-parts systems are commonly the order picker walks or drives along the aisles to pick items [8]. In parts-to-picker system, an automatic device brings unit loads from the storage area to the picking stations (sometimes also called picking bays), where the pickers select the required amount of each item. Cart picking systems are particularly popular in case for e-commerce companies, which deal with many customer order lines, must be picked in a short time window. For small items, a well-managed cart picking system can result in around 500 picks on average per order picker hour. Wave picking is used if orders by splitting and/or batching for a common destination or common truck loading time are released simultaneously for picking in multiple warehouse areas. It is a picking system that considers customer orders handling, storage locations, and/or picking equipment concurrently.

Algorithms for generating picking orders are: Rim and Park [22] proposed a linear programming approach to assign the inventory to the orders to maximize the order fill rate, Lu et al. [20] developed an algorithm for dynamic order picking that allows for changes to pick-lists during a pick cycle, Füßle and Boysen [18] aimed at a synchronization between the batches of picking orders concurrently assembled and the sequence of SKUs moved along the line, such that the number of lines passing to be accomplished by the picker is minimized, Giannikas et al. [9] introduced an interventionist order picking strategy that aims to improve the responsiveness of order picking systems, Ho and Lin [15] improved order-picking performance by converting a sequential zone-picking line into a zone-picking network, Schwerdfeger and Boysen [19] proposed a multi-objective approach to solve order picking along a crane-supplied pick face, etc. With all these algorithms, none of them considered truck loading due date into their models.

1.3. The Operational Definition of Wave Planning

From WMS perspective, wave picking is one of the most complex functions of the system, because it includes the following planning functions:

- Batching or slitting the customer orders into appropriate picking lists (also called picking orders)
- Locating SKUs in the warehouse
- Creating sequences of picking tasks by considering routing distances
- Release picking tasks to the pickers/equipment to be fulfilled

1.4. The Objective of This Paper

The objective of this research is to develop a mixed integer programming model for wave planning. The model is inspired by a real-world case study of an online bookseller. The model aims to help the bookseller to generate optimal wave picking lists in a given time window. It also can serve as a simulation with the purpose to determine what is the best time window of a wave picking.

2. The Case Study

In Taiwan, 24-h express shipping domestically is one of the promises of an online bookstore. When the promise comes to warehousing, the quality and trustworthiness of the supplier is important. There is a 2-h express warehousing company, located in north of Taiwan, fully supporting the 24-h express shipping for the online bookstore company. The online bookstore company transfers their customer orders, which could be up to around 10,000 orders, to the warehousing company every hour. Based on the customer orders, the warehousing company generates picking orders every 2 h. Currently, picking orders are planned and separated into two categories. One is (type I) picking orders for those customer orders, that only purchase one book. Type I picking orders are also called discrete picking. The other is (type II) picking orders for those customer orders, that purchase two or more books, and type II picking orders are also called batch picking. Carts are devices for those type II picking orders and carts with trays are devices for those type I picking orders. For those type I picking orders, packing processes are as follow placing a box onto a table scale, checking the contents, inserting a pack slip, void filling, and taping. For those type II picking orders, additional sorting orders and packing process are required. Sorting orders is another time-consuming process for warehousing operators. Operators must pick up books from carts and dispatch them into customer-oriented trays as shown in Figure 2. To live up to its reputation of being a 2-h express warehousing company, all the picking and packing processes for around 20,000 customer orders must be done in a 2-h time window.

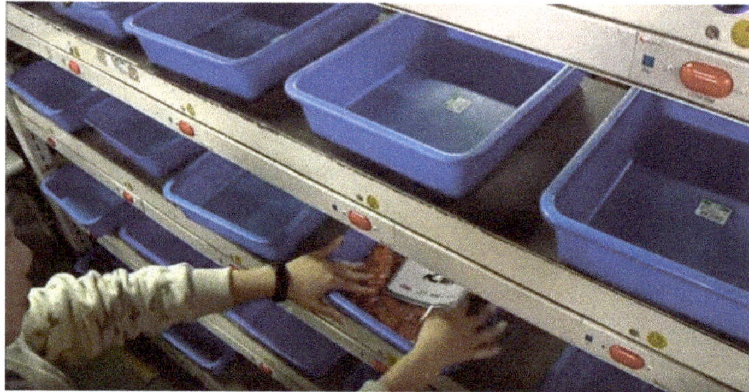

Figure 2. Order sortation process.

There are six delivery routes per day in the 2-h express warehousing company. The first group of the delivery routes is called to-door delivery. There is to-door delivery in the morning, to-door delivery in the afternoon, and to-door delivery in the evening. The second group of the delivery routes is called to-store delivery. There is to-store delivery in northern Taiwan, to-store delivery in central Taiwan, and to-store delivery in southern Taiwan. The packed delivery boxes, which are completed every 2 h, will be stacked in a large buffer space and wait for loading to delivery trucks six times per day according to the six delivery routes.

3. Problem Description and Assumptions

3.1. Problem Description

Despite the fact that warehouses have been around for hundreds of years, they continue to evolve because of dramatic changes in customer demands, along with the constant advancements in technology. This is leading to a revolution in warehouse capabilities. According to [16], the global trends of warehousing businesses include (1) higher workforce expectations, (2) continued focus on regulatory compliance, (3) next-generation postponement strategies, (4) increased use of automation and robotics, (5) geographic expansion, (6) increased presence in urban area, (7) compressed order processing times, (8) grater internal and external collaboration, and (9) smart optimization. Basically, to conquer these global trends of warehousing businesses, an enterprise can tackle them by developing new supply chain collaboration strategies, by developing new logistics automation devices, or by developing intelligent mathematic models or algorithms to optimize the warehousing/logistics resources. In this study, we concentrate on tackling the challenge by smart optimization.

This trend of growing online shopping sales has significantly influenced the goods supply chains throughout its international, regional, and last mile segments. Kang [23] pointed trade and logistics businesses have restructured operational and system-wide aspects of their supply chain to accommodate the ever-growing demand for just-in-time (JIT) production, online shopping, and instant shipment/delivery.

To apply JIT concept to order picking process, a number of researches have been tackled jointly in previous studies to provide a comprehensive solution to order picking [6,10–13,24,25], fulfilling a specified objective function to reduce operational costs [11,12,26–29], and enhance customer service [10,21,30,31]. We summarized previous studies, that provide integration solutions for order picking processes, as shown in Table 2.

Table 2. The integrated studies in recent years.

Reference	Integration	Benefit/Goal
Kulak et al. [12] Hong and Kim [32] Choy et al. [33]	Order batching, Picker routing Order picking, Sequencing	Improve order picking
Schubert et al. [34]	Order picking, Vehicle routing	Provide high-quality solutions.
Zhang et al. [4]	Order batching, Delivery	Deliver numerous orders within the shortest time
Won and Olafsson* [6]	Order batching, Order picking	Optimize customer response time.
Chen et al. [10] Henn [31]	Order batching, Sequencing, Routing problem Order batching, Sequencing	Minimum total tardiness of customer orders.

Recall the case study, stocking packed delivery boxes generated every 2 h is not an efficient way of managing a warehouse. It is not only increasing total travelling paths of all picking operators, but also might waste warehouse buffering space. The idea to improve warehousing efficiency of this case study is to apply JIT concept to order picking process in the 2-h express warehousing company.

Try to think about this. Does it make sense to stack packed delivery boxes generated every 2 h (i.e., 2-h based wave planning)? If we can plan order picking waves to as close to as their loading due time as possible, the time with the packed delivery boxes stacking at the buffering space will be reduced, the buffering space for stacking packed delivery boxes will be reduced, and the processing time window from customer orders to picking orders can be expanded, too. With the expanding time windows, more similar order list can then be scheduled into this time frame. The efficiency of the picking operator can be improved.

In this study, we developed a wave planning model to integrate order batching, sequencing, picker routing, packing, and delivery (as green blocks illustrated in Figure 3) for improving warehouse capabilities in terms of reduce buffer spaces and deliver numerous orders within the shortest time.

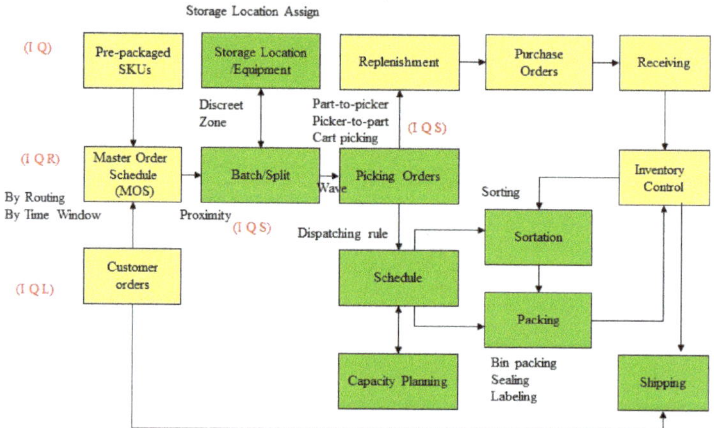

Figure 3. Wave planning model that integrates those green blocks.

3.2. Basic Assumption

- The picking area is not zoning.
- The efficiency of pickers is the same.
- The truck loading time is fixed and known.
- Items are cuboid and the length, width and high are known.
- The items, quantity, and correlation of each order are known.
- The picking distance between the storage locations of items are known.
- Candidate containers are cuboid and the cost and the capacity are known.
- The number of pickers, the constrain of picking quantity and the volume capacity of order picker truck are known.
- The distribution center is the kind that can deal with a great amount of orders that are small-volume with large-variety.

4. The Wave Planning Algorithm

In this study, we present a wave planning mathematic model by integrating order picking with packing planning and scheduling by simultaneously consider three major factors: (1) by time window from MOS, (2) random storage SKUs, and (3) picker-to-box.

This model is proposed to the distribution center which focuses on 24-h delivery and the way of picking is by manual picking. Nowadays, the order type of e-commerce is often low-volume and high-mix, and this makes the processes more complicated in the distribution center. How they deal with the e-orders and send the packages to customers within 24 h by the most efficiency way has become the most important goal.

This study extends from Shiau and Liao [17], and takes more factors into consideration (as shown in Figure 4). First, taking the orders as the center, and establish relations with truck loading time, items, batches, containers, and schedule. For example, according to the truck loading time of orders to schedule the batches, and the batch orders to include items and containers for the particular package.

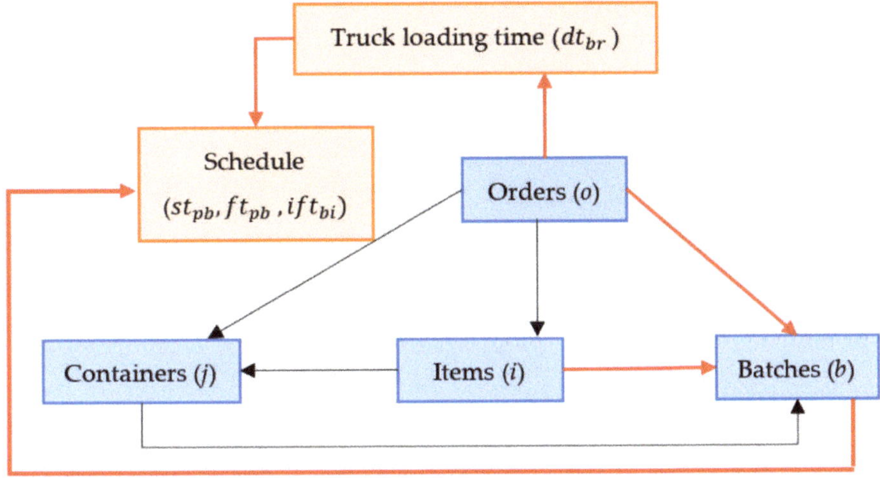

Figure 4. The relation of considered factors.

After the distribution center accumulates numerous customers' orders for 24 h, it should confirm the environment parameters whether should be adjusted. The next step is to consider the relation between picking route, order batching, loading configuration, containers, departure time, and items. The system will calculate the distance from one storage to another. Moreover, it will restrict the picking quantity of each picking order, and compute the shortest picking route and waiting time of containers in the buffer area.

Sets

$O = \{1,2,...,OR\}$	set of customer orders
$R = \{1,2,...,NR\}$	set of items
$BS = \{1,2,...,BM\}$	set of batches
$AP = \{0,1,...,N\}$	set of picking points included pickup and deposit (P/D) points
$V = \{1,2,...,m\}$	set of candidate containers
$PS = \{1.....PE\}$	set of pickers
$DR = \{1.....ND\}$	set of the truck loading time

Relationship matrix

$RS_{oi} = \{RS_{oi},...,RS_{on}\}^T$: the relation between orders and items

Parameters

OR	: Number of customer orders
NR	: Number of items in customer orders
m	: Number of candidate containers
B	: Batches
BM	: Number of batches
o	: The customer orders
i	: The items customer ordered
j	: The candidate container

P_{oi}	: The length of item i in customer order o
q_{oi}	: The width of item i in customer order o
r_{oi}	: The height of item i in customer order o
L_j	: The length of container j
W_j	: The width of container j
H_j	: The height of container j
C_j	: The cost of container j
t_{ik}	: The storage distance between item i and item k
sd_b	: The processing efficiency of batch
RS_{oi}	: 1 if item i is in order o; 0 otherwise
PCP_b	: The amount of picking items of each batch b
PCB_p	: The capacity of the assigned batches by picker p
ECP_b	: The volume capacity of the order picker cart for batch b
MAV	: A very large number
MAC	: The maximum cost of packing
MIC	: The minimum cost of packing
MAD	: The maximum travel distance between items
MID	: The minimum travel distance between items
MAW	: The maximum waiting time for truck loading
MIW	: The minimum waiting time for truck loading

Input variables

TL	: The limited of total operating time of all pickers
T_{ir}	: The truck loading time r of item i
dt_{br}	: The truck loading time r of batch b

Output variables

u_{bi}	: The picking sequence of item i in batch b
x_{oi}	: x-axis position of the front-left bottom corner of item i in customer order o be assigned
y_{oi}	: y-axis position of the front-left bottom corner of item i in customer order o be assigned
z_{oi}	: z-axis position of the front-left bottom corner of item i in customer order o be assigned
sp_b	: The number of total picking points of batch b
st_{pb}	: The start time of batch b by picker p
ft_{pb}	: The finish time of batch b by picker p
ift_{bi}	: The finish time of item i in batch b
wt_{pb}	: The waiting time for departure of batch b by picker p
mt_b	: The operation time of batch b
ad	: The total distance of all picking route
aw	: The total wait time for truck loading of batches

Decision variables

S_{oij}	: 1 if item i is put into box j; 0 otherwise
n_{oj}	: 1 if box j is used; 0 otherwise
v_{bi}	: 1 if item i is in batch b; 0 otherwise
q_{bij}	: 1 if item i of batch b is put into box j; 0 otherwise
lx_{oi}	: 1 if the length of item i is parallel to x-axis of the box; 0 otherwise
ly_{oi}	: 1 if the length of item i is parallel to y-axis of the box; 0 otherwise
lz_{oi}	:1 if the length of item i is parallel to z-axis of the box; 0 otherwise
wx_{oi}	: 1 if the width of item i is parallel to x-axis of the box; 0 otherwise
wy_{oi}	: 1 if the width of item i is parallel to y-axis of the box; 0 otherwise
wz_{oi}	: 1 if the width of item i is parallel to z-axis of the box; 0 otherwise
hx_{oi}	: 1 if the height of item i is parallel to x-axis of the box; 0 otherwise
hy_{oi}	: 1 if the height of item i is parallel to y-axis of the box; 0 otherwise
hz_{oi}	: 1 if the height of item i is parallel to z-axis of the box; 0 otherwise
a_{oik}	: 1 if item i is on the left side of item k in customer order o; 0 otherwise
b_{oik}	: 1 if item i is on the right side of item k in customer order o; 0 otherwise
c_{oik}	: 1 if item i is in front of item k in customer order o; 0 otherwise

d_{oik}	: 1 if item i is behind item k in customer order o; 0 otherwise
e_{oik}	: 1 if item i is under item k in customer order o; 0 otherwise
f_{oik}	: 1 if item i is above item k in customer order o; 0 otherwise
g_{oik}	: 1 if the picking sequence of item i is before item k in batch B; 0 otherwise
w_{pb}	: 1 if batch b is picking by picker p; 0 otherwise

Formulation

- *Objective Function*

The objective function (1) is to minimize the picking route, the cost of containers and waiting time for truck loading of packaged containers heaping in the buffer area. Since the units of the three factors are different, we have to normalize the three objects.

$$\text{Min}\left(\frac{a\,d - MID}{MAD - MID} + \frac{\sum_{o=1}^{OR} \sum_{j=1}^{m} C_j \cdot n_{oj} - MIC}{MAC - MIC} + \frac{a\,w - MIW}{MAW - MIW} \right) \tag{1}$$

- *Subject to*

Constraints (2)–(4) ensure the relation between items and containers. The length, width, and height of each items must be parallel with one axis of the container. For instance, the length of item i is only parallel to x-axis, y-axis, or z-axis of the container (i.e., Equation (2)), where as lx_{oi}, ly_{oi}, and lz_{oi} are binary variables.

$$lx_{oi} + ly_{oi} + lz_{oi} = 1 \qquad \forall o \in O, i \in R \tag{2}$$

$$wx_{oi} + wy_{oi} + wz_{oi} = 1 \qquad \forall o \in O, i \in R \tag{3}$$

$$hx_{oi} + hy_{oi} + hz_{oi} = 1 \qquad \forall o \in O, i \in R \tag{4}$$

In other words, Constraints (5)–(7) ensure what parallel with some axis is one of the lengths, width, or height of item i in order o.

$$lx_{oi} + wx_{oi} + hx_{oi} = 1 \qquad \forall o \in O, i \in R \tag{5}$$

$$ly_{oi} + wy_{oi} + hy_{oi} = 1 \qquad \forall o \in O, i \in R \tag{6}$$

$$lz_{oi} + wz_{oi} + hz_{oi} = 1 \qquad \forall o \in O, i \in R \tag{7}$$

Constraints (8)–(13) ensure the relative position how pickers put into the container between item and item in order o. For instance, if item i is on the left side of item k, x-axis position of the front-left bottom corner of item k (i.e., the right-hand side of Equation (8)) must be larger than x-axis position of the front-left bottom corner of item i plus its size of item i is parallel to x-axis of the box (i.e., the left-hand side of Equation (8)).

$$x_{oi} + p_{oi} \cdot lx_{oi} + q_{oi} \cdot wx_{oi} + r_{oi} \cdot hx_{oi} \le x_{ok} + (1 - a_{oik}) \cdot MAV$$
$$\forall o \in O, i \in R, k \in R; i \ne k \tag{8}$$

$$x_{oi} + p_{ok} \cdot lx_{ok} + q_{ok} \cdot wx_{ok} + r_{oi} \cdot hx_{ok} \le x_{oi} + (1 - b_{oik}) \cdot MAV$$
$$\forall o \in O, i \in R, j \in V \tag{9}$$

$$y_{oi} + p_{oi} \cdot ly_{oi} + q_{oi} \cdot wy_{oi} + r_{oi} \cdot hy_{oi} \le y_{ok} + (1 - c_{oik}) \cdot MAV$$
$$\forall o \in O, i \in R, j \in V \tag{10}$$

$$y_{ok} + p_{ok} \cdot ly_{ok} + q_{ok} \cdot wy_{ok} + r_{oi} \cdot hy_{ok} \le y_{oi} + (1 - d_{oik}) \cdot MAV$$
$$\forall o \in O, i \in R, j \in V \tag{11}$$

$$z_{oi} + p_{oi} \cdot lz_{oi} + q_{oi} \cdot wz_{oi} + r_{oi} \cdot hz_{oi} \leq z_{ok} + (1 - e_{oik}) \cdot MAV$$
$$\forall o \in O, i \in R, j \in V \tag{12}$$

$$z_{ok} + p_{ok} \cdot lz_{ok} + q_{ok} \cdot wz_{ok} + r_{oi} \cdot hz_{ok} \leq z_{oi} + (1 - f_{oik}) \cdot MAV$$
$$\forall o \in O, i \in R, j \in V \tag{13}$$

Constraints (14) ensures the items of the order o are put into the same container. For instance, if in order o, item i ($RS_{oi} = 1$) and item k ($RS_{ok} = 1$) are put into container j ($s_{oij} = 1$ and $s_{okj} = 1$), then there exists some correlation between item i and item k.

$$a_{oik} + b_{oik} + c_{oik} + d_{oik} + e_{oik} + f_{oik} \geq s_{oij} \cdot RS_{oi} + s_{okj} \cdot RS_{ok} - 1$$
$$\forall o \in O, i \in R, k \in R, j \in V; i \neq k \tag{14}$$

Constraints (15)–(16) ensure items of order o only put into one container. Only one container j (s_{oij}) could be equal to one in Equation (15), and in Equation (16) if container j is used ($n_{oj} = 1$), item can be put in container j ($s_{oij} >= 0$).

$$\sum_{j=1}^{m} s_{oij} \cdot RS_{oi} = 1 \qquad \forall o \in O, i \in R \tag{15}$$

$$\sum_{i=1}^{N\,R} s_{oij} \cdot RS_{oi} \leq MAV \cdot n_{oj} \qquad \forall o \in O, j \in V \tag{16}$$

Constraints (17)–(18) ensure one order can use several containers and the items of different order cannot be put into the same container. In Equation (17), at least one container j must be used for order o. In Equation (18), one container j can only be used in one customer order o.

$$\sum_{j=1}^{m} n_{oj} \geq 1 \qquad \forall o \in O \tag{17}$$

$$\sum_{o=1}^{OR} n_{oj} \leq 1 \qquad \forall j \in V \tag{18}$$

Constraints (19)–(21) ensure the total size of items putting in the same container is not bigger than the container. For instance, if item i is placed in container j, the size of item i is parallel to x-axis of the box (i.e., the left-hand side of Equation (19)) must be smaller than the size of length of box j (i.e., the right-hand side of Equation (19)).

$$x_{oi} + p_{oi} \cdot lx_{oi} + q_{oi} \cdot wx_{oi} + r_{oi} \cdot hx_{oi} \leq L_j + \left(1 - s_{oij} \cdot RS_{oi}\right) \cdot MAV$$
$$\forall o \in O, i \in R, j \in V \tag{19}$$

$$y_{oi} + p_{oi} \cdot ly_{oi} + q_{oi} \cdot wy_{oi} + r_{oi} \cdot hy_{oi} \leq W_j + \left(1 - s_{oij} \cdot RS_{oi}\right) \cdot MAV$$
$$\forall o \in O, i \in R, j \in V \tag{20}$$

$$z_{oi} + p_{oi} \cdot lz_{oi} + q_{oi} \cdot wz_{oi} + r_{oi} \cdot hz_{oi} \leq H_j + \left(1 - s_{oij} \cdot RS_{oi}\right) \cdot MAV$$
$$\forall o \in O, i \in R, j \in V \tag{21}$$

Constraints (22)–(24) ensure items putting in the same container must be in the same batch. In Equation (22) only one batch b is used for a container j (i.e., only one q_{bij} could be equal to 1). In Equation (23) if item i and item k are in the same container j ($s_{oij} = 1$ and $s_{okj} = 1$), they must be in the same batch b ($q_{bij} = 1$ and $q_{bkj} = 1$). Equation (24) determines those container j for item i ($q_{bij} = 1$) if item i is assigned to batch b ($v_{bi} = 1$).

$$\sum_{j=1}^{m}\left(\sum_{b=1}^{B\,M}q_{bij}\right)\cdot s_{oij}\cdot RS_{oi}=1 \qquad \forall o\in O, i\in R \tag{22}$$

$$q_{bij}\cdot s_{oij}\cdot RS_{oi}-q_{bkj}\cdot s_{okj}\cdot RS_{ok}=0 \qquad \forall b\in BS, o\in O, i\in R, j\in V; ik \tag{23}$$

$$\sum_{j=1}^{m}q_{bij}=v_{bi} \qquad \forall b\in BS, i\in R \tag{24}$$

Constraints (25) ensures every picker starts with P/D point.

$$v_{b0}=1 \qquad \forall b\in BS \tag{25}$$

Constraints (26)–(27) ensure each order picker cart (ECP_b) does not overload and the picking quantities do not exceed the capability of each picking operator (PCP_b).

$$\sum_{i=1}^{N\,R}\sum_{j=1}^{m}L_j\cdot W_j\cdot H_j\cdot q_{bij}\cdot s_{oij}\cdot RS_{oi}\le ECP_b \qquad \forall b\in BS, o\in O \tag{26}$$

$$\sum_{i=1}^{N\,R}v_{bi}\le PCP_b \qquad \forall b\in BS \tag{27}$$

Constraints (28)–(30) calculate the total distant (ad) and ensure pickers will not walk to the repeated route. For instance, if item 1 is picked before item 2, and item 2 is picked before item 3, then it is not possible for item 3 to be picked before item 1.

$$\sum_{b=1}^{B\,M}\sum_{j=1}^{m}\sum_{i=1}^{N\,R}\sum_{k=0}^{N\,R}t_{ik}\cdot g_{bik}\cdot v_{bi}\cdot v_{bk}=ad \qquad i\ne k \tag{28}$$

$$\sum_{k=1}^{N\,R}g_{b0k}\cdot v_{b0}\cdot v_{bk}=1 \qquad \forall b\in BS \tag{29}$$

$$\sum_{b=1}^{B\,M}\sum_{i=0}^{N\,R}g_{bik}\cdot v_{bi}\cdot v_{bk}=1 \qquad \forall k\in AP; i\ne k \tag{30}$$

Constraints (31)–(33) ensure pickers will walk from one picking point to another and finally back to P/D point. In Equation (31) each batch b, only one item i can be picked after P/D point (i.e., only one g_{bi0} could be equal to 1). In Equation (32), for each batch b, only one item k can be picked after item i (i.e., only one g_{bi0} could be equal to 1). In Equation (33), the number of total picking points of batch b is calculated.

$$\sum_{i=1}^{N\,R}g_{bi0}\cdot v_{bi}\cdot v_{b0}=1 \qquad \forall b\in BS \tag{31}$$

$$\sum_{b=1}^{B\,M}\sum_{k=0}^{N\,R}g_{bik}\cdot v_{bi}\cdot v_{bk}=1 \qquad \forall i\in AP; i\ne k \tag{32}$$

$$\sum_{i=1}^{N\,R}v_{bi}=sp_b \qquad \forall b\in B \tag{33}$$

Constraints (34) determines picking sequence of each item i (u_{bi}) and item k (u_{bk}).

$$u_{bi} - u_{bk} + (sp_b + 1) \cdot g_{bik} \leq sp_b + (1 - v_{bi} \cdot v_{bk}) \cdot MAV$$
$$\forall b \in BS, i \in AP, k \in R; i \neq k \tag{34}$$

If items in the batches, the picking sequence of items (u_{bi}) should be more than or equal to zero by Constraints (35)–(36).

$$u_{bi} \leq sp_b + (1 - v_{bi}) \cdot MAV \qquad \forall b \in BS, i \in AP \tag{35}$$

$$u_{bi} \geq 0 - (1 - v_{bi}) \cdot MAV \qquad \forall b \in BS, i \in AP \tag{36}$$

Constraints (37)–(38) ensure the items placing in the lower position of containers will be picked first. If item i is placed in lower layer to item k (i.e., $e_{oik} = 1$ and $f_{oik} = 0$), then Equations (37) and (38) are held.

$$e_{oik}(u_{bi} - u_{bk}) \leq 0 + (1 - v_{bi} \cdot v_{bk}) \cdot MAV$$
$$\forall b \in BS, o \in O, i \in R, k \in R; i \neq k \tag{37}$$

$$f_{oik}(u_{bi} - u_{bk}) \geq 0 - (1 - v_{bi} \cdot v_{bk}) \cdot MAV$$
$$\forall b \in BS, o \in O, i \in R, k \in R; i \neq k \tag{38}$$

Constraints (39)–(40) determine how to assign the batch to pickers. In Equation (39), a picker p can be only assigned to one batch b. Equation (40) calculates the capacity of the assigned batches (PCB_p) by picker p.

$$\sum_{p=1}^{P E} W_{pb} = 1 \qquad \forall b \in BS \tag{39}$$

$$\sum_{b=1}^{B M} W_{pb} = PCB_p \qquad \forall p \in PS \tag{40}$$

Constraint (41) calculates the operation time of batch (mt_b).

$$\sum_{i=1}^{N R} \sum_{k=1}^{N R} t_{ik} \cdot g_{bik} \cdot v_{bi} \cdot v_{bk} \cdot sd_b = mt_b \qquad i \neq k \tag{41}$$

Constraint (42) limits the total operation time.

$$\sum_{b=1}^{B M} mt_b \cdot W_{pb} \leq TL \qquad \forall b \in BS \tag{42}$$

Constraint (43) calculates the start time of batch (st_{pb}).

$$st_{pb} + mt_b * W_{pb} \leq ft_{pb} \qquad \forall p \in PS; \forall b \in BS \tag{43}$$

Constraint (44) ensures the sequence of batches (i.e., finish time of batch b must less than start time of batch k).

$$ft_{pb} \leq st_{pk} \qquad b \neq k; \forall p \in PS; \forall b \in BS; \forall k \in BS \tag{44}$$

Constraint (45) calculates the waiting time for truck loading (wt_{pb}) of batches.

$$ft_{pb} + wt_{pb} = dt_{br} \qquad \forall p \in PS; \forall b \in BS; \forall r \in DR \tag{45}$$

Constraint (46) is the finish time of batch b to be equal to the finish time of items in batch b. In Constraint (47), all finish time of items in batch b (ift_{bi}) must be earlier than the given due time (T_{ir}).

$$ift_{bi} = ft_{pb} \cdot v_{bi} \qquad \forall p \in PS; \forall b \in BS; \forall i \in AP \tag{46}$$

$$ift_{bi} \leq T_{ir} \qquad \forall b \in BS; \forall i \in AP; \forall r \in DR \tag{47}$$

Constraint (48) calculates the total waiting time of batches (aw).

$$\sum_{p=1}^{PE} \sum_{b=1}^{BM} wt_{pb} = aw \tag{48}$$

$$x_{oi}, y_{oi}, z_{oi}, ft_{pb}, ift_{bi}, wt_{pb} \geq 0 \tag{49}$$

Constraints (50)–(51) declare the 0/1 variables and nonnegative variables in this model.

$$lx_{oi}, ly_{oi}, lz_{oi}, wx_{oi}, wy_{oi}, wz_{oi}, hx_{oi}, hy_{oi}, hz_{oi} = 0, 1 \tag{50}$$

$$a_{oik}, b_{oik}, c_{oik}, d_{oik}, e_{oik}, f_{oik}, g_{bik}, S_{oij}, n_{oj}, q_{bij}, v_{bi}, W_{pb} = 0, 1 \tag{51}$$

5. The Example Sets

This study takes the distribution center of the e-retailer of a bookstore for example. We suppose the working time will be from 6:00 a.m. to 10:00 p.m. and they will fulfill customers' orders within 24 h. If customers place the orders before 12:00 p.m., they will receive the items on the following day before 12:00 p.m. However, if they place the orders after 12:00 p.m., they will receive the items until the day after tomorrow before 12:00 p.m.

The e-commerce will send a batch of orders to the distribution center every two hours and the distribution center will start to deal the orders rapidly before the next batch arrives. The problem with this is that it will mean that many packages will be waiting for delivery in the buffer area. Another problem is that the pickers may travel to the same locations every two hours.

By wave planning, the distribution center can accumulate the batches of orders which the e-commerce sends every two hours. After 24 h, the distribution center will transfer the customer orders to picking orders at 6:00 a.m. and 12:00 p.m. according to the truck loading time and items of the orders. Before pickers start to pick, they need to set up the environment parameters. Then, use the optimization modeling software to solve the problem.

First, the setup of WMS including the storage location of items, the type of containers, the capacity of pickers, and the volume capacity of order picker trucks. There are 32 items, numbered from 1 to 32, and the storage locations are transferred as a distance matrix. Containers have three kinds of sizes showed in Table 3. To assign the jobs to pickers equally, the capacity of pickers is the same (let us say 4 units/each) and the volume capacity of order picker trucks are also the same (let us say 149,903 cm^3/unit). According to the difference delivery place, there are five truck loading times, which are (1) 9 a.m. for south shop delivery, (2) 3 p.m. for central shop delivery, (3) 10 p.m. for north shop delivery, (4) 12 p.m. for to-door delivery, and (5) 5 p.m. for another to-door delivery.

After inputting the environment parameters, it can start to check the information of customer orders including item number, quantities, and item size (see Table 4). Then the system would generate the relation between customer orders and items RS_{oi} (see Table 5) and the item number i is the picking point i. The start and finish place of pickers are P/D point, and the number of picking points is showed in Table 6. Compared to the five truck loading schedules, an associated truck loading time is assigned to each item which ensures the items will be picked before the truck loading time.

Table 7 is the distance between picking point. Table 8 is the performance of picking before using the model. The calculation of operation time is the distance multiplied by the processing efficiency and the waiting time starts from the final of that processing time window. For example, if the processing time is 2–4 p.m., trucking loading time is 3 p.m. of the next day. The waiting time will start from 4 p.m. to 3 p.m. of the next day. The total waiting time is 13,320 min and the total operation time is 207 min.

Table 3. The sizes and price of three kinds of containers.

Container Number (j)	Container Graph	Container Size (cm)			Price (C)
		Length (L_j)	Width (W_j)	Height (H_j)	
Size 1 ($j = 1\sim7$)		23	14	13	55
Size 2 ($j = 8\sim20$)		23	18	19	70
Size 3 ($j = 21\sim27$)		39.5	27.5	23	100

Table 4. The customer orders.

Place Order Time	Customer Order (o)	Item Number (i)	Quantity	Item Size (cm)			Truck Loading Time (r) (The Next Day)
				Length (p_{oi})	Width (q_{oi})	Height (r_{oi})	
12–14	1	3	1	13	5	6	3 p.m.
		8	1	20	10	10	3 p.m.
14–16	2	9	1	15	13	10	9 a.m.
		10	1	17	12	13	9 a.m.
	3	11	1	14	13	5	9 a.m.
		17	1	5	5	10	9 a.m.
	4	12	1	7	14	7	9 a.m.
		13	1	16	4	5	9 a.m.
16–18	5	1	1	5	3	6	9 a.m.
		2	1	16	19	7	9 a.m.
	6	4	1	8	4	7	3 p.m.
		14	1	13	5	5	3 p.m.
	7	6	1	14	8	5	5 p.m.
		7	1	8	4	4	5 p.m.
	8	15	1	5	5	10	5 p.m.
		23	1	15	10	6	5 p.m.
	9	5	1	13	5	6	12 p.m.
		18	1	20	10	10	12 p.m.

Table 4. *Cont.*

Place Order Time	Customer Order (o)	Item Number (i)	Quantity	Item Size (cm)			Truck Loading Time (r) (The Next Day)
				Length (p_{oi})	Width (q_{oi})	Height (r_{oi})	
18–20	10	19	1	15	13	10	12 p.m.
		20	1	17	12	13	12 p.m.
20–22	11	21	1	14	13	5	12 p.m.
		24	1	5	5	10	12 p.m.
22–24	12	16	1	7	14	7	12 p.m.
		22	1	16	4	5	12 p.m.
4–6	13	25	1	13	5	6	10 p.m.
		26	1	20	10	10	10 p.m.
6–8	14	31	1	15	13	10	10 p.m.
		32	1	17	12	13	10 p.m.
8–10	15	27	1	14	13	5	10 p.m.
		29	1	5	5	10	10 p.m.
10–12	16	28	1	7	14	7	10 p.m.
		30	1	16	4	5	10 p.m.

Table 5. The relation matrix of customer orders and items (RS_{oi}).

Item (i) \ Order (o)	1	2	3	4	5	6	7	8	9	10	11	12	13	14	15	16
1	0	0	0	0	1	0	0	0	0	0	0	0	0	0	0	0
2	0	0	0	0	1	0	0	0	0	0	0	0	0	0	0	0
3	1	0	0	0	0	0	0	0	0	0	0	0	0	0	0	0
4	0	0	0	0	0	1	0	0	0	0	0	0	0	0	0	0
5	0	0	0	0	0	0	0	0	1	0	0	0	0	0	0	0
6	0	0	0	0	0	0	1	0	0	0	0	0	0	0	0	0
7	0	0	0	0	0	0	1	0	0	0	0	0	0	0	0	0
8	1	0	0	0	0	0	0	0	0	0	0	0	0	0	0	0
9	0	1	0	0	0	0	0	0	0	0	0	0	0	0	0	0
10	0	1	0	0	0	0	0	0	0	0	0	0	0	0	0	0
11	0	0	1	0	0	0	0	0	0	0	0	0	0	0	0	0
12	0	0	0	1	0	0	0	0	0	0	0	0	0	0	0	0
13	0	0	0	1	0	0	0	0	0	0	0	0	0	0	0	0
14	0	0	0	0	0	1	0	0	0	0	0	0	0	0	0	0
15	0	0	0	0	0	0	0	1	0	0	0	0	0	0	0	0
16	0	0	0	0	0	0	0	0	0	0	0	1	0	0	0	0
17	0	0	1	0	0	0	0	0	0	0	0	0	0	0	0	0
18	0	0	0	0	0	0	0	0	1	0	0	0	0	0	0	0
19	0	0	0	0	0	0	0	0	0	1	0	0	0	0	0	0
20	0	0	0	0	0	0	0	0	0	1	0	0	0	0	0	0

Table 5. *Cont.*

Order (o) / Item (i)	1	2	3	4	5	6	7	8	9	10	11	12	13	14	15	16
21	0	0	0	0	0	0	0	0	0	0	1	0	0	0	0	0
22	0	0	0	0	0	0	0	0	0	0	0	1	0	0	0	0
23	0	0	0	0	0	0	0	1	0	0	0	0	0	0	0	0
24	0	0	0	0	0	0	0	0	0	0	1	0	0	0	0	0
25	0	0	0	0	0	0	0	0	0	0	0	0	1	0	0	0
26	0	0	0	0	0	0	0	0	0	0	0	0	1	0	0	0
27	0	0	0	0	0	0	0	0	0	0	0	0	0	0	1	0
28	0	0	0	0	0	0	0	0	0	0	0	0	0	0	0	1
29	0	0	0	0	0	0	0	0	0	0	0	0	0	0	1	0
30	0	0	0	0	0	0	0	0	0	0	0	0	0	0	0	1
31	0	0	0	0	0	0	0	0	0	0	0	0	0	1	0	0
32	0	0	0	0	0	0	0	0	0	0	0	0	0	1	0	0

Table 6. The item numbers and picking point numbers.

Item Number (i)	-	1	2	3	4	5	6	7	8	9	10
Storage location	P/D	F9	G6	D14	I13	K7	J4	I9	E2	D7	E2
Picking Point (i)	33	1	2	3	4	5	6	7	8	9	10
Item number (i)	11	12	13	14	15	16	17	18	19	20	21
Storage location	F1	F5	G12	H8	I3	J6	F14	J11	K13	K14	H14
Picking Point (i)	11	12	13	14	15	16	17	18	19	20	21
Item number (i)	22	23	24	25	26	27	28	29	30	31	32
Storage location	J15	K10	H3	J12	B14	B12	E12	D11	B10	C13	I11
Picking Point (i)	22	23	24	25	26	27	28	29	30	31	32

Table 7. The distance between picking point.

	1	2	3	4	5	6	7	8	9	10	11	12	13	14	15	16	17	18	19	20	21	22	23	24	25	26	27	28	29	30	31	32	33
1	0	3	12	13	22	19	17	18	19	15	8	3	3	18	14	21	5	18	16	15	12	14	19	15	17	15	17	14	15	19	16	15	7
2	3	0	15	16	19	16	18	15	16	12	5	1	6	17	11	18	8	21	19	18	15	17	22	12	20	18	20	17	18	22	19	18	10
3	12	15	0	11	20	23	15	18	7	11	18	16	9	12	22	21	7	16	14	13	10	12	17	21	15	7	9	2	3	11	8	13	5
4	13	16	11	0	15	18	4	25	18	22	17	17	10	5	11	16	8	11	9	8	1	7	12	10	10	14	16	13	14	18	15	2	6
5	22	19	20	15	0	3	19	22	23	19	14	18	19	20	12	1	17	4	5	7	14	8	3	13	5	23	25	22	23	27	24	17	15
6	19	16	23	18	3	0	16	19	20	16	11	15	22	15	9	2	20	7	9	10	17	11	6	10	8	26	28	25	24	26	27	18	18
7	17	18	15	4	19	16	0	21	22	18	13	17	14	1	7	18	12	15	13	12	5	11	16	6	14	18	20	17	18	22	19	2	10
8	18	15	18	25	22	19	21	0	13	9	10	14	22	20	14	21	21	30	28	27	25	26	24	26	27	12	10	17	16	8	11	22	19
9	19	16	7	18	23	20	22	13	0	5	11	15	16	23	15	22	14	23	21	20	17	19	24	16	22	14	16	5	4	18	15	20	12
10	15	12	11	22	19	16	18	9	5	0	7	11	18	17	11	18	18	23	25	24	22	24	21	11	24	19	17	10	9	15	20	19	16
11	8	5	18	17	14	11	13	10	11	7	0	4	11	12	6	13	13	18	20	21	18	22	17	7	19	21	19	16	15	17	20	15	15
12	3	1	16	17	18	15	17	14	15	11	4	0	7	16	10	17	9	22	20	19	16	18	21	11	21	19	21	18	19	21	20	19	11
13	3	6	9	10	19	22	14	22	16	18	11	7	0	15	21	20	2	15	13	12	9	11	16	18	14	12	14	11	12	16	13	12	4
14	18	17	16	5	20	15	1	20	23	17	12	16	15	0	6	17	13	16	14	13	6	12	17	5	15	19	21	18	19	23	20	3	11
15	14	11	22	11	12	9	7	14	15	11	6	10	21	6	0	11	19	16	20	19	12	18	15	1	17	25	23	20	19	21	24	9	17
16	21	18	21	16	1	2	18	21	22	18	13	17	20	17	11	0	18	5	7	8	15	9	4	12	6	24	26	23	24	28	25	18	16
17	5	8	7	8	17	20	12	21	14	18	13	9	2	13	19	18	0	13	11	10	7	9	14	18	12	10	12	9	10	14	11	10	2
18	18	21	16	11	4	7	15	30	23	23	18	22	15	16	16	5	13	0	2	3	10	4	1	17	1	19	21	18	19	23	20	13	11
19	16	19	14	9	6	9	13	28	21	25	20	20	13	14	20	7	11	2	0	1	8	2	3	19	1	17	19	16	17	21	18	11	9
20	15	18	13	8	7	10	12	27	20	24	21	19	12	13	19	8	10	3	1	0	7	1	4	18	2	16	18	15	16	20	17	10	8
21	12	15	10	1	14	17	5	25	17	22	18	16	9	6	12	15	7	10	8	7	0	6	11	11	9	13	15	12	13	17	14	3	5
22	14	17	12	7	8	11	11	26	19	24	22	18	11	12	18	9	9	4	2	1	6	0	5	17	3	15	17	14	15	19	18	9	7
23	19	22	17	12	3	6	16	24	24	21	17	21	16	17	15	4	14	1	3	4	11	5	0	16	2	20	22	19	20	24	23	14	12
24	15	12	21	10	13	10	6	14	16	11	7	11	18	5	1	12	18	17	19	18	11	17	16	0	18	24	24	23	24	22	25	18	16

Table 7. *Cont.*

	1	2	3	4	5	6	7	8	9	10	11	12	13	14	15	16	17	18	19	20	21	22	23	24	25	26	27	28	29	30	31	32	33
25	17	20	15	10	5	8	14	27	22	24	19	21	14	15	17	6	12	1	1	2	9	3	2	18	0	18	20	17	18	22	19	12	10
26	15	18	7	14	23	26	18	12	14	19	21	19	12	19	25	24	10	19	17	16	13	15	20	24	18	0	2	9	10	4	1	16	8
27	17	20	9	16	25	28	20	10	16	17	19	21	14	21	23	26	12	21	19	18	15	17	22	24	20	2	0	11	12	2	1	18	10
28	14	17	2	13	22	25	17	17	5	10	16	18	11	18	20	23	9	18	16	15	12	14	19	23	17	9	11	0	1	13	10	15	7
29	15	18	3	14	23	24	18	16	4	9	15	19	12	19	19	24	10	19	17	16	13	15	20	24	18	10	12	1	0	14	11	16	8
30	19	22	11	18	27	26	22	8	18	15	17	21	16	23	21	28	14	23	21	20	17	19	24	22	22	4	2	13	14	0	3	20	12
31	16	19	8	15	24	27	19	11	15	20	20	20	13	20	24	25	11	20	18	17	14	18	23	25	19	1	1	10	11	3	0	17	9
32	15	18	13	2	17	18	2	22	20	19	15	19	12	3	9	18	10	13	11	10	3	9	14	8	12	16	18	15	16	20	17	0	8
33	7	10	5	6	15	18	10	19	12	16	15	11	4	11	17	16	2	11	9	8	5	7	12	16	10	8	10	7	8	12	9	8	0

Table 8. The performance of picking before using the module.

Placing Order Time	Processing Time	Batch (b)	Customer Order (o)	Operation Time	Waiting Time (wt_{pb})	Truck Loading Time (r)
12–14	14–16	1	1	21	1380 min.	3 p.m. of the next day
14–16	16–18	2	2	20	900 min.	9 a.m. of the next day
			3		900 min.	9 a.m. of the next day
		3	4	11	900 min.	9 a.m. of the next day
16–18	18–20	4	5	19	780 min.	9 a.m. of the next day
			6		1140 min.	3 p.m. of the next day
		5	7	22	1200 min.	5 p.m. of the next day
			8		1200 min.	5 p.m. of the next day
		6	9	15	960 min.	12 p.m. of the next day
18–20	20–22	7	10	9	840 min.	12 p.m. of the next day
20–22	The next day 6–8	8	11	22	240 min.	12 p.m. of the next day
22–24			12		240 min.	12 p.m. of the next day
4–6 of the next day		9	13	18	840 min.	10 p.m. of the next day
6–8 of the next day	The next day 8–10	10	14	17	720 min.	10 p.m. of the next day
8–10 of the next day	The next day 10–12	11	15	15	600 min.	10 p.m. of the next day
10–12 of the next day	The next day 12–14	12	16	18	480 min.	10 p.m. of the next day

6. The Computational Results

This study accumulates the customer orders for 24 h from 12:00 p.m. to 12:00 p.m. of the next day. If the customer orders were placed within the period of 12:00 p.m. to 6 a.m. of the next day, and to be delivered to the south and home before 12:00 p.m., the wave planning module would generate the picking schedule at 6:00 a.m. on the next day. If orders were placed within the period of 12:00 p.m. to 12:00 p.m. of the next day and to be delivered to the north shop, central shop, and home before 17:00 p.m., the wave planning module would generate the picking schedule at 12:00 p.m. of the next day.

There are 16 orders assign to two pickers and dividing into eight batches (Table 9). In this case, the result is a feasible solution (Figure 5). The planned waves are showed in Figure 6. The total waiting time is 65 min and operation time is 173 min. The total picking route is 346 m and the total cost of container price is 1030 dollars (Table 10).

Table 9. The improvement results.

Batch (*b*)/Picker (p)	Placing Order Time	Customer Order (*o*)	Start Time (st_{pb})	Operation Time (min.) (mt_b)	Finish Time (ft_{pb})	Waiting Time (wt_{pb})	Truck Loading Time (*r*)
1/1	14–16	2	8:23 a.m.	26	8:49 a.m.	11 min.	12:00 p.m.
	14–16	3				11 min.	12:00 p.m.
2/1	14–16	4	8:49 a.m.	11	9:00 a.m.	0	12:00 p.m.
	16–18	5					12:00 p.m.
3/1	16–18	9	11:22 a.m.	16	11:38 a.m.	22 min.	12:00 p.m.
	22–24	12				22 min.	12:00 p.m.
4/2	18–20	10	11:38 a.m.	22	12:00 p.m.	0	12:00 p.m.
	20–22	11					12:00 p.m.
5/1	12–14	1	2:33 p.m.	27	3:00 p.m.	0	5:00 p.m.
	16–18	6					5:00 p.m.
6/2	16–18	7	4:34 p.m.	22	5:00 p.m.	0	5:00 p.m.
	16–18	8					5:00 p.m.
7/2	8–10	15	9:11 p.m.	17	9:28 p.m.	32 min.	10:00 p.m.
	10–12	16				32 min.	10:00 p.m.
8/2	4–6	13	9:28 p.m.	32	10:00 p.m.	0	10:00 p.m.
	6–8	14					10:00 p.m.

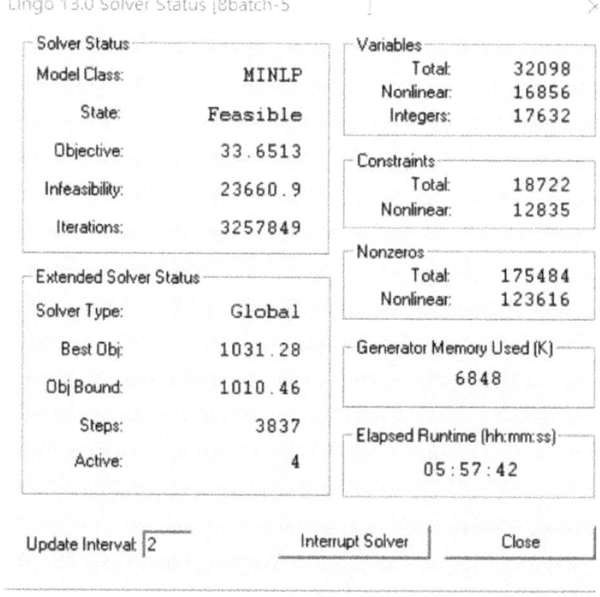

Figure 5. The Screenshot of Lingo solver status.

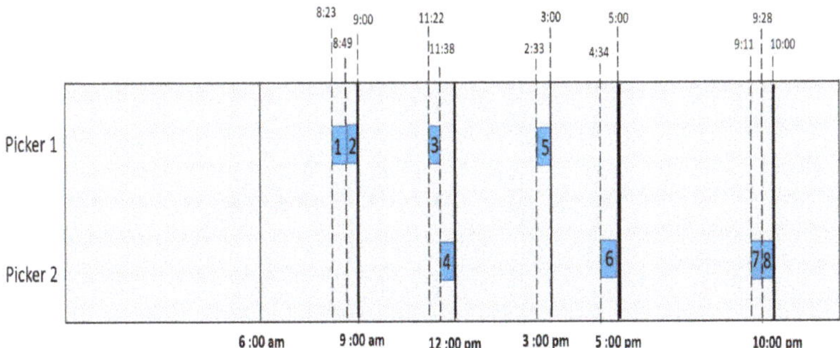

Figure 6. The schedule of batches.

Table 10. The packaged container of customer orders.

Batch (b)	Customer Orders (o)	Container (j)	Container size	Container Price (C_j)
1	2	14	2	70
	3	2	1	55
2	4	13	2	70
	5	8	2	70
3	9	15	2	70
	12	11	2	70
4	10	10	2	70
	11	7	1	55
5	1	9	2	70
	6	5	1	55
6	7	6	1	55
	8	20	2	70
7	15	3	1	55
	16	1	1	55
8	13	16	2	70
	14	19	2	70

The system will compute the most suitable container and the sequence of putting the items into the containers is done according to the picking sequence (Table 11). The loading configuration is also generated.

Before Lingo 13.0 was used to find the model solution, the limit was increased to 50,000 through the general memory limit under options to facilitate a smooth solution-finding process.

(1) Model output: The model solution was found using the Lingo 13.0 optimization program. Figure 5 presents the Lingo solution-finding result screen.

(2) Numbers of decision variables and constraint equations:

Number of decision variables: 32,098;
Number of constraint equations: 18,722;
Model efficiency:
Number of solution-finding cycles: 3,257,849;
Number of solution-finding steps: 3837;

Time used to solve the model: 5 h 57 min and 42 s.

Table 11. The picking sequence of items in batch b.

Picking Sequence of Items (u_{bi}) \\ Batch (b)	1	2	3	4	5	6	7	8
1	0	2	0	0	0	0	0	0
2	0	4	0	0	0	0	0	0
3	0	0	0	0	1	0	0	0
4	0	0	0	0	4	0	0	0
5	0	0	4	0	0	0	0	0
6	0	0	0	0	0	3	0	0
7	0	0	0	0	0	1	0	0
8	0	0	0	0	2	0	0	0
9	2	0	0	0	0	0	0	0
10	3	0	0	0	0	0	0	0
11	1	0	0	0	0	0	0	0
12	0	3	0	0	0	0	0	0
13	0	1	0	0	0	0	0	0
14	0	0	0	0	3	0	0	0
15	0	0	0	0	0	2	0	0
16	0	0	3	0	0	0	0	0
17	4	0	0	0	0	0	0	0
18	0	0	2	0	0	0	0	0
19	0	0	0	2	0	0	0	0
20	0	0	0	1	0	0	0	0
21	0	0	0	3	0	0	0	0
22	0	0	1	0	0	0	0	0
23	0	0	0	0	0	4	0	0
24	0	0	0	4	0	0	0	0
25	0	0	0	0	0	0	0	2
26	0	0	0	0	0	0	0	1
27	0	0	0	0	0	0	4	0
28	0	0	0	0	0	0	1	0
29	0	0	0	0	0	0	2	0
30	0	0	0	0	0	0	3	0
31	0	0	0	0	0	0	0	4
32	0	0	0	0	0	0	0	3

After calculating the performance, waiting time of the original picking planning is 13,320 min, operation time of the original picking planning is 207 min, and the total distance traveled for the original picking planning is 414 m. The calculated waiting time of the wave planning is 130 min, operation time of the wave planning is 173 min, and the total distance traveled for the wave planning logic is 346 m.

Figure 6 illustrates the Gantt chart of the eight batches for the two pickers. For example, the picking route of batch one is showed in Figure 7. Picker 1 will start to pick from P/D point at 8:23 a.m., and the picking sequence is 11→9→10→17. After picking the four items, Picker 1 would go back to P/D point at 8:49 a.m. The load configuration of order 2 and order 3 are showed in Figure 8. The container size of order 2 is size two and order 3 is size one.

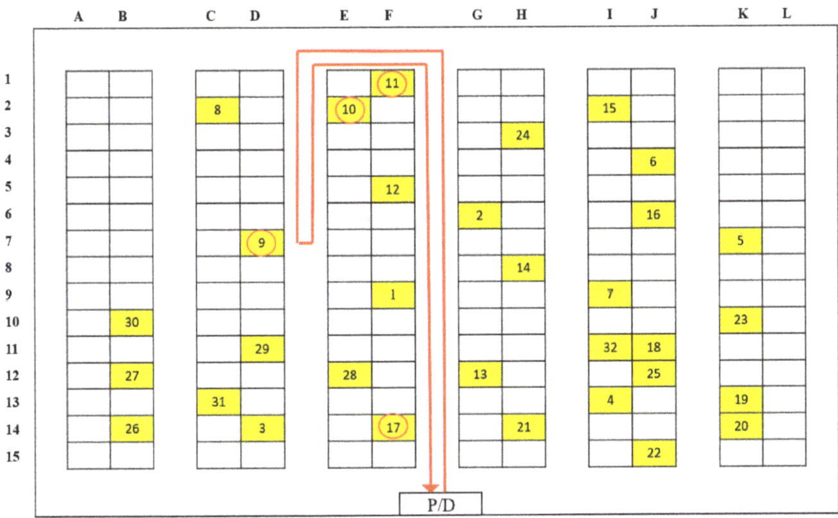

Figure 7. The picking route of batch 1.

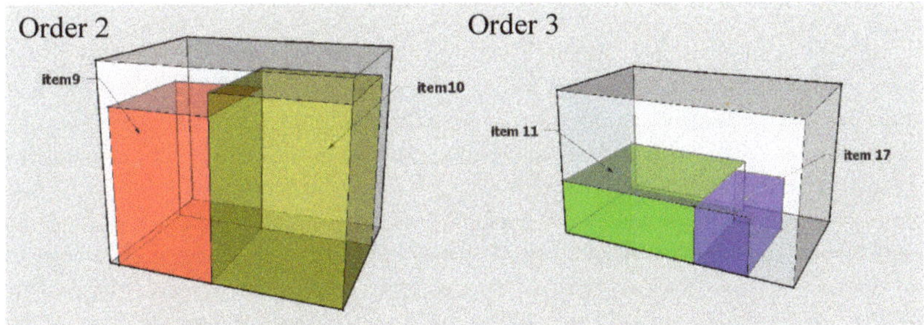

Figure 8. The load configuration of order 2 and order 3.

After fulfilling the batch one, Picker 1 will start to deal batch two continually. Picker 1 will start to pick from P/D point at 8:49 am, and the picking sequence is 13→1→12→2. After picking the four items, Picker 1 would back to P/D point at 9:00 am. The container size of order 4 and 5 is size two.

Picker 1 will start to pick from P/D point at 11:22 am, and the picking sequence is 22→18→16→5. After picking the four items, Picker 1 would back to P/D point at 11:38 am. The container size of order 9 and 12 is size two.

At 2:33 pm, the Picker 1 will start to deal with batch five. Picker 1 will start to pick from P/D point at 2:33 pm, and the picking sequence is 3→8→14→4. After picking the four items, Picker 1 would back to P/D point at 3:00 pm. The container size of order 1 is size two and order 6 is size one.

Comparing the waiting times, operation times, total travelling distances between original picking planning, and proposed wave planning, the results shown that the wave planning model can reduce the waiting time for truck loading of packages significantly, and also can reduce the time packages heaping in buffer area. The operation time and total distance of picking route also have better results when tested with wave planning.

The characteristics of the SKUs being handled, total number of transactions, total number of orders, picks per order, quantity per pick, total number of SKUs, value-added processing activities such as kitting or private labeling, and handling of piece pick, case pick, or full-pallet loads are all factors that affect the method for order picking. Comparing with previous studies as shown in Table 2, our model demonstrated a successful integration of batching and splitting customer orders, sequencing picking orders, planning picker routes, scheduling picking waves, assigning packing configurations, and eliminating packing operations. Moreover, it also can increase capacity utilization rate and reduce the packaging cost.

7. Conclusions

Order picking policy is a case-oriented design and planning task. In this paper, we reviewed an order picking problem from forms and workflow of a WMS perspectives. By using them to observe an order picking problem of a 24-h express shipping online bookstore, the considering parameters of wave planning for designing the optimization programming model were identified.

To the best of our knowledge, there was no wave order picking planning algorithm been revealed so far. The wave planning model we developed was a mixed integer nonlinear programming model. It takes multiple customer orders, SKUs' sizes, volumes of mailing boxes, locations of SKUs, and planned delivering truck loading times as input parameters, then calculates the configuration of containers (i.e., bin packing problem), the groups of picking SKUs (i.e., order batch problem), and the groups of pickers (i.e., batch assignment problem), and, finally, outputs a set of order picking lists with their associated schedules.

The past studies showed that picking orders in a short time window is a better picking policy in the random storage warehouse [5]. However, having our proposed model, distribution centers can adjust the frequency of order batching. In other words, the distribution center can accumulate orders for a longer time window which may collect more similar SKUs. The combined picking of several orders can bring about a reduction of approximately 60% in walking time [2]. Moreover, this can avoid order pickers to travel to the same storage locations in the short time. To sum up, the object of the model is that distribution centers can not only achieve the goal of fast delivering but also increase the capacity utilization rate which can make distributions fulfill more orders within the same time period and can completed orders before departure time.

The proposed model is a nonlinear mixed integer programming model. The limitation to this kind of model is that the derivation of an optimal solution is not always guaranteed. Therefore, we could only use an example to prove the capability of the conceptual design of such an order picking system. The calculating time of the example is around 6 h, which also makes the practical use of this model limited. This brings the need of developing a heuristic algorithm for finding the near optimal solution of our model in a relative short computing time as our future work. Our current model could be used as a benchmark reference model for comparing solution results while developing such heuristic algorithms.

Author Contributions: Conceptualization, J.-Y.S.; resources, J.-Y.S.; methodology, J.-A.H.; formal analysis, J.-A.H.; writing—original draft preparation, J.-A.H.; writing—review and editing, J.-Y.S.; supervision, J.-Y.S.; funding acquisition, J.-Y.S. All authors have read and agreed to the published version of the manuscript.

Funding: This research is partially supported by the Ministry of Science and Technology, Taiwan under the grant MOST 109-2637-E-992-015.

Conflicts of Interest: The authors declare no conflict of interest.

References

1. Chabot, T.; Lahyani, R.; Coelho, L.C.; Renaud, J. Order picking problems under weight, fragility and category constraints. *Int. J. Prod. Res.* **2016**, *55*, 6361–6379. [CrossRef]
2. Brynzér, H.; Johansson, M. Design and performance of kitting and order picking systems. *Int. J. Prod. Econ.* **1995**, *41*, 115–125. [CrossRef]
3. Sharp, G.P.; Il-Choe, K.; Yoon, C.S.; Graves, R.J.; Wilhelm, M.R.; McGinnis, L.F.; Ward, R.E. Small Parts Order Picking: Analysis Framework and Selected Results. In *Material Handling'90*; Springer: Berlin/Heidelberg, Germany, 1991; pp. 317–341.
4. Zhang, J.; Wang, X.; Huang, K. Integrated on-line scheduling of order batching and delivery under B2C e-commerce. *Comput. Ind. Eng.* **2016**, *94*, 280–289. [CrossRef]
5. De Koster, R.; Le-Duc, T.; Roodbergen, K.J. Design and control of warehouse order picking: A literature review. *Eur. J. Oper. Res.* **2007**, *182*, 481–501. [CrossRef]
6. Won, J. Joint order batching and order picking in warehouse operations. *Int. J. Prod. Res.* **2005**, *43*, 1427–1442. [CrossRef]
7. Petersen, C.G.; Aase, G. A comparison of picking, storage, and routing policies in manual order picking. *Int. J. Prod. Econ.* **2004**, *92*, 11–19. [CrossRef]
8. Çelik, M.; Süral, H. Order picking in a parallel-aisle warehouse with turn penalties. *Int. J. Prod. Res.* **2016**, *54*, 4340–4355. [CrossRef]
9. Giannikas, V.; Lu, W.; Robertson, B.; McFarlane, D. An interventionist strategy for warehouse order picking: Evidence from two case studies. *Int. J. Prod. Econ.* **2017**, *189*, 63–76. [CrossRef]
10. Chen, T.-L.; Cheng, C.-Y.; Chen, Y.-Y.; Chan, L.-K. An efficient hybrid algorithm for integrated order batching, sequencing and routing problem. *Int. J. Prod. Econ.* **2015**, *159*, 158–167. [CrossRef]
11. Cheng, C.-Y.; Chen, Y.-Y.; Chen, T.-L.; Yoo, J.J.-W. Using a hybrid approach based on the particle swarm optimization and ant colony optimization to solve a joint order batching and picker routing problem. *Int. J. Prod. Econ.* **2015**, *170*, 805–814. [CrossRef]
12. Kulak, O.; Sahin, Y.; Taner, M.E. Joint order batching and picker routing in single and multiple-cross-aisle warehouses using cluster-based tabu search algorithms. *Flex. Serv. Manuf. J.* **2012**, *24*, 52–80. [CrossRef]
13. Scholz, A.; Schubert, D.; Wäscher, G. Order picking with multiple pickers and due dates—Simultaneous solution of Order Batching, Batch Assignment and Sequencing, and Picker Routing Problems. *Eur. J. Oper. Res.* **2017**, *263*, 461–478. [CrossRef]
14. Dallari, F.; Marchet, G.; Melacini, M. Design of order picking system. *Int. J. Adv. Manuf. Technol.* **2008**, *42*, 1–12. [CrossRef]
15. Ho, Y.-C.; Lin, J.-W. Improving order-picking performance by converting a sequential zone-picking line into a zone-picking network. *Comput. Ind. Eng.* **2017**, *113*, 241–255. [CrossRef]
16. Harner, L.; Hagen, N.; Zickert, S.; Grimm, D. 2016 Top Warehouse Trends for the Decade Ahead. Available online: http://media.logistique-agroalimentaire.com/Presentation/future_series_wp_top_warehouse_trends_digital_818921.pdf (accessed on 27 October 2020).
17. Shiau, J.-Y.; Liao, T.-C. Developing an order picking policy for economical packing. In Proceedings of the 2013 IEEE International Conference on Service Operations and Logistics, and Informatics; Institute of Electrical and Electronics Engineers (IEEE), Dongguan, China, 28–30 July 2013; pp. 387–392.
18. Füßler, D.; Boysen, N. Efficient order processing in an inverse order picking system. *Comput. Oper. Res.* **2017**, *88*, 150–160. [CrossRef]
19. Schwerdfeger, S.; Boysen, N. Order picking along a crane-supplied pick face: The SKU switching problem. *Eur. J. Oper. Res.* **2017**, *260*, 534–545. [CrossRef]
20. Lu, W.; McFarlane, D.C.; Giannikas, V.; Zhang, Q. An algorithm for dynamic order-picking in warehouse operations. *Eur. J. Oper. Res.* **2016**, *248*, 107–122. [CrossRef]
21. Ballestín, F.; Perez, A.; Lino, P.; Quintanilla, S.; Valls, V. Static and dynamic policies with RFID for the scheduling of retrieval and storage warehouse operations. *Comput. Ind. Eng.* **2013**, *66*, 696–709. [CrossRef]
22. Rim, S.-C.; Park, I.-S. Order picking plan to maximize the order fill rate. *Comput. Ind. Eng.* **2008**, *55*, 557–566. [CrossRef]
23. Kang, S. Relative logistics sprawl: Measuring changes in the relative distribution from warehouses to logistics businesses and the general population. *J. Transp. Geogr.* **2020**, *83*, 102636. [CrossRef]

24. Tsai, C.-Y.; Liou, J.J.H.; Huang, T.-M. Using a multiple-GA method to solve the batch picking problem: Considering travel distance and order due time. *Int. J. Prod. Res.* **2008**, *46*, 6533–6555. [CrossRef]

25. Valle, C.A.; Beasley, J.E.; Da Cunha, A.S. Optimally solving the joint order batching and picker routing problem. *Eur. J. Oper. Res.* **2017**, *262*, 817–834. [CrossRef]

26. Henn, S. Algorithms for on-line order batching in an order picking warehouse. *Comput. Oper. Res.* **2012**, *39*, 2549–2563. [CrossRef]

27. Hsieh, L.-F.; Huang, Y.-C. New batch construction heuristics to optimise the performance of order picking systems. *Int. J. Prod. Econ.* **2011**, *131*, 618–630. [CrossRef]

28. Hsu, C.-M.; Chen, K.-Y.; Chen, M.-C. Batching orders in warehouses by minimizing travel distance with genetic algorithms. *Comput. Ind.* **2005**, *56*, 169–178. [CrossRef]

29. Rubrico, J.; Higashi, T.; Tamura, H.; Nikaido, M.; Ota, J.; Systems, L.M. A Fast Scheduler for Multiagent in a Warehouse. *Int. J. Autom. Technol.* **2009**, *3*, 165–173. [CrossRef]

30. Henn, S.; Schmid, V. Metaheuristics for order batching and sequencing in manual order picking systems. *Comput. Ind. Eng.* **2013**, *66*, 338–351. [CrossRef]

31. Henn, S. Order batching and sequencing for the minimization of the total tardiness in picker-to-part warehouses. *Flex. Serv. Manuf. J.* **2015**, *27*, 86–114. [CrossRef]

32. Hong, S.; Kim, Y. A route-selecting order batching model with the S-shape routes in a parallel-aisle order picking system. *Eur. J. Oper. Res.* **2017**, *257*, 185–196. [CrossRef]

33. Choy, K.L.; Ho, G.T.S.; Lam, H.Y.; Lin, C.; Ng, T.W. A sequential order picking and loading system for outbound logistics operations. In Proceedings of the 2014 Portland International Conference on Management of Engineering & Technology (PICMET 2014), Kanazawa, Japan, 27–31 July 2014; pp. 507–513.

34. Schubert, D.; Scholz, A.; Wäscher, G. Integrated order picking and vehicle routing with due dates. *OR Spectr.* **2018**, *40*, 1109–1139. [CrossRef]

Publisher's Note: MDPI stays neutral with regard to jurisdictional claims in published maps and institutional affiliations.

Article

Solving Order Planning Problem Using a Heuristic Approach: The Case in a Building Material Distributor

Chia-Nan Wang [1,*], Ngoc-Ai-Thy Nguyen [1,*] and Thanh-Tuan Dang [1,2]

[1] Department of Industrial Engineering and Management, National Kaohsiung University of Science and Technology, Kaohsiung 80778, Taiwan; i108143106@nkust.edu.tw

[2] Department of Logistics and Supply Chain Management, Hong Bang International University, Ho Chi Minh 723000, Vietnam

* Correspondence: cn.wang@nkust.edu.tw (C.-N.W.); i109143106@nkust.edu.tw (N.-A.-T.N.)

Received: 15 October 2020; Accepted: 11 December 2020; Published: 15 December 2020

Abstract: For building material distributors, order planning is a key process as a result of the increase in construction projects' scale and complexity. In this paper, the integration of simulation modeling and the response surface methodology (RSM) is presented to solve an order planning problem in the construction supply chain. The interactions of various factors are examined to observe their effects on key system measurements, and a combination of factor levels is determined to achieve the optimal performance. RSM is applied to find the possible values of the optimal setting for system responses, which consists of three main steps: central composite design (CCD), Box–Behnken design (BBD), and a comparison of both designs. The model is tested with a realistic case study of a building material distributor in Vietnam to demonstrate its effectiveness. Controllable factors (independent variables), which are the review period (T), order quantity (Q), and safety stock (SS), are found to significantly affect system responses, which are the total cost (TC) and customer service level (CSL). The results provide the best settings of factor levels that produce the possible minimum TC and maximum CSL. The developed framework could be applied as a useful reference for decision-makers, purchasing managers, and warehouse managers to obtain the most suitable order policy for a robust order planning process.

Keywords: building material distributors; central composite design (CCD); Box–Behnken design (BBD); optimal cost; customer service level; forecasting; order planning; inventory management

---

1. Introduction

Buildings are becoming increasingly complex because of the ongoing rise in the size and scope of construction projects. Following that, the rising segmentation of the construction industry goes hand-in-hand with the growth of specialist suppliers or contractors, and the diversity of products, designs, and control activities [1]. Relationships in the construction industry are short-term and normally informal/ad-hoc, which focus on the projects, not the business. Construction demand is inherently unstable as the industry is project-based with defined start and endpoints, and a conventional separation between design and construction. Demand is viewed by temporary coalitions as a series of competitively tendered prototypes. Competencies between construction projects vary from time to time and are distinguished from each other. In the supply chain of the construction industry, order planning is the key process, especially for building material distributors. Adequate order planning is quintessential to a more comprehensive integrated supply chain solution, by which businesses can achieve real-time capabilities to act as an intermediary between manufacturing, sales, and customer service concerns, in order to guarantee on-time, effective and reliable processes of order fulfillment and delivery [2]. For building material distributors, the most common problems are inaccurate forecasts of annual projects received and the overloading in stock due to the delayed construction work progress. Therefore, demand forecasting and inventory management are crucial tasks in the order planning process, as the progress of construction projects is often not always as smooth as scheduled. Accurately forecasting spikes in demand and appropriate levels of safety stock can give a competitive advantage in the industry.

In the context of demand variability, sufficient order planning policies are crucial [3]. Some common challenges of order planning problems in all different types of industries are accurate monitoring of all stock-keeping units (SKUs) and stock levels, availability of historic sales forecasting data, calculating correct supplier lead-times, new product introductions, and centralizing stock control [4]. In this context, building material distributors often deal with serious stock-outs (unknown/unaware projects) or extreme overstock (delayed projects). Consequently, additional costs are incurred due to extra shipments, transportation and transferring activities, inventory holding, and handling costs, to name a few. In addition, the expectation of customers for timely and reliable products and service is ever-increasing, so that a more efficient ordering operation is indispensable and is a key measure for customer service. Incomplete and delayed orders are the most common causes of worsening customer satisfaction. Hence, strategies in ordering planning problems, namely which order to fulfill and in what sequence that strongly affects customer satisfaction and businesses' profits, as it decides how to divide the inventory between orders and how to attend to orders to provide an efficient process [5].

Some strategies are used to mitigate risk in the ordering process: classifying stock [6], drop-shipping [3], prepurchasing stocks [7], using hybrid strategies [8,9], and setting stock minimization policies [10]. Many studies developed stock control policies to solve the problem of managing inventory. They applied lot sizing, stockpiling, and period length instead of controlling on-hand inventory, at the same time examining certain stochastic variables in the ability to respond with high demand fluctuations and shortages in the multistage supply chain [11,12]. In practice, demands vary from time to time, they are more of fuzzy numbers than deterministic ones. Problems concerning production and inventory decisions are regarded as a joint economic lot-sizing problem (JELP), of which a vendor and a buyer with stochastic demand and variable lead time are considered [13]. Jauhari and Laksono [14] solved the JELP with the demand changing over time, imperfect production emission cost, and defective production rate. By considering the fuzzy annual demand, their mathematical model contributed to the existing inventory literature. Chiang Kao and Wen-Kai Hsu [15] provided a framework for the development of the fuzzy total inventory cost with uncertain demand and the used Yager's ranking method to minimize the total cost and to prevent lost-sales. Kim et al. [16] introduced a multiperiod newsvendor problem, which formulated integer recourse decisions as a multistage stochastic programming model. The proposed method offered novel knowledge to optimize the trade-off between delivery, transshipment, shortage, and holding costs, which is the gap to other studies that suggested stochastic inventory models. Additionally, most studies suggested different considerations for customer service's perception, assumed that quality products, sales price, corporate social responsibility, and recycling commitment are key factors to the customer service level [17–22].

In this paper, simulation modeling is developed to explain the ordering process for building materials in a distributor. For this case study, key system performance measures are the total cost (TC) and the customer service level (CSL). The total cost includes transportation cost, holding cost, over-storage cost, and shortage cost. On the other hand, CSL simplifies by the accumulation of total demand satisfied. Parameters of products include unit weight, unit holding cost, unit shortage cost, and unit over-storage cost. Controllable factors of the ordering process are the monthly demand, review period (T), order quantity (Q), and safety stock (SS), which are treated as independent variables. Using RSM, the best settings, which are appropriate combinations of T, Q, and SS are determined as an indicator to measure the performance of each possible setting, in which TC and CSL are response variables. The solution is to place the right order quantities at the right time, with the appropriate safety stock to achieve customer satisfaction levels between 90 and 95 percent at minimum costs.

The contributions of our paper are three-fold. First, the paper proposed a computational framework to solve the order planning problem by a heuristics approach using RSM, which includes the central composite design (CCD), Box–Behnken design (BBD), and variance dispersion graph (VDG), in order to offer the best possible time-efficient and resource-saving solution in terms of key performances. Most exact methods cannot solve the problem within a satisfactory amount of time, especially for large-scale problems. In contrast, most heuristics allows decision-makers to obtain a feasible and timelier solution. For order planning problems in the construction industry, managing the demand, duration, and progress between projects that are supposed to be implemented simultaneously by a distributor is a complicated task. Thus, to facilitate the practical process and available resources, heuristics can provide decision-makers feasible short-term solutions within a reasonable amount of time. Second, the proposed heuristic model is validated by a realistic case study of a distributor in Vietnam to illustrate the effectiveness of the model's result. Third, the managerial implications of the paper could be a beneficial guide for decision-makers, purchasing managers, and warehouse managers to obtain the most suitable order policy for a robust order planning process. In addition, the practical contribution of this study is the comprehensive insight into order planning problems contributed by the case study of the building material distributor in Vietnam.

The paper is divided into five sections. The introduction and relevant studies are discussed in the first section. In the second section, the research procedure and related methodology are included. This paper proposed a computational framework to solve the order planning problem. Additionally, the case study of a distributor is presented in the third section. In the fourth section, the empirical results are shown, including the results of excel spreadsheet simulation, and experimental design of central composite design, and Box–Behnken design. Moreover, discussions, conclusions, and recommendations are given in the last section.

2. Literature Review

This paper combined simulation modeling and the response surface methodology (RSM) to solve an order planning problem. A simulation and/or computational experiment are used to model a comprehensive supply chain network that corresponds with a number of supply chain operational elements and management levels [23]. Some studies that proposed simulation-based optimization models to explain supply chain systems are discussed in [24,25]. Supply chain dynamics (response) are typically quantified in terms of order and inventory variance ratios. To experiment with the stochastic supply chain model, a simulation modeling approach was adopted. As a result of modifying the controllable variables, the simulation model was run to generate the various supply chain responses, and thus the functional relationship could be fitted and its response surface defined. RSM is a feasible alternative to modeling, on the other hand, which optimizes stochastic, dynamic, and complex systems such as supply chains. This approach is used to find a functional link between the complex responses of the supply chain and significant controllable variables that influence them [26]. Giddings et al. [27] used RSM for optimality analysis of the cost coefficients in mixed integer linear programming in facility location problems. The study implemented the design of experiments and applied least squares regression to determine cost coefficients that significantly affect the optimal total cost surface within setup coefficient ranges. In a problem of finding the suitable capacity for a factory, Shang et al. [28] aimed at integrating simulation, Taguchi techniques, and RSM. Findings of this hybrid approach is a useful reference for businesses to analyze the dynamic relations among various factors, so as to determine the best combination of their levels that optimizes the impact of demand uncertainty on the performance of the supply chain. Buchholz et al. [29] developed a hybrid model of a process-based simulation ProC/B toolset and RSM for the optimization of the process chain models. In this paper, the experiment of the central composite design was applied to find the optimal values of response variables. The "bullwhip" effect, known also as order variance amplification, is a major cause of supply chain deficiencies. Hassanzadeh et al. [30] studied a three-stage simulation with a single retailer, a single wholesaler, and a singer-producer under both centralized and decentralized chains, in order to analyze the causes of the bullwhip effect from two dimensions of order and inventory variance using RSM. As an extension, a multiobjective minimization problem in which the order and inventory variance ratios are separated into two objective functions was developed by Devika et al. [31]. RSM was used in a hybrid evolutionary approach to analyze interactions between the variance ratios and their corresponding effects, in order to optimize supply chain systems. For measuring design parameters on the bullwhip effect and dynamic responses, Tang et al. [32] analyzed its influences on the supply chain's performance using a hybrid Taguchi and dual RSM. This paper offered practical solutions to the supply chain managers and designers in the trade-off between customer service level and inventory holding cost under an uncertain environment.

To provide an overview of recent contributions, several studies that presented relevant problem characteristics are shown in Table 1. Concerning order planning problems, many studies approached exact methods such as NLP [33], MOLP [34,35], and mathematical models [36–39]. The widely used method is applying inventory policies such as the economic order quantity (EOQ), economic production quantity (EPQ), (S, T), or (R, Q) model [36,39,40]. The order planning problem is actually the flow shop (FS) problem, which is an NP-hard problem [41]. Most exact methods cannot solve large-scale problems within an acceptable amount of time. Therefore, heuristics [33,42] have widely been applied currently, which produce good feasible solutions timely and effectively. In this paper, to optimize an order planning problem, namely to minimize total relevant costs and maximize customer satisfaction level, RSM was used to determine the best combination of the system parameters. Independent variables, which are the review period (T), order quantity (Q), and safety stock (SS), were treated as factors for response variables, which are the total cost (TC) and customer service level (CSL). Analytical approaches have many limitations to simultaneously solve two or three factors [23]. It is difficult to observe the factor interactions and predict their effects on the overall objective when multiple factors are considered at a time. By using RSM, our research aims to determine the best parameter settings of considered factors to find the possible optimum values of the responses. Findings are expected to offer purchasing managers or decision-makers to obtain the optimal combination of the levels of independent factors, which strongly affect the key system performance measures, i.e., total relevant costs and customer satisfaction level.

Table 1. List of common problem characteristics for several previous studies.

Authors	Independent Variables	Response Variables	Methods/Techniques
Zhang et al. [33], 2015	Number of orders Planning horizon Production capacity	Penalty for delivery Penalty for order Inventory cost Production cost	NLP PSO algorithm
Guo et al. [34], 2015	Due date of order Production workload Capacity efficiency	Completion time Production capacity Processing time	MOLP DOE
Lim et al. [35], 2017	Demand variability Lead time	Holding cost Supply cost Safety stock	Simulation modeling MOLP
Singha et al. [36], 2017	Reorder point Order quantity Inventory policies	Ordering cost Shortage cost Holding cost Over-ordering cost	EOQ (R, Q) model Mathematical model
Jansen et al. [37], 2019	Planned lead time Production plan	Holding cost Penalty cost	Mathematical model
Thinakaran et al. [40], 2019	Time-varying Stock demand Lot size	Backorder Lost sales	EOQ EPQ

Table 1. *Cont.*

Authors	Independent Variables	Response Variables	Methods/Techniques
Wang et al. [42], 2020	Ordering quantity Reorder point Target stock Inventory policies	Holding cost Penalty cost	Simulation modeling DOE
Kokuryo et al. [38], 2020	Order quantity Production schedule Order period	Delivery time Profit	Mathematical model
Tai et al. [39], 2020	Order quantity Lead time Inventory policies	Over-storage amount Inventory cost	EOQ (S, T) model Mathematical model
This paper, 2020	Review period Order quantity Safety stock	Holding cost Shortage cost Over-storage cost Transportation cost	Simulation modeling RSM

Note: NLP: nonlinear programming, PSO: particle swarm optimization, MOLP: multi-objective linear programming, DOE: design of experiment, EOQ: economic order quantity, EPQ: economic production quantity, RSM: response surface methodology.

3. Materials and Methods

3.1. Response Surface Methodology (RSM)

Response surface methodology (RSM) was used to design and perform an experimental design via statistical and mathematical approaches. RSM was applied to determine the possible optimal values of factors (i.e., independent variables) to minimize and/or maximize the outputs (response variables). RSM was assumed that there is no correlation among independent variables. There are many stages to perform the process of optimization using RSM. The following parts present the simple and polynomial regression model [43–46].

The simple regression model is presented in Equation (1) as follows.

$$Y = \beta_o + \beta_1 X_1 + \beta_2 X_2 + \ldots + \beta_k X_k + \varepsilon \tag{1}$$

where Y denotes response variables, $X_1, X_2, \ldots X_k$ denotes independent variables, k is the total of independent variables, β_o is the intercept, $\beta_1, \beta_2, \ldots \beta_k$ is the slope, and ε is the random error.

Equation (2) shows the quadratic polynomial regression model.

$$Y = \beta_o + \sum \beta_1 X_1 + \sum \beta_2 X_2{}^2 + \sum \beta_3 X_1 X_2 + \ldots + \varepsilon \tag{2}$$

where Y denotes response variables, X denotes independent variables, β_o is the intercept, β_1 is the coefficient for the linear terms, β_2 is the coefficient for the square terms, β_3 is the coefficient for the interaction among terms, and ε is the random error.

In the polynomial regression model, CCD and BBD are two types of experimental designs often referred to by the researchers. CCD is an experimental design with two-level. Minitab provides the rotatability value (α) to ensure the design displays the desirable properties. Meanwhile, BBD is the experiment with a three-level design. BBD does not have the rotatability value (α) and embedded factorial design. Points on the diagrams represent the experimental runs of CCD and BBD are shown in Figure 1 as follows [47].

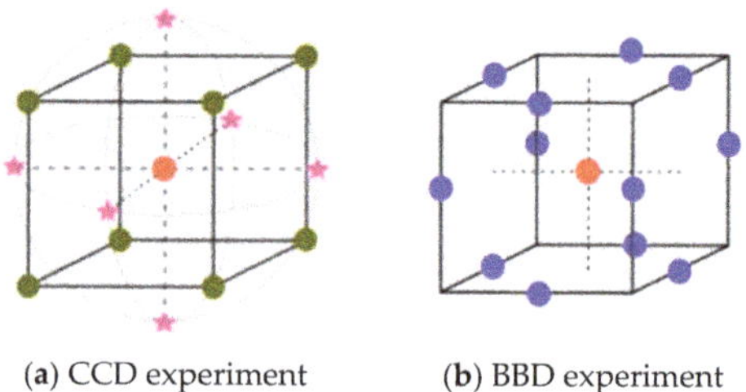

(**a**) CCD experiment   (**b**) BBD experiment

Figure 1. The experimental runs of the central composite design (CCD) and Box–Behnken design (BBD).

3.2. A Computational Framework

In this paper, a computational framework was built (Figure 2), which consists of two parts, described as follows: (1) an Excel spreadsheet was used to develop the order planning model, and calculate the response variables, i.e., total cost and customer service level. Then, (2) Minitab was used to find the feasible values of the optimal setting for those responses using RSM, CCD, and BBD. Besides, a VDG was applied to validate these experimental designs. Three independent variables were being considered to have effects on the response variables in this experiment, consisting of the review period (T, month), order quantity (Q, unit), and safety stock (SS, unit). The combination in the total cost (TC, USD) that includes the holding cost, shortage cost, over-storage cost, transportation cost, and customer service level (CSL, %) were chosen as response variables of the order planning model.

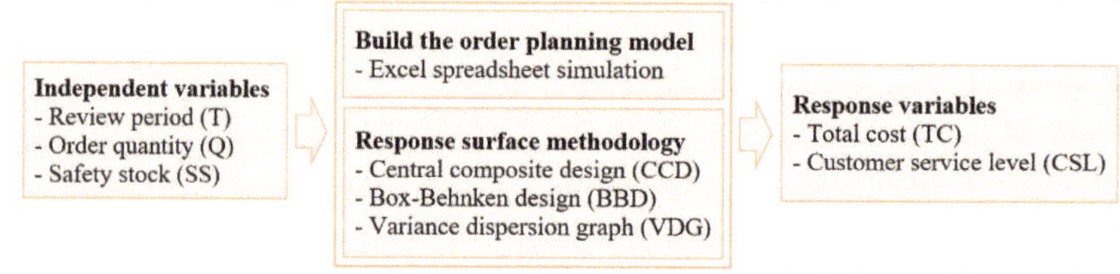

Figure 2. A computational framework.

4. A Case Study

4.1. Problem Description

The description of the primary building materials of a distributor in Vietnam is shown in Figure 3. There are four main material products for coating in construction works: Spectite CW100 (a), Spectite WS (b), Spectite HP600 (c), and Speccoat PE145 (d). The statistical summary of historical sale data (2017–2019), and the component costs of materials are demonstrated in Tables 2 and 3, respectively.

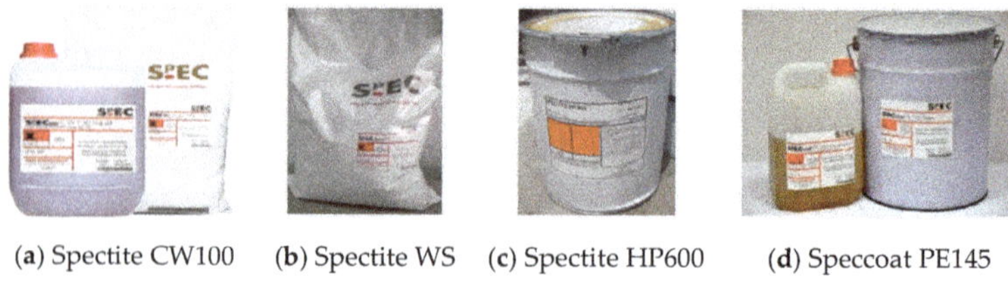

(**a**) Spectite CW100   (**b**) Spectite WS   (**c**) Spectite HP600   (**d**) Speccoat PE145

Figure 3. List of the main materials group of the distributor.

Table 2. Statistical summary of historical sale data added up to 36 months (2017–2019).

2017	Spectite CW100	Spectite WS	Spectite HP600	Speccoat PE145
Max	1663	131	224	40
Min	215	0	0	1
Avg	1044.58	45	36.33	5.83
SD	470.18	47.13	63.58	11.81

2018	Spectite CW100	Spectite WS	Spectite HP600	Speccoat PE145
Max	2140	965	100	114
Min	411	0	0	2
Avg	1093.50	198.17	21.42	31.83
SD	538.73	265.93	35.64	34.91

2019	Spectite CW100	Spectite WS	Spectite HP600	Speccoat PE145
Max	2290	359	75	83
Min	219	0	0	0
Avg	1387.75	142.42	32.67	30.17
SD	649.19	124.85	30.79	24.88

Note: calculated by the researchers.

Table 3. Component costs of materials.

Materials	Spectite CW100	Spectite WS	Spectite HP600	Speccoat PE145	Unit
Packing size	22 kg/set	25 kg/bag	20 L/pail	15 L/pail	-
Wholesale price	11	20	78	49.5	USD/unit
Receipt cost	12.93	22.12	84.73	54.70	USD/unit
Selling price	18.42	32.89	116.22	87.72	USD/unit
Holding cost	0.075	0.129	0.494	0.319	USD/unit/month
Shortage cost	0.553	0.987	3.487	2.632	USD/unit
Over-storage cost	0.162	0.277	1.059	0.684	USD/unit/month

Note: calculated by the researchers.

Currently, the company places orders from a supplier every three or four months. Afterward, the transportation freight charge for 20 ft and 40 ft containers that are used for shipments directly to a warehouse. For inventory management, each unit of products (set, pail, or bag) is charged for holding cost per unit per month. Over-storage cost per unit is charged when ending inventory excesses the warehouse capacity. Shortage cost per unit is charged if satisfied demand is less than the actual demand. Key performances are measure periodically, which are the total cost, calculated by the sum of holding inventory, shortage cost, over-storage cost, and transportation cost, as presented in Equation (3); and the customer service level, which is determined by the accumulation of the total demand satisfied, as shown in Equation (4). Subsequently, the company is facing a problem in which should setup different factors (i.e., referred to independent variables in this paper) that affect the performance of the inventory system. Therefore, designing a computational framework to find the feasible values of the optimal setting of levels for independent variables that strongly affect the minimum total cost of holding cost, shortage cost, over-storage cost, transportation cost, and customer service level is needed.

Total cost (TC) can be calculated from the following Equation (3):

$$TC = C_h + C_s + C_o + C_t \tag{3}$$

where C_h is holding cost, C_s is shortage cost, C_o is over-storage cost, and C_t is transportation costs.

The customer service level (CSL) in this model simplifies by the accumulation of the total demand satisfied, and it can be calculated from the following Equation (4):

$$CSL = \frac{Demand - Shortage\ Amount}{Total\ Demand} \times 100\% \qquad (4)$$

4.2. Numerical Example

In order to demonstrate the model's effectiveness, the authors chose "Spectite CW100", the key product of the company. Compared to other products, Spectite CW100 is always indispensable for construction projects, yet often running out of stock. Additionally, this product was selected based on ABC analysis (i.e., highest sale volume and high order frequency) [48,49]. The ABC classification divides inventory stock-keeping units (SKUs) into three groups based on their annual monetary purchases. These SKUs are high-value items, i.e., the 15–20% of the total inventory that accounts for 75–80% of the total annual inventory value. The product is distributed to customers and given inventory carrying cost, shortage cost, and over-storage cost of each unit. Based on product behaviors and the current order planning process, the interval ranges of the variables were set by the decision-makers as shown in Table 4 and were chosen as potential values in the factorial design for this problem.

Table 4. The interval ranges of the variables.

No.	Variables	Notation	Interval Range	Unit
1	Review period (categorical factor)	T	(3, 4)	Month
2	Order quantity (continuous factor)	Q	(3500, 4000)	Unit
3	Safety stock (continuous factor)	SS	(1000, 1300)	Unit
4	Total cost (response)	TC	TC (C_h, C_s, C_o, C_t)	USD
5	Customer service level (response)	CSL	CSL (90% and 95%)	%

Note: calculated by the researchers.

5. Empirical Results

5.1. Excel Spreadsheet Simulation

First, the authors pivoted the quantity of historical sale data to count monthly order frequency to generate the probability of demand. Next, the authors used the random function in excel to simulate customer demand and get the values response variables (TC and CSL). The experimental design was created using a central composite design and Box–Behnken design to find the possible independent variables settings for solving the order planning problem. The excel spreadsheet simulation template for the experimental design is described in Figure 4.

In this paper, a case study of a distributor is presented to demonstrate the effectiveness of the developed computational framework to solve the current order planning problem in the company.

The model's parameters setup is described as follows.

- Unit holding cost, $C_h = 0.075$ USD/unit/month;
- Unit shortage cost, $C_s = 0.553$ USD/unit;
- Unit over-storage cost, $C_o = 0.162$ USD/unit/month;
- Warehouse capacity = 5000 units;
- Weight (CW100) = 22 kg/unit;
- Maximum capacity of container type 20 ft = 21,000 kg;
- Maximum capacity of container type 40 ft = 42,000 kg;
- Freight charge of container type 2 0ft = 1500 USD/trip;
- Freight charge of container type 40 ft = 2700 USD/trip.

Formula bar: `=IF(AND(D10>0,B10<$E$2),$E$2-B10,IF(AND(D10>0,B10=$E$2,B10>$E$2),0,0))`

Excel Spreadsheet Simulation

Month	Beginning Inventory	Total Order	Quantity Order	Safety Stock	Total Weight	Available Weight	No. of 20ft	No. of 40ft	Trans Cost	Arrived Order	Demand	Satisfied Demand	No. of Shortage	Cumulative Shortage	Ending Inventory	Over Storage
0	0	4500	3500	1000	99,000	105,000	1	2	6900	0	0	0	0	0	0	0
1	0	0	0	0	0	0	0	0	0	0	0	0	0	0	0	0
2	4500	0	0	0	0	0	0	0	0	4500	1461	1461	0	0	3039	0
3	3039	3500	3500	0	77,000	84,000	0	2	5400	0	1431	1431	0	0	1608	0
4	1608	0	0	0	0	0	0	0	0	0	1667	1608	59	59	0	0
5	3500	0	0	0	0	0	0	0	0	3500	432	432	0	59	3068	0
6	3068	3500	3500	0	77,000	84,000	0	2	5400	0	1485	1485	0	59	1583	0
7	1583	0	0	0	0	0	0	0	0	0	2292	1583	709	768	0	0
8	3500	0	0	0	0	0	0	0	0	3500	1662	1662	0	768	1838	0
9	1838	3500	3500	0	77,000	84,000	0	2	5400	0	815	815	0	768	1023	0
10	1023	0	0	0	0	0	0	0	0	0	652	652	0	768	371	0
11	3871	0	0	0	0	0	0	0	0	3500	454	454	0	768	3417	0
12	3417	3500	3500	0	77,000	84,000	0	2	5400	0	510	510	0	768	2907	0
13	2907	0	0	0	0	0	0	0	0	0	1436	1436	0	768	1471	0
14	4971	0	0	0	0	0	0	0	0	3500	237	237	0	768	4734	0
15	4734	3500	3500	0	77,000	84,000	0	2	5400	0	1216	1216	0	768	3518	0
16	3518	0	0	0	0	0	0	0	0	0	1944	1944	0	768	1574	0
17	5074	0	0	0	0	0	0	0	0	3500	1572	1572	0	768	3502	0
...	...	...	...	...	...	...	...	...	...	...	...	...	...	...	...	...
43	7093	0	0	0	0	0	0	0	0	3500	1565	1565	0	768	5528	528
44	5528	3500	3500	0	77,000	84,000	0	2	5400	0	2161	2161	0	768	3367	0
45	3367	0	0	0	0	0	0	0	0	0	1593	1593	0	768	1774	0
46	5274	0	0	0	0	0	0	0	0	3500	1648	1648	0	768	3626	0
47	3626	3500	3500	0	77,000	84,000	0	2	5400	0	957	957	0	768	2669	0
48	2669	0	0	0	0	0	0	0	0	0	1988	1988	0	768	681	0
49	4181	0	0	0	0	0	0	0	0	3500	517	517	0	768	3664	0
50	3664	3500	3500	0	77,000	84,000	0	2	5400	0	1434	1434	0	768	2230	0

T	Q	SS
3	3500	1000

Model's Parameters Setup

Unit holding cost	0.075	USD/unit/month
Unit shortage cost	0.553	USD/unit
Unit overstorage cost	0.162	USD/unit/month
Warehouse capacity	5000	units
Weight (CW100)	22	kg/unit
20ft capacity	21,000	kgs
40ft capacity	42,000	kgs
20ft freight charge	1500	USD/trip
40ft freight charge	2700	USD/trip

Holding Cost	6492	USD/month
Shortage cost	474	USD/month
Over-storage cost	85	USD/month
Transportation cost	66,300	USD/month

Total cost	73,351	USD/month
Customer service level	98%	%CSL

Figure 4. Excel spreadsheet simulation.

5.2. Statistical Analysis

In the initial step, the statistical analysis using the full factorial design was performed to test the curvature of the response variables (TC and CSL) and independent variables (T, Q, and SS). Then, the data was analyzed in Minitab using stepwise, and the result shows curvature. The analysis of variance of factorial regression and the Pareto chart of the standardized effects are presented in Figures 5 and 6, respectively.

Factorial Regression: Total Cost versus T, Q, SS, CenterPt

```
Analysis of Variance

Source                   DF       Adj SS         Adj MS    F-Value  P-Value
Model                     7   19173065223     2739009318   24655.43    0.000
  Linear                  3   16051048270     5350349423   48161.63    0.000
    T                     1   15338631444    15338631444  138072.00    0.000
    Q                     1     701813796      701813796    6317.44    0.000
    SS                    1      10603030       10603030      95.44    0.002
  2-Way Interactions      3     706255141      235418380    2119.14    0.000
    T*Q                   1     698427462      698427462    6286.95    0.000
    T*SS                  1       5073995        5073995      45.67    0.007
    Q*SS                  1       2753684        2753684      24.79    0.016
  Curvature               1    2415761812     2415761812   21745.69    0.000
Error                     3        333275         111092
  Lack-of-Fit             1        127000         127000       1.23    0.383
  Pure Error              2        206275         103137
Total                    10   19173398497


Model Summary

      S      R-sq   R-sq(adj)   R-sq(pred)
333.304   100.00%     99.99%       99.96%
```

Figure 5. Analysis of variance of factorial regression.

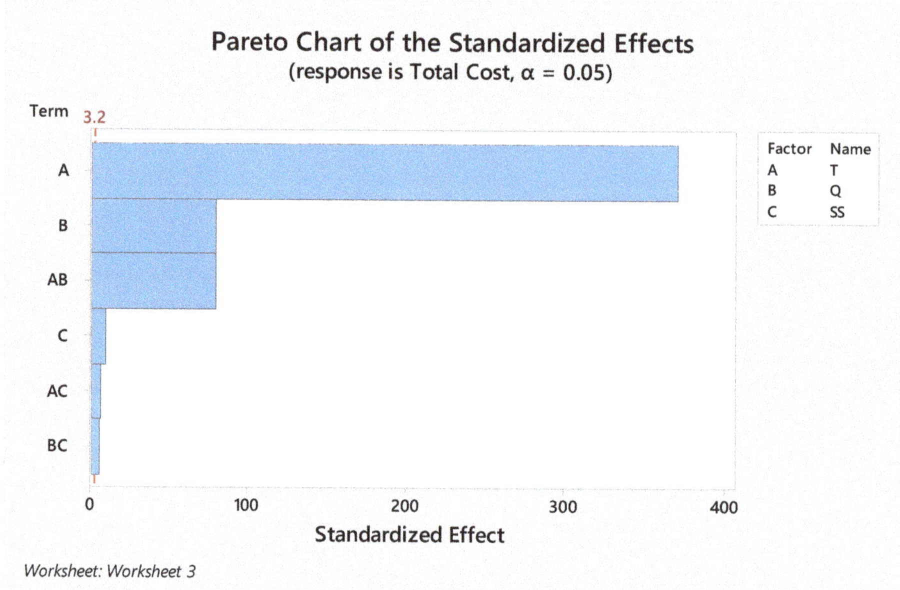

Figure 6. Pareto chart of the standardized effects.

In Figure 5, the analysis of variance of factorial regression, the results suggested that the model had significant curvature because the *p*-value (*p*-value = 0.000) was less than the confidence level at 95%. The presence of curvature usually indicates that the factor settings were near a feasible response value. Therefore, two response surface designs, central composite design, and Box–Behnken design were considered afterward to find the possible parameter settings for this case. In the step of full factorial design, for the starting center point, T = 3, Q = 3500, and SS = 1000, were considered. Figure 6 shows the Pareto chart of the standardized effects from the results and factors. It can be seen that their interaction significantly affected the model's results.

5.3. Central Composite Design (CCD)

In this section, a central composite design was performed to find the effects of independent variables on the response variables. In this paper, the CCD experiment (one continuous factor and two categorical factors) used the 20-run design with two blocks. Minitab provides the rotatability value (α) to ensure the design displays the desirable properties. The rotatability value of the CCD experiment was chosen based on factorial run and factorial point, $\alpha = 1.41$ was applied in this experimental design. The CCD experiment running setup and randomized design table are presented in Figure 7 and Table 5 below. Figure 7 shows all set-up parameters for the experimental CCD in the research. From the results of Table 5, TC and CSL were calculated based on independent variables, which T, Q, and SS. Note: StdOrder (standard order), RunOrder (run order), and PtType (center point type).

The initial experimental model was built for the response variables. Figure 8 shows the response surface regression of total cost versus T, Q, and SS. In the first running, the results suggest that the terms SS, the square terms SS*SS, and the interaction among Q*SS, Q*T, and SS*T were not significant (*p*-value > 0.05), and the paper considered the confidence level of the experimental design at 95%. In the next steps, these terms were eliminated from the CCD model.

Central Composite Design

```
Factors:            3      Replicates:        1
Base runs:         20      Total runs:       20
Base blocks:        2      Total blocks:      2

Two-level factorial: Full factorial

Cube points:                      8
Center points in cube:            2
Axial points:                     8
Center points in axial:           2

α: 1.41421
```

Figure 7. Central composite design running setup.

Table 5. Central composite design (one continuous factor and two categorical factors).

StdOrder	RunOrder	PtType	Blocks	T	Q	SS	TC	CSL
1	1	1	1	1	−1	−1	76,990.53	86
2	2	1	1	1	1	−1	78,115.85	93
3	3	1	1	1	−1	1	76,789.66	87
4	4	1	1	1	1	1	79,264.98	94
5	5	0	1	1	0	0	76,829.35	91
6	6	1	1	2	−1	−1	66,298.47	69
7	7	1	1	2	1	−1	64,918.34	78
8	8	1	1	2	−1	1	65,582.80	74
9	9	1	1	2	1	1	67,053.43	81
10	10	0	1	2	0	0	65,334.14	75
11	11	−1	2	1	−1.41	0	77,047.57	84
12	12	−1	2	1	1.41	0	96,797.76	94
13	13	−1	2	1	0	−1.41	76,971.17	90
14	14	−1	2	1	0	1.41	78,037.13	91
15	15	0	2	1	0	0	76,826.26	92
16	16	−1	2	2	−1.41	0	66,215.61	70
17	17	−1	2	2	1.41	0	74,732.52	81
18	18	−1	2	2	0	−1.41	65,732.97	73
19	19	−1	2	2	0	1.41	69,031.07	78
20	20	0	2	2	0	0	63,287.22	75

Note: calculated by the researchers.

Response Surface Regression: Total Cost versus T, Q, SS

```
Coded Coefficients

Term         Coef  SE Coef  T-Value  P-Value   VIF
Constant    70569     1834    38.49    0.000
Q            2729      917     2.98    0.014  1.00
SS            534      917     0.58    0.573  1.00
T            6274      820     7.65    0.000  1.00
Q*Q          3141     1213     2.59    0.027  1.23
SS*SS          13     1213     0.01    0.991  1.23
Q*SS          525     1297     0.40    0.694  1.00
Q*T          1212      917     1.32    0.216  1.00
SS*T         -227      917    -0.25    0.810  1.00

Regression Equation in Uncoded Units

T
3  Total Cost = 665705 - 344 Q - 28.6 SS + 0.0503 Q*Q + 0.0002 SS*SS + 0.0084 Q*SS

4  Total Cost = 685286 - 354 Q - 26.8 SS + 0.0503 Q*Q + 0.0002 SS*SS + 0.0084 Q*SS
```

Figure 8. CCD—response surface regression of total cost (TC) versus the period (T), order quantity (Q), and safety stock (SS).

After excluded these terms, the experiment was run once more. The results are shown in Figure 9, the reduced model. From the results, term T and the square term Q*Q were highly significant

(p-value < 0.05), while term Q was barely significant. Thus, term Q was kept for the next analysis to see the effects of these independent variables on the objective functions.

Response Surface Regression: Total Cost versus T, Q, SS (reduced model)

```
Coded Coefficients

Term         Coef  SE Coef  T-Value  P-Value   VIF
Constant    70585     1167    60.48    0.000
Q            2729      891     3.06    0.007  1.00
T            6274      797     7.87    0.000  1.00
Q*Q          3135     1065     2.94    0.010  1.00

Regression Equation in Uncoded Units

T
3   Total Cost = 653172 - 340 Q + 0.0502 Q*Q

4   Total Cost = 640623 - 340 Q + 0.0502 Q*Q
```

Figure 9. CCD—response surface regression of TC versus T, Q, and SS (reduced model).

For the second objective function, CSL (%), the CCD experiment was performed consequentially. The response surface regression of the CSL versus T, Q, and SS is presented in Figure 10. In the first running, the results display that the square terms SS*SS, and the interaction among Q*SS and Q*T was not significant (p-value > 0.05). Then, these terms were eliminated from the CCD model.

Response Surface Regression: CSL versus T, Q, SS

```
Coded Coefficients

Term        Coef  SE Coef  T-Value  P-Value   VIF
Constant  83.250    0.371   224.69    0.000
Q          3.731    0.185    20.14    0.000  1.00
SS         1.155    0.185     6.24    0.000  1.00
T          7.400    0.166    44.66    0.000  1.00
Q*Q       -0.469    0.245    -1.91    0.082  1.23
SS*SS     -0.094    0.245    -0.38    0.709  1.23
Q*SS      -0.250    0.262    -0.95    0.360  1.00
Q*T       -0.213    0.185    -1.15    0.274  1.00
SS*T      -0.729    0.185    -3.93    0.002  1.00

Regression Equation in Uncoded Units

T
3   CSL % = -67.7 + 0.0706 Q + 0.0187 SS - 0.000008 Q*Q - 0.000001 SS*SS - 0.000004 Q*SS

4   CSL % = -94.3 + 0.0723 Q + 0.0245 SS - 0.000008 Q*Q - 0.000001 SS*SS - 0.000004 Q*SS
```

Figure 10. CCD—response surface regression of the customer service level (CSL; %) versus T, Q, and SS.

After excluded these terms, the CCD model was run again, the reduced model is shown in Figure 11. The results proposed that only the square term Q*Q was not significant because of the p-value = 0.068 was out of the confidence level at 95%. Meanwhile, the terms Q, SS, T, and the interaction of SS*T dramatically affected the CCD model (p-value < 0.05).

Response Surface Regression: CSL versus T, Q, SS (reduced model)

```
Coded Coefficients

Term        Coef  SE Coef  T-Value  P-Value   VIF
Constant  83.143    0.237   350.58    0.000
Q          3.731    0.181    20.60    0.000  1.00
SS         1.155    0.181     6.38    0.000  1.00
T          7.400    0.162    45.68    0.000  1.00
Q*Q       -0.429    0.216    -1.98    0.068  1.00
SS*T      -0.729    0.181    -4.02    0.001  1.00

Regression Equation in Uncoded Units

T
3   CSL % = -47.4 + 0.0629 Q + 0.00171 SS - 0.000007 Q*Q

4   CSL % = -68.0 + 0.0629 Q + 0.00754 SS - 0.000007 Q*Q
```

Figure 11. CCD—response surface regression of CSL (%) versus T, Q, and SS (reduced model).

After reducing the CCD model into the final form, the residual plots for TC and CSL are displayed in Figure 12, including normal probability plot, histogram, versus fits, and versus order. The results show that the assumption of the model had no concern about the violation of the acceptable outliner. In detail, the normal probability plot nearly followed a straight line, then the residuals could assume from a normal distribution. The versus fits were used to verify that the residuals were scattered randomly about zero. From the plots, the non-constant variance of the residuals was fit. The histogram in the CCD model approximately followed a distribution. The versus order shows the residuals in the order of data collection. As can be seen, the versus order did not show any pattern, hence, there was no time correlation in the residuals.

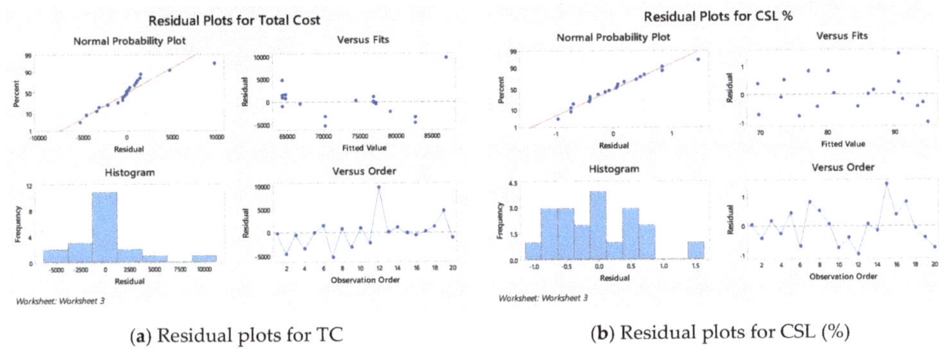

(**a**) Residual plots for TC
(**b**) Residual plots for CSL (%)

Figure 12. CCD—residual plots for TC and CSL.

Figure 13 shown the response optimizer of the CCD experiment to find the feasible response variables with CSL at 90% and 95%. From the result, the CSL set to target that was 90% with the possible minimum total cost 76,330 of the possible settings were T = 3, Q = 3428, and SS = 1336. Meanwhile, when the CSL was increased up to 95%, the possible minimum total cost was 77,090 of the optimal settings that were T = 3, Q = 3520, and SS = 1354.

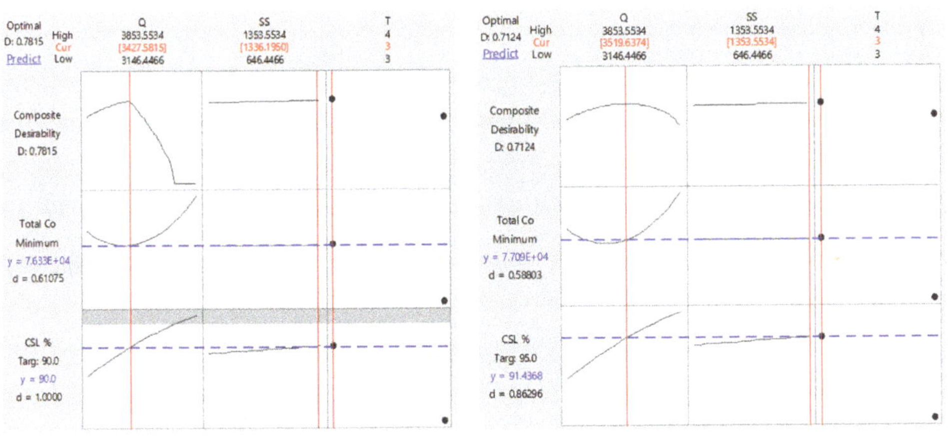

(**a**) Response optimizer for TC and CSL (90%)
(**b**) Response optimizer for TC and CSL (95%)

Figure 13. CCD—response optimizer for TC and CSL.

Besides, the surface and contour plots for CSL are displayed in Figure 14. According to the contour plot, the CSL located in an area of dark green color was more satisfying. The trade-off existed in the objective function of the model. The more the customer service level is satisfied, the more of the total cost will be paid by the distributor.

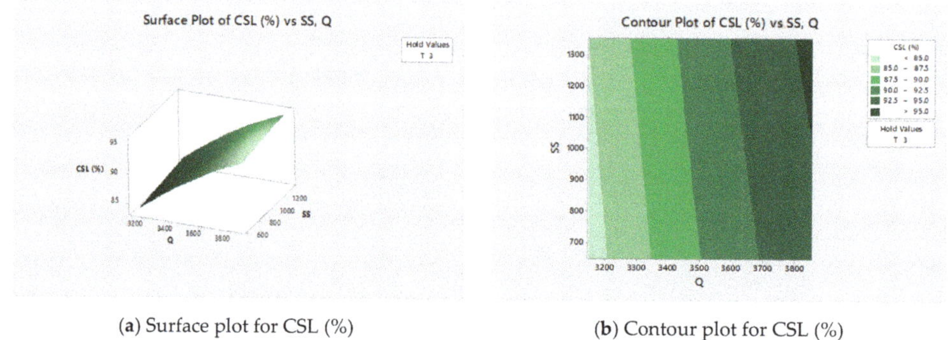

(**a**) Surface plot for CSL (%)   (**b**) Contour plot for CSL (%)

Figure 14. CCD—surface and contour plots for CSL (%).

5.4. Box–Behnken Design (BBD)

Following the results from the experiment of the central composite design, Box–Behnken design was performed to compare the effects of independent variables on the response variables. In this experiment, BBD with three continuous factors used the total 15-run design with one block. Compared to the central composite design, the BBD experiment did not have an embedded factorial design and extreme points, and the BBD experiment also did not have the rotatability value (α) in the experimental design. The advantage of the BBD model was that the design usually had fewer design points, hence, they were often less expensive to run than the CCD design with the same number of independent variables. The BBD experiment running setup and randomized design table is presented in Figure 15 and Table 6 as follows. Figure 15 shows all set-up parameters for the experimental BBD. From the results of Table 6, TC and CSL were calculated based on considered independent variables using an excel spreadsheet.

The initial model was obtained for the response variables (TC and CSL). Figure 16 shows the response surface regression of total cost versus T, Q, and SS. In the first running, the results suggest that the square terms Q*Q, SS*SS (p-values were 0.976 and 0.915 respectively), and the interaction among T*SS and Q*SS (p-values were 0.721 and 0.902, respectively) were not significant (all p-value > 0.05), while the paper considered the confidence level of the BBD at 95%. Then, these terms were eliminated from the BBD model to obtain the following final model.

After excluded insignificant terms, the experimental BBD was run again, the results of the reduced model are shown in Figure 17. The results display that term SS was statistically significant while the terms T, Q, square term T*T, and the interaction of T*Q significantly affected the model (p-value < 0.05). The authors decided to keep the term SS for further analysis to see the effects of these independent variables on the objective functions.

Box-Behnken Design

```
Factors:         3      Replicates:       1
Base runs:      15      Total runs:      15
Base blocks:     1      Total blocks:     1

Center points:   3
```

Figure 15. Box–Behnken design running setup.

Table 6. Box–Behnken design (three continuous factors).

StdOrder	RunOrder	PtType	Blocks	T	Q	SS	TC	CSL
1	1	2	1	2	3250	1000	134,828.00	100
2	2	2	1	4	3250	1000	65,946.84	71
3	3	2	1	2	3750	1000	171,568.82	100
4	4	2	1	4	3750	1000	64,635.88	79
5	5	2	1	2	3500	750	151,097.12	100
6	6	2	1	4	3500	750	65,616.60	73
7	7	2	1	2	3500	1250	154,401.55	100
8	8	2	1	4	3500	1250	64,982.51	77
9	9	2	1	3	3250	750	76,990.53	86
10	10	2	1	3	3750	750	78,115.85	93
11	11	2	1	3	3250	1250	76,789.66	87
12	12	2	1	3	3750	1250	79,264.98	94
13	13	0	1	3	3500	1000	76,826.26	92
14	14	0	1	3	3500	1000	80,511.79	100
15	15	0	1	3	3500	1000	77,198.46	87

Note: calculated by the researchers.

Response Surface Regression: Total Cost versus T, Q, SS

```
Coded Coefficients

Term         Coef  SE Coef  T-Value  P-Value   VIF
Constant    78179     3004    26.03    0.000
T          -43839     1839   -23.83    0.000  1.00
Q            4879     1839     2.65    0.045  1.00
SS            452     1839     0.25    0.816  1.00
T*T         31150     2708    11.50    0.000  1.01
Q*Q           -84     2708    -0.03    0.976  1.01
SS*SS        -305     2708    -0.11    0.915  1.01
T*Q         -9513     2601    -3.66    0.015  1.00
T*SS         -985     2601    -0.38    0.721  1.00
Q*SS          337     2601     0.13    0.902  1.00


Regression Equation in Uncoded Units

Total Cost = 6125 - 93620 T + 138 Q + 4 SS + 31150 T*T - 0.0013 Q*Q - 0.0049 SS*SS
 - 38.1 T*Q - 3.9 T*SS + 0.0054 Q*SS
```

Figure 16. BBD—response surface regression of TC versus T, Q, and SS.

Response Surface Regression: Total Cost versus T, Q, SS (reduced model)

```
Coded Coefficients

Term         Coef  SE Coef  T-Value  P-Value   VIF
Constant    77957     1491    52.29    0.000
T          -43839     1395   -31.43    0.000  1.00
Q            4879     1395     3.50    0.007  1.00
SS            452     1395     0.32    0.753  1.00
T*T         31178     2042    15.27    0.000  1.00
T*Q         -9513     1972    -4.82    0.001  1.00

Regression Equation in Uncoded Units

Total Cost = 20419 - 97725 T + 133.7 Q + 1.81 SS + 31178 T*T - 38.05 T*Q
```

Figure 17. BBD—response surface regression of TC versus T, Q, and SS (reduced model).

For the second objective function related to CSL (%), the BBD was performed consequentially. The response surface regression of the CSL versus T, Q, and SS is presented in Figure 18. In the first running, the results display that the term SS (p-value was 0.585) and the square terms Q*Q and SS*SS (p-values were 0.461 for both) were not significant (p-value > 0.05). In this case, the square terms Q*Q and SS*SS were eliminated from the BBD model in priority, the term SS was kept for further analysis purposes.

Response Surface Regression: CSL versus T, Q, SS

```
Coded Coefficients

Term          Coef  SE Coef  T-Value  P-Value   VIF
Constant     93.00     2.15    43.24    0.000
T           -12.50     1.32    -9.49    0.000  1.00
Q             2.75     1.32     2.09    0.070  1.00
SS            0.75     1.32     0.57    0.585  1.00
T*T          -4.00     1.94    -2.06    0.073  1.01
Q*Q          -1.50     1.94    -0.77    0.461  1.01
SS*SS        -1.50     1.94    -0.77    0.461  1.01

Regression Equation in Uncoded Units

CSL (%) = -265 + 11.5 T + 0.179 Q + 0.0510 SS - 4.00 T*T - 0.000024 Q*Q - 0.000024 SS*SS
```

Figure 18. BBD—response surface regression of CSL (%) versus T, Q, and SS.

After excluded these terms, the BBD experiment was run again, the reduced model is shown in Figure 19. The results show that only the term Q, SS, and the square term T*T (*p*-values were 0.054, 0.564, and 0.067, respectively) were slightly significant because the *p*-value was slightly out of the confidence level at 95%. Otherwise, only the terms T was highly significant to the BBD experiment (*p*-value < 0.05).

Response Surface Regression: CSL versus T, Q, SS (reduced model)

```
Coded Coefficients

Term          Coef  SE Coef  T-Value  P-Value   VIF
Constant     91.29     1.34    67.92    0.000
T           -12.50     1.26    -9.94    0.000  1.00
Q             2.75     1.26     2.19    0.054  1.00
SS            0.75     1.26     0.60    0.564  1.00
T*T          -3.79     1.84    -2.06    0.067  1.00

Regression Equation in Uncoded Units

CSL (%) = 53.2 + 10.2 T + 0.01100 Q + 0.00300 SS - 3.79 T*T
```

Figure 19. BBD—response surface regression of CSL (%) versus T, Q, and SS (reduced model).

After reducing the BBD model into the final form, the residual plots (normal probability plot, histogram, versus fits, and versus order plot) for the response variables (TC and CSL) are displayed in Figure 20. The results show that all residual plots satisfied the normality and constant variance assumptions, which means the simulated data was random and followed normal distribution.

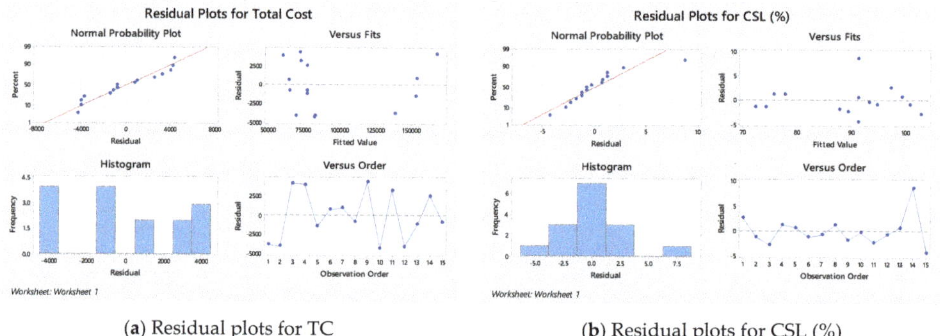

(**a**) Residual plots for TC

(**b**) Residual plots for CSL (%)

Figure 20. BBD—residual plots for TC and CSL.

Figure 21 shows the response optimizer of BBD to determine the feasible response variables, i.e., TC and CSL (CSL, 90%, 95%). From the result, the CSL set to target was 90% with the possible minimum total cost 70,870 of the possible settings were T = 3, Q = 3603, and SS = 1250. Otherwise, when the CSL was at 95%, the possible minimum total cost was 85,920 of the optimal settings were T = 3, Q = 3690, and SS = 1250.

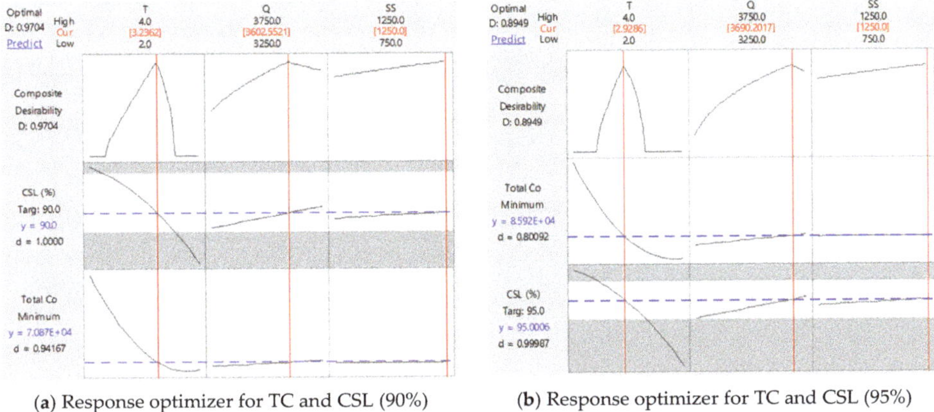

(**a**) Response optimizer for TC and CSL (90%)   (**b**) Response optimizer for TC and CSL (95%)

Figure 21. BBD—response optimizer for TC and CSL.

Moreover, the surface and contour plots for CSL are shown in Figure 22. The results suggest that the possible minimum total cost was in an area of light green (<74,000), the area of SS less than 1000, and the area of Q was less than 3500 according to the contour plot. The trade-off existed in the objective function of the model, which means, the more the customer service level is satisfied, the more of the total cost will be paid by the distributor.

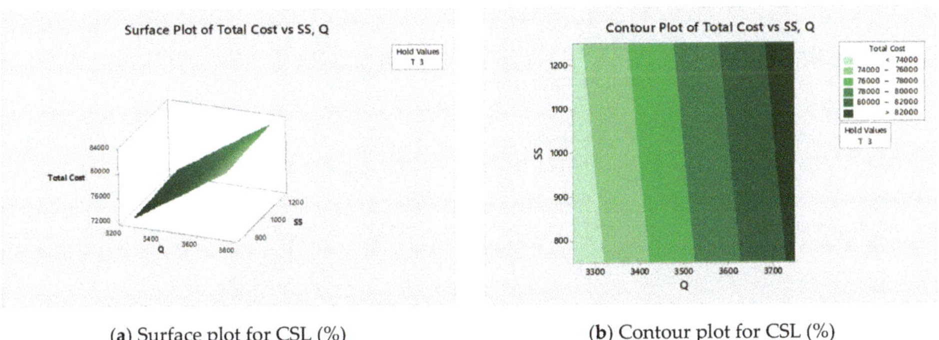

(**a**) Surface plot for CSL (%)   (**b**) Contour plot for CSL (%)

Figure 22. BBD—surface and contour plots for the total cost.

5.5. The Comparision of the CCD and BBD Experiment

This section performs the comparisons of the CCD and BBD experiments. The summary of the response optimizer results of CCD and BBD are displayed in Table 7 below. Next, the overlaid variance dispersion graph for CCD and BBD are drawn as in Figure 23. RSM was applied to build the prediction models, hence, prediction variance was significant. For more than two independent variables in the model, the variance dispersion graph (VDG) was one of the useful tools to validate the surface's results. VDG shows the value of minimum, maximum, and average of the scaled prediction variance (SPV) versus the distance of the design point from the center point [50].

Table 7. The summary of the response optimizer results of CCD and BBD.

	Central Composite Design (CCD)		Box-Behnken Design (BBD)	
	CSL (90%)	**CSL (95%)**	**CSL (90%)**	**CSL (95%)**
T	3	3	3	3
Q	3428	3520	3603	3690
SS	1336	1354	1250	1250
TC	76,330	77,090	70,870	85,920

Note: calculated by the researchers.

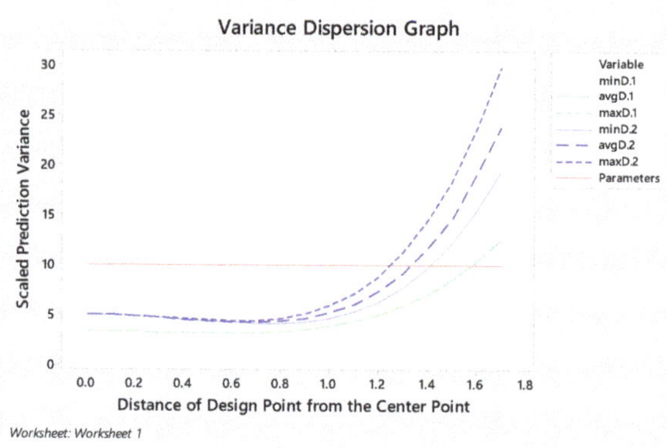

Figure 23. Variance dispersion graph for the CCD and BBD experiment.

According to the terms of the variance in Figure 23, the blue lines show the BBD experiment while the green lines display the CCD experiment. Both of these designs almost had significant differences in the total cost results with 2000 difference, which means, the distance of design point from the center point, and the scaled prediction variance (SPV) shows the CCD design had better SPV (smallest values) than the BBD design, therefore, CCD experiment can be considered as the best model in this paper. On the other hand, based on the results of the response optimizer, BBD design gave the smallest total cost. In other words, this paper provided useful insights for reference decision-makers in the related industry. Based on the objective of the industry, the managers or policymakers can consider the possible experimental designs that give the best performance toward the strategy of sustainable development.

6. Discussions and Conclusions

This paper proposed a computational framework for the order planning problem in a distributor. In the proposed model, values of response variables were simulated and calculated, and Minitab was used to determine the possible optimal setting for those responses using the central composite design and Box–Behnken design. In the case study, a combination of the review period (T), order quantity (Q), and safety stock (SS) was selected as independent variables of the model. Additionally, the total cost of the holding cost, shortage cost, over-storage cost, and transportation cost, and customer service level was selected as response variables. The objective of the paper was to minimize the total cost by planning on placing the right order quantities, in the right periods, with the possible safety stock in order to reach customer satisfaction around 90–95 percent using RSM.

For solving an order planning problem of Spectite CW100, the statistical result from the factorial design shows that the curvature had a significant effect on the responses, therefore, two response surface designs, central composite design, and Box–Behnken design, were considered afterward to find the possible parameter settings. Based on the results in the response optimizer, the following values for the response surface design were as follows:

1. Central composite design (CCD), customer service level CSL was at 90%, the minimum total cost was 76,330 with the possible optimal settings were the time period (T) = 3, order quantity (Q) = 3428, safety stock (SS) = 1336; (2) customer service level CSL was at 95%, the minimum total cost was 77,090 with the possible optimal settings were T = 3, Q = 3520, and SS = 1354.

2. Box–Behnken design (BBD), customer service level CSL was at 90%, the minimum total cost was 70,870 with the possible optimal settings were the time period (T) = 3, order quantity (Q) = 3603, and safety stock (SS) = 1250; (2) customer service level CSL was at 95%, the minimum total cost was 85,920 with the possible optimal settings were T = 3, Q = 3690, and SS = 1250.

According to the scaled prediction variance (SPV), the CCD design had better SPV (smallest values) than the BBD design, therefore, the CCD experiment could be considered as the best model in this paper. On the other hand, based on the results of the response optimizer, BBD design gave the smallest total cost. When the CSL increased from 90% to 95%, the total cost also increased in both designs. The proposed computational framework can be applied to other products of the distributor (Spectite WS, Spectite HP600, and Speccoat PE145), or any ordering process in businesses within the supply chain. In summary, our contributions include (1) the paper proposed a computational framework to solve the order planning problem by a heuristics approach using RSM, (2) the model was tested with a case study of a distributor to show the model's effectiveness, and (3) The insights implication of the model's results was to provide a useful reference for purchasing managers or decision-makers to obtain the most suitable order policy for a robust order planning process, which includes responses that were considered in monetary values. Besides, this could be used as a guideline for other related industries. However, there were some limitations to this paper. To get more robust solutions, future studies should consider considerable data on the products and address this problem under stochastic demand, lead time uncertainty, and unstable markets.

Author Contributions: Conceptualization, N.-A.-T.N. and C.-N.W.; Data curation, N.-A.-T.N.; Formal analysis, T.-T.D.; Funding acquisition, C.-N.W.; Investigation, N.-A.-T.N. and T.-T.D.; Methodology, C.-N.W. and T.-T.D.; Project administration, C.-N.W.; Software, N.-A.-T.N. and T.-T.D.; Validation, N.-A.-T.N. and C.-N.W.; Writing—original draft, N.-A.-T.N. and T.-T.D.; Writing—review and editing, C.-N.W. and T.-T.D. All authors have read and agreed to the published version of the manuscript.

Funding: This research was partly supported by the National Kaohsiung University of Science and Technology, and MOST 109-2622-E-992-026 from the Ministry of Sciences and Technology in Taiwan.

Acknowledgments: The authors appreciate the support from the National Kaohsiung University of Science and Technology, Ministry of Sciences and Technology in Taiwan.

Conflicts of Interest: The authors declare no conflict of interest.

References

1. Designing Building Wiki. Supply Chain Management in Construction. Available online: https://www.designingbuildings.co.uk/wiki/Supply_chain_management_in_construction#The_need_for_SCM_in_construction (accessed on 31 November 2020).

2. Right Place, Right Time: The Importance of Order-Slotting Solutions. Available online: https://blog.flexis.com/right-place-right-time-the-importance-of-order-slotting-solutions (accessed on 31 November 2020).

3. Cavalcante, C.A. Order planning policies for business-to-consumer e-tail stores. *Comput. Ind. Eng.* **2019**, *136*, 106–116.

4. Chien, C.F.; Lin, Y.S.; Lin, S.K. Deep reinforcement learning for selecting demand forecast models to empower Industry 3.5 and an empirical study for a semiconductor component distributor. *Int. J. Prod. Res.* **2020**, *58*, 2784–2804. [CrossRef]

5. Geunes, J.; Akçali, E.; Pardalos, P.M.; Romeijn, H.E.; Shen, Z.J.M. (Eds.) *Applications of Supply Chain Management and e-Commerce Research*; Springer Science & Business Media: Berlin, Germany, 2006; Volume 92.

6. Partovi, F.Y.; Anandarajan, M. Classifying inventory using an artificial neural network approach. *Comput. Ind. Eng.* **2002**, *41*, 389–404. [CrossRef]

7. Yao, X.; Huang, R.; Song, M.; Mishra, N. Pre-positioning inventory and service outsourcing of relief material supply chain. *Int. J. Prod. Res.* **2018**, *56*, 6859–6871. [CrossRef]
8. Diabat, A.; Abdallah, T.; Le, T. A hybrid tabu search based heuristic for the periodic distribution inventory problem with perishable goods. *Ann. Oper. Res.* **2016**, *242*, 373–398. [CrossRef]
9. Diabat, A.; Deskoores, R. A hybrid genetic algorithm based heuristic for an integrated supply chain problem. *J. Manuf. Syst.* **2016**, *38*, 172–180. [CrossRef]
10. Clark, A.J.; Scarf, H. Optimal policies for a multi-echelon inventory problem. *Manag. Sci.* **1960**, *6*, 475–490. [CrossRef]
11. Gharaei, A.; Pasandideh, S.H.R.; Khamseh, A.A. Inventory model in a four-echelon integrated supply chain: Modeling and optimization. *J. Model. Manag.* **2017**, *12*, 739–762. [CrossRef]
12. Hoseini Shekarabi, S.A.; Gharaei, A.; Karimi, M. Modelling and optimal lot-sizing of integrated multi-level multi-wholesaler supply chains under the shortage and limited warehouse space: Generalised outer approximation. *Int. J. Syst. Sci. Oper. Logist.* **2019**, *6*, 237–257. [CrossRef]
13. Ben-Daya, M.; Hariga, M. Integrated single vendor single buyer model with stochastic demand and variable lead time. *Int. J. Prod. Econ.* **2004**, *92*, 75–80. [CrossRef]
14. Jauhari, W.A.; Laksono, P.W. A joint economic lot-sizing problem with fuzzy demand, defective items and environmental impacts. *Mater. Sci. Eng.* **2017**, *273*, 12–18. [CrossRef]
15. Kao, C.; Hsu, W.K. A single-period inventory model with fuzzy demand. *Comput. Math. Appl.* **2002**, *43*, 841–848. [CrossRef]
16. Kim, G.; Wu, K.; Huang, E. Optimal inventory control in a multi-period newsvendor problem with non-stationary demand. *Adv. Eng. Inform.* **2015**, *29*, 139–145. [CrossRef]
17. Modak, N.M.; Panda, S.; Sana, S.S.; Basu, M. Corporate social responsibility, coordination and profit distribution in a dual-channel supply chain. *Pac. Sci. Rev.* **2014**, *16*, 235–249. [CrossRef]
18. Modak, N.M.; Panda, S.; Sana, S.S. Pricing policy and coordination for a two-layer supply chain of duopolistic retailers and socially responsible manufacturer. *Int. J. Logist. Res. Appl.* **2016**, *19*, 487–508. [CrossRef]
19. Modak, N.M.; Panda, S.; Sana, S.S. Pricing policy and coordination for a distribution channel with manufacturer suggested retail price. *Int. J. Syst. Sci. Oper. Logist.* **2016**, *3*, 92–101. [CrossRef]
20. Modak, N.M.; Panda, S.; Sana, S.S. Three-echelon supply chain coordination considering duopolistic retailers with perfect quality products. *Int. J. Prod. Econ.* **2016**, *182*, 564–578. [CrossRef]
21. Modak, N.M.; Panda, S.; Sana, S.S. Two-echelon supply chain coordination among manufacturer and duopolies retailers with recycling facility. *Int. J. Adv. Manuf. Technol.* **2016**, *87*, 1531–1546. [CrossRef]
22. Modak, N.M.; Modak, N.; Panda, S.; Sana, S.S. Analyzing structure of two-echelon closed-loop supply chain for pricing, quality and recycling management. *J. Clean. Prod.* **2018**, *171*, 512–528. [CrossRef]
23. Wang, C.-N.; Dang, T.-T.; Le, T.Q.; Kewcharoenwong, P. Transportation Optimization Models for Intermodal Networks with Fuzzy Node Capacity, Detour Factor, and Vehicle Utilization Constraints. *Mathematics* **2020**, *8*, 2109. [CrossRef]
24. Chilmon, B.; Tipi, N.S. Modelling and simulation considerations for an end-to-end supply chain system. *Comput. Ind. Eng.* **2020**, *150*, 106870. [CrossRef]
25. Tordecilla, R.D.; Juan, A.A.; Montoya-Torres, J.R.; Quintero-Araujo, C.L.; Panadero, J. Simulation-optimization methods for designing and assessing resilient supply chain networks under uncertainty scenarios: A review. *Simul. Model. Pract. Theory* **2020**, *106*, 102166. [CrossRef]
26. Shaban, A.; Shalaby, M.A. Modeling and optimizing of variance amplification in supply chain using response surface methodology. *Comput. Ind. Eng.* **2018**, *120*, 392–400. [CrossRef]
27. Giddings, A.P.; Bailey, T.G.; Moore, J.T. Optimality analysis of facility location problems using response surface methodology. *Int. J. Phys. Distrib. Logist. Manag.* **2001**, *31*, 38–52. [CrossRef]
28. Shang, J.S.; Li, S.; Tadikamalla, P. Operational design of a supply chain system using the Taguchi method, response surface methodology, simulation, and optimization. *Int. J. Prod. Res.* **2004**, *42*, 3823–3849. [CrossRef]
29. Buchholz, P.; Müller, D.; Thümmler, A. Optimization of process chain models with response surface methodology and the ProC/B toolset. In *Supply Chain Management und Logistik*; Physica-Verlag HD: Heidelberg, Germany, 2005; pp. 553–575.
30. Hassanzadeh, A.; Jafarian, A.; Amiri, M. Modeling and analysis of the causes of bullwhip effect in centralized and decentralized supply chain using response surface method. *Appl. Math. Model.* **2014**, *38*, 2353–2365. [CrossRef]

31. Devika, K.; Jafarian, A.; Hassanzadeh, A.; Khodaverdi, R. Optimizing of bullwhip effect and net stock amplification in three-echelon supply chains using evolutionary multi-objective metaheuristics. *Ann. Oper. Res.* **2016**, *242*, 457–487. [CrossRef]

32. Tang, L.; Yang, T.; Ma, Y.; Wang, J. Analysis of bullwhip effect and the robustness of supply chain using a hybrid Taguchi and dual response surface method. In Proceedings of the 2017 International Conference on Service Systems and Service Management, Dalian, China, 16–18 June 2017; pp. 1–6.

33. Zhang, T.; Zheng, Q.P.; Fang, Y.; Zhang, Y. Multi-level inventory matching and order planning under the hybrid Make-To-Order/Make-To-Stock production environment for steel plants via Particle Swarm Optimization. *Comput. Ind. Eng.* **2015**, *87*, 238–249. [CrossRef]

34. Guo, Z.X.; Yang, C.; Wang, W.; Yang, J. Harmony search-based multi-objective optimization model for multi-site order planning with multiple uncertainties and learning effects. *Comput. Ind. Eng.* **2015**, *83*, 74–90. [CrossRef]

35. Lim, L.L.; Alpan, G.; Penz, B. A simulation-optimization approach for sales and operations planning in build-to-order industries with distant sourcing: Focus on the automotive industry. *Comput. Ind. Eng.* **2017**, *112*, 469–482. [CrossRef]

36. Singha, K.; Buddhakulsomsiri, J.; Parthanadee, P. Mathematical model of inventory policy under limited storage space for continuous and periodic review policies with backlog and lost sales. *Math. Probl. Eng.* **2017**, *2017*, 1–9. [CrossRef]

37. Jansen, S.; Atan, Z.; Adan, I.; de Kok, T. Setting optimal planned leadtimes in configure-to-order assembly systems. *Eur. J. Oper. Res.* **2019**, *273*, 585–595. [CrossRef]

38. Kokuryo, D.; Yamashita, K.; Kaihara, T.; Fujii, N.; Umeda, T.; Izutsu, R. A proposed production decision method for order planning considering decision criteria of multiple organizations. *Procedia CIRP* **2020**, *93*, 933–937. [CrossRef]

39. Thinakaran, N.; Jayaprakas, J.; Elanchezhian, C. Survey on inventory model of EOQ & EPQ with partial backorder problems. *Mater. Today Proc.* **2019**, *16*, 629–635.

40. Tai, P.; Huyen, P.; Buddhakulsomsiri, J. A novel modeling approach for a capacitated (S,T) inventory system with backlog under stochastic discrete demand and lead time. *Int. J. Ind. Eng. Comput.* **2020**, *12*, 1–14. [CrossRef]

41. Hunsucker, J.L.; Shah, J.R. Performance of priority rules in a due date flow shop. *Omega* **1992**, *20*, 73–89. [CrossRef]

42. Wang, C.-N.; Dang, T.-T.; Nguyen, N.-A.-T. A computational model for determining levels of factors in inventory management using response surface methodology. *Mathematics* **2020**, *8*, 1210. [CrossRef]

43. Myers, R.H.; Montgomery, D.C.; Cook, C.M.A. *Response Surface Methodology: Process and Product Optimization Using Designed Experiments*, 3rd ed.; John Wiley & Sons: Hoboken, NJ, USA, 2009; Volume 39.

44. Khuri, A.I. Response surface methodology and its applications in agricultural and food sciences. *Int. J. Biom. Biostat.* **2017**, *5*, 1–11. [CrossRef]

45. Goh, T.N. The role of statistical design of experiments in six sigma: Perspectives of a practitioner. *Qual. Eng.* **2002**, *14*, 659–671. [CrossRef]

46. Shi, W.; Shang, J.; Liu, Z.; Zuo, X. Optimal design of the auto parts supply chain for JIT operations: Sequential bifurcation factor screening and multi-response surface methodology. *Eur. J. Oper. Res.* **2014**, *236*, 664–676. [CrossRef]

47. Widodo, E. A review of response surface methodology approach in supply chain management. In Proceedings of the 3rd Asia Pacific Conference on Research in Industrial and Systems Engineering, Depok, Indonesia, 16–17 June 2020; pp. 322–327.

48. Radasanu, A.C. Inventory management, service level and safety stock. *J. Public Adm. Financ. Law* **2016**, *9*, 145–153.

49. Coleman, B.J. Determining the correct service level target. *Prod. Inventory Manag. J.* **2000**, *41*, 19.

50. Montgomery, D.C. *Design and Analysis of Experiments*; John Wiley & Sons: Hoboken, NJ, USA, 2017.

Publisher's Note: MDPI stays neutral with regard to jurisdictional claims in published maps and institutional affiliations.

Article

Strategic Supply Chain Planning for Food Hubs in Central Colombia: An Approach for Sustainable Food Supply and Distribution

Gonzalo Mejía [1,*], Daniela Granados-Rivera [1], Jairo Alberto Jarrín [2], Alejandra Castellanos [1], Natalia Mayorquín [1] and Erika Molano [1]

[1] Sistemas Logísticos Research Group, Faculty of Engineering, Campus Universitario Puente del Común, Universidad de La Sabana, Km. 7 Autopista Norte de Bogotá, Chía 250001, Colombia; danielagrri@unisabana.edu.co (D.G.-R.); alejandracagu@unisabana.edu.co (A.C.); lizethmacu@unisabana.edu.co (N.M.); erikamobo@unisabana.edu.co (E.M.)

[2] Operations & Supply Chain Management Research Group, International School of Economics and Administrative Sciences, Campus Universitario Puente del Común, Universidad de La Sabana, Km. 7 Autopista Norte de Bogotá, Chía 250001, Colombia; jairojq@unisabana.edu.co

* Correspondence: gonzalo.mejia@unisabana.edu.co; Tel.: +57-1-861-5555 (ext. 25223)

Featured Application: This work considers a real-life problem posed by public institutions of Colombia responsible for the design and implementation of food supply master plans. As such, the results presented in this paper will likely have a direct impact in public policies that consider both ends of a fresh food supply chain: Farmers in the countryside and end consumers at the major demand centers.

Abstract: This paper investigates the problem of sustainable rural supply and urban distribution of fresh food products in central Colombia. Paradoxically, while farmers in the countryside suffer from poverty due to the low profitability of the agricultural activity, inhabitants at urban centers pay high prices for fresh and nutritious foods. In this work, we propose a supply chain system and a business model based on food hubs located on existing (and often abandoned) public facilities in the central region of Colombia. There are many examples in which the hub strategy has facilitated trade and logistics in supply chains. However, few studies consider the particularities of the presented case. We study a business strategy through a mathematical model which considers both the sustainable and efficient operation of the food hubs and better trading conditions for farmers. We propose a variant of the competitive hub location problem adapted to this case study. We tested the model under different scenarios such as changes in the attractiveness parameters, operation costs, and profit margins. The results suggest that if hubs are able to attract farmers, the model can be both sustainable for the hub concessionaires and for the farmers.

Keywords: competitive hub location problem; network design; food systems; rural development; mathematical programming

check for
updates

Citation: Mejía, G.; Granados-Rivera, D.; Jarrín, J.A.; Castellanos, A.; Mayorquín, N.; Molano, E. Strategic Supply Chain Planning for Food Hubs in Central Colombia: An Approach for Sustainable Food Supply and Distribution. *Appl. Sci.* **2021**, *11*, 1792. https://doi.org/10.3390/app11041792

Academic Editor: Farouk Yalaoui
Received: 31 December 2020
Accepted: 1 February 2021
Published: 18 February 2021

1. Introduction

Modern supply chains must not only be cost efficient and guarantee quick responses to their customers, but must also guarantee the alignment of the interests of its stakeholders and the equitable distribution in costs and margins. The location of facilities is a strategic decision that has a significant influence in supply chain performance that can create a competitive advantage in commercial supply chains and improve coverage in public supply chains. A good facility location strategy can also significantly decrease the transportation and emission costs, and mitigate its environmental impact.

This research addresses a real problem of a fresh food supply chain in rural areas of Central Colombia. This problem is currently under study by RAP-E-(Administrative Office

279

for Special Regional Planning—in Spanish), a region-wide public institution responsible for designing and implementing public policies for its food supply master plan. As such, this paper not only fed its models with real data, but also received valuable inputs from the senior RAP-E-management.

The fresh food supply chain studied in this paper is typical in emerging economies. In general, a large number of fragmented and disperse farmers supply fresh products to urban areas. These farmers rely heavily on intermediation that increases the prices paid by the final customers and create handling problems and food losses [1]. In central Colombia, this issue of intermediation has been long discussed, and several actions have been devised by local and central governments to reduce or eliminate its impact [2]. The most prominent was the construction of urban food hubs in Bogotá, the country capital [2]. This strategy was designed to directly connect farmers and shop owners with the support of an information system. The main idea was that shop owners placed orders directly to farmers through an Internet platform and a third-party logistics (3-PL) company were responsible for the hub operation and for the transportation of the agricultural products from the countryside to the city hubs. Important efforts were made by the mayor office of Bogotá to create networks of both shop owners and farmers and to educate them in the best commercialization and technology practices. Unfortunately, this strategy failed due to several reasons that included lack of Internet connectivity and poor logistics and business considerations [1]. No major 3-PL company participated in the concession bid, and the farmers' and shop owners' networks disbanded after a few attempts. A recent research [1] showed that the best strategy in terms of profits was intermediation.

More recently, the COVID 19 pandemic has seen a number of apps supported by Industry 4.0 technologies (e.g., [3,4]) which link farmers directly with both store owners and final consumers. As of today, these efforts have been modest at best and the vast majority of products is still traded through middlemen. There are several reasons, as pointed out in [5]: Middlemen play a significant role in this type of supply chain by matching supply and demand and shortening transaction times [6]. They also have a close relationship with farmers, as they, farmers, and middlemen, often live in the same region and share traditions and culture [5,7]. This fragmented farming system is also very inefficient in logistics and environmental terms: For example, an estimate of 12,000 trucks arrive daily to Bogotá with fresh food products. Most of these trucks are small and old 10-ton vehicles that travel half-loaded from locations within a 200 Km radius [1,8].

A common strategy to mitigate intermediation and to help small and mid- size farmers access larger markets is the development of regional food hubs. This strategy has been implemented in both first world [9,10] and in emerging countries [11,12]. The U.S. Department of Agriculture (USDA) defines a food hub as "a centrally located facility with a business management structure facilitating the aggregation, storage, processing, distribution, and/or marketing of locally/regionally produced food products" [13]. Hubs act both as supply consolidators and distributors. As such, hubs can also facilitate access to healthy foods to vulnerable populations by redirecting food flows to underserved areas that could hardly be reached out to with direct links [11]. In Colombia, the CONPES 3982—National Policy of Logistics of January 2020 document, establishes as one of its strategic pillars, the adoption of specialized logistics infrastructures—(ILEs in Spanish), which can play a prominent role as articulators between cargo and transport modes and providing value-added logistics services.

Food hub location problems have attracted researchers from different disciplines such as operations research [9], economics [14,15], and social sciences [16]. This paper focuses on the hub location problem adapted to the food systems in the central region of Colombia. Accordingly, we propose the following research questions:

- RQ1: "How rural food hubs in emerging economies can become a sustainable strategy to improve farmer's access to urban markets?"
- RQ2: "Where should a network of food hubs be located with the goal of both maximizing farmer coverage and minimize logistics costs?"

- RQ3: "How the transportation and emissions costs are impacted by the hub strategy?"

This paper addresses the above three questions. The main contributions of this work can be summarized as follows:

- A new mathematical model of competitive hub locations that incorporates constraints specific of the situation of farmers in emerging countries with provisions for a sustainable operation.
- A combination of heuristics with the mathematical model aimed at solving realistic size problems.
- An extensive analysis of scenarios with recommendations that can be used for public policies.

2. Context

The central region of Colombia encompasses 5 administrative departments: Boyacá, Cundinamarca, Meta, Tolima, and Huila, and the Bogotá Capital District as a special administrative unit. Table 1 shows the composition according to population (according to figures from the most recent population census in 2018, Colombia has over 48 M habitants [17]):

Table 1. The central region in figures.

Department	Population	% Colombia	% Region
Bogotá D.C	7,181,469	14.9	50.3
Boyacá	1,135,698	2.4	8.0
Cundinamarca	2,792,877	5.8	19.6
Meta	919,129	1.9	6.4
Tolima	1,228,763	2.5	8.6
Huila	1,009,548	2.1	7.1
Totals	14,267,484	29.6%	100%

Figure 1 illustrates the population distribution according to the areas of the region and provides recent figures of the agricultural production. Bogotá is the most densely populated municipality, with more than 200,000 habitants per square kilometer, followed by Cundinamarca with between 100.1 to 200 inhabitants per square kilometer, and, Boyacá, Meta, Tolima, and Huila with between 50.1 to 100 inhabitants per square kilometer.

The population of the Central Region of Colombia is estimated in approximately 2.1 million inhabitants (14% of the country), of which 75% are rural. The region concentrates 26% of the agricultural production of the country (agriculture, livestock, hunting, forestry, and fishing).

The central region has a wide variety of climates. The Andes cordillera that runs south-north splits the central region in three sub-regions: The hot and humid valleys in the west, the mountainous area in the central part, and the semi-arid plains in the east. The mountainous region itself presents a wide variety of ecosystems: Dense cloud forests, high plains, and semi-deserts to name a few.

Table 2. Hectares cultivated and production of the main products of the Central Region.

Department	Production (Ha)	% Colombia
Rice	185,688	15.9%
Corn	167,086	22.0%
Potato	121,632	4.4%
Sugar cane	66,074	16.1%
Plantain	52,025	9.0%

These climates create an excellent environment for agricultural activities. The central region of Colombia, in terms of the agricultural economic activity, contributes with approximately 30% to the GDP of Colombia. This contribution has remained relatively constant in the last 15 years. The central region produces 26% of Colombia's food. The main agricultural food products are potato, plantain, rice, sugar cane, and corn. Table 2 shows the total hectares planted and the agricultural production in tons in the central region. This information was compiled from [18].

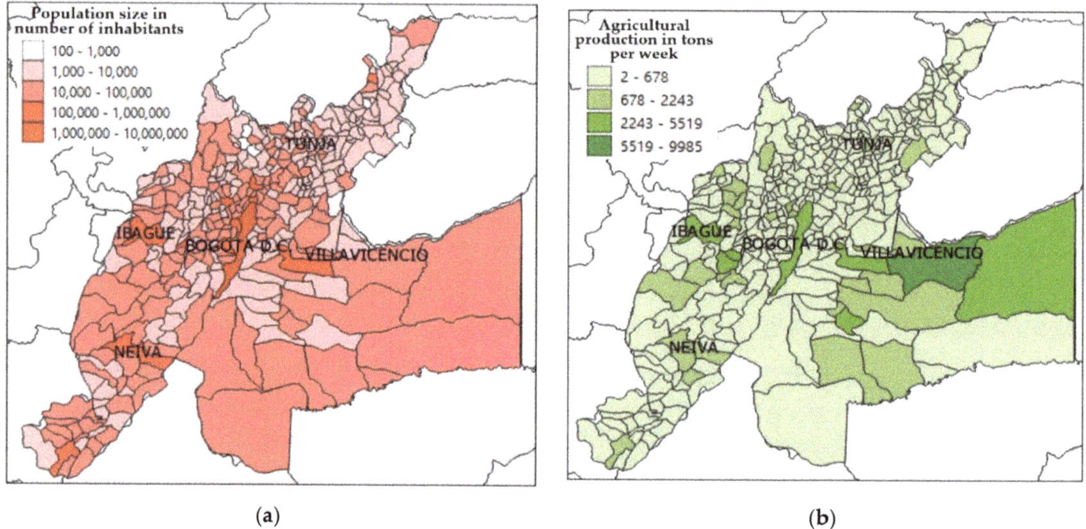

(a)   (b)

Figure 1. Characterization of the Central Region in terms of (**a**) population size and (**b**) production quantities of all five agricultural chains considered in this paper.

According to data from the latest available National Logistics Survey (2018) [19], the operating and logistics costs of the agricultural industry in Colombia account for 12.8% of the final price (below the national average which is 13.5%), where the main components of such operation costs are storage (35.9%), transportation (33.3%), overhead (27.7%), and customer service and other costs (3.1%).

A factor that contributes to the reduction of storage costs is to improve inventory turnover (in days), according to the same survey [19]. The agricultural industry reports 11.8 days of inventory in raw materials and agricultural supplies and 24.4 days in finished product, while supply days are 20.5, and in distribution, 16.7. Likewise, in terms of transportation costs, reducing the traveled distance can clearly contribute to the reduction of operating costs. In the Colombian agricultural industry, trucks and other cargo vehicles average 2744 km/month on intercity journeys and 2066 km/month on urban journeys. In relation to the quality of deliveries (deliveries on time, error-free documentation, and so forth), for the agricultural industry the performance rate is around 79.8%, which can be considered very low [19]. The main reported problems were damage to the goods (food loss), transport delays and customer-related delivery problems [19].

The above figures show that a food hub strategy specialized in the supply, storage, and distribution of agricultural products in the central region of Colombia, can be an opportunity to reduce (i) the number of intervening actors, (ii) the days of inventory, and (iii) the traveled distance.

3. Literature Review

In this section, we explore the mathematical foundations of hub location problems. This is a particular case of the facility location problem (FLP) that consists of both establishing the location of one or more facilities and the assignment of customers to such facilities [20]. Most problems studied in the literature derive from the classical *p*-medians, fixed charge and set covering problems [20]. The p-medians consist of establishing the location of exactly p facilities among a set of $n > p$ possible sites with the objective of minimizing the total weighted distance. The fixed charge cost is a natural extension of the p-medians in which the number of facilities is a decision variable itself. In this model, in addition to transportation costs, there are fixed charge costs associated with the opening/operation of the facilities. The set covering problem (SCP) differs from the above two in the sense that the objective is minimizing the number of facilities subject to the constraint that all customers must be served [21]. Facilities can only serve customers located within a predefined radius of influence (i.e., customers are covered). A different approach to coverage are the Huff's gravity [22] and the logistics regression (logit) [23] models. These models, instead of having the sharp coverage distinction of the SCP (i.e., covered or not), establish the probability that a customer patronizes a facility. This probability is related to some measure of attractiveness of the facility. The first papers considered only retail stores [22]. Later models included other services [24].

There are many variants of the FLP and it is not the purpose of this paper to cover them all. For a complete review, please see [25,26]. An extension of the FLP used in this paper is the competitive facility location problem in which the decision maker does not have control on the choices of the customers, but rather has estimations of their shopping behavior [27–29]. In general, these choices are modeled through either the Huff or the logit formulas. Most papers consider Mixed Integer Linear Programming (MILP) formulations and use either standard MILP solvers [30] or decomposition algorithms [31,32]. Other approaches use metaheuristics algorithms [33–35].

Facility location models are also common for fresh food and agriculture supply chain design [36,37]: This topic of has attracted many researchers that have addressed multiple problems [38,39]. Some examples related to emerging countries are: Villegas et al. [12] proposed a bi-level algorithm for the location of purchasing centers in the coffee belt region of Colombia. The two objectives considered were the cost minimization and the coverage maximization. Amorim et al. [40] presented a MILP model for supply chain design with JIT planning and scheduling with perishability constraints. Orjuela-Castro [8] proposed a MILP model for the location of fruit processing centers in the central region of Colombia. This model included humidity and temperature related constraints of the potential sites. Suraraksa and Shin [41] introduced a MILP model for the network design of urban distribution centers of fruits and vegetables in Bankok (Thailand). They investigated not only location, but also vehicle routing extensions. Later, Granillo [42] presented a real-life application model for the location of a catering distribution problem in urban schools in Mexico considering waiting times. In most of these papers, the objective function was the cost minimization including fixed, transportation, and other costs related to quality, and/or delivery. Several extensions in food supply chain combine facility location with combinatorial another problem: Examples are the location and routing [43,44] and the inventory location [45] problems.

A special case is the hub location problem: In this problem, entities travel between predefined origin and destination (OD) nodes. These entities can either travel directly between OD nodes or through a hub. These hubs may facilitate the flow of entities, and reduce transportation costs due to economies of scale [9]. Clearly, the hub location is not restricted to supply chains. Many applications exist in the telecommunication, airline, and urban transportation sectors [46,47]. The hub location involves several decisions: (1) Selecting a subset of hubs among a set of candidate locations, (2) determining inflows and outflows at the selected hubs, and (3) determine the transportation modes. A variant, closely related to this research, is the competitive hub location problem [24,46,48]. In the competitive hub location problem, a company intends to establish new hubs where there are already other competitors including the company's own hubs. The objective is generally to maximize profit, which involves two terms: (1) The company's revenue, which is proportional to the attracted demand, and (2) fixed and variable costs [46].

The mathematical problem of perishable food hubs has been little studied: Among the few works, Etemadnia et al. [9] proposed a MILP formulation for regional food hubs in the United States, which is solved in two phases with a linear relaxation in the first phase to filter our non-promising nodes and a second phase that included integer and binary variables. Later, Musavi et al. [49] presented a multi-objective perishable food hub location and scheduling problem for sustainable supply chains. Finally, Gharehyakheh et al. [50] presented a multi-objective formulation for the case of distribution of perishable foods considering emissions and shelf life. The last two references use Multi Objective Evolutionary Algorithms (MOEAs).

Most works on food hub and food supply chains consider decisions based on centralized control of all links: Suppliers, hubs, and retailers [44,49–51]. This is not realistic in the studied case: First, farmers in central Colombia have choices to sell and distribute their products. Second, in the literature of supply chains, the demand must be fully or partially satisfied [49,52]. In this paper, we assume that the hub management makes decisions as to where to deliver the products based on some parameter of attractiveness of the trading point (i.e., the city wholesaler) as in [26,27,46] Normally, this decision has to do with several factors such as familiarity with the trading point, or with the probability of finding customers. We make the reasonable assumption that the demand of each city is far larger than the quantities that the hubs can deliver. After all, these regional hubs in the central region of Colombia are facilities located in small-to-medium size towns with limited capacity (the largest with around 600 tons) and for instance, the total demand of the country capital Bogotá alone is over 40,000 tons per day.

In the problem studied in this paper, we have a combination of centralized and decentralized decisions. The purchasing center has control on the operation of the facilities and of the distribution to the cities (i.e., outbound logistics); however, since farmers are not contractually bound by the food hubs, one goal is to attract farmers and therefore the agricultural production in the most profitable and sustainable way. As such, farmers are customers of the food supply chain system and therefore a portion of their yield is expected to be captured by the food hubs.

Our work has both obvious similarities and important differences with the competitive hub location problem: On the similarities side, farmers may choose to travel to the food hubs or sell their products directly to middlemen, wholesalers, or to final customers as in [1,6]. The decision will be made based again on some measure of attractiveness [7]. The objective function (maximizing profit) presented in this paper is consistent with those of the hub facility location literature (e.g., [27,46,53]). In terms of differences, the proposed food hub model is more than a consolidation and storage facility where the product arrives and then it is re-distributed. In our proposal, the hub company would also purchase the product directly to the farmers (as suggested in [2]). Many successful cases in emerging countries use this business model (e.g., [11,12] to name a few) because of culture and traditions of the social network. Second, there is not a predefined destination for the fresh food products as this is a management decision as explained above. Finally, in our model, farmers do not bear the cost of traveling from the hub to the cities. This cost is borne by the hub concessionaire.

4. Materials and Methods

The problem studied in this paper can be summarized as follows: Given a set of famers, a set of candidate hub locations, and a set of customers (represented as cities or towns), locate the best subset of hubs in order to maximize profits. On one hand, the hubs' location must be attractive to farmers (i.e., close to them) and on the other hand, these hubs should also be located close to the demand points to reduce transportation costs. Attracting farmers also has the benefit of economies of scale on transportation and on fixed operation costs.

4.1. Data Preparation

In this section we provide the details of the proposed mathematical model. We study the five largest agricultural chains defined by RAP-E-. In this context, an agricultural chain is defined as a set of agricultural food products which share some special characteristics such as temperature requirements during handling/transportation and packing. As a result, products of the same agricultural chain can be transported in the same vehicle. Appendix A shows the description of the agricultural chains.

Table 3 shows the data used in the models and their sources. A description follows next:

Figure 2 shows the location of the 30 selected potential hub locations in the central region.

Table 3. Data collection for modeling the problem.

Entity	Data	Source
Farmers	Annual yield of 54 agricultural products grouped in 5 agricultural chains Approximate geographical location [1] Distances to logistics hubs	RAP-E-
Demand points	Annual intake of fresh foods per person-day Geographical location Population	RAP-E-, DANE [17]
Logistics operators	Capacity in tons Monthly cost of operation/opening Geographical location Distances to farmers/demand points Cost of transportation (fixed & variable) Emissions	RAP-E- Ministry of Transportation [54] Ministry of Environment [55] Corabastos (bulletin of prices) [56]

[1] Georeferenced location and area of the production zones.

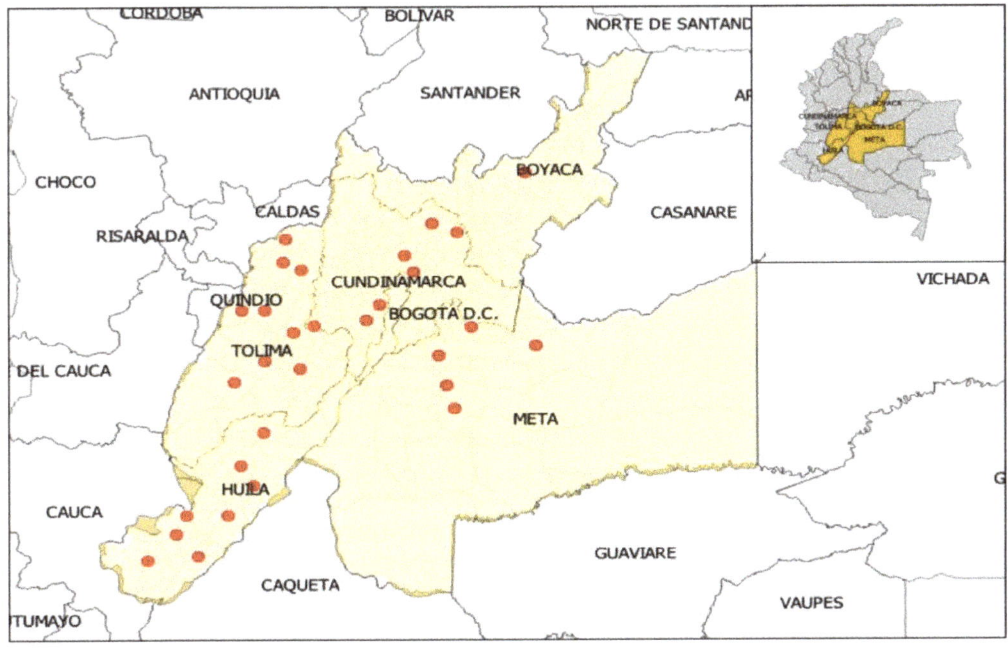

Figure 2. Location of 30 potential hubs within the central region's area of Colombia.

4.2. Model Formulation

We developed a model based on the Hub Location Competitive model formulated in [46]. Our model considers that farmers can choose a buyer not only among available hubs' potential locations, but among the demand points to sell their products. In this paper, the probability of a farmer choosing the products' destination is given by the Huffs gravity model that involves population size and distance as the main factors. The original Huff's formula establishes the probability that a customer selects a market. In this context, we made the reasonable assumptions that the main criteria for market selection are size (i.e., the probability of finding buyers is greater) and distance (i.e., the greater the distance, the more costly and the less familiarity with the market).

As such, and according to the Huff's formula, the probability $y_{ij'}$ that a seller i select a market j' is given by Equation (1):

$$y_{ij'} = \frac{\dfrac{w_{j'}^{\delta}}{d_{ij'}^{\gamma}}}{\sum_{j \in J} \dfrac{w_j^{\delta}}{d_{ij}^{\gamma}}} \qquad \forall h \in H; j \in J \tag{1}$$

where $d_{ij'}$ is the distance between seller i and market j', $w_{j'}$ is the size of market j' and γ and δ are exponents greater or equal to 0 to be determined.

In addition, we used this choice model to determine the quantities to be delivered by a hub to its potential customers. The same rationale can be applied to the hub management when choosing markets to sell their products. It means we use a double selection, one in each echelon of the supply chain. We regarded a set of production points (farmers' locations) $i \in I$ that have to choose between sending directly to a demand point $j \in J$ or use an available intermediary (set of hubs) $h \in H$. The selection is made with a probability y_{ih} that depends on the distance between the farmer and the hub d'_{ih} or the demand point $\hat{d}_{ij}$ and the population size of hub municipality $\hat{w}_h$ or demand point city w'_j. Both factors, distance and population, define corresponding sensitivity exponents γ and δ, respectively. With the probability variable y_{ih}, we defined the quantities of each product $k \in K$ to send to each available open hub. The other important aspect that we considered in this model is the selection of transportation vehicles. In the central Colombian case, agricultural products are transported in mainly three types of trucks: Trailer trucks (32 ton), 2-axle (16 ton) trucks, and "turbo" trucks (smaller 4.5 ton) trucks. According to the information of the Ministry of Transportation [57], most trucks travel around 50% loaded. For this reason, we set a minimum occupancy of 50% for the larger trucks in our model. Tables 4 and 5 show the definition of sets, parameters, and variables.

The objective function is to maximize the hubs' net profit considering a gross profit $margin_k$ for each product, fixed costs of operating a hub c_h, fixed $\hat{c}_v$ and variable c_v costs of transportation, and emission costs of carbon dioxide c'_v. The costs of transportation and emissions are associated with each type of vehicle $v \in V$. Equation (2) corresponds to the objective function formulation.

$$Max \sum_{k \in K} \sum_{h \in H} \sum_{j \in J} \sum_{v \in V} margin_k s'_{khjv} - \left(\sum_{h \in H} c_h x_h + \sum_{l \in L} \sum_{h \in H} \sum_{j \in J} \sum_{v \in V} \hat{c}_v truck_{vhjl} + \sum_{k \in K} \sum_{h \in H} \sum_{j \in J} \sum_{v \in V} c_v d_{hj} s'_{khjv} \right.$$
$$\left. + \sum_{l \in L} \sum_{h \in H} \sum_{j \in J} \sum_{v \in V} c'_v e_v d_{hj} truck_{vhjl} \right) \tag{2}$$

The constraints are defined as follows. Constraint set Equation (3) determines a farmer's probability to select a hub or a demand point according to the Huff's gravity model. Constraint set Equation (4) establishes the quantities of product k delivered by farmer i to hub h. Constraint set Equation (5) corresponds to the balance of quantities of product k traded by farmer i. Constraints set Equation (6) is the hub capacity constraint. Constraint set Equation (7) guarantees that the quantities delivered from hubs to the demand point are not greater than the quantities received by the hub. Constraint sets

Equations (8) and (9) ensure the minimum and maximum occupancy levels per vehicle type. Constraint set Equation (10) determines the quantities sent from hub h to each demand point j according to the gravity model's probability. Constraint sets Equations (11) and (12) refer to the non-negativity and binary constraints of the decision variables.

Table 4. Definition of indices and parameters.

Sets	
I	Farmers' locations.
H	Potential location of hubs.
J	Demand points' locations.
K	Products.
L	Agricultural chains' classification.
V	Type of vehicle.
Parameters	
c_h	Operation fixed cost of hub h.
q_h	Capacity of hub h.
$margin_k$	Unit profit margin of product k.
$\hat{c}_v$	Operation fixed cost for each vehicle type v.
c_v	Variable cost per kilometer and ton of each vehicle type v.
e_v	Average carbon dioxide emission in tons per kilometer for vehicle v.
c'_v	Unit cost of carbon dioxide emission per kilometer by vehicle v.
q_v	Capacity of vehicle v.
min_v	Minimum occupancy of vehicle v (%).
$\hat{w}_h$	Population of hubs' location h.
w'_j	Population of demand point j.
d'_{ih}	Distance between farmer i and hub h.
$\hat{d}_{ij}$	Distance between farmer i and demand point j.
d_{hj}	Distance between hub h and demand point j.
o_{ki}	Weekly quantities of product k by sold by farmer i.
a_{kl}	Binary parameter that establishes if product k belongs to agricultural chain l.
γ	Sensitivity exponent for the distance between farmers and hubs.
$\hat{\gamma}$	Sensitivity exponent for the distance between hubs and cities.
δ	Sensitivity exponent for population factor.
p_{hj}	Probability of hub h of selecting demand point j.

Table 5. Definition of decision variables.

Decision Variables	
x_h	Binary variable if hub location h is open
$\hat{s}_{ki}$	Weekly quantities (tons) of product k sent from farmer i to demand points
y_{ih}	Probability of farmer i of choosing to go to hub h
s_{kih}	Weekly quantities (tons) of product k sent from farmer i to hub h
s'_{khjv}	Weekly quantities (tons) of product k sent from hub h to demand point j in vehicle v
$truck_{vhjl}$	Number of vehicles v with products of agricultural chain l sent from hub h to demand point j

$$y_{ih} = \frac{\frac{\hat{w}_h^{\delta}}{d'^{\gamma}_{ih}} x_h}{\sum_{h' \in H} \frac{\hat{w}_{h'}^{\delta}}{d'^{\gamma}_{ih'}} x_{h'} + \sum_{j \in J} \frac{w'^{\delta}_j}{\hat{d}^{\hat{\gamma}}_{ij}}} \qquad \forall i \in I; h \in H \tag{3}$$

$$o_{ki} y_{ih} = s_{kih} \qquad \forall k \in K; i \in I; h \in H \tag{4}$$

$$\sum_{h \in H} s_{kih} + \hat{s}_{ki} = o_{ki} \qquad \forall i \in I; k \in K \tag{5}$$

$$\sum_{k \in K} \sum_{i \in I} s_{kih} \leq q_h x_h \qquad \forall h \in H \tag{6}$$

$$\sum_{i \in I} s_{kih} \geq \sum_{j \in J} \sum_{v \in V} s'_{khjv} \qquad \forall k \in K; h \in H \tag{7}$$

$$\sum_{k \in K} a_{kl} s'_{khjv} \leq q'_v truck_{vhjl} \quad \forall v \in V; h \in H; l \in L; j \in J \tag{8}$$

$$\sum_{k \in K} a_{kl} s'_{khjv} \geq q'_v \left(truck_{vhjl} - (1 + min_v) \right) \quad \forall v \in V; h \in H; l \in L; j \in J \tag{9}$$

$$\sum_{v \in V} s'_{khjv} \leq p_{hj} \sum_{i \in I} s_{kih} \quad \forall k \in K; h \in H; j \in J \tag{10}$$

$$x_h \in \{0, 1\} \quad \forall h \in H \tag{11}$$

$$truck_{vhjl}, \hat{s}_{ki}, y_{ih}, s_{kih}, s'_{khjv} \geq 0 \quad \forall v \in V; h \in H; l \in L; j \in J; k \in K \tag{12}$$

As Equation (3) is not linear, we adapted the linear reformulation proposed in [23] for multinomial logit choice probabilities to the gravity model. In this work, we set the parameter φ_{ih} shown in Equation (13).

$$\varphi_{ih} = \frac{\dfrac{w_h^{\delta}}{d'^{\gamma}_{ih}}}{\sum_{j \in J} \dfrac{w'^{\delta}_j}{d'^{\gamma}_{ij}}} \quad \forall i \in I; h \in H \tag{13}$$

The above linear reformulation introduces a non-negative variable $\overline{y}_i$, which is the accumulative choice probability of farmer $i \in I$. This reformulation adds the constraints sets defined in Equations (14)–(16) to our model, and replaces Equation (3).

$$\overline{y}_i + \sum_{h \in H} y_{ih} \leq 1 \quad \forall i \in I \tag{14}$$

$$y_{ih} - \frac{\varphi_{ih} x_h}{1 + \varphi_{ih}} \leq 0 \quad \forall i \in I; h \in H \tag{15}$$

$$y_{ih} - \varphi_{ih} \overline{y}_i \leq 0 \quad \forall i \in I; h \in H \tag{16}$$

Notice that the value of the parameter $\widehat{p_{hj}}$ for a given hub is independent of the other open hubs since hubs are not competing among each other to supply cities.

5. Results

5.1. Description of Computer Tests

This section presents the experimental results of our tests. We coded our model in GAMS™ and ran the experiments using the CPLEX ™ solver on a computer with an Intel® Core ™ i7-6500U chipset, 8 Mb RAM memory, and 2.5 GHz microprocessor. Due to the huge size of the model (98 hub locations, 54 products, 342 farming locations, 5 chains, and 3 truck types), the model could not run in a reasonable time (after 5 h, no feasible solution was provided). Therefore, we implemented the classical "add" and "drop" heuristics for facility location [25], and according to the results, we short-listed the number of potential candidate hubs to 30 locations. First, we dropped those facilities that were not suitable for fresh food trading (e.g., slaughtering houses, or very small facilities). Next, we ran 24 instances with combinations of values of δ, γ and fixed costs and filtered out those locations that were never open or that handled minimum quantities. We also kept those facilities that were always present in the solution open. The selected market cities were (i) those with over 500,000 inhabitants for cities outside the central region, and (ii) the provincial capitals of the central region regardless of their population. A similar approach to handle real-life models was proposed in [9]. Figure 3 shows the location of the selected market cities. Still, the number of possible combinations was large, but could be handled in a reasonable time. We set a time limit of 3600 s for each run.

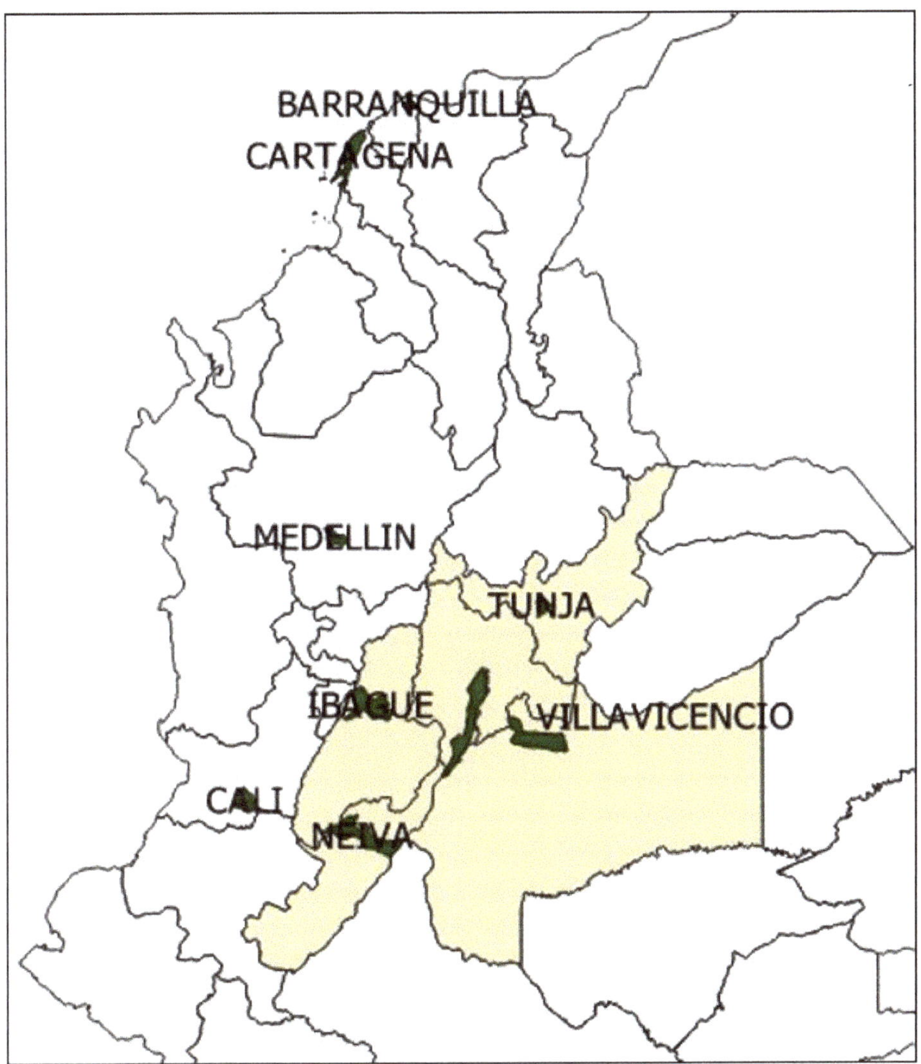

Figure 3. Selected demand cities for distributing the agricultural supply of the three agricultural chains.

The parameter calibration consisted of determining the values of the exponents of the gravity model. The RAP-E-officers estimated that 65% of the total fresh agricultural food supply is commercialized outside the central region. With this in mind, we ran the model in the situation of no open hubs, with all production traded through middlemen located at the production sites. As a result, we set the values of δ (population exponent) and $\hat{\gamma}$ (exponent of distance from hubs to cities) to 1.25 and 0.1, respectively.

Preliminary tests showed that the factor that affected most of the decisions and, hence the profitability of the model, was the ability to capture "market share" from farmers. Since, at the present pandemic time, it was not possible to conduct surveys to farmers to have accurate information about their choices, we varied the exponent γ in the Huff formula of Equation (3). We did some preliminary validations with officers of RAP-E- in which essentially, they told us that the majority of farmers would not travel long distances to sell their products, and most likely they would only trade at the nearest town. We also corroborated the exponent values with those proposed in [58]. In the experiments, we varied this exponent and observed its effects on the objective function. The other uncertain parameter is the fixed operation hub costs. As said above, we established this value based on estimates made for the Food Security and Supply Master Plan of Bogotá (PMASAB in Spanish). However, as these estimates may change, we varied the original estimates in ±20%.

We carried out a sensitivity analysis in which we varied the hub capacities and the gross profit margins. Although, the RAP-E- officers said that there were no plans for new infrastructure, the increase in capacity can be seen as adding days of operation. In this case, the fixed operation costs were adjusted in the same proportion. We did not vary the transportation costs for the following reasons: First, the sources we used were reliable, and second, the major components of such costs are unlikely to vary: Salaries, taxes, and insurance in the term of fixed transportation cost and gas and tolls in the term of variable transportation cost. Last, we varied the gross margins, and the original figures were reduced in 10%, 25%, and 50%. The idea is that reduced margins for hubs may eventually result in better prices paid to the farmers. Table 6 shows the parameter values used in the experiments.

Table 6. Combination of parameter values. The letter x is referred to the initial value of the parameter to vary.

Parameter Values	
γ (distance exponent (farmers to hubs)	0.1, 0.5, 1.0, 2.0, 3.0
fixed_cost_factor	Changes in fixed operation costs: $0.8\times, 1.0\times, 1.2\times$
hub_capacity_factor	Increase in hub capacity: $1.5\times, 2\times, 3\times$ *

* denotes 3 times the current capacity of all hubs.

In this paper, we assume that farmers are paid the same at both hubs and demand points, and therefore, the attractiveness of a hub/demand point does not depend on prices. This assumption may not be true in most cases, but that will create a different decision problem that will (i) add too much complexity to our model, and (ii) be very difficult to calibrate due to reliable information. In the experiments in which we varied gross margin values, the purpose was to investigate how the profitability of the hub can be transferred to the farmers. The topic of price related attractiveness will be the subject of future research.

5.2. Baseline Scenario

We performed several preliminary tests, we set all the exponents of the Huff formula with values close to 0.0. In this case, no hubs were open in the optimal solution. This implies that hubs would not be able to attract farmers and that the business would not be profitable. We propose a baseline scenario which, based on our validations with the RAP-E-officers, is the most likely to occur. In this scenario, we considered a distance exponent γ of 2.0 and 0% of variation in hubs' costs.

Figure 4 shows the hub locations and their links with the farmers in this baseline. The solid lines are links that account for more than two (2) weekly tons of delivered products.

Figure 5a shows the geographical location of the 23 open hubs and the total tons supplied in the baseline scenario. We can observe that the larger supplied quantities are concentrated to the west of the central region where more and larger hubs are open. In this part of the central region, most hubs operate at full capacity. This area concentrates the larger portion of total agricultural production (see Figure 1b). Additionally, the population of the cities in this area, indicated in Figure 1a, and of its neighbors make the area attractive for the food hub business. The largest open hubs are located in the provincial capital cities of Villavicencio (east in the Meta department) and Ibagué (west in the Tolima department) and in the mid-size city of Duitama (Boyacá, north). The first two hubs would be supplied by the rich agricultural regions of the neighbor lowlands whereas the Duitama hub would be supplied by many small farmers of the Andean northern region. The other larger open hubs would be located around the high plain areas of Cundinamarca, the valleys of Huila in the south, and the fertile valleys of the Magdalena river in Tolima.

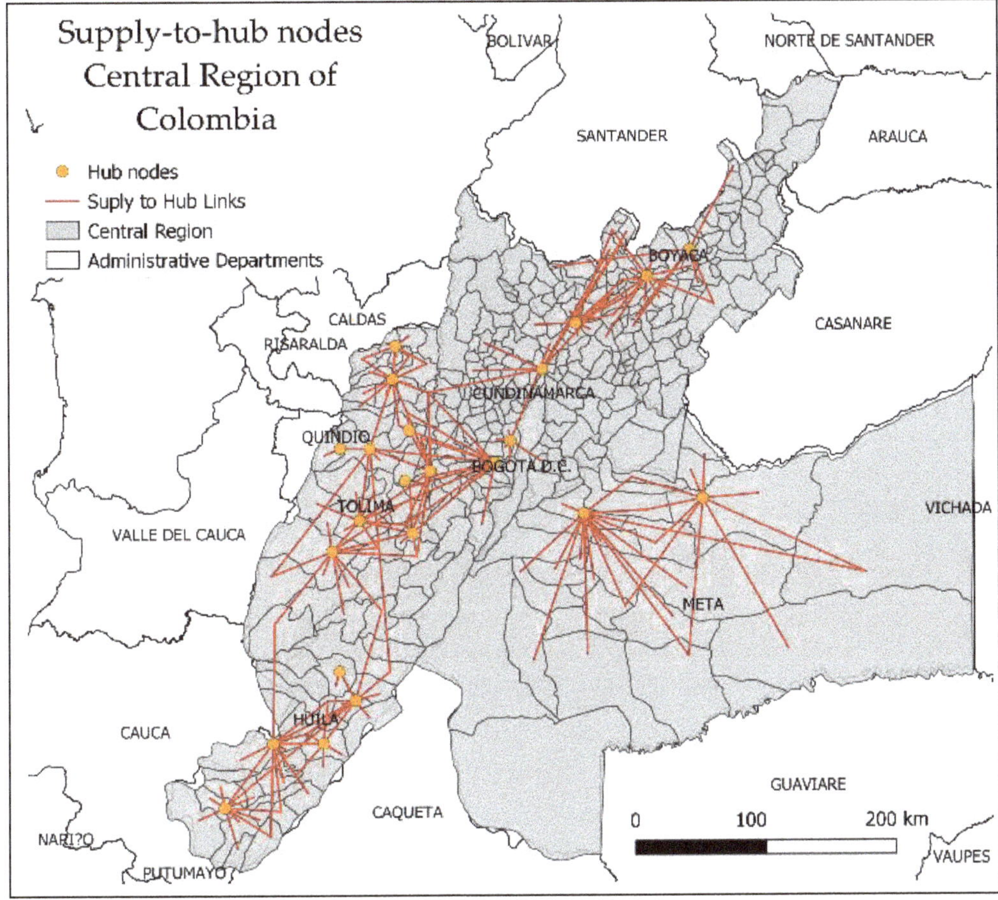

Figure 4. Supply-to-hubs main links.

The participation of the agricultural chains at each hub is shown in Figure 5b. We observe that these results are consistent with the agricultural vocation of each area of the region: Hot and humid flat areas of the western and eastern part of the central region are ideal for rice and other grains, whereas potato, some fruits, and vegetables are mostly cultivated in the mountainous areas of the central region.

We also analyze the quantities delivered to the main cities. Figure 6a shows the quantities from hub-to-demand. As expected, the city that receives the largest quantities is Bogotá, the country capital (about 2800 weekly tons). Considering the hubs in the western part of the central region, we can observe that the proximity to Bogotá clearly facilitates the business activity. Bogotá is by far the largest urban center of the country that also hosts the largest wholesaler (Corabastos) in which, as of today, over 12,000 tons of food are traded daily [1]. Following Bogotá, the other two larger demand points are Medellín (northwest of the region in the Antioquia department) and Cali (west of the region in the Valle del Cauca department). These two cities receive around 400 weekly tons. These cities are out of the Central Region, but their size, population, and proximity with several hubs near the borders of the central region, are a reason for their business. The other smaller cities (Ibagué, Neiva, Villavicencio, Tunja, Cartagena, and Barranquilla) commercialize between 5 and 65 weekly tons. Although, the area around Ibagué in Tolima (west) is a large producer of foods, a large portion of them is delivered to other big cities. In general, Figure 6b indicates that the most traded products are grains, fruits and vegetables (F&V), and "other products". Potato and plantain chains are smaller chains because they only contain two or three products. Notice that production quantities of these two chains are virtually not delivered to the northern cities of Cartagena and Barranquilla (circles in the upper part of Figure 6b). The cost and profits analysis follows next.

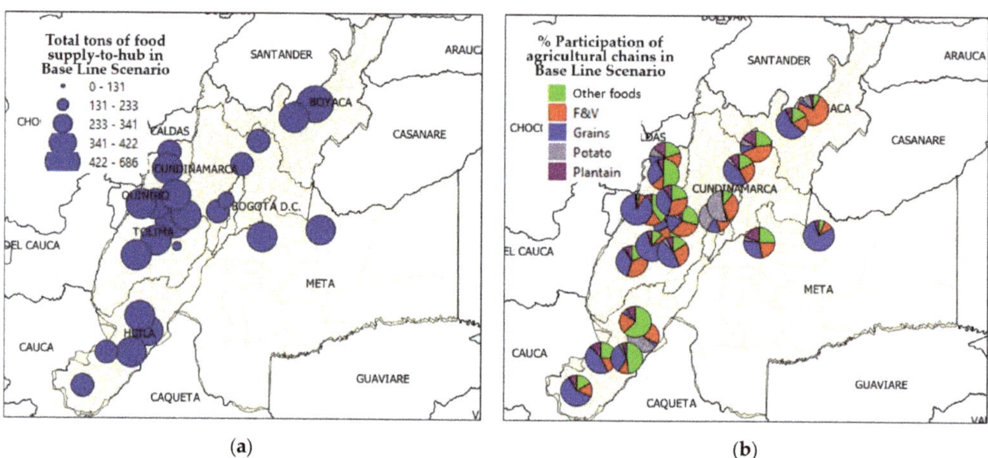

Figure 5. (**a**) Total tons of food supplied to the food hubs. (**b**) Participation of the agricultural chains in the baseline scenario.

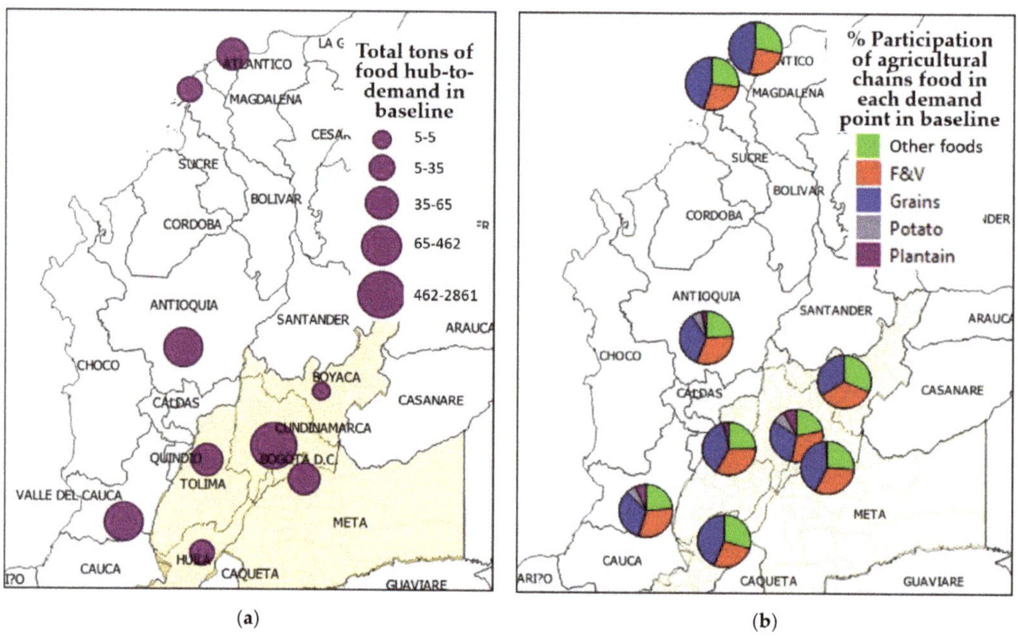

Figure 6. (**a**) Total tons delivered to each point of demand. (**b**) Distribution of delivered products by supply chains.

We also compare the baseline in relation of the current situation of no hubs in which products are delivered directly to the cities. We can observe that all transportation and emission costs decrease with the hubs' network. This is achieved mainly due to reductions in the vehicles' operation fixed costs and emissions' costs (up to 95% reduction). This cost savings can be explained as fewer vehicles are required when cargo is consolidated at the hubs. In the scenario of no hubs, the number of vehicles used is approximately 8000 per week, considering that each farmer (or middlemen) travels to the cities in small trucks. With the hubs' network, the number of vehicles required to transport food is around 800, which represents a 90% reduction. The additional benefit is that fewer vehicles also reduce the total emissions of CO_2, improving the carbon print. Figure 7 presents this comparison.

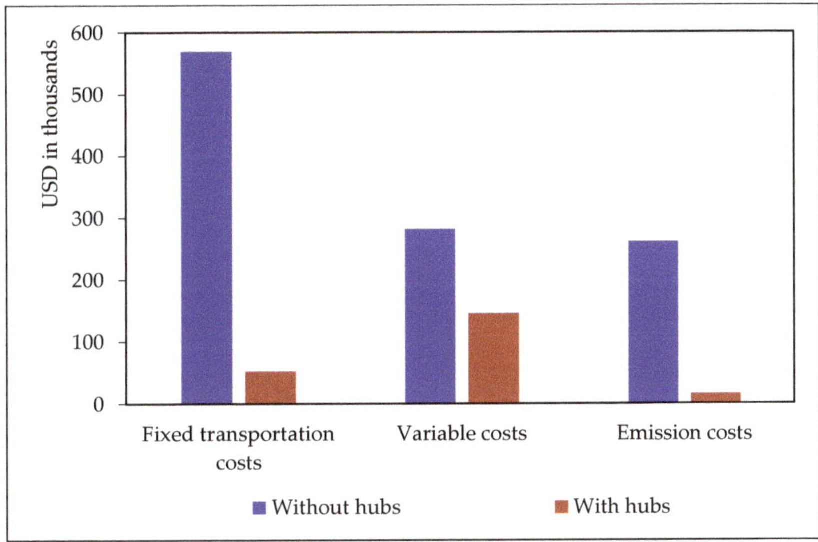

Figure 7. Transportation costs' variation in the baseline scenario with and without hubs.

5.3. Variation in the Exponent γ and in Hubs' Costs

Table 7 shows the results of scenarios presented in Table 6. Analyzing the results, we can observe that with low values of the sensitivity distance exponent γ, fewer hubs are open as smaller quantities are delivered from farmers to hubs; the opposite is true for larger values of γ. The differences are important: For $\gamma = 0.1$ (low distance attractiveness), the number of open hubs ranged between 12 and 16, whereas for $\gamma = 3.0$, the number of hubs ranged between 24 and 25.

Figure 8 illustrates the effects on the gross profit and the total cost of these parameter combinations. We can notice that the variation in hubs' costs represents a less significant change in the number of open hubs and on the traded quantities in comparison with changes in the exponent γ. This behavior corroborates the results shown in Table 7. With values of γ less between 0.0 and 1.0, the variations are small; however, if the exponent γ is greater than 1.0, the gross profits rise steeply. On the other hand, Figure 8 also shows that net profit in all cases is greater than 0. This means that the hubs' network might be a cost-effective business for concessionaires.

Table 7. Summary of the results in 14 scenarios with variation on the baseline scenario.

Scenario	γ	Hub Fixed Costs Variation	Average Quantities Handled by Hubs (Tons)	Total Quantities (Weekly Tons)	Number of Hubs	Net Profit
Baseline	2.0	0%	77	8866	23	951
1	0.1	−20%	58	4601	16	413
2	0.1	0%	60	4486	15	413
3	0.1	+20%	69	4157	12	372
4	0.5	−20%	54	4866	18	440
5	0.5	0%	58	4661	16	435
6	0.5	+20%	67	4328	13	389
7	1	−20%	54	5446	20	512
8	1	0%	58	5237	18	506
9	1	+20%	65	4519	14	379
10	2	−20%	77	8866	23	982
11	2	+20%	79	8695	22	895
12	3	−20%	84	10,100	24	1293
13	3	0%	81	10,089	25	1201
14	3	+20%	81	10,089	24	1204

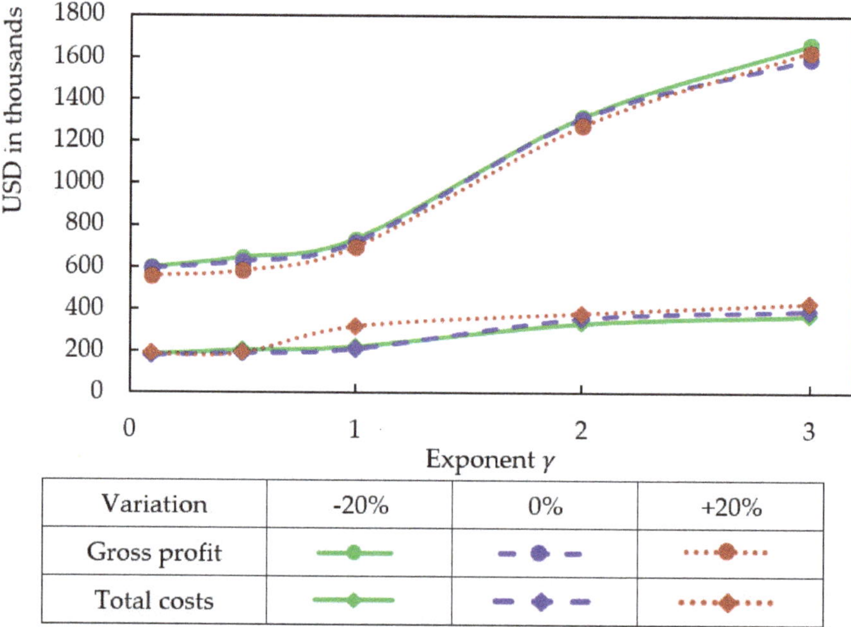

Figure 8. Gross profits and total costs with changes on the exponent γ and hubs' costs ±20% (the number at the right in the conventions represents this last variation).

For the purpose of comparison, we selected a scenario with low sensitivity to distance $(\gamma) = 0.5$ and 0% fixed cost variation. This is scenario 5 of Table 7. The results are illustrated in Figure 9a. The solution shows that 16 hubs are open, and that the pattern of hub locations is similar to the baseline scenario. However, in most cases, the traded quantities are smaller, as fewer farmers are attracted in this scenario. Only in the case of the Duitama hub (the large circle in the Boyacá department), the quantities handled are similar in comparison with the baseline. The explanation is that the strategic location and the lack of competitors of this hub make it the only trading option in this region for the small and medium size farmers of northern Boyacá. In the department of Huila in this scenario, only two (as opposed to five in the baseline) hubs are open, and the corresponding trading quantities are smaller.

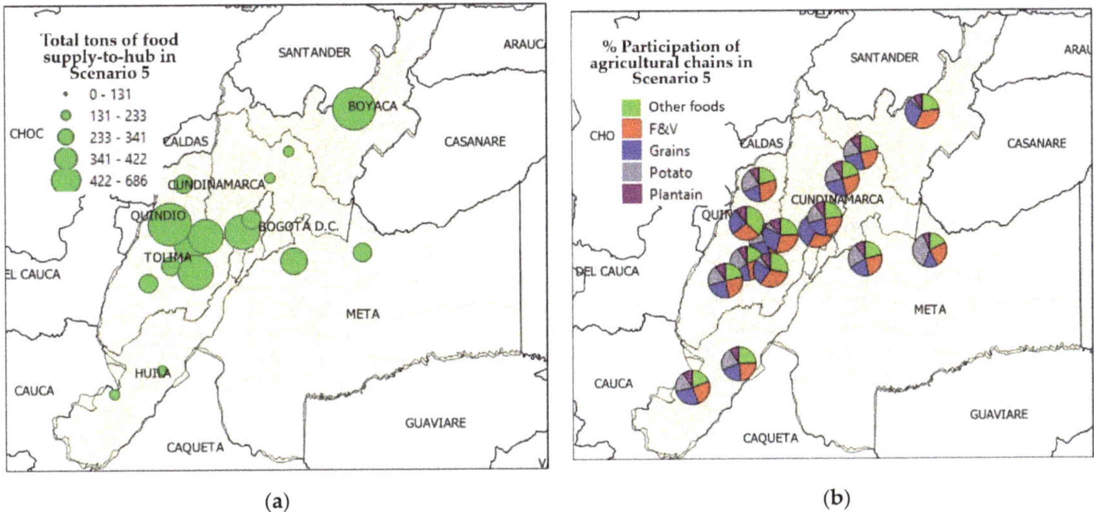

(a)   (b)

Figure 9. Representation of supply-to-hub quantities in scenario 5 shown in Table 7. Figure (**a**) shows the total sent tons, and (**b**) shows the classification of those delivered tons in the five food supply chains.

5.4. Results of Capacity Variations

This section shows the results of the experiments with the variation in hubs' capacity in comparison with the baseline scenario. Table 8 shows these results.

Table 8. Summary of results of hubs' capacity variations. The weekly gross profit, total costs, and net profit are presented in thousands of USD. The letter $\times$ represents the capacity's initial value.

Increase in Hubs' Capacity	Gross Profit	Transportation Costs	Emission Costs	Fixed Hubs' Operation Costs	Net Profit	No Hubs	Total Handled Quantities (tons)
Baseline 1×	1310	198	16	146	950	23	8866
1.5×	1617	232	17	206	1162	21	11,095
2×	1884	265	19	274	1326	21	13,255
3×	2244	313	21	412	1498	21	16,488

As expected, the greater the capacity, the greater the quantities are traded, and the greater the net profits are. In the three new scenarios, the number of open hubs is 21. The trend, however, is of diminishing returns. Doubling the capacity in relation to the baseline produces a net profit of USD 1,162,000 (39% increase) whereas tripling the capacity produces a net profit of USD 1,498,000 (57% increase). In terms of quantities, the relation with the capacity increments is approximately linear: An addition of one percentage point in capacity represents an increase of approximately 37 handled tons. The fixed operation costs also increase in a linear fashion in relation to the baseline. An increment of one percentage point in capacity represents on average an increase of around USD 1360 in fixed costs.

We can also observe that the transportation costs also increase compared to the baseline as hubs handle larger quantities of products, which in turn require more trucks. However, the transportation costs do not increase that steeply: For example, tripling the hubs' capacity results in an increase of 58% in the transportation cost. This implies that the greater the hubs' capacity, the better use the vehicles' capacity. In terms of emission costs, there is a moderate increase. For example, tripling the hub's capacity increases the emission costs in 23%.

5.5. Results of Margin Variations

The last proposed experiment is the variations in the gross profit margin. As said above, this business is meant not only to be profitable, but also to help farmers to obtain better prices. Thus, a reduction in the hubs' margin can be translated to the farmers as in [11]. Table 9 illustrates these results.

Table 9. Summary of margin variations' results. The gross profit, total costs, and net profit are presented in thousands of USD per week. The letter × represents the margin's initial value.

Variation in Margin of Gross Profit	Gross Profit	Transport Costs	Emission Costs	Fixed Operation Costs	Net Profit	No Hubs	Total Handled Quantities (Tons)
Baseline 1×	1310	198	16	146	950	23	8866
0.9×	1147	187	15	134	812	21	8405
0.75×	941	174	13	134	621	21	8405
0.5×	593	138	9	131	315	20	8345

We observe that with the reduction in the gross margins, the number of hubs reduces as well. Likewise, the delivered quantities are smaller. The net profit also reduces accordingly. However, analyzing the net profit, even with a 50% gross margin reduction, the profits can be still attractive for concessionaires.

The model shows that a reduction in margins has a moderate impact on the overall logistics operation: Although fewer hubs are open, the reduction in the number of hubs is not significant (23 in the baseline vs. 20 in the 0.5× scenario and a reduction of approximately 500 tons per week or 166 tons/hub-week). A similar analysis can be performed for the transport costs, emissions, and fixed operation costs of the hubs. The greatest impact is on profits: Reducing the margins in 50%, the profits would decrease in around 66% (950,000 vs. 315,000). This situation may be extreme, but shows the potential savings that can be achieved at both ends of the supply chain (farmers and end costumers).

6. Conclusions

The results illustrate many interesting things: First and most noteworthy is that the hub strategy can be a profitable activity if the hubs are put into operation. In the experiments, we varied everything that could reasonably be changed, and even in scenarios with low attractiveness (low quantities handled by the hubs) and low margins, the business was still profitable. For example, in the baseline, the estimate weekly net profit is USD 951,000 which corresponds to an average weekly net profit of around USD 40,000 per hub. The total costs in this baseline scenario, calculated from Figure 5, are USD 360,000. In this case, the weekly net profits are almost three times the total costs. In the extreme case of the lowest value of sensitivity to distance (γ) and the highest fixed costs, the weekly net profit is USD 372,000 and an average weekly net profit of around USD 31,000 per hub. In terms of factors that affect the profitability, the sensitivity exponent appears to be the most important as illustrated in Figure 8. For instance, keeping the fixed costs variation constant at 0%, we can observe that the weekly net profits ranged from USD 413,000 to USD 1,201,000 (almost three times). It can also be noticed, in the same figure, that varying the fixed costs produced nearly no variations in the gross profits.

Second, as said above, there were scenarios in which the optimal solution resulted in zero (0) open hubs. These were scenarios in which hubs failed to attract farmers (low values of γ and δ). This implies that the private (hub concessionaires) and the public institutions (e.g., RAP-E-) have the challenging task of promoting the usage of the hubs (i.e., increase the values of γ). This can be done through education and better trading conditions for farmers. This is not easy for people who have been isolated for years and are reluctant to change their culture and traditions. In our point of view, major attractors would be offering

(i) better margins in comparison with those of the middlemen, (ii) technical assistance and farm inputs, and (iii) on-the-spot cash payments. The E-choupal initiative [11] is an example of success in the transformation of rural India with these three ideas.

Third, increasing the capacity of the hubs does increase the net profits, but there can be limitations. Although the fixed costs and transportation costs also increase, the larger quantities that these hubs can handle result in greater profits as presented in Section 5. The question is if the hubs' concessionaires can increase their operation capability. As said above, increasing the capacity can be seen as adding days of operation, which seems plausible. A potential limitation for the extended operation is the availability of trucks. For example, a given truck cannot be used on a daily basis and will only be available at the hub in three to five days, depending on the route.

Lastly, there are difficulties to overcome before the successful implementation of a model like this: First, middlemen would not easily give up their business. Intermediation in Colombia is a big business, and a big hub competitor would not be welcome. A potential strategy is creating middlemen memberships to the hub concessionaires and therefore they would participate of the profits. Second, if the hubs are too distant from the farmers, hubs would be of no use. In this sense, hubs would only be useful for farmers located in relatively well-connected areas, and those farmers in remote areas would still rely on middlemen. Finally, the investment to refurbish the hubs can be significant, and at the present pandemic time, the central government has other priorities and limited budget to execute projects like this.

7. Final Thoughts and Further Research

This paper has presented a mathematical model for food hub locations in the central region of Colombia. The model considers the special characteristics of the supply and distribution of the fresh food supply chain in the region, namely dispersed farmers in the countryside that make choices as to where they sell their products and food hubs that distribute such products to urban centers. To the best of our knowledge, this problem has not been studied in the literature. This problem is of special relevance because public institutions in Colombia, responsible for the design and implementation of the master plans for food supply and food security, are closely following the results of this model.

This research uses extensions and adaptations of several competitive hub and facility location formulations to create a realistic model that can be used as a first input for policy making. The results show that the model can be viable and sustainable in the long term under several conditions, as described throughout the paper. This proposed model, of realistic size, was tested with real data. Extensive experiments suggest that the ability of the food concessionaires to attract farmers is the main success factor for the business model.

Based on the results presented throughout this paper, managers and decision makers of the central region can devise an articulated set of tax and legal incentives to attract private investment for the creation and/or strengthening of logistics strategies with technical and operational capacities to efficiently deal with the operation of the hubs. At the same time, education and sensibilization campaigns must be carried out with all stakeholders in order to guarantee a smooth operation of the hub network.

Future research will be carried out with extensions such as logistic regression models that capture additional decision factors such as prices and social network, to name a few. Multi-objective extensions considering environmental factors are also foreseen.

Author Contributions: Conceptualization, G.M., D.G.-R., and J.A.J.; methodology, G.M. and D.G.-R.; software, D.G.-R., N.M., E.M., and A.C.; validation, J.A.J., G.M., and D.G.-R.; formal analysis, G.M. and D.G.-R.; investigation, G.M., D.G.-R., N.M., E.M., and A.C.; resources, G.M.; data curation, D.G.-R., N.M., E.M., and A.C.; writing—original draft preparation, G.M., J.A.J., and D.G.-R.; writing—review and editing, All; visualization, D.G.-R.; supervision, G.M.; project administration, G.M.; funding acquisition, G.M. and J.A.J. All authors have read and agreed to the published version of the manuscript.

Funding: This research was funded by Dirección de Investigación de la Universidad de la Sabana, grant number ING-210-2018. The APC was funded by the same grant.

Institutional Review Board Statement: Not applicable.

Informed Consent Statement: Not applicable.

Data Availability Statement: Data sharing is not applicable to this article.

Acknowledgments: Special thanks to the officers of RAP-E-. Eng. Dayana Rico and Jorge Aya, senior officers of RAP-E- and José Antonio Meza, senior consultant of RAP-E-, who provided valuable information and supervised the conceptualization of the mathematical model. The authors also want to thank the Universidad de La Sabana for providing the funds and means to undertake this research.

Conflicts of Interest: The authors declare no conflict of interest. The funders had no role in the design of the study; in the collection, analyses, or interpretation of data; in the writing of the manuscript, or in the decision to publish the results.

Appendix A

This appendix shows the products that belong to each of the analyzed agricultural chains.

Table A1. the products and each of the analyzed agricultural chains.

Agricultural Chain	Product
F&V	Collard greens, avocado, crookneck pumpkin, green peas, banana, pumpkin and field pumpkin, big headed onion, green onion, plum, cauliflower, curuba (Andean passion fruit), peach, spinach, strawberry, sweet granadilla, soursop, guava, bean, lettuce, lemon, lulo (passion fruit), tangerine, mango, apple, passion fruit, blackberry, orange, papaya, cucumber, pear, pineapple, beet, cabbage, tomato, tree tomato, carrot
Grains	Dried rice, dried bean, chickpea
Other products	Oil—palm oil, oatmeal, sugar—sugarcane, cacao—chocolate, coffee, crackers, dried grain corn—corn flour, bread, panela, pasta, salt, dry grain wheat—wheat flour
Plantain	Plantain
Potato	Potato, yellow small potato

References

1. Mejía, G.; García-Díaz, C. Market-level effects of firm-level adaptation and intermediation in networked markets of fresh foods: A case study in Colombia. *Agric. Syst.* **2018**, *160*, 132–142. [CrossRef]
2. Secretaría Distrital de Desarrollo Económico, Subdirección de Abastecimiento Alimentario. "Mejoramiento de la eficiencia del Sistema de Abastecimiento y Seguridad Alimentaria de Bogotá". 2016. Available online: http://www.desarrolloeconomico.gov.co/sites/default/files/planeacion/1020_seguridad_alimentaria.pdf (accessed on 4 February 2021).
3. Saurabh, S.; Dey, K. Blockchain technology adoption, architecture, and sustainable agri-food supply chains. *J. Clean. Prod.* **2021**, *284*, 124731. [CrossRef]
4. Lezoche, M.; Hernandez, J.E.; Díaz, M.D.M.E.A.; Panetto, H.; Kacprzyk, J. Agri-food 4.0: A survey of the supply chains and technologies for the future agriculture. *Comput. Ind.* **2020**, *117*, 103187. [CrossRef]
5. Hernández, J.M.; Pedroza, C. The influence of the network topology on the agility of a supply chain. *arXiv* **2016**, arXiv:1611.10094.
6. Arya, A.; Löffler, C.; Mittendorf, B.; Pfeiffer, T. The middleman as a panacea for supply chain coordination problems. *Eur. J. Oper. Res.* **2015**, *240*, 393–400. [CrossRef]
7. Abebe, G.K.; Bijman, J.; Royer, A. Are middlemen facilitators or barriers to improve smallholders' welfare in rural economies? Empirical evidence from Ethiopia. *J. Rural. Stud.* **2016**, *43*, 203–213. [CrossRef]
8. Orjuela-Castro, J.A.; Sanabria-Coronado, L.A.; Peralta-Lozano, A.M. Coupling facility location models in the supply chain of perishable fruits. *Res. Transp. Bus. Manag.* **2017**, *24*, 73–80. [CrossRef]
9. Etemadnia, H.; Goetz, S.J.; Canning, P.; Tavallali, M.S. Optimal wholesale facilities location within the fruit and vegetables supply chain with bimodal transportation options: An LP-MIP heuristic approach. *Eur. J. Oper. Res.* **2015**, *244*, 648–661. [CrossRef]
10. Krejci, C.; Beamon, B.M. Modeling Food Supply Chain Sustainability Using Multi-Agent Simulation. *Int. J. Sustain. Econ. Soc. Cult. Context* **2013**, *8*, 143–157. [CrossRef]
11. Anupindi, R.; Sivakumar, S. Supply Chain Reengineering in Agri-Business: A Case Study of ITC's e-Choupal. In *International Series in Operations Research & Management Science*; Price, C.C., Ed.; Springer: Boston, MA, USA, 2007; pp. 265–307.

12. Villegas, J.G.; Palacios, F.; Medaglia, A.L. Solution methods for the bi-objective (cost-coverage) unconstrained facility location problem with an illustrative example. *Ann. Oper. Res.* **2006**, *147*, 109–141. [CrossRef]
13. U.S. Department of Agriculture. "Getting to Scale with Regional Food Hubs", Food and Nutrition Farming, 2017. Available online: https://www.usda.gov/media/blog/2010/12/14/getting-scale-regional-food-hubs#:~{}:text=Ourworkingdefinitionofa, valuechain%2Cfoodhubsare (accessed on 20 December 2020).
14. Cleary, R.; Goetz, S.J.; McFadden, D.T.; Ge, H. Excess competition among food hubs. *J. Agric. Resour. Econ.* **2019**, *44*, 141–163.
15. Ge, H.; Canning, P.; Goetz, S.J.; Perez, A. Effects of scale economies and production seasonality on optimal hub locations: The case of regional fresh produce aggregation. *Agric. Econ.* **2018**, *49*, 157–169. [CrossRef]
16. Morganti, E.; Gonzalez-Feliu, J. City logistics for perishable products. The case of the Parma's Food Hub. *Case Stud. Transp. Policy* **2015**, *3*, 120–128. [CrossRef]
17. Departamento Administrativo Nacional de Estadística [DANE]. "Resultados del Censo Nacional de Población y Vivienda 2018". 2018. Available online: https://www.dane.gov.co/index.php/estadisticas-por-tema/demografia-y-poblacion/censo-nacional-de-poblacion-y-vivenda-2018 (accessed on 16 December 2019).
18. Ministerio de Agricultura. "Agronet". 2021. Available online: https://www.agronet.gov.co/Paginas/inicio.aspx (accessed on 25 January 2021).
19. Departamento Nacional de Planeación. "Encuesta Nacional Logística". 2018. Available online: https://onl.dnp.gov.co/es/Publicaciones/SiteAssets/Paginas/Forms/AllItems/Informe%20de%20resultados%20Encuesta%20Nacional%20Log%C3%ADstica%202018.pdf (accessed on 20 December 2020).
20. Owen, S.H.; Daskin, M.S. Strategic facility location: A review. *Eur. J. Oper. Res.* **1998**, *111*, 423–447. [CrossRef]
21. Church, R.L.; Roberts, K.L. Generalized coverage models and public facility location. *Pap. Reg. Sci. Assoc.* **1983**, *53*, 117–135. [CrossRef]
22. Huff, D.L. Defining and Estimating a Trading Area. *J. Mark.* **1964**, *28*, 34. [CrossRef]
23. Haase, K.; Müller, S. A comparison of linear reformulations for multinomial logit choice probabilities in facility location models. *Eur. J. Oper. Res.* **2014**, *232*, 689–691. [CrossRef]
24. Khosravi, S.; Jokar, M.R.A. Facility and hub location model based on gravity rule. *Comput. Ind. Eng.* **2017**, *109*, 28–38. [CrossRef]
25. Farahani, R.Z.; Asgari, N.; Heidari, N.; Hosseininia, M.; Goh, M. Covering problems in facility location: A review. *Comput. Ind. Eng.* **2012**, *62*, 368–407. [CrossRef]
26. Farahani, R.Z.; Fallah, S.; Ruiz, R.; Hosseini, S.; Asgari, N. OR models in urban service facility location: A critical review of applications and future developments. *Eur. J. Oper. Res.* **2019**, *276*, 1–27. [CrossRef]
27. Ahmadi, Z.; Ghezavati, V. Developing a new model for a competitive facility location problem considering sustainability using Markov chains. *J. Clean. Prod.* **2020**, *273*, 122971. [CrossRef]
28. Drezner, T.; Drezner, Z.; Kalczynski, P.J. A cover-based competitive location model. *J. Oper. Res. Soc.* **2011**, *62*, 100–113. [CrossRef]
29. Drezner, T.; Drezner, Z.; Kalczynski, P.J. Strategic competitive location: Improving existing and establishing new facilities. *J. Oper. Res. Soc.* **2012**, *63*, 1720–1730. [CrossRef]
30. Xu, J.; Murray, A.T.; Wang, Z.; Church, R. Challenges in applying capacitated covering models. *Trans. GIS* **2020**, *24*, 268–290. [CrossRef]
31. Cordeau, J.-F.; Furini, F.; Ljubić, I. Benders decomposition for very large scale partial set covering and maximal covering location problems. *Eur. J. Oper. Res.* **2019**, *275*, 882–896. [CrossRef]
32. Miranda, P.A.; Garrido, R.A. Incorporating inventory control decisions into a strategic distribution network design model with stochastic demand. *Transp. Res. Part E Logist. Transp. Rev.* **2004**, *40*, 183–207. [CrossRef]
33. Sarker, B.R.; Wu, B.; Paudel, K.P. Optimal number and location of storage hubs and biogas production reactors in farmlands with allocation of multiple feedstocks. *Appl. Math. Model.* **2018**, *55*, 447–465. [CrossRef]
34. Vafaeinejad, A.; Bolouri, S.; Alesheikh, A.A.; Panahi, M.; Lee, C.-W. The Capacitated Location-Allocation Problem Using the VAOMP (Vector Assignment Ordered Median Problem) Unified Approach in GIS (Geospatial Information Systam). *Appl. Sci.* **2020**, *10*, 8505. [CrossRef]
35. Casisi, M.; Buoro, D.; Pinamonti, P.; Reini, M. A Comparison of Different District Integration for a Distributed Generation System for Heating and Cooling in an Urban Area. *Appl. Sci.* **2019**, *9*, 3521. [CrossRef]
36. Luo, J.; Ji, C.; Qiu, C.; Jia, F. Agri-Food Supply Chain Management: Bibliometric and Content Analyses. *Sustainability* **2018**, *10*, 1573. [CrossRef]
37. Paciarotti, C.; Torregiani, F. The logistics of the short food supply chain: A literature review. *Sustain. Prod. Consum.* **2021**, *26*, 428–442. [CrossRef]
38. Ahumada, O.; Villalobos, J.R. Application of planning models in the agri-food supply chain: A review. *Eur. J. Oper. Res.* **2009**, *196*, 1–20. [CrossRef]
39. Soto-Silva, W.E.; Nadal-Roig, E.; González-Araya, M.C.; Plà-Aragonès, L.M. Operational research models applied to the fresh fruit supply chain. *Eur. J. Oper. Res.* **2016**, *251*, 345–355. [CrossRef]
40. Amorim, P.; Günther, H.O.; Almada-Lobo, B. Multi-objective integrated production and distribution planning of perishable products. *Int. J. Prod. Econ.* **2012**, *138*, 89–101. [CrossRef]
41. Suraraksa, J.; Shin, K.S. Shin Urban Transportation Network Design for Fresh Fruit and Vegetables Using GIS–The Case of Bangkok. *Appl. Sci.* **2019**, *9*, 5048. [CrossRef]

42. Granillo-Macías, R. Socio-Economic Planning Sciences Logistics optimization through a social approach for food distribution. *Socioecon. Plann. Sci.* **2020**, 100972, in press. [CrossRef]

43. Govindan, K.; Jafarian, A.; Khodaverdi, R.; Devika, K. Two-echelon multiple-vehicle location—Routing problem with time windows for optimization of sustainable supply chain network of perishable food. *Int. J. Prod. Econ.* **2014**, *152*, 9–28. [CrossRef]

44. Patidar, R.; Venkatesh, B.; Pratap, S.; Daultani, Y. A Sustainable Vehicle Routing Problem for Indian Agri-Food Supply Chain Network Design. In Proceedings of the 2018 International Conference on Production and Operations Management Society (POMS) 2018, Peradeniya, Sri Lanka, 14–16 December 2018; pp. 1–5. [CrossRef]

45. Gholami-Zanjani, S.M.; Jabalameli, M.S.; Klibi, W.; Pishvaee, M.S. A robust location-inventory model for food supply chains operating under disruptions with ripple effects. *Int. J. Prod. Res.* **2021**, *59*, 301–324. [CrossRef]

46. Lüer-Villagra, A.; Marianov, V. A competitive hub location and pricing problem. *Eur. J. Oper. Res.* **2013**, *231*, 734–744. [CrossRef]

47. Mahmoodjanloo, M.; Tavakkoli-Moghaddam, R.; Baboli, A.; Jamiri, A. A multi-modal competitive hub location pricing problem with customer loyalty and elastic demand. *Comput. Oper. Res.* **2020**, *123*, 105048. [CrossRef]

48. Mahmutogullari, A.I.; Kara, B.Y. Hub location under competition. *Eur. J. Oper. Res.* **2016**, *250*, 214–225. [CrossRef]

49. Musavi, M.; Bozorgi-Amiri, A. A multi-objective sustainable hub location-scheduling problem for perishable food supply chain. *Comput. Ind. Eng.* **2017**, *113*, 766–778. [CrossRef]

50. Gharehyakheh, A.; Krejci, C.; Cantu, J.; Rogers, K.J. A Multi-Objective Model for Sustainable Perishable Food Distribution Considering the Impact of Temperature on Vehicle Emissions and Product Shelf Life. *Sustainability* **2020**, *12*, 6668. [CrossRef]

51. Meneghetti, A.; Monti, L. Greening the food supply chain: An optimisation model for sustainable design of refrigerated automated warehouses. *Int. J. Prod. Res.* **2015**, *53*, 6567–6587. [CrossRef]

52. Mogale, D.G.; Cheikhrouhou, N.; Tiwari, M.K. Modelling of sustainable food grain supply chain distribution system: A bi-objective approach. *Int. J. Prod. Res.* **2019**, *58*, 5521–5544. [CrossRef]

53. Tammy, D.; Zvi, D.K.P.J.; Kalczynski, P.J. The multiple markets competitive location problem. *Kybernetes* **2016**, *45*, 854–865.

54. Ministerio de Transporte. Sistema de Información de Costos Eficientes para el Transporte Automotor de Carga SICE-TAC. 2021. Available online: https://www.mintransporte.gov.co/publicaciones/4462/sice-tac/ (accessed on 25 January 2021).

55. Ministerio de Medio Ambiente. *Impuesto Nacional al Carbono*; Ministerio de Medio Ambiente y Sostenibilidad: Bogotá, Colombia, 2017.

56. Corabastos, "Boletín de precios". 2021. Available online: https://precios.corabastosonline.co/#/tendencia/grupos (accessed on 25 January 2021).

57. Ministerio de Transporte. "Transporte en Cifras", Transporte en Cifras. Estadísticas 2019; 2019. Available online: https://plc.mintransporte.gov.co/Estadísticas/Transporte-en-Cifras (accessed on 25 January 2021).

58. Drezner, T.; Drezner, Z.; Zerom, D. Competitive facility location with random attractiveness. *Oper. Res. Lett.* **2018**, *46*, 312–317. [CrossRef]

applied sciences

MDPI

Article

Lot-Sizing and Scheduling for the Plastic Injection Molding Industry—A Hybrid Optimization Approach

Nathalie Klement [1], Mohamed Amine Abdeljaouad [2,*], Leonardo Porto [3] and Cristóvão Silva [3]

[1] Arts et Métiers Institute of Technology, LISPEN, HESAM Université, 59000 Lille, France; nathalie.klement@ensam.eu
[2] CEA Tech Hauts-de-France, 59000 Lille, France
[3] CEMMPRE, Department of Mechanical Engineering, University of Coimbra, 3030-790 Coimbra, Portugal; leonardo-rocha-porto@hotmail.com (L.P.); cristovao.silva@dem.uc.pt (C.S.)
* Correspondence: mohamed-amine.abdeljaouad@cea.fr

Abstract: The management of industrial systems is done through different levels, ranging from strategic (designing the system), to tactical (planning the activities and assigning the resources) and operational (scheduling the activities). In this paper, we focus on the latter level by considering a real-world scheduling problem from a plastic injection company, where the production process combines parallel machines and a set of resources. We present a scheduling algorithm that combines a metaheuristic and a list algorithm. Two metaheuristics are tested and compared when used in the proposed scheduling approach: the stochastic descent and the simulated annealing. The method's performances are analyzed through an experimental study and the obtained results show that its outcomes outperform those of the scheduling policy conducted in a case-study company. Moreover, besides being able to solve large real-world problems in a reasonable amount of time, the proposed approach has a structure that makes it flexible and easily adaptable to several different planning and scheduling problems. Indeed, since it is composed by a reusable generic part, the metaheuristic, it is only required to develop a list algorithm adapted to the objective function and constraints of the new problem to be solved.

Keywords: heuristic; metaheuristics; scheduling; injection molding

check for
updates

Citation: Klement, N.; Abdeljaouad, M.A.; Porto, L.; Silva, C. Lot-Sizing and Scheduling for the Plastic Injection Molding Industry—A Hybrid Optimization Approach. *Appl. Sci.* **2021**, *11*, 1202. https://doi.org/10.3390/app11031202

Academic Editor: Farouk Yalaoui
Received: 29 December 2020
Accepted: 23 January 2021
Published: 28 January 2021

Publisher's Note: MDPI stays neutral with regard to jurisdictional claims in published maps and institutional affiliations.

1. Introduction

Industry is changing rapidly and is facing the fourth industrial revolution. In this new context, industrial systems need to be more flexible and intelligent to deal with an ever-growing data availability, real time decisions and increasing customization. With Internet of Things, which allows various objects in the company or on the shop-floor to communicate with each other, industrial systems will become more intelligent and able to anticipate shortage of materials, the need for maintenance operations or to respond to urgent commands. This will imply the use of more efficient tools for planning and re-planning shop-floor operations.

Such decision support tools will be important at all decision levels: For example, at a strategic level, they can be used to decide which activities must be performed during a given planning horizon and to estimate the resources needed. At this level they can also be useful to model and test new shop-floor configurations and anticipate the impact of these changes on the system performance. Considering the tactical level, decision tools will allow to allocate resources to each activity planned to occur during a given time horizon. Finally, at the operational level they can be used to schedule the activities, determining the time allocation of each resource and to re-schedule them to respond to emergencies (new activities that have to be done as soon as possible). Table 1 summarizes the problems encountered at the different levels.

Table 1. Levels of decision.

Decision Level	Strategic	Tactical	Operational
Planning Horizon	Years	Months	Weeks/Days
Typical Problems	Determine the number of resources	Plan activities Assign resources	Schedule activities Consider emergencies

In this paper, we present a hybrid optimization method to solve a lot sizing and scheduling problem encountered in a plastic injection facility that produces components for electronic products. The proposed method was first introduced in [1] and concisely used to solve optimization problems from different areas: healthcare systems, as well as textile and plastic injection industries, with promising preliminary results. In the current work, the method's application to plastic injection molding is further developed and its performance analyzed with a comparison to optimal and heuristic solutions.

Injection molding can be considered as a single-stage manufacturing process where parts are produced by injecting a molten material, typically a thermoplastic polymer, like in our case study company, into a custom-made mold. In the injection machine, the material is molten, mixed and injected into the mold, acquiring, after cooling, the shape imposed by the configuration of the cavity and being then ejected. The obtained part usually have the finished shape, not requiring further operations. In general, the shop-floor is composed by several injection machines which are shared by different products necessitating different molds. A setup time occurs each time a new product has to be manufactured. This setup time consists in the time required to dismount the mold used for the previous part and to mount the new mold on the machine. Obtaining a good schedule for such a constrained problem can be a real challenge.

The above described problem can be stated as follows: n jobs have to be processed on parallel machines, using their needed mold. These jobs have to be scheduled, considering their given processing time and due date before they have to be completed. When two jobs consecutively processed on one machine need two different molds, a sequence dependent setup time is recorded. Jobs have to be processed at once (non-preemptive constraint). To save setup costs, jobs using the same mold may be scheduled one after another. There are compatibility constraints between the molds and the machines: Each mold can only be assigned to some specific machines. Each mold exists in a single copy, thus the same mold cannot be used by more than one machine during the same time period. The objective is to find the suitable allocation of the jobs to the machines as well as the schedule of jobs on each machine so that minimizing the total tardiness.

This paper is structured as follows—Section 2 provides a literature review on the problems that present some analogies with the scheduling issue considered in this paper, such as batching and lot-sizing. A mathematical formulation modelling the problem and its constraints is detailed in Section 3. Then, in Section 4, we present an optimization method to solve the problem, consisting in a list algorithm and a metaheuristic. The different results of the experimental study are then presented in Section 5. The article ends with a Conclusion and some proposals for further work.

2. Literature Review

Several papers from the literature addressed the injection molding planning and scheduling problems. Dastidar and Nagi [2] studied a parallel machine scheduling problem in an injection molding facility, where different products require different types of resources. The authors provide a mathematical formulation and a decomposition based approach to minimize the setup and the inventory holding costs. A decomposition strategy is also used in [3,4] where the compatibility constraints between injection molds and machines are taken into account. In both articles, the problem is decomposed into sub-problems by grouping the machines by sets according to their compatible molds. Each sub-problem is then dealt with as a Capacitated Lot-Sizing Problem (CLSP). Van Wassenhove and De

Bodt [3] used two heuristic procedures while [4] provides a goal programming formulation and a heuristic to minimize the production's costs. The obtained results are shown to outperform the planning and scheduling procedures currently used in their case-study injection plants.

The problem we study in this paper also shares some features with the CLSP. Indeed, the lot-sizing issue reflects the fact that companies usually have to manufacture several types of products with limited production capacity. It is a very popular subject in the field of combinatorial optimization and has been defined since the early 1900s with the EOQ Economic Order Quantity [5]. Lot-sizing problems have been widely studied since then; a recent literature review can be found in [6]. Different lot-sizing models have been developed, depending on the characteristics of the production process. Although most of the lot-sizing studies focus on "product lots" (i.e., determining the right period and the quantity to produce for each type of product), some of them can also be related to "process lots"; in other words, to the decision of how many times or how long each process should be used and how the processes should be scheduled, given that each process can lead to a set of products. Such a feature appears in several industrial problems [7–9], including injection molding [10], for which a mixed integer programming model is proposed and tested on real data from a Brazilian molded pulp plant.

CLSP has many extensions, including the consideration of parallel production machines per period and setup times between products of different types. However, the scheduling is not part of this issue and is traditionally made as a second step. Therefore there is a significant difference between CLSP and the problem we study. Indeed, separating assignment and scheduling or decomposing a problem can immediately eliminate several good solutions from the searching space. In this work, we take the problem as a whole and this issue may thus be referred to as a batching problem. Batching is indeed the decision of whether or not to schedule similar jobs contiguously, to avoid setup times or setup costs [11].

Drexl and Kimms [12] draw an analogy between the batching and lot-sizing problems. In the continuous time lot sizing and scheduling problem, each demand is characterized by its deadline and its size. Demands are interpreted as jobs and the demand size determines the processing time of a job. An important assumption is that the capacity (e.g., the speed of the machine) is constant over time and thus the processing time of a job does not depend on the schedule. Another fundamental assumption is that jobs are not allowed to be split, which means that a certain demand must always be processed from the beginning to the end. Of course, several demands for the same item may be grouped together and processed by lots to save setup costs. Problems with these assumptions can thus be referred to as a batching and scheduling problem (BSP) rather than a lot sizing and scheduling problem. The relationship between batching and lot-sizing is also analyzed in [13], where the studied lot sizing problem is solved as a batch sequencing problem.

Potts and Kovalyov [14] review the literature on scheduling with batching, in the particular case of parallel machines. In this paper, parallel machines are classified as: identical, when the processing times depend only of the jobs, regardless of the machine on which they are processed; uniform, if the processing times depend on the jobs and the machines' speed; or unrelated, when the processing time is a function of both the job and the machine where it is allocated. One of the specificities of our problem is that the processing time of the jobs depends only on the jobs. But because of the incompatibility between some molds and some machines, not all the machines can process all the jobs. To the best of our knowledge, this aspect has not been considered in batching problems but [15,16] solved some scheduling problems with eligibility restrictions using constructive heuristics. The eligibility aspect can also be found in [17] but in the flowshop case. Nevertheless, our problem can be seen as a problem with parallel unrelated machines, where the processing time will be: (1) the time required to process the job if the mold is compatible with the selected machine or (2) infinite if the mold is not compatible with the selected machine. A batch-scheduling problem with parallel unrelated machines is dealt with in [18], where the authors compare the performances of four heuristics to minimize the total weighted tardiness. Their study showed the superiority of a two-level batching heuristic and a simulated annealing approach over heuristics that use priority rules such as earliest weighted due date or shortest weighted processing time. Reference [19] also addressed a parallel unrelated batch scheduling problem with sequence dependent setup times and developed a GRASP metaheuristic to minimize the makespan. Reference [20] provide a constructive algorithm and a simulated annealing to portion and schedule batches in a multi-stages semi-conductor manufacturing plant where the parallel machines per stage are non-identical.

Because the weighted tardiness minimization in an unrelated parallel machine scheduling problem, $m\left|\left|\sum_{j=1}^{n} w_j T_j\right.\right.$, is already NP-hard [21], our problem is also NP-hard. Indeed, the $Rm\left|\left|\sum_{j=1}^{n} w_j T_j\right.\right.$ is a particular case of our problem when the jobs have equal weights, the setup-times are null and each mold is required by only one job. As seen in this literature review, whether for solving the lot-sizing or batching problems, the developed methods are quite sophisticated constructive approaches or optimization approaches. Our aim in this paper is to provide a high performing optimization method which is easy to develop and use. Although in this article the proposed method will only be tested in a case study company, injection molding is a common industrial activity and several plants may share the same features. Our solving method will thus not be limited to a single use-case and this work may be considered as a first step toward a final objective which is to generalize the optimization tool to other scheduling problems with new characteristics.

In the following section, we provide a mathematical model that will be useful to spot the limit of the exact resolution for this NP-hard problem in terms of solving times and to ensure the quality of our method for the small-sized instances. The method itself is described and implemented in Section 4 and then experimented in a real plastic injection case.

3. Mathematical Model

The mathematical model of the problem we intend to solve, described in this section, was inspired by a model proposed by [22], although there are some significant differences between them. In the original model both earliness and tardiness were to be minimized. In our case study company, earliness is not considered as an issue, thus only tardiness is considered leading to a difference in the objective function. Furthermore, in the original model changeover occur whenever two jobs from different families are sequenced one after the other but this time is constant. In our case, when a setup time occur, this time depends of the sequence of jobs i and j, considering the time required to dismount the mold to produce i plus the time necessary to mount the mold to process j. Thus, unlike the original model we consider a sequence dependent setup time, which lead to a difference in the way the setups are considered in the model, specifically in the equations used to determine the completion time of a given job. Finally and this is probably the main difference between the models, in [22] the model is developed for identical parallel machines, which means that any job can be processed in any machine. In our case study company, a job can only be processed in a subset of the available machines. This has led to the introduction of a new constraint (3) to ensure that a job is always allocated to a compatible machine.

Below, we provide the inputs of our model:

- m : Number of molds
- n : Number of jobs
- r : Number of machines
- f_j : Mold for job j
- d_j : Due date of job j
- p_j : Processing time of job j
- A_j : Time to mount the mold of job j on the machine
- D_j : Time to dismount the mold of job j

- M_{jk}: The setup time when job k follows job j on a machine. This setup time is equal to the time to dismount the mold used by j (D_j) plus the time to mount the mold used by k (A_k), if j and k require different molds. Thus: $M_{jk} = A_k + D_j$ if $f_j \neq f_k$ 0 otherwise
- CF_{jb}: Binary variable denoting if job j is compatible with machine b:
- $CF_{jb} = \begin{cases} 1 \text{ if job } j \text{ is compatible with machine } b \\ 0 \text{ otherwise} \end{cases}$
- G: "Big M," sufficiently large number to make the problem solvable by linear programming. In our case, G should be much higher than the ending time of any job.

The decision variables of our model are the following:

- C_j : Completion time of job j
- T_j : Tardiness of job j
- α_{jb}: Binary variable denoting if j is the first job to be processed on machine b:
  $\alpha_{jb} = \begin{cases} 1 \text{ if job } j \text{ is the first to be processed on machine } b \\ 0 \text{ otherwise} \end{cases}$
- θ_{jk}: Binary variable denoting if job k is scheduled right after job j:
  $\theta_{jk} = \begin{cases} 1 \text{ if job } k \text{ is immediately processed after job } j \\ 0 \text{ otherwise} \end{cases}$
- β_{jb}: Binary variable denoting if job j is processed on machine b but not in the first place:
  $\beta_{jb} = \begin{cases} 1 \text{ if job } j \text{ is processed on machine } b \text{ but not on the first place} \\ 0 \text{ otherwise} \end{cases}$

The mathematical model can be stated as follows:
Minimize:

$$\sum_{j=1}^{n} T_j. \tag{1}$$

Restricted to:

$$\sum_{b=1}^{r} \left(\alpha_{jb} + \beta_{jb} \right) = 1, \ \forall j = 1, 2, \ldots n \tag{2}$$

$$\sum_{b=1}^{r} CF_{jb} \left(\alpha_{jb} + \beta_{jb} \right) = 1, \ \forall j = 1, 2, \ldots n \tag{3}$$

$$\alpha_{jb} + \beta_{jb} \leq \beta_{kb} + 1 - \theta_{jk}, \ \forall j = 1, 2, \ldots n; \ k = 1, 2, \ldots n; \ b = 1, 2, \ldots r \tag{4}$$

$$\sum_{j=1}^{n} \alpha_{jb} \leq 1, \ \forall b = 1, 2, \ldots r \tag{5}$$

$$\sum_{b=1}^{r} \alpha_{jb} + \sum_{k=1}^{n} \theta_{kj} = 1, \ \forall j = 1, 2, \ldots n \tag{6}$$

$$\sum_{k=1}^{n} \theta_{jk} \leq 1, \ \forall j = 1, 2, \ldots n \tag{7}$$

$$C_k \geq C_j + M_{jk} + \sum_{b=1}^{r} p_k \beta_{kb} + G \left(\theta_{jk} - 1 \right), \ \forall j = 1, 2, \ldots n; \ k = 1, 2, \ldots n \tag{8}$$

$$C_j \geq p_j + A_j \sum_{b=1}^{r} \alpha_{jb}, \ \forall j = 1, 2, \ldots n; \tag{9}$$

$$C_j \leq d_j + T_j, \ \forall j = 1, 2, \ldots n; \tag{10}$$

$$C_j, T_j \geq 0, \ \forall j = 1, 2, \ldots n; \tag{11}$$

Equation (1) represents the objective function, which is to minimize the sum of all tardiness. A job is considered tardy if its completion time is higher than its due date; and in that case, the difference between its completion time and its due date is added to the sum.

Equation (2) ensures that each job must be assigned to one machine. Equation (3) checks compatibility, which means that a job can only be scheduled on a compatible machine. Equation (4) enforces a job and its direct successor in the processing sequence to both be produced on the same machine. Equation (5) ensures that, for each machine, there can only be one job scheduled first. Equation (6) states that if a job is not the first to be scheduled on any machines, then it must have a predecessor. Equation (7) says that for each job, there can only be one other scheduled immediately before it or none in case it is the first job. Equation (8) forces the completion time of a certain task to be bigger than its processing time plus the setup and the completion time of its predecessor.

Equation (9) computes the completion time of the first job processed by each machine as its processing time plus the setup time. Equation (10) provides the completion time of a job smaller or equal to the sum of its due date and the tardiness. Equation (11) is the non-negativity constraint.

4. Proposed Solution Approach

Since the studied problem is NP-hard, solving our model with a mathematical solver is not expected to be efficient for the large-sized industrial problems. Therefore, we provided a solving algorithm that can quite rapidly reach good solutions. We propose a hybrid method, which combines a constructive heuristic (a list algorithm) with a metaheuristic. The problem characteristics are specified in the list algorithm while the metaheuristic is more generic and can be adapted to other scheduling problems without changes. The principle of the method is given in Figure 1. It can be seen as a master/slave resolution scheme, where the metaheuristic handles the searching process and the list-algorithm performs the quality assessment of the solutions [23,24].

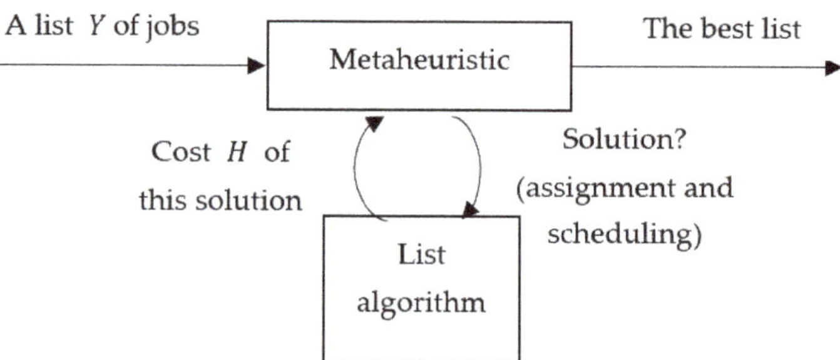

Figure 1. Hybridization metaheuristic—List algorithm.

The encoding scheme of the proposed metaheuristic is a list Y of jobs, which is the order in which they will be selected by the algorithm to be assigned to the machines. The move between neighbor solutions is performed by jobs' permutation. The jobs are selected one by one by list algorithm L, following their order as given in List Y and assigned to the requested resources, building a scheduling that takes into account the problem constraints. From this scheduling emerges Solution X, which will be evaluated by objective function H. On the basis of this assessment, the metaheuristic decides if the solution should be selected or not. At the end of its run, the proposed method gives the scheduling sequence that gives the best value of objective function.

The general encoding scheme can be represented by Equation (12). Ω is the set of all lists Y; S the set of all admissible assignments X of the jobs to the machines. An assignment is considered admissible if all constraints are respected. More details about the encoding are given in [25].

$$Y \in \Omega \underset{Heuristic\ L}{\rightarrow} L(Y) = X \in S \underset{Criterion\ H}{\rightarrow} H(X). \qquad (12)$$

4.1. List Algorithm

List algorithms are greedy heuristics that are widely used to solve scheduling problems. The standard list algorithm iteratively constructs a schedule by considering the jobs in a listed order and assigning each one of them to the first machine that becomes idle [26]. Examples of list algorithms applied to and activity planning and resource allocation in a hospital are presented in [27].

In the injection molding problem described in this paper, we propose a list scheduling algorithm to schedule jobs and assign them to the machines and dates considering tool compatibility. One of the advantages of using a list algorithm is to provide a simple neighborhood structure for metaheuristics, which can browse the set of solutions by permutations between two items in a list, as we will see in the next subsection. The details of the list algorithm is described by Algorithm 1. The algorithm takes as input the number of machines, the number of molds, as well as a sorted list of N jobs $(\sigma_i)_{i \in N}$ which states the order in which they will be selected to be assigned to the machines. Of course, the jobs' durations and their associated molds are part of the input. The algorithm gives as output the assignation of the jobs to the machines.

Algorithm 1 List algorithm for the injection problem

Data: List of jobs $(\sigma_i)_{i \in N}$
forall *Job* σ_i **do**
    Order the machines according to their release date: $R[i]$
    $j: = 0$
    while *Job* σ_i *not assigned AND* $j < Nb_{machines}$ **do**
        if *Job* σ_i *and machine* $R[j]$ *compatible with the mold* **then**
            if *Needed mold available* **then**
                Release date machine $R[j]: =$ Release date machine $R[j]$ + processing time of
Job σ_i + setup time
                Release date needed mold: $=$ Release date machine $R[j]$
                Assign *Job* σ_i to machine $R[j]$
            else
                Find on which machine the needed mold is used: *MachineUsing*
                If the mold currently used in *MachineUsing* is the one needed by *Job* σ_i **then**
                    Release date *MachineUsing*: $=$ Release date *MachineUsing* + processing
time of *Job* σ_i
                else
                    Release date *MachineUsing*: $=$ Release date *MachineUsing* + processing
time of *Job* σ_i + setup time
                Release date needed mold: $=$ Release date machine *MachineUsing*
                Assign *Job* σ_i to machine *MachineUsing*
        else
            Next machine: $j(j + 1)\% Nb_{machines}$

Figure 2 shows an example of the algorithm's assignment rule for a 3-machines instance. At iteration u, the mold required by the next job i on the list Y is not being used on any machine. i is thus assigned to the less loaded machine with a setup time. Next, job j needs a mold that is currently used on machine $M3$ that has the lowest release time. j will therefore be assigned to $M3$ without setup. For job k however, the required mold is currently used on the most-loaded machine $M1$ and therefore the job has to wait for its release time before being processed. k is assigned to $M1$ without setup time.

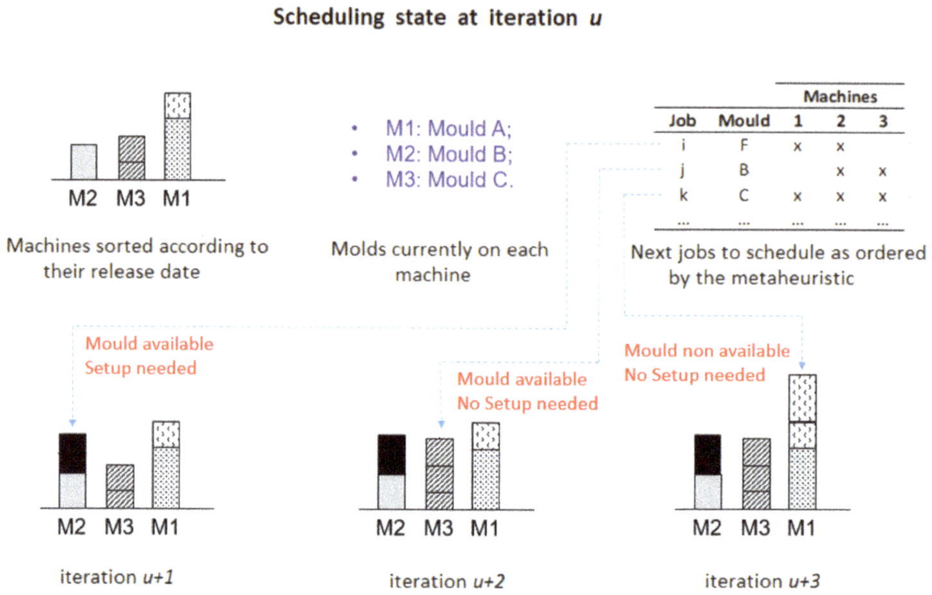

Figure 2. Illustration of the List Algorithm's assignment rule.

4.2. Metaheuristic

The proposed list algorithm is combined with a metaheuristic. We chose to use two single solution-based metaheuristics: stochastic descent and simulated annealing.

4.2.1. Tuning

The metaheuristic performs in the set Ω of all jobs' permutations. Given that n is the number of jobs, cardinal of Ω is $n!$. One solution $Y \in \Omega$ is a list of jobs.

Solutions are compared according to an objective function which characterizes their quality. In our case, the objective function is the total tardiness: the sum of the tardiness of all jobs. After having assigned the jobs to the molds and to the machines using the list algorithm described in the previous section, the value of the objective function is computed as follows: for each job, it is the difference between the completion time and the due date. All positive values are added.

The neighborhood system V is the exchange of two jobs' positions in List Y: job at position i permutes with the one at position j. V checks accessibility and reversibility properties. An initial solution is randomly computed, based on a randomly sorted list of jobs.

4.2.2. Stochastic Descent

Stochastic descent is one of the oldest metaheuristics. Its principle is to compute a solution Y' which is a neighbor of the current solution Y, according to the neighborhood system V. After applying the list algorithm L, a solution $X' = L(Y')$ is obtained. If the value of the objective function $H(X')$ is lower or equal to the current one $H(X)$, then the solution X' of the list Y' is accepted. Stochastic descent converges to a local minimum. This method is easy to implement but the quality of the results may be insufficient. Algorithm 2 describes stochastic descent.

Algorithm 2 Principle algorithm of stochastic descent

Data: Initial solution Y
$XL(Y)$
while *necessary* **do**
    choose uniformly and randomly $Y' \in V(Y)$
    $X'L(Y')$
    if $H(X') \leq H(X)$ **then**
    YY'

4.2.3. Simulated Annealing

This metaheuristic comes from process used in metallurgy. That process alternates cycles of slow cooling and heating [28] used inhomogeneous simulated annealing to emulate this process. Applied to the optimization field, the algorithm consists of executing a local search with a computed probability to choose a worse solution than the current one, allowing thus to escape local optimum. This probability progressively decreases throughout the running process of the algorithm. If neighborhood system V satisfies the accessibility and reversibility properties, simulated annealing converges in probability to the set of optimal solutions [29]. Algorithm 3 describes the principle algorithm of simulated annealing.

Algorithm 3 Principle algorithm of simulated annealing

Data: Initial solution Y, Temperature T_0, Decreasing factor α
TT_0
$XL(Y)$
while *necessary* **do**
    choose uniformly and randomly $Y' \in V(Y)$
    $X'L(Y')$
    if $H(X') \leq H(X)$ **then**
        YY'
    else
        if *rand[0, 1]* $\leq e^{-\frac{H(Y')-H(Y)}{T}}$ **then** YY'
    Compute the new temperature $T\ \alpha * T$

Two parameters have to be chosen:

- The initial temperature T_0 is chosen to be big enough so that most of the transitions are accepted at the beginning: $e^{-\frac{H(Y')-H(Y)}{T_0}} \approx 1\ \forall(Y,\ Y')$.
- The decreasing factor α is chosen in such a way that the final temperature would be close to zero.

5. Experiments

In this section, we carry out an experimental study to assess the performance of the proposed method. The outcomes are compared with an exact resolution performed by a mathematical solver for small-sized instances and with the results obtained with a heuristic presented in [30] for bigger ones. This benchmark heuristic is the scheduling method currently used in the molding company and it consists of a two-phase algorithm: first it assigns molds to machines and then it schedules jobs on each machine. Each mold is thus assigned to only one machine throughout the scheduling, with no possibility to move to another one. This method has thus he advantage to ensure a small number of setups throughout the production process.

5.1. Instances Generation

The case study company can produce more than 500 different parts, each one requiring its own mold. 25 injection machines are available in the shop-floor. An average of 200 production orders are received each week. The production planner needs to generate a weekly plan for the received orders, considering the items to be produced, their required quantities and the due date negotiated with the costumer. To execute this plan, the production planner must decide which job will be processed in each machine and the sequence by which they will be produced.

To test the proposed approach, 11 instances were randomly generated. The characteristics of these instances (number of machines available, number of molds required and number of jobs to be planned) are presented in Table 2. To correctly replicate the case study company problem, real data (jobs processing times and setup times) were collected on the shop-floor. The following statistical distributions fitting the collected data were used to generate the instances:

(1) Jobs' processing times: Exponential with an average of 10.75 h.
(2) Due dates: Uniform with a minimum of 24 and a maximum of 312 h.
(3) Setup times:

    ○ Time required to dismount a mold: Uniform with a minimum of 15 and a maximum of 45 min

    ○ Time required to mount a mold: Uniform with a minimum of 20 and a maximum of 60 min.

Table 2. Generated instances.

Instance number	1	2	3	4	5	6	7	8	9	10	11
Machines		2			3			5		10	
Molds	3	6	12	16	18	20	26	26	25	59	63
Jobs	10	15	32	47	53	57	79	80	81	177	191

The *big M* value of our mathematical model (i.e., variable G) has been chosen as equal to 10.000. Let us note that this value has no impact on the model's solving times, since there is no loop or sum that depends on it.

5.2. Results

The computer that was used in the experiments is powered by an i7 CPU running at 2.6 GHz. First, to validate the quality of our method, we compared its results to those of the mathematical model. The solver used to solve the mathematical model is CPLEX 12.6.3. Table 3 summarizes results obtained by both methods. Results represent tardiness. Computational times are also given. In case the solver does not find any solution in a reasonable computational time, the upper and lower bounds are reported.

Table 3. Comparison of results between the solver and the hybrid method.

		Solver		Our Method	
Instance	**Number of Jobs**	**Time**	**Tardiness**	**Time**	**Tardiness**
1	10	19 s	38.00	19 s	38.00
2	15	12 min	18.21	7 s	18.21
3	32	30 h	[37.65, 80.76]	96 s	80.76

As we can see in Table 3, for the 10 and 15 jobs instances, our method finds an optimal solution in less time than the solver. For larger instances, the solver does not reach any optimal solution for hours while our method finds a good solution (actually the upper bound found by the solver) in a few seconds. Since the CPLEX solver ran for more than 30 h without finding a solution for the 32 jobs instance, we decided not to continue the exact resolution for larger instances, for which we propose a comparison between our method and the scheduling heuristic currently used by the company.

Table 4 presents the results obtained for the big-sized instances with this benchmark heuristic and with our method. The comparison considers the tardiness and the number of setups. Over several replication, the shortest time needed to find one best solution is reported in the last column of the table.

Table 4. Comparison of results between the benchmark method and the proposed approach.

| | | Current Approach | | Proposed Approach | | | | |
| | | | | Stochastic Descent | | Simulated Annealing | | Time |
Instance	Jobs	Tardiness	Setup	Tardiness	Setup	Tardiness	Setup	
4	47	8.32	16	0	33	0	31	4 min
5	53	37.76	26	24.37	40	17.98	42	37 min
6	57	159.6	23	136.29	33	66.16	27	3 min
7	79	395	37	241.26	49	148.63	44	6 min
8	80	77.8	33	0.98	49	0	41	1 h
9	81	165.9	28	52.56	39	46.17	43	1 h
10	177	1020.4	82	963.95	114	343.82	101	5 h
11	191	580.2	83	786.18	122	579.5	124	4 h

The results show that besides being easy to develop, the proposed method is effective, giving good results in reasonable computational time. In comparison with the benchmark method proposed by [30], an average reduction of 35% is achieved in tardiness. For some instances, the reduction of tardiness is as high as 65%. This can be explained by the fact that the proposed method is less constraining, as the molds are not fixed and can switch from a machine to another throughout the scheduling. Nevertheless, the proposed method leads to an increase in the number of setups. It is important to note that, since the main objective of the company was to reduce the tardiness, the number of setups was not considered in the objective function of the proposed method but for other problems or future studies, an objective function considering tardiness and number of setups, with different weights, can be considered.

From the experimental study conducted in this section, we can thus conclude that our proposed approach brings a significant improvement in terms of tardiness when compared to the company current approach. Therefore, it is expected that the proposed approach will be implemented by the company as part of their production planning and scheduling system.

6. Conclusions and Perspectives

In this paper, a real world scheduling problem from a plastic injection company has been studied and solved. This problem considers sequence dependent setup time, parallel machines and compatibility constraints between machines and resources (molds). The considered objective is to minimize the jobs' tardiness (i.e., the difference between the completion time of the jobs and their due date). First, a mathematical formulation of the problem has been provided. Then, an optimization method aiming to solve the big-sized instances has been developed. The proposed tool consists of a hybridization of a list-algorithm and a metaheuristic and can be seen as a master/slave approach. The obtained results show that the method provides high quality results for the proposed problem, outperforming a two-stage heuristic previously developed for the same problem [30] and currently used by the case-study injection plant.

This hybridization has been proposed instead of a classical metaheuristic because it is easier to develop. Indeed, using a classical metaheuristic needs to define a neighborhood system to browse the solution space. This neighborhood system can be complex: the neighbor solution needs to respects all constraints of the specific problem. In our approach, the neighbor is a swap between two jobs in the list of jobs and then the application of the list algorithm ensure the constraints respect. This has two interests. Our tool aims to solve industrial applications, so an easy, friendly tool may have more chance to be used by a company. Plus, we intend to solve others industrial applications, using the same tool, only adapting the list algorithm and the objective function (see [1] for other implementation of this tool). Ultimately, a generic tool that could be used to solve several different operational management problems could be developed. A strong point of the proposed method is that the metaheuristic can be used without any change for a large portfolio of problems. The intended tool would thus have a generic part composed of this metaheuristic and a library of list algorithms, which will allow rapid implementation by different companies, facing different planning, scheduling and assignment issues.

Several ways of improvement are considered to go deeper in the development of our tool. Within the hybridization, new metaheuristics could be used for the proposed method, including other single solution-based methods as iterated local search or population-based metaheuristics, such as particle swarm optimization using parallel computation with GPU. This will make it possible to assess the effectiveness of the proposed tool under different metaheuristic approaches. New list algorithms are currently proposed for different industrial problems. We are currently collecting data in companies with lot-sizing, scheduling and assignment problems. These problems take into account new constraints, as precedencies between jobs. Moreover, new objective function are considered: for instance improving the ergonomic and not only the economic aspect. Thus, our tool could face more realistic industrial problems from Industry 4.0, which deals with new markets and new technologies.

Author Contributions: Methodology, L.P. and N.K.; Supervision, C.S.; Writing—review & editing, N.K. and M.A.A. All authors have read and agreed to the published version of the manuscript.

Funding: This research was partially funded by FEDER Hauts de France and CEA Tech.

Institutional Review Board Statement: Not applicable.

Informed Consent Statement: Not applicable.

Data Availability Statement: The data presented in this study were generated on the basis of information provided by a case-study company and are available on request from the corresponding author.

Conflicts of Interest: The authors declare no conflict of interest.

References

1. Klement, N.; Silva, C. A generic decision support tool to planning and assignment problems: Industrial applications and industry 4.0. In *Scheduling in Industry 4.0 and Cloud Manufacturing*; Sokolov, B., Ivanov, D., Dolgui, A., Eds.; Springer International Publishing: Cham, Switzerland, 2020; pp. 167–192. [CrossRef]
2. Dastidar, S.G.; Nagi, R. Scheduling injection molding operations with multiple resource constraints and sequence dependent setup times and costs. *Comput. Oper. Res.* **2005**, *32*, 2987–3005. [CrossRef]
3. Van Wassenhove, L.N.; De Bodt, M.A. Capacitated lot sizing for injection moulding: A case study. *J. Oper. Res. Soc.* **1983**, *34*. [CrossRef]
4. Nagarur, N.; Vrat, P.; Duongsuwan, W. Production planning and scheduling for injection molding of pipe fittings a case study. *Int. J. Prod. Econ.* **1997**, *53*. [CrossRef]
5. Erlenkotter, D. Ford whitman harris and the economic order quantitymodel. *Oper. Res.* **1990**, *38*, 937–946. [CrossRef]
6. Copil, K.; Wörbelauer, M.; Meyr, H.; Tempelmeier, H. Simultaneous lotsizing and scheduling problems: A classification and review of models. *OR Spectr.* **2016**, 1–64. [CrossRef]
7. Gramani, M.C.N.; França, P.M. The combined cutting stock and lot-sizing problem in industrial processes. *Eur. J. Oper. Res.* **2006**, *174*, 509–521. [CrossRef]
8. Luche, J.R.D.; Morabito, R.; Pureza, V. Combining process selection and lot sizing models for production scheduling of eletrofused grains. *Asia-Pac. J. Oper. Res.* **2009**, *26*, 421–443. [CrossRef]
9. Gaudreault, J.; Frayret, J.M.; Rousseau, A.; Amours, S.D. Combined planning and scheduling in a divergent production system with co-production:A case study in the lumber industry. *Comput. Oper. Res.* **2011**, *38*, 1238–1250. [CrossRef]
10. Martinez, K.Y.P.; Toso, E.A.V.; Morabito, R. Production planning in the molded pulp packaging industry. *Comput. Ind. Eng.* **2016**, *98*, 554–566. [CrossRef]
11. Potts, C.N.; Van Wassenhove, L. Integrating scheduling with batching and lot-sizing: A review of algorithms and complexity. *J. Oper. Res. Soc.* **1992**, 395–406. [CrossRef]
12. Drexl, A.; Kimms, A. Lot sizing and scheduling: Survey and extensions. *Eur. J. Oper. Res.* **1997**, *99*, 221–235. [CrossRef]

13. Jordan, C. Discrete lot-sizing and scheduling by batch sequencing. In *Batching and Scheduling*; Springer: Berlin/Heidelberg, Germany, 1996; pp. 95–119. [CrossRef]
14. Potts, C.N.; Kovalyov, M.Y. Scheduling with batching: A review. *Eur. J. Oper. Res.* **2000**, *120*, 228–249. [CrossRef]
15. Centeno, G.; Armacost, R.L. Parallel machine scheduling with release time and machine eligibility restrictions. *Comput. Ind. Eng.* **1997**, *33*, 273–276. [CrossRef]
16. Centeno, G.; Armacost, R.L. Minimizing makespan on parallel machines with release time and machine eligibility restrictions. *Int. J. Prod. Res.* **2004**, *42*, 1243–1256. [CrossRef]
17. Ruiz, R.; Maroto, C. A genetic algorithm for hybrid flowshops with sequence dependent setup times and machine eligibility. *Eur. J. Oper. Res.* **2006**, *169*, 781–800. [CrossRef]
18. Kim, D.-W.; Na, D.-G.; Chen, F.F. Unrelated parallel machine scheduling with setup times and a total weighted tardiness objective. *Robot. Comput.-Integr. Manuf.* **2003**, *19*, 173–181. [CrossRef]
19. Sáenz-Alanís, C.A.; Jobish, V.D.; Salazar-Aguilar, M.A.; Boyer, V. A parallel machine batch scheduling problem in a brewing company. *Int. J. Adv. Manuf. Technol.* **2016**, *87*, 65–75. [CrossRef]
20. Yugma, C.; Dauzère-Pérès, S.; Artigues, C.; Derreumaux, A.; Sibille, O. A batching, and scheduling algorithm for the diffusion area in semiconductor manufacturing. *Int. J. Prod. Res.* **2012**, *50*, 2118–2132. [CrossRef]
21. Lenstra, J.K.; Rinnooy Kan, A.H.G.; Brucker, P.J. Complexity of machine scheduling problems. *Ann. Discret. Math.* **1977**, *1*, 343–362. [CrossRef]
22. Omar, M.K.; Teo, S.C. Minimizing the sum of earliness/tardiness in identical parallel machines schedule with incompatible job families: An improved MIP approach. *Appl. Math. Comput.* **2006**, *181*, 1008–1017. [CrossRef]
23. Toutouh, J.; Nesmachnow, S.; Alba, E. Fast energy-aware olsr routing invanets by means of a parallel evolutionary algorithm. *Clust. Comput.* **2012**, *16*. [CrossRef]
24. Afsar, H.M.; Lcomme, P.; Ren, L.; Prodhon, C.; Vigo, D. Resolution of a job-shop problem with transportation constraints: A master/slave approach. In Proceedings of the 8th IFAC Conference on Manufacturing Modelling, Management and Control MIM, Troyes, France, 28 June 2016. [CrossRef]
25. Gourgand, M.; Grangeon, N.; Klement, N. An analogy between bin packing problem and permutation problem: A new encoding scheme. In *Advances in Production Management Systems. Innovative and Knowledge-Based Production Management in a Global-Local World*; Springer: Berlin/Heidelberg, Germany, 2014; Volume 438, pp. 572–579. [CrossRef]
26. Zhu, X.; Wilhelm, W.E. Scheduling and lot sizing with sequence-dependent setup: A literature review. *IIE Trans.* **2006**, *38*, 987–1007. [CrossRef]
27. Klement, N.; Gourgand, M.; Grangeon, N. Medical imaging: Exams planning and resource assignment: Hybridization of a metaheuristic and a list algorithm. In Proceedings of the 10th International Conference on Health Informatics, Porto, Portugal, 21–23 February 2017. [CrossRef]
28. Metropolis, N.; Rosenbluth, A.W.; Rosenbluth, M.N.; Teller, A.H.; Teller, E. Equation of state calculations by fast computing machines. *J. Chem. Phys.* **1953**, *21*, 1087–1092. [CrossRef]
29. Aarts, E.; van Laarhoven, P. *Simulated Annealing: Theory and Applications*; Kluwer Academic Publishers: Dordrecht, The Netherlands, 1987. [CrossRef]
30. Silva, C.; Ferreira, L.M. Microplano—A scheduling support system for the plastic injection industry. In *E-Manufacturing: Business Paradigms and Supporting Technologies*; Springer: Berlin/Heidelberg, Germany, 2004; pp. 81–89. [CrossRef]

 *applied sciences*

Article

An Optimization Model for Operational Planning and Turnaround Maintenance Scheduling of Oil and Gas Supply Chain

Ahmed M. Ghaithan

Construction Engineering and Management Department, College of Environmental Design,
King Fahd University of Petroleum and Minerals, Dhahran 31261, Saudi Arabia; ahmedgh@kfupm.edu.sa;
Tel.: +966-138603212

Received: 1 October 2020; Accepted: 22 October 2020; Published: 26 October 2020

Abstract: Hydrocarbon supply chain (HCSC) is a complex network that extends from oil and gas fields to demand nodes. Integrating operation and maintenance activities along this complex network is crucial since the hydrocarbon industry is the most influential sector in the world economy, and any disruptions or variations in hydrocarbon product supply will affect the whole world economy. Therefore, effective and thoughtful maintenance extends the life of an asset and enhances its reliability. To prevent huge losses in production and ultimately satisfy customer needs, the maintenance jobs are preferred to be performed during times of low demand. Thus, operation planning and maintenance scheduling decisions are dependent and should be optimized simultaneously. Therefore, the aim of this study is to develop an integrated mathematical model for the operation and maintenance planning of the oil and gas supply chain. The utility of the proposed model has been demonstrated using the Saudi Arabian HCSC. The proposed model effectively produces optimal operation and maintenance schedule decisions. A sensitivity analysis was conducted to study the effect of critical parameters on the obtained results.

Keywords: supply chain optimization; oil and gas supply chain; maintenance scheduling; operation planning; energy

---

1. Introduction

The hydrocarbon supply chain (HCSC) comprises oil and gas networks. The oil network involves oil fields, oil and gas separation plants (GOSPs), primary storage facilities, oil processing plants, refinery plants, secondary storage facilities, and demand nodes. In comparison, the gas network consists of gas fields, storages facilities, gas plants, fractionation plants, secondary storage facilities, and demand nodes. The two networks overlap in some entities and share some products.

Few researchers have studied the optimization of operational planning under the assumption of plant availability with full capacity, yet this assumption is practical. It is crucial to consider interruptions or plants' sudden shutdown when planning for operation. In the real world, plants might be totally or partially unavailable due to periodic maintenance, shutdown, etc. Thus, production capacities will be reduced or reach zero. To restore the plants to their full capacities and keep them in good condition, plants should be maintained regularly and routinely to prevent unplanned failures and increase production rates. Maintenance activities reduce the sudden breakdown and prolong the unit's life. Thus, operation planning models need to be more realistic and should consider the maintenance schedule. To prevent huge losses in production and eventually satisfy customer needs, maintenance jobs should be performed during times of low demands. Besides, connections between upstream and downstream plants need to be considered before deciding to perform maintenance activities, especially in complex networks such as HCSC.

Thus, operation planning models must be more realistic and take into account maintenance tasks. Several studies have addressed the integration of operation and maintenance planning activities within the scope of a single machine in different areas and applications, such as the inventory control area. According to the author's knowledge, no work has been done to integrate operation planning and maintenance scheduling in the supply chain.

The study aimed to develop a mixed integer programming (MIP) model to coordinate the operation and maintenance activities of the oil supply chain plants, taking into account the best use of available human resources and maintenance resources, while satisfying the maintenance and operational constraints. The expected outputs of the proposed model are the production volumes, flow quantity, exportation volume, inventory level, as well as maintenance schedules for influential HCSC plants while maximizing profit. Maximizing profit is necessary for any organization to meet financial obligations and support the country's budget.

The utility of the model is demonstrated using a real case (Saudi Arabia's oil and gas supply chain). The model is expected to have a positive impact on the future management of this important component of the energy sector. A valuable and practical sensitivity analysis was conducted to study the effect of the key parameters on the obtained results. Some managerial insights will be derived and are expected to help and guide decision-makers as they build their decisions.

The integration of operation and maintenance activities have not been studied adequately, specifically in the supply chain context. In the area of the hydrocarbon supply chain, no work has tackled production planning and maintenance scheduling simultaneously. Therefore, the maintenance scheduling models of petroleum plants and integrated models in the area of inventory control will be reviewed in this section. In the HSCS, operation and maintenance schedules have been treated independently. For example, [1] developed a multi-objective model for the operational planning of the Hydrocarbon supply chain under the assumption of the entities available during the planning horizon. Few studies have studied the maintenance scheduling of single entities of a petroleum network such as the refinery plant. For instance, [2] developed a Mixed Integer Linear Programming (MILP) model for scheduling the preventive maintenance of units of a refinery plant. The proposed model aims to maximize the utilization level of units while satisfying maintenance constraints.

Mendez et al. [3] developed an MILP model for the optimization of the short term blending and scheduling problem in applications of an oil refinery. The model aims to optimize the profit and satisfy the quality requirements. The authors developed a novel method to solve the proposed model. Göthe-Lundgren et al. [4] proposed an MILP model for the production scheduling of the oil-refinery problem. In the proposed model, the production process consisted of two hydro-treatment units and one unit of distillation. The aim of the scheduling was to decide which model of operation to use in each processing unit at each point in time in order to satisfy the demand while minimizing the production cost and considering storage capacities. Antomarioni et al. [5] developed a data-driven model for determining the type of maintenance that should be performed to a refinery plant. Bertolini et al. [6] addressed an approach based on risk inspection and maintenance for maintaining activities in an oil refinery to minimize the risk. A heuristic algorithm was utilized to solve the proposed model, considering the time, human resources, and budget constraints.

Aissani et al. [7] presented a multi-agent approach for the scheduling of dynamic maintenance activities for the petroleum industry using a reinforcement approach. Amaran et al. [8] proposed an MILP model for planning the turnaround activities of interconnected plants in a chemical site. The aim of the proposed model was to maximize the net present value. Awad and Ertem [9] developed a stochastic model to schedule preventive maintenance for an oil refinery plant taking into consideration the uncertainty in maintenance time. The model aimed to minimize the total adjusted tardiness times while satisfying the amount of multiple-skill-level maintenance crew and budget limits. Xu et al. [10] developed a proactive scheduling model for maintaining refinery plants under the uncertainty of crude shipping delays. A mixed-integer nonlinear programming model was developed to optimize crude unloading, transferring, and processing at the refinery plant. Recently, [11] extended the [10] work by integrating proactive and reactive scheduling.

Several studies have tackled the issue of coordinating production and maintenance activities, especially in the inventory control area. The first pioneering work in this field was performed by Rosenblatt and Lee [12], who formulated a model to optimize economic production quantity and at the same time, generate a maintenance plan for a production system. Joshi and Gupta [13] employed the production plans along with a history of machine failure to optimize maintenance activities. In his dissertation, Pintelon [14] proposed an integrated model for production and maintenance planning simultaneously. The aim of the developed model was to find out the period between two consecutive preventive maintenance tasks. Ashayeri et al. [15] proposed an MILP model for scheduling production jobs and the preventive maintenance of a multi-line production system. The aim of the model is to examine the possibility of performing preventive maintenance before a new job or run the job without performing preventive maintenance until a breakdown happened then repair the production system. Rahim and Ben-Daya [16] reviewed inventory models that integrated production planning, quality, and maintenance activities. Iravani and Duenyas [17] developed a policy for a production inventory system operated by a single machine subjected to deterioration. The maintenance and production decisions were optimized by solving the model using the Markov decision process.

Cassady and Kutanoglu [18] developed an integrated preventive maintenance and production scheduling model. The aim of the model was to decrease the overall weighted job slowness. Cassady and Kutanoglu [19] investigated an integrated model for maintenance and production scheduling with the aim to minimize the overall anticipated weighted job completion time. Darwish and Ben-Daya [20] introduced an inventory model combining preventive maintenance and production and taking into account an inspection error. Aghezzaf et al. [21] developed a hybrid production planning and maintenance model for a deteriorating production system. To renew and restore the system, preventive maintenance was carried out on a regular basis, and minimal repair was carried out. Production and maintenance plans have been developed to reduce overall production and maintenance costs. Aghezzaf and Najid [22] extended the [21] work for a system consisting of more than one production line which operates in parallel.

Zied et al. [23] formulated a model for a manufacturing system that was subjected to random failure and assuming the demand was stochastic. The failure rate was assumed to be a function of machine working time and production rate. The output of the model was an optimal production plan combined with a preventive maintenance policy, which minimized the total cost subject to a target service level. Berrichi et al. [24] focused on the problem of maintenance and production planning in a parallel machine case by developing a bi-objective integrated model which enabled the manager to decide the production assignment to machine and plan the maintenance for each machine at the same time. Thus, machines not in duty can be taken for maintenance as per a pre-identified plan. Ayed et al. [25] developed a model for optimizing a manufacturing system that faces random failures. Due to failure, a random shortage demand was substituted by subcontracting from a different production system. It was assumed that the manufacturing system failure rates are both functions of time and production rate. A preventive maintenance schedule was determined to minimize manufacturing system decay.

Fitouhi and Nourelfath [26] formulated an integrated production and maintenance planning model for a single machine that was subjected to random failure. Preventive maintenance is applied periodically, and minimal repair was performed to renew the machine to its normal state. Yalaoui et al. [27] developed a model similar to [22] work for multi-machines. Assid et al. [28] developed a simulation model that integrated production, setup, and maintenance activities simultaneously.

Khosrojerdi et al. [29] developed a robust mathematical model for designing a power supply chain. The model considers planning the power generation and scheduling of preventive and corrective maintenance due to possible failures in the electricity transmission system. Cheng et al. [30] proposed a model for a machining system in which the production rate affects both the cutting tool age and the reliability of the machine. The developed model takes into consideration the addition to production scheduling, a preventive maintenance of the machine, and the tool replacement. Ettaye et al. [31] explored the aspect of integrated production operations and maintenance planning activities at the tactical level of multi-line systems. The policy of maintenance provides preventive substitutions at the start of every cycle as well as minimal repairs in the event random failures occur. The aim of the model was to minimize the total costs and concurrently determine the maximum production plan alongside the replacement time.

Géhan et al. [32] presented a joint optimization model for integrating master production schedule and maintenance strategy for a manufacturing system. The proposed policy allows minimizing the impacts of potential random failures. A heuristic algorithm was invented for solving the proposed model. Liu and Lv [33] discussed a maintenance optimization strategy for a multi-component system with the use of the degradation information of the system. They argued that combining the maintenance activities was cheaper than having each component of a system planned separately. Zhao et al. [34] proposed a model for the integration of maintenance and production planning considering the production losses due to preventive maintenance, corrective maintenance and operation dependent failures. Hnaien et al. [35] discussed the problem of integrating production planning and preventative maintenance for a single machine in order to maximize its production capacity. A mixed-integer linear programming model was proposed to tackle the problem.

Askri et al. [36] proposed a preventive maintenance and production policy for a manufacturing system consisting of parallel machines considering the lease contract. The aim of the proposed model was to determine the economical production quantity and the preventive maintenance interval for the machines while minimizing the total cost of production and maintenance. Amiri and Honarvar [37] investigated a bi-objective mathematical model which integrated both tasks; preventive maintenance planning and energy hub components repair under the assumption of stochastic demand. Khatab et al. [38] discussed the integration of production quality and the condition-based maintenance of a single-machine production system. The problem tackled assumes that a machine will have some product defects after a certain level of degradation. The model considers the inspection time and preventive maintenance to reduce off-specification products.

The oil and gas supply chain includes many petrochemical plants such as sweetening plants, refinery plants, gas plants, and fractionation plants which constitute connected and advanced equipment to help maximize the value of the petroleum product. To reduce the sudden breakdown, prolong their lives, and minimize the lifetime cost of energy, both operations and maintenance costs need to be optimized simultaneously. Similarly, renewable energy plants such as onshore/offshore wind energy plants involve many connected units that need to be maintained regularly. Wind power is often considered as an expensive source of energy. Therefore, coordinating operations and maintenance decisions will reduce the costs for operations and maintenance which indicates that there is large potential in cost-savings. Several studies have optimized operation and maintenance scheduling for renewable energy power plants applications. For example, Rinaldi et al. [39] developed an optimization model to coordinate the operation and maintenance activities for offshore renewable energy farm. Martin et al. [40] studied the integration of operation and maintenance scheduling of an offshore wind farm by conducting a sensitivity analysis to identify the most significant factors affecting availability and operation and maintenance costs.

Rinaldi et al. [41] optimized the operation and maintenance activities of an offshore wind farm using genetic algorithms in order to minimize operation and maintenance costs of the offshore farm while maximizing both the availability and reliability. Helsen et al. [42] studied the optimization of the operation and maintenance strategy of a wind farm using a big data approach with the aim to understand the failure mechanisms of wind turbines. Nielsen and Sørensen [43] developed a mathematical model for operation planning and maintenance scheduling of a single wind turbine. Both condition-based and corrective maintenance are compared taking into consideration constraints of design and inspection. Dinwoodie [44] developed a mathematical model for coordinating the operation and maintenance decisions for an offshore wind farm. This model minimizes the lifetime cost of energy by optimizing both the operations and maintenance costs of offshore wind farm. Seyr and Muskulus [45] conducted a comprehensive review on the area of operation and maintenance of offshore wind farms.

The above literature review revealed that the integration of operation and maintenance activities in the oil and gas supply chain area has not been tackled; however, several studies have addressed the integration of operation and maintenance activities in the scope of a single entity in different areas and applications such as inventory control. To the author's knowledge, no work has appeared in the literature to integrate operation planning and maintenance scheduling for HCSC. Therefore, this paper fills the aforementioned gaps.

2. Materials and Methods

The focus of this research is on developing an optimization model for integrating operation planning and the turnaround maintenance of oil and gas supply. To achieve the research objectives and answer the research question, a systematic approach is followed as: first, the problem is formulated, and an optimization model is developed. Second, the actual case study employed to study the utility of the proposed model is presented. Finally, the output findings are discussed, and a sensitivity analysis is conducted to study the impact of changes in input parameters on the optimal findings.

2.1. Problem Statement

Hydrocarbon supply chain (HCSC) is a complex and long network extending from the production of crude oil and natural gas to marketing the oil and gas products. The entities along the HCSC constitute oil and gas fields, oil and gas separation plants (GOSPs), primary storage facilities, oil and gas processing plants, refinery plants, fractionation plants, secondary storage facilities, and demand nodes. The crude oil from oil fields and/or GOSPs is pumped to oil processing plants for removing sulfur. After removing the sulfur, the oil is shipped to international and local terminals. Locally, the crude oil is further transformed in refinery plants into its oil by-products (i.e., diesel, naphtha, gasoline, Liquefied Petroleum Gas (LPG), fuel oil, asphalt, and kerosene). The oil by-products are then used to satisfy international and local demands. The associated natural gas produced at gas–oil separation plants (GOSPs) and the non-associated natural gas produced directly from pure gas fields are streamed to gas plants in which they are processed into natural gas liquid (NGL), methane and hydrogen sulfide. The methane is used as a raw material by industry, and the NGL is fractionated in fractionation plants into its gas products (i.e., ethane, propane, butane, and natural gasoline) used to satisfy the local and international demand. A general representation of the entities of the HCSC are shown in Figure 1.

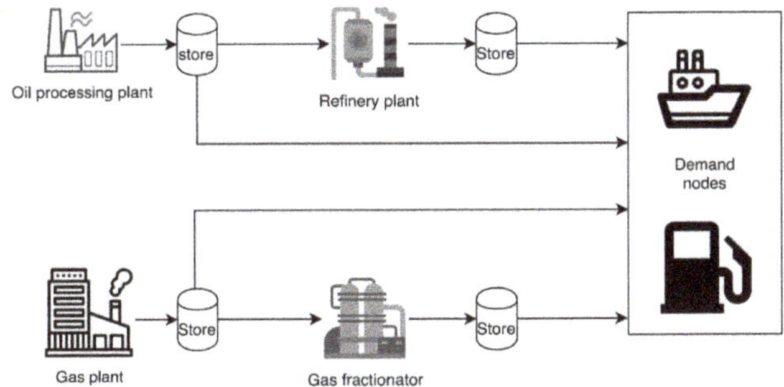

Figure 1. Schematic representation of downstream oil and gas supply chain.

In the real-world, HCSC plants might be totally or partially unavailable due to periodic maintenance and sudden shutdown. Thus, their production capacities will be reduced or reach zero. To restore these plants to their full capacities and keep them in good condition, plants should be maintained regularly and routinely to prevent unplanned failures, increase plant production rates, and prolong their life. Thus, operation planning models need to be more realistic and should consider maintenance scheduling. The maintenance cycle and duration of petroleum plants such as oil processing plants, refinery plants, gas plants, and fractionation plants must be identified before model formulation.

The aim of the proposed MIP model is to coordinate the operation and maintenance activities of the HCSC plant network, taking into account the best use of available manpower and maintenance resources, while satisfying maintenance and operational constraints. The objective of the proposed model is to maximize the HCSC profit. The model will help to determine the optimal production, inventory and transport volumes and at the same time, generate an optimal maintenance schedule. Each HCSC plant is assumed to have its own storage facility to satisfy the needs during planned and/or unexpected failures, as well as a plant shutdown. In the hydrocarbon industry, inventory is an important and expensive asset; therefore, it is important to manage the inventory level and specify their capacities precisely.

2.2. Mathematical Model

In this section, an integrated operation and maintenance scheduling model was developed. The sets, indices, variables, and parameters used in the proposed model are summarized in Table 1.

After defining the notations required for model development, the constraints and objective functions will be formulated below:

The following constraint limits the number of times a plant *i* has to be maintained during the planning horizon:

$$\sum_{t \in T} St_{it} = u_i \quad \forall \ i \in \{N1, N2\} \tag{1}$$

The efficiency of the maintenance at time-period intervals is implemented by equating the corresponding binary variables using the following equation where fi represents the length of time between each maintenance activity for a plant *i*. For instance, if a particular plant *i* has been maintained at time *t*, the second maintenance task will be performed at time $t + f$:

$$St_{it} = St_{i\,t+f_i} \quad \forall \ i \in \{N1, N2\}, t \in T \tag{2}$$

The size of the crew needed to perform maintenance jobs in plant *i* should not exceed the available manpower. The following constraint is added to ensure that there should be no maintenance task for the same crew while they are maintaining other plants. In addition, no more than a fixed number of plants can be served at the same time due to the limited number of crew available. This constraint can be represented mathematically as follows:

$$\sum_{\tau=t-n_i+1}^{t} St_{i\tau} R_{it} \leq W_{it} \quad \forall \ i \in \{N1, N2\}, t \in T \tag{3}$$

The material balance of the plants belong to the first level of HCSC (oil processing plant and gas plants) are represented by Equation (4) where the material flow from the upstream is equal to the the production volumes at each plant *i*. The processed quantities are then sent to the next level of the HCSC (refinery plants; N2, fractionation plant; N2 and demand node; N3) to satisfy the local and international demands:

$$\sum_{r} w_{it}^{p} Y_{it}^{r} = \sum_{j \in N2} X_{ijt}^{p} + \sum_{j \in N3} X_{ijt}^{p} \quad \forall \ t \in T, p \in P, i \in N1 \tag{4}$$

Table 1. Sets, subsets, variable and parameters notations.

	Sets	
$i, j \in I$	:	All nodes;
$N1 \subseteq I$	:	Set of petroleum plants belong to the first level of the supply chain;
$N2 \subseteq I$	:	Set of petroleum plants belong to the second level of the supply chain;
$N3 \subseteq I$	:	Set of demand nodes;
$S1 \subseteq I$	:	Set of storage tanks belong to the first level of the supply chain;
$S2 \subseteq I$	:	Set of storage tanks belong to the second level of the supply chain;
R	:	Set of raw material imported from upstream of the supply chain to the first level;
P	:	Set of products processed at level 1;
B	:	Set of products processed at level 2;
T	:	Time horizon.

	Indices	
t	:	Time periods in weeks, $t = \{1, 2, \ldots, T\}$;
r	:	Raw material imported from upstream, $r = \{1, 2, \ldots, R\}$;
p	:	Product produced at level 1, $p = \{1, 2, \ldots, P\}$;
b	:	Product produced at level 2, $b = \{1, 2, \ldots, B\}$.

	Decision Variables	
Y^r_{it}	:	Supply of raw material r from upstream to plants belong to first level $i \in N1$ at time t;
X^p_{ijt}	:	Flow of product p from plant i to customer/plant j at time t;
X^b_{ijt}	:	Flow of by-product b from plant i to customer/plant j at time t;
I^p_{it}	:	Inventory level of product p at time t at plants of level 1;
I^b_{it}	:	Inventory level of product p at time t at plants of level 2;
x^{p+}_{it}, x^{p-}_{it}	:	Amount of product $p \in P$ produced above, below the demand of demand node $i \in N3$ during time $t \in T$;
x^{b+}_{it}, x^{b-}_{it}	:	Amount of product $b \in B$ produced above, below the demand of demand node $i \in N3$ during time $t \in T$;
St_{it}	:	$\begin{cases} 1 \text{ if plant } i \text{ starts its maintenance at period } t, \\ 0 \text{ otherwise} \end{cases}$

	Parameters	
Dt_i	:	The duration of maintenance for the plant i.
R_{it}	:	The manpower required for maintaining plant i during time t.
W_{it}	:	The total manpower available for plant i during time t.
Ca_i	:	Capacity of plant i.
CS_i	:	Capacity of storage tank i.
Cr_{ij}	:	Capacity of route connecting plant with plant/customer j.
D^p_{jt}	:	Demand quantities of product p at demand node $j \in N3$ at time t.
D^b_{jt}	:	Demand quantities of by-product b at demand node $j \in N3$ at time t.
Ls^p_{jt}	:	Lost sales of products p at demand node $j \in N3$ at time t;
Ls^b_{jt}	:	Lost sales of by-products b at demand node $j \in N3$ at time t;
Cm_{it}	:	Cost of maintaining plant i at time t;
Cp^r_{it}	:	Cost of processing raw material r at plant i at time t;
Cp^p_{ijt}	:	Cost of processing product stream X^p_{ijt} coming from plant i and processed at plant j at time t;
Ct^p_{ijt}	:	Cost of shipping product p from plant i to customer/plant j at time t;
Ct^b_{ijt}	:	Cost of shipping by-product b from plant i to customer/plant j at time t;
w^{p+}_{it}, w^{p-}_{it}	:	Cost incurred due to producing product $p \in P$ more than, less than the demand of node $i \in N3$ at time $t \in T$;
w^{b+}_{it}, w^{b-}_{it}	:	Cost incurred due to producing product $b \in B$ more than, less than the demand of node $i \in N3$ at time $t \in T$;
Hc^p_{it}	:	Inventory holding cost of product p per period t at plants i of level 1;
Hc^b_{it}	:	Inventory holding cost of product b per period t at plants i of level 2;
Sp^p_{jt}	:	Selling price of product p at demand node $j \in N3$ at time t;
Sp^b_{jt}	:	Selling price of by-product b at demand node $j \in N3$ at time t;
w^p_{it}	:	Composition of product p obtained from raw materials streaming from upstream to the plant i at time t;
w^b_{ijt}	:	Composition of by-product b obtained from product p streaming from plant i to plant j at time t;
$OPEC_t$	:	Market share of crude oil specified to an oil production country at time t.

The inlet to the second level of the HCSC (refinery plants and fractionation plant) is then processed and fractionated to its by-products based on the inlet compositions. The by-product flow is then forced to the third level of the HCSC to satisfy the local and international demands N3, Equation (5):

$$\sum_p \sum_{i \in N1} w_{ijt}^b X_{ijt}^p = \sum_{k \in N3} X_{jkt}^b \quad \forall \ t \in T, b \in B, \ j \in N2, \ k > i \tag{5}$$

The material balance for the storage tanks belongs to the first and the second levels of the HCSC which are denoted by the two constraints:

$$\sum_{i \in N1} X_{ijt}^p + I_{it-1}^p = \sum_{j \in N2} X_{jkt}^p + \sum_{j \in N3} X_{jkt}^p + I_{it}^p \quad \forall \ t \in T, p \in P, \ j \in S1, \ k > i \tag{6}$$

The inlet flows from the upstream plants to each store tanks plus inventory left from the previous period $(t-1)$ equal to the outlet from the store tank plus inventory level at time period t:

$$\sum_{i \in N2} X_{ijt}^b + I_{it-1}^b = + \sum_{j \in N3} X_{jkt}^b + I_{it}^b \quad \forall \ t \in T, b \in B, \ j \in S2, \ k > i \tag{7}$$

The inlet flows from plants located at the second level to each store tanks plus inventory left from the previous period $(t-1)$ equal to the outlet from store tank plus inventory level at time period t.

The inlet to any petroleum plant is limited to the plant's capacity. If a plant is in maintenance, the capacity is dropped to zero. Equations (8) and (9) represent the capacity constraints for the plants belong to the first and the second levels of HCSC, respectively:

$$\sum_{r \in R} Y_{it}^r \le \left(1 - \sum_{\tau = t-n_i+1}^{t} St_{i\tau}\right) Ca_{it} \quad \forall \ i \in N1, t \in T \tag{8}$$

$$\sum_{i \in N1} \sum_{p \in P} X_{ijt}^p \le \left(1 - \sum_{\tau = t-n_j+1}^{t} St_{j\tau}\right) Ca_{jt} \quad \forall \ j \in N2, t \in T \tag{9}$$

The storage tanks' capacities are restricted by the following two constraints:

$$CS_j^{min} \le I_{jt}^p \le CS_j^{max} \quad \forall \ j \in S1, \ p \in P, \ t \in T \tag{10}$$

$$CS_j^{min} \le I_{jt}^b \le CS_j^{max} \quad \forall \ j \in S2, b \in B, t \in T \tag{11}$$

Exportation volumes from each crude oil type should not exceed the quota specified by OPEC to each member in the Organization of the Petroleum Exporting Countries, Equation (12). In this study, maintenance was carried out on a weekly basis, while demand is on a monthly basis. Therefore, Dp_t represents the set of time periods included in the demand period starting at time t, and dt denotes the duration of the demand period:

$$\sum_{i \in N1} \sum_{j \in N3} \sum_{p \in P} \sum_{t \in Dp_t} X_{ijt}^p \le OPEC_t \quad \forall \ Dp_t = \{t, \ldots, t+d_t\} \tag{12}$$

Equations (13) and (14) denote that the amount of sales plus shortages minus excess equals demand at any time period t:

$$\sum_{i \in N1} \sum_{t \in Dp_t} X_{ijt}^p + X_{it}^{p+} - X_{it}^{p-} = D_{jt}^p \quad \forall \ j \in N3, p \in P, Dp_t = \{t, \ldots, t+d_t\} \tag{13}$$

$$\sum_{i \in N2} \sum_{t \in Dp_t} X_{ijt}^b + X_{it}^{b+} - X_{it}^{b-} = D_{jt}^b \quad \forall \ j \in N3, b \in B, Dp_t = \{t, \ldots, t+d_t\} \tag{14}$$

The profit Equation (15) involves two main parts: the revenue obtained minus expenses, which includes maintenance costs, processing costs, transportation costs, holding costs, and the penalty costs of producing over and under the specified demand.

$$
\begin{aligned}
Max f = & \left[\sum_{i \in N1} \quad \sum_{j \in N3} \sum_{p \in P} \sum_{t \in T} Sp_{jt}^{p} X_{ijt}^{p} + \sum_{i \in N2} \sum_{j \in N3} \sum_{b \in B} \sum_{t \in T} Sp_{jt}^{b} X_{ijt}^{b} \right] \\
& - \left[\sum_{i \in N1,N2} \sum_{t \in T} St_{it} R_{it} \sum_{\tau}^{t+n_i-1} Cm_{i\tau} + \sum_{i \in N1} \sum_{r \in R} \sum_{t \in T} Cp_{it}^{r} Y_{it}^{r} \right. \\
& + \sum_{i \in N1} \sum_{j \in N2} \sum_{p \in P} \sum_{t \in T} Cp_{ijt}^{p} X_{ijt}^{p} + \sum_{i \in N1} \sum_{j \in N2} \sum_{p \in P} \sum_{t \in T} Ct_{ijt}^{p} X_{ijt}^{p} \\
& + \sum_{i \in N2} \sum_{j \in N3} \sum_{b \in B} \sum_{t \in T} Ct_{ijt}^{b} X_{ijt}^{b} + \sum_{i \in S1} \sum_{p \in P} \sum_{t} Hc_{it}^{p} I_{it}^{p} + \sum_{i \in S2} \sum_{b \in B} \sum_{t \in T} Hc_{it}^{b} I_{it}^{b} \\
& + \sum_{i \in N3} \sum_{p \in P} \sum_{t} W_{it}^{p+} X_{it}^{p+} + \sum_{i \in N3} \sum_{p \in P} \sum_{t} W_{it}^{p-} X_{it}^{p-} + \sum_{i \in N3} \sum_{b \in B} \sum_{t} W_{it}^{b+} X_{it}^{b+} \\
& \left. + \sum_{i \in N3} \sum_{b \in B} \sum_{t} w_{it}^{b-} X_{it}^{b-} \right]
\end{aligned}
\tag{15}
$$

2.3. Case Study

The Saudi Arabian oil and gas network was used to validate the proposed model. The entities of the oil and gas supply chain with the corresponding units and relevant data are reported below. The planning horizon was four years and discretized over 192 weeks. To simplify the development of a complex HCSC network, the supply chain was divided into two subnetworks shown in Figures 2 and 3. The HCSC network is a representation of Figure from [46].

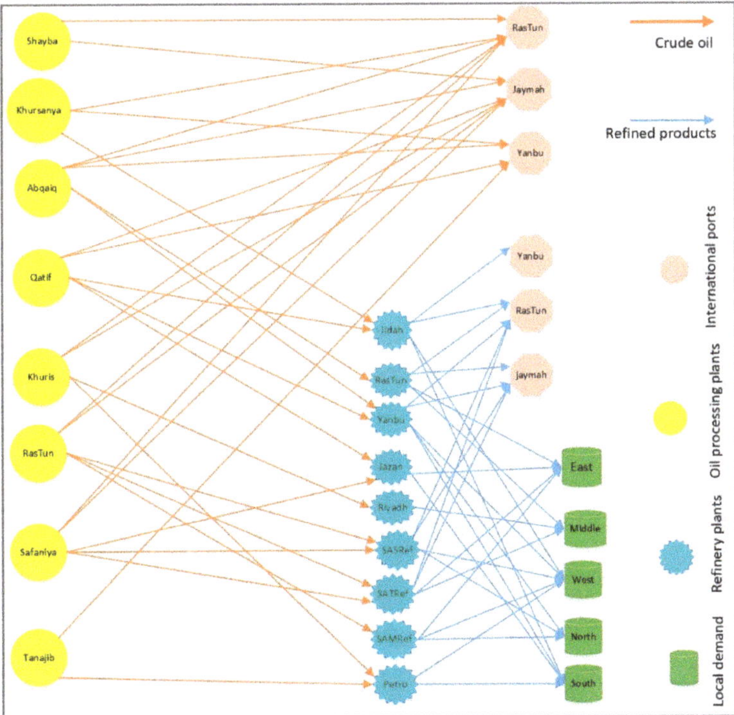

Figure 2. The oil network.

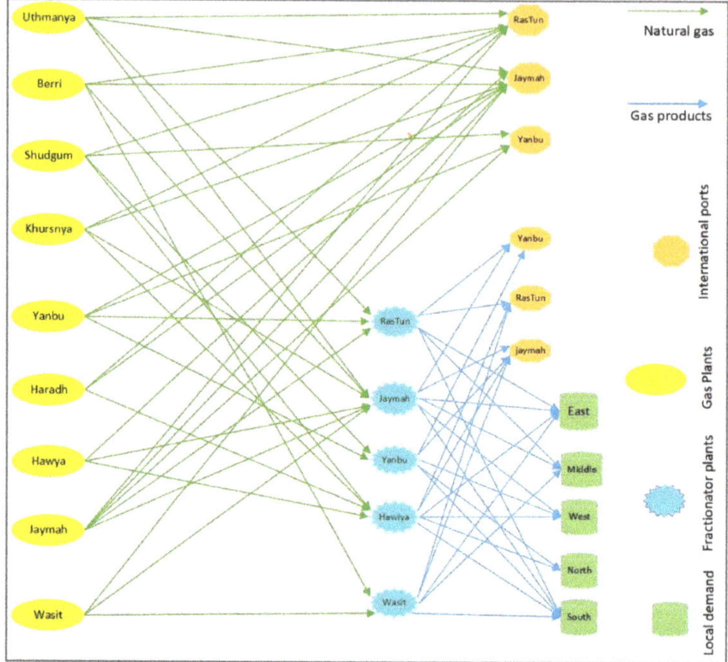

Figure 3. The natural gas network.

The flow and operation of the HCSC is as follows: wellhead oil of different grades; Arabian light, Arabian extra light, Arabian medium, and Arabian heavy produced in upstream units of the HCSC are transported through pipelines to eight (8) oil processing plants for the processing and removal of sulfur. The crude oil is then transported to three (3) major international terminals and (9) local refineries where crude oil is transformed into seven (7) by-products (LPG, naphtha, gasoline, diesel, kerosene, fuel oil, and asphalt) based on their yield. Oil by-products are then distributed to meet the needs of local customers in the main five (5) regions of the Kingdom and three (3) major industrial cities. Parts of LPG, naphtha, and fuel oil are used to meet international demand through terminals.

On the other side, the associated and non-associated natural gas streams produced at gas–oil separation plants (GOSPs) and gas fields, respectively, are transported to nine (9) gas plants for the impurity removal and recovery of hydrogen sulfide. The output gases are NGL and methane. The NGL is supplied to five (5) fractionation plants for the further separation and fractionation into gas products (ethane, butane, propane, and natural gasoline). Part of NGL is exported through international terminals to meet international demand, while methane and ethane are used to meet local demand. At the same time, propane, butane, and natural gasoline meet the demands of international and local customers.

Entities such as oil processing plants, refinery plants, gas plants, etc. will be totally or partially unavailable due to failures or planned maintenance. Thus, their production capacity may be reduced or reached zero. Maintenance is used to restore these entities to their full capacity and keep them in good condition. Each HCSC plant is assumed to have its own storage facility to satisfy needs during planned and/or unexpected failures. In HCSC, inventory is an expensive asset that needs to be considered. Therefore, it is essential to manage inventory levels and capacities. Therefore, when optimizing the HCSC, there is a need to consider operation and maintenance activities in an integrated manner. In this study, a proposed MILP model will be developed and solved. The model will jointly produce a maintenance schedule for each HCSC plant and operational plans such as production volumes, transported volumes, inventory levels, shortages, and excess volumes, etc.

To develop a proposed model, the maintenance cycle and duration for oil processing plants, refinery plants, gas plants, and fractionation plants must be specified, and logical relationships between plants need to be considered. For example, dependent plants are preferred to be stopped for maintenance simultaneously. In addition, similar operating units cannot be taken for maintenance at the same time in order to prevent a huge loss in production. As per consultation with people working in the oil industry, the average maintenance duration for the HCSC plant is between 3–4 weeks. It is assumed that every HCSC plant needs to be maintained twice over the four-year planning horizon. The maintenance crew required to perform maintenance activity in each HCSC plant is ten workers, and the total number of crews available is 20. Every worker earns, on average $120 a week.

Real data were collected from [1,47,48] and were used as input to the proposed model to evaluate its utility. To represent the above stated problem, the MIP model has been developed and solved using the CPLEX 13.3 commercial solver [49]. All computations were done on a PC type intel (R) Core (TM)2 Quad CPU processor with 2.67 GHz and 4 GB RAM using the GAMS (General Algebraic Modeling System) [50]. The model output will be discussed and analyzed in the following sections.

3. Results

To investigate the utility of the proposed model, the Saudi Arabian oil and gas supply chain was employed. The entities of the HCSC with their associated units and relevant data are reported in Section 4. The planning horizon is four years with a one week planning period. In this study, maintenance tasks are carried out on a weekly basis, while demand is on a monthly basis.

The maintenance schedule for all HCSC entities over the planning horizon is displayed in Figure 4. The Gantt chart shows a schedule displaying which each unit will be taken for maintenance and in which weeks. The width of the bar indicates the duration of maintenance in weeks.

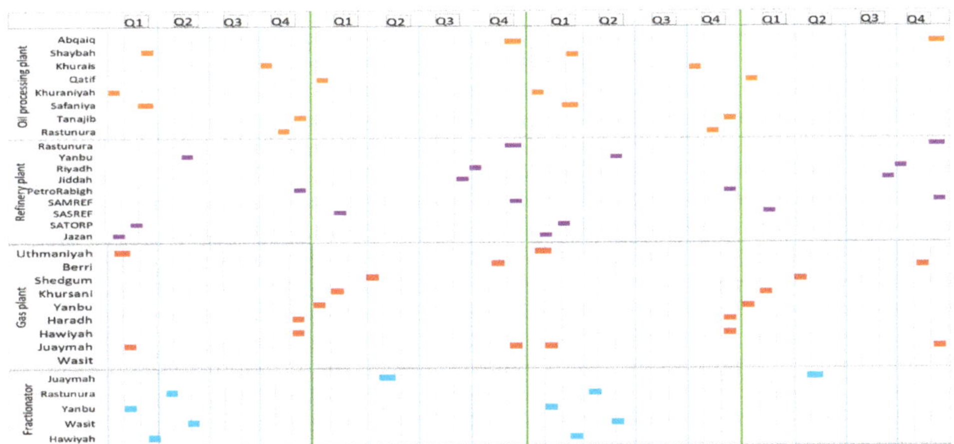

Figure 4. Maintenance schedule for all hydrocarbon supply chain (HCSC) plants (operations and maintenance model); Color code: Orange: oil processing plants; Purple: Refinery plants; Red: Gas plants; Light Blue: Fractionator plants.

The following conclusions can be derived to describe the Gantt chart:

- It was observed that the proposed model distributes maintenance activities uniformly and perfectly among the HCSC plants along the time horizon of four years.

- In addition, it is shown that the model minimizes the amount of lost sales by forcing most of HCSC units to be maintained during the months of low demand. Almost all HCSC units are taken for maintenance during the quarters of low demand (quarters 1, 2, and 4). The third quarter is the summer period, which represents the months of high demand.
- The HCSC is a complex network that involves multi-connections where each plant belongs to the upstream is connected to more than one downstream plant and even to more than one demand node. Similarly, each plant has tanks with enough capacity to feed the associated downstream plants during the maintenance period. This situation provides flexibility for choosing maintenance timing. Therefore, the majority of HCSC plants were taken for maintenance randomly at different times.
- Few connected plants have been taken for maintenance at the same time. For instance, Abqaiq oil processing plant and RasTunura refinery plant are maintained simultaneously. Similarly, both the Tanajib oil processing plants and PetroRabigh refinery plant are maintained at the same time.
- For natural gas plants, maintenance tasks are decoupled; almost all connected upstream and downstream plants are maintained randomly because each downstream plant has connections with more than one upstream unit, and the storage capacity is enough during the maintenance time. In the sensitivity analysis section, the impact of upstream storage capacities on maintenance schedule and arrangement will be investigated.

For the deep investigation, a sample of output for some selected plants will be displayed. For instance, the maintenance schedule, utilization, and production for gas plants during the first year are shown in Figure 5. The third quarter represents the months of the highest demand. Therefore, almost all gas plants have been taken for maintenance during the months of the demand (Quarters 1, 2 and 4).

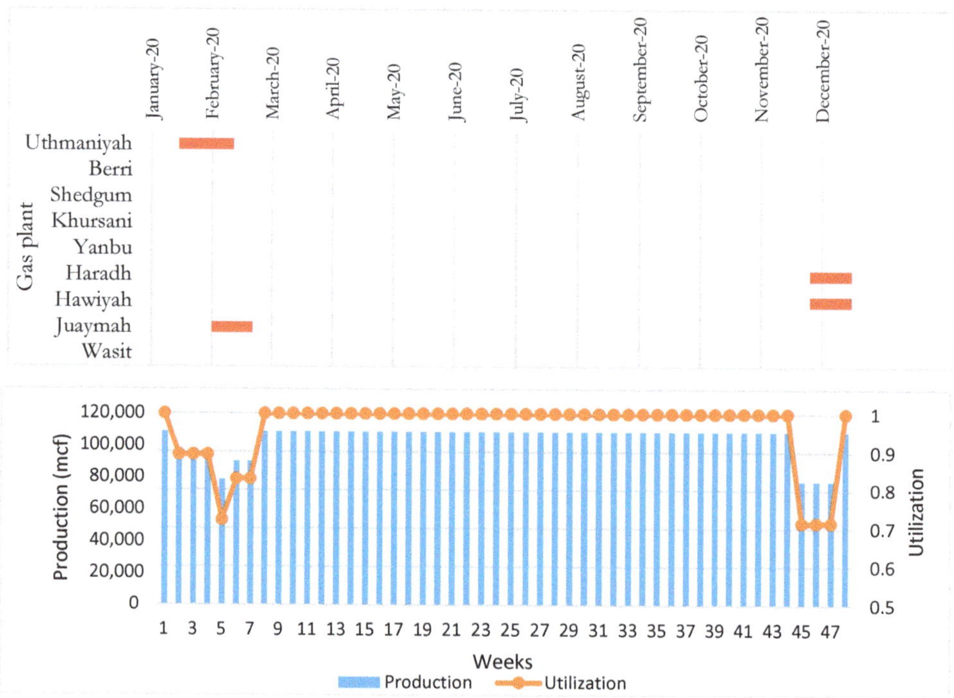

Figure 5. Gas plants maintenance scheduling, production plans, and utilization (year 1).

Similarly, Figure 6 shows the oil processing plants' maintenance schedule, production, and utilization for the first year.

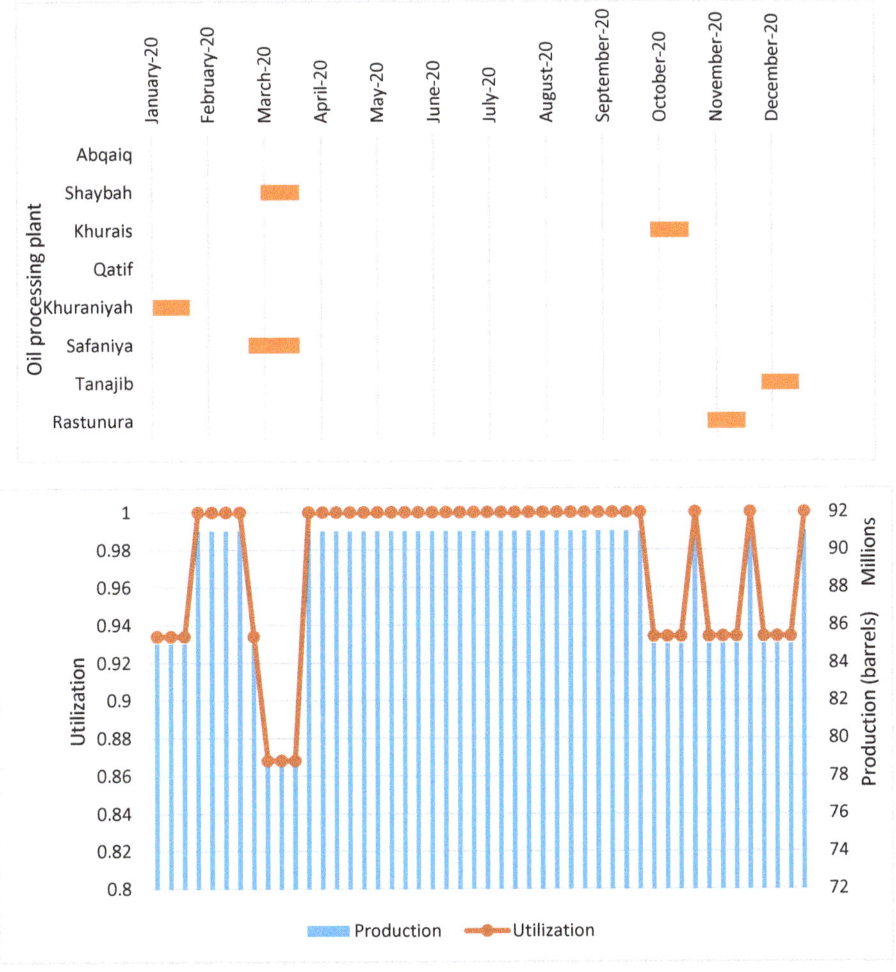

Figure 6. Oil processing plants maintenance scheduling, production, and utilization (year 1).

Figure 7 indicates the fractionation plant storage capacities and inventory levels fluctuations across the planning horizon. Production and maintenance are carried out on a weekly basis, while demand is on a monthly basis. As a result, the inventory level accumulated over the first four weeks of a month, then decreased as demand occurred at the end of the last week of a month.

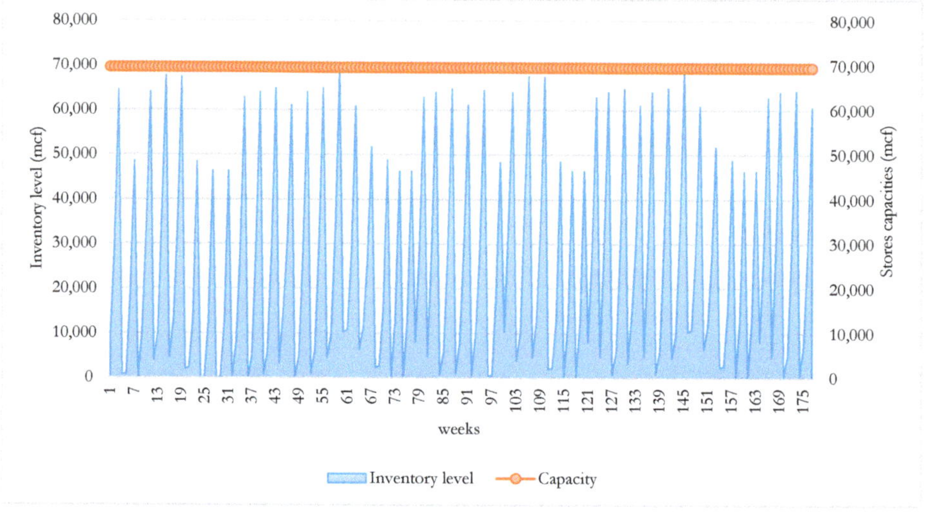

Figure 7. Fractionation plants inventory level and storage capacity.

Sensitivity Analysis

A valuable sensitivity analysis was conducted to show how the model was sensitive to the changes in the key parameters. In this study, the impact of changes in selling prices, demand, and store capacity on model output will be evaluated. Figure 8 shows the variation in profit with changes in selling price. It is noted that as the selling price increases, the total profit linearly increases.

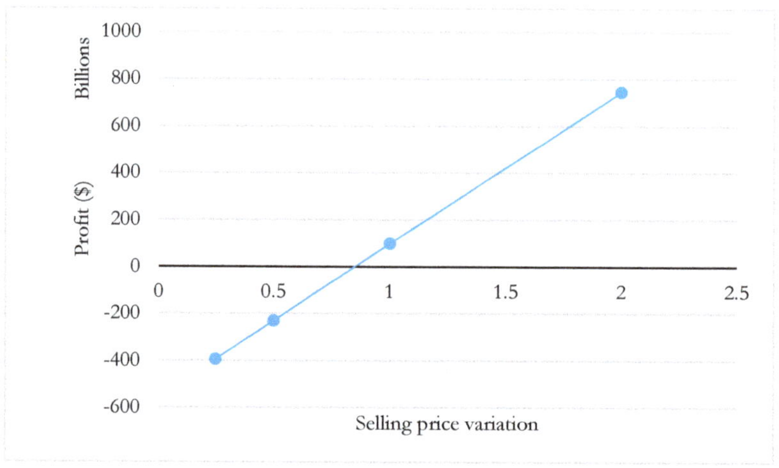

Figure 8. Profatibility changes with variation in demand.

Figure 9 shows the fluctuations in profit with variability in demand. As demand increases, profit decreases because plant and storage capacities are kept constant; therefore, lost sales increase, which leads to customer dissatisfaction and ultimately lower profitability.

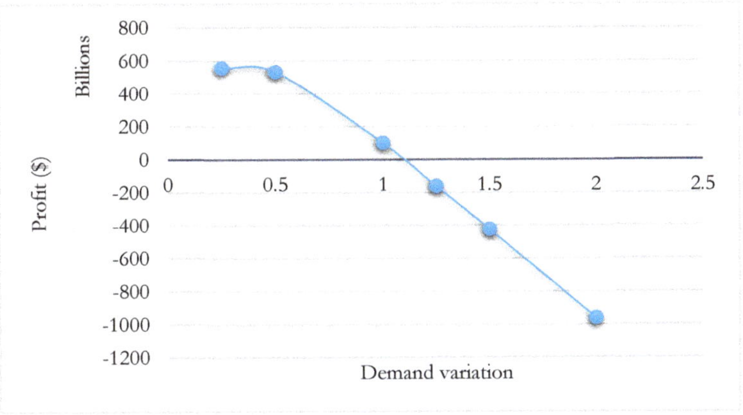

Figure 9. Profitability changes with the variation in demand.

The impact of storage capacity variability on profitability is demonstrated in Figure 10. It is revealed that as storage capacity increases, profit increases due to a decline in lost sales and the satisfaction of most customer needs. Reducing store capacities by more than 55% will generate non-profitable plans.

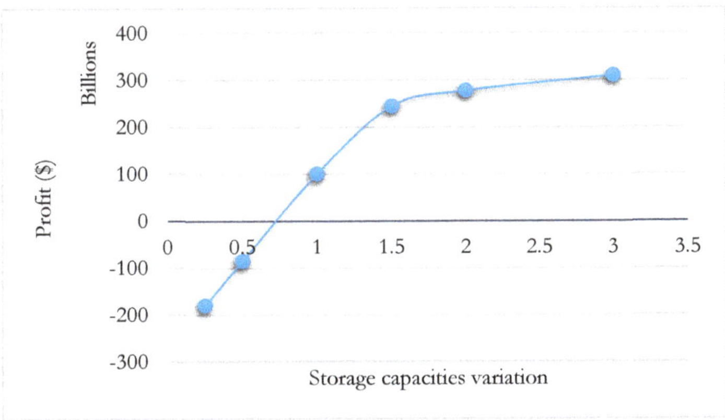

Figure 10. Profitability changes with variation in store capacity.

The impact of upstream storage capacities on the maintenance schedule and arrangement is investigated in Figure 11.

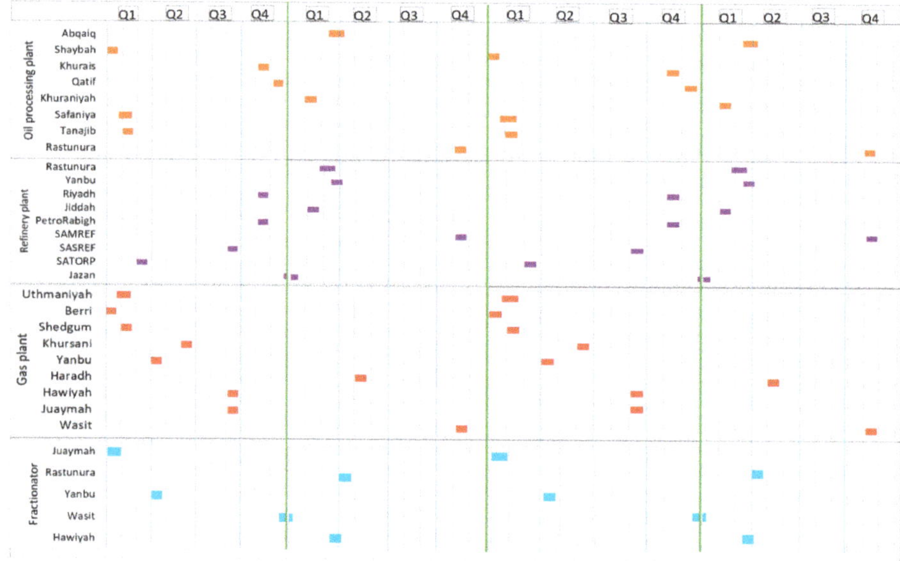

Figure 11. Maintenance schedule for all HCSC plants (reducing the storage capacity of upstream petroleum plants by 50%); Color code: Orange: oil processing plants; Purple: Refinery plants; Red: Gas plants; Light Blue: Fractionator plants.

The subsequent investigations can be derived from the above Gantt chart after reducing the storage capacity of upstream petroleum plants by 50%:

- The maintenance activities were distributed uniformly and perfectly among HCSC plants and over the study period.
- The majority of petroleum plants are taken for maintenance during the quarters of low demands.
- Due to the reducing storage capacities of upstream plants, most of the connected plants have been taken for maintenance at the same time because the storage capacity of the upstream units is not sufficient to feed downstream units. For instance, the Jiddah refinery plant and Khurasaniya oil processing plant have been maintained in the same period. Similarly, both Riyadh and PetroRabigh refinery plants are maintained with Khurais oil processing plant.
- Likewise, RasTunura and Yanbu refinery plants were maintained exchangeably; Rastunura is maintained during the first two weeks of the Abqaiq oil processing plant's maintenance time, and then Yanbu is maintained during the last two weeks of the Abqaiq oil processing plant's maintenance period because during the first two weeks of maintaining the Abqaiq plant, inventory capacity is sufficient to feed only RasTunura refinery plant and vice versa for Yanbu refinery plant.
- For natural gas petroleum plants, Jaymah fractionation plant is recommended to be maintained with Uthmaniya and Berri gas plants. Similarly, the Yanbu fractionation plant and Yanbu gas plants are maintained at the same time.

To determine the significance of integration between operations and scheduling activities during modelling, the model was resolved without integration (only maintenance model). The obtained maintenance schedule is depicted in Figure 12.

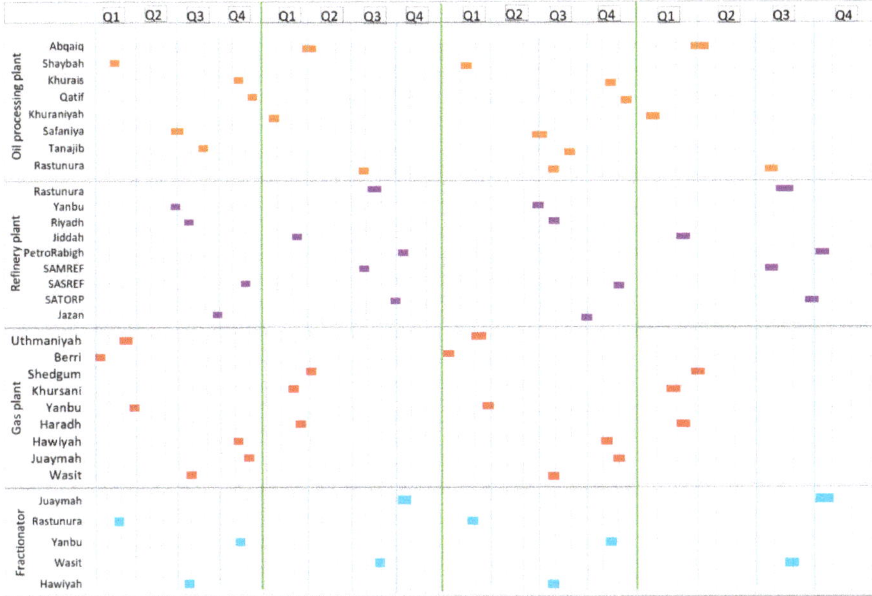

Figure 12. Maintenance schedule for all HCSC plants (maintenance model only); Color code: Orange: oil processing plants; Purple: Refinery plants; Red: Gas plants; Light Blue: Fractionator plants.

It appeared that the HCSC units were taken for maintenance on a random basis during the planning period without taking into account the seasons of high or low demand, which would inevitably result in a huge loss of production and ultimately does not satisfy most customer needs.

4. Discussion

In this study, an integrated operation and maintenance model for both the oil and gas supply chain was developed. The model integrated and generated operation plans and maintenance schedules for the HCSC plants jointly. The integration between the operation and maintenance activities was studied and compared to a model without integration. The obtained results justified the importance of the proposed model. The results of the model were promising in which the model forced all the HCSC plants to be maintained during the months of low demand to prevent huge losses in the production and satisfy most of the demand and minimize lost sales as much as possible. The output of the model shows the role of storage in satisfying needs during plant maintenance, as well as the impact of storage capacities on coupling or decoupling maintenance tasks for connected plants. The model can be extended by addressing the uncertainty of input parameters such as the demand, price, etc.

Funding: This research received no external funding.

Acknowledgments: The author would like to acknowledge the financial support provided by the Deanship of Scientific Research (DSR) at King Fahd University of Petroleum and Minerals (KFUPM) for funding this work through project NO. SR181022.

Conflicts of Interest: The author declares no conflict of interest.

References

1. Ghaithan, A.M.; Attia, A.; Duffuaa, S.O. Multi-objective optimization model for a downstream oil and gas supply chain. *Appl. Math. Model.* **2017**, *52*, 689–708. [CrossRef]
2. Alkhamis, T.M.; Yellen, J. Refinery units maintenance scheduling using integer programming. *Appl. Math. Model.* **1995**, *19*, 543–549. [CrossRef]
3. Mendez, C.A.; Grossmann, I.E.; Harjunkoski, I.; Kaboré, P. A simultaneous optimization approach for off-line blending and scheduling of oil-refinery operations. *Comput. Chem. Eng.* **2006**, *30*, 614–634. [CrossRef]
4. Göthe-Lundgren, M.; Lundgren, J.T.; Persson, J.A. An optimization model for refinery production scheduling. *Int. J. Prod. Econ.* **2002**, *78*, 255–270. [CrossRef]
5. Antomarioni, S.; Bevilacqua, M.; Potena, D.; Diamantini, C. Defining a data-driven maintenance policy: An application to an oil refinery plant. *Int. J. Qual. Reliab. Manag.* **2019**. [CrossRef]
6. Bertolini, M.; Bevilacqua, M.; Ciarapica, F.E.; Giacchetta, G. Development of risk-based inspection and maintenance procedures for an oil refinery. *J. Loss Prev. Process Ind.* **2009**, *22*, 244–253. [CrossRef]
7. Aissani, N.; Beldjilali, B.; Trentesaux, D. Dynamic scheduling of maintenance tasks in the petroleum industry: A reinforcement approach. *Eng. Appl. Artif. Intell.* **2009**, *22*, 1089–1103. [CrossRef]
8. Amaran, S.; Sahinidis, N.V.; Sharda, B.; Morrison, M.; Bury, S.J.; Miller, S.; Wassick, J.M. Long-term turnaround planning for integrated chemical sites. *Comput. Chem. Eng.* **2015**, *72*, 145–158. [CrossRef]
9. Awad, M.; Ertem, M. Stochastic scheduling of workforce-constrained preventive maintenance activities in petroleum plants. In Proceedings of the 2017 Annual Reliability and Maintainability Symposium (RAMS), Orlando, FL, USA, 23–26 January 2017; pp. 1–5.
10. Xu, J.; Qu, H.; Wang, S.; Xu, Q. A new proactive scheduling methodology for front-end crude oil and refinery operations under uncertainty of shipping delay. *Ind. Eng. Chem. Res.* **2017**, *56*, 8041–8053. [CrossRef]
11. Qu, H.; Wang, S.; Xu, Q. Integrated Proactive and Reactive Scheduling for Refinery Front-end Crude Movement with Consideration of Unit Maintenance. *Ind. Eng. Chem. Res.* **2019**, *58*, 12192–12206. [CrossRef]
12. Rosenblatt, M.J.; Lee, H.L. Economic production cycles with imperfect production processes. *IIE Trans.* **1986**, *18*, 48–55. [CrossRef]
13. Joshi, S.; Gupta, R. Scheduling of routine maintenance using production schedules and equipment failure history. *Comput. Ind. Eng.* **1986**, *10*, 11–20. [CrossRef]
14. Pintelon, L. Performance reporting and decision tools for maintenance management. Ph.D. Thesis, Department of Industrial Management, Catholic University of Leuven, Leuven, Belgium, 1990. (Unpublished).
15. Ashayeri, J.; Teelen, A.; Selenj, W. A production and maintenance planning model for the process industry. *Int. J. Prod. Res.* **1996**, *34*, 3311–3326. [CrossRef]
16. Rahim, M.A.; Ben-Daya, M. *Integrated Models in Production Planning, Inventory, Quality, and Maintenance*; Springer Science & Business Media: Berlin, Germany, 2012.
17. Iravani, S.M.; Duenyas, I. Integrated maintenance and production control of a deteriorating production system. *IIE Trans.* **2002**, *34*, 423–435. [CrossRef]
18. Cassady, C.R.; Kutanoglu, E. Integrating preventive maintenance planning and production scheduling for a single machine. *IEEE Trans. Reliab.* **2005**, *54*, 304–309. [CrossRef]
19. Cassady, C.R.; Kutanoglu, E. Minimizing job tardiness using integrated preventive maintenance planning and production scheduling. *IIE Trans.* **2003**, *35*, 503–513. [CrossRef]
20. Darwish, M.A.; Ben-Daya, M. Effect of inspection errors and preventive maintenance on a two-stage production inventory system. *Int. J. Prod. Econ.* **2007**, *107*, 301–313. [CrossRef]
21. Aghezzaf, E.H.; Jamali, M.A.; Ait-Kadi, D. An integrated production and preventive maintenance planning model. *Eur. J. Oper. Res.* **2007**, *181*, 679–685. [CrossRef]
22. Aghezzaf, E.-H.; Najid, N.M. Integrated production planning and preventive maintenance in deteriorating production systems. *Inf. Sci.* **2008**, *178*, 3382–3392. [CrossRef]
23. Zied, H.; Sofiene, D.; Nidhal, R. An optimal production/maintenance planning under stochastic random demand, service level and failure rate. In Proceedings of the 2009 IEEE International Conference on Automation Science and Engineering, Bangalore, India, 22–25 August 2009; pp. 292–297.
24. Berrichi, A.; Amodeo, L.; Yalaoui, F.; Châtelet, E.; Mezghiche, M. Bi-objective optimization algorithms for joint production and maintenance scheduling: Application to the parallel machine problem. *J. Intell. Manuf.* **2009**, *20*, 389. [CrossRef]

25. Ayed, S.; Sofiene, D.; Nidhal, R. Joint optimisation of maintenance and production policies considering random demand and variable production rate. *Int. J. Prod. Res.* **2012**, *50*, 6870–6885. [CrossRef]
26. Fitouhi, M.-C.; Nourelfath, M. Integrating noncyclical preventive maintenance scheduling and production planning for multi-state systems. *Reliab. Eng. Syst. Saf.* **2014**, *121*, 175–186. [CrossRef]
27. Yalaoui, A.; Chaabi, K.; Yalaoui, F. Integrated production planning and preventive maintenance in deteriorating production systems. *Inf. Sci.* **2014**, *278*, 841–861. [CrossRef]
28. Assid, M.; Gharbi, A.; Hajji, A. Joint production, setup and preventive maintenance policies of unreliable two-product manufacturing systems. *Int. J. Prod. Res.* **2015**, *53*, 4668–4683. [CrossRef]
29. Khosrojerdi, A.; Zegordi, S.H.; Allen, J.K.; Mistree, F. A method for designing power supply chain networks accounting for failure scenarios and preventive maintenance. *Eng. Optim.* **2016**, *48*, 154–172. [CrossRef]
30. Cheng, G.Q.; Zhou, B.H.; Li, L. Joint optimisation of production rate and preventive maintenance in machining systems. *Int. J. Prod. Res.* **2016**, *54*, 6378–6394. [CrossRef]
31. Ettaye, G.; El Barkany, A.; El Khalfi, A. Integrating production and maintenance for a multi-lines system. *Int. J. Perform. Eng.* **2017**, *13*, 29. [CrossRef]
32. Géhan, M.; Castanier, B.; Lemoine, D. Integration of maintenance in the tactical production planning process under feasibility constraint. In *IFIP International Conference on Advances in Production Management Systems*; Springer: Berlin/Heidelberg, Germany, 2014; pp. 467–474.
33. Liu, Q.; Lv, W. Multi-component manufacturing system maintenance scheduling based on degradation information using genetic algorithm. *Ind. Manag. Data Syst.* **2015**, *115*, 1412–1434. [CrossRef]
34. Zhao, S.; Wang, L.; Zheng, Y. Integrating production planning and maintenance: An iterative method. *Ind. Manag. Data Syst.* **2014**, *114*, 162–182. [CrossRef]
35. Hnaien, F.; Yalaoui, F.; Mhadhbi, A.; Nourelfath, M. A mixed-integer programming model for integrated production and maintenance. *IFAC-Pap.* **2016**, *49*, 556–561. [CrossRef]
36. Askri, T.; Hajej, Z.; Rezg, N. Jointly production and correlated maintenance optimization for parallel leased machines. *Appl. Sci.* **2017**, *7*, 461. [CrossRef]
37. Amiri, S.; Honarvar, M. Providing an integrated Model for Planning and Scheduling Energy Hubs and preventive maintenance. *Energy* **2018**, *163*, 1093–1114. [CrossRef]
38. Khatab, A.; Diallo, C.; Aghezzaf, E.-H.; Venkatadri, U. Integrated production quality and condition-based maintenance optimisation for a stochastically deteriorating manufacturing system. *Int. J. Prod. Res.* **2019**, *57*, 2480–2497. [CrossRef]
39. Rinaldi, G. An integrated Operation and Maintenance Framework for Offshore Renewable Energy. Ph.D. Thesis, University of Exeter, Exeter, UK, 2019.
40. Martin, R.; Lazakis, I.; Barbouchi, S.; Johanning, L. Sensitivity analysis of offshore wind farm operation and maintenance cost and availability. *Renew. Energy* **2016**, *85*, 1226–1236. [CrossRef]
41. Rinaldi, G.; Pillai, A.C.; Thies, P.R.; Johanning, L. Multi-objective optimization of the operation and maintenance assets of an offshore wind farm using genetic algorithms. *Wind Eng.* **2020**, *44*, 390–409. [CrossRef]
42. Helsen, J.; Peeters, C.; Doro, P.; Ververs, E.; Jordaens, P.J. Wind farm operation and maintenance optimization using big data. In Proceedings of the 2017 IEEE Third International Conference on Big Data Computing Service and Applications (BigDataService), San Francisco, CA, USA, 6–9 April 2017; pp. 179–184.
43. Nielsen, J.J.; Sørensen, J.D. On risk-based operation and maintenance of offshore wind turbine components. *Reliab. Eng. Syst. Saf.* **2011**, *96*, 218–229. [CrossRef]
44. Dinwoodie, I. Modelling the Operation and Maintenance of Offshore Wind Farms. Ph.D. Thesis, University of Strathclyde, Glasgow, UK, 2014.
45. Seyr, H.; Muskulus, M. Decision support models for operations and maintenance for offshore wind farms: A review. *Appl. Sci.* **2019**, *9*, 278. [CrossRef]
46. Murray, S. *Energy to the World: The Story of Saudi Aramco*; Aramco Services Company: Houston, TX, USA, 2011.
47. Attia, A.M.; Ghaithan, A.M.; Duffuaa, S.O. Data on upstream segment of a hydrocarbon supply chain in Saudi Arabia. *Data Brief* **2019**, *27*, 104804. [CrossRef]
48. Attia, A.M.; Ghaithan, A.M.; Duffuaa, S.O. A multi-objective optimization model for tactical planning of upstream oil & gas supply chains. *Comput. Chem. Eng.* **2019**, *128*, 216–227.

49. Manual, C.U. IBM ILOG CPLEX Optimization Studio. *Version* **1987**, *12*, 1987–2018.
50. Brook, A.; Kendruck, D.; Meeraus, A. *GAMS—General Algebraic Modeling System: A User's Guide;* Scientific Press: Redwood City, CA, USA, 1988.

Publisher's Note: MDPI stays neutral with regard to jurisdictional claims in published maps and institutional affiliations.

MDPI

St. Alban-Anlage 66

4052 Basel

Switzerland

Tel. +41 61 683 77 34

Fax +41 61 302 89 18

www.mdpi.com

Applied Sciences Editorial Office

E-mail: applsci@mdpi.com

www.mdpi.com/journal/applsci